DEPOSITION TECHNOLOGIES FOR FILMS AND COATINGS

MATERIALS SCIENCE SERIES

Series Editor:

Rointan F. Bunshah, University of California at Los Angeles

Deposition Technologies for Films and Coatings

Developments and Applications

by

Rointan F. Bunshah

John M. Blocher, Jr.
Thomas D. Bonifield
John G. Fish
P.B. Ghate
Birgit E. Jacobson

Donald M. Mattox
Gary E. McGuire
Morton Schwartz
John A. Thornton
Robert C. Tucker, Jr.

np NOYES PUBLICATIONS
Park Ridge, New Jersey, U.S.A.

Copyright © 1982 by Noyes Publications
   No part of this book may be reproduced in any form
   without permission in writing from the Publisher.
Library of Congress Catalog Card Number: 82-7862
ISBN: 0-8155-0906-5
Printed in the United States

Published in the United States of America by
Noyes Publications
Mill Road, Park Ridge, New Jersey 07656

10 9 8 7 6

Library of Congress Cataloging in Publication Data
Main entry under title:

Deposition technologies for films and coatings.

   Bibliography: p.
   Includes index.
   1. Coating processes. I. Blocher, J. M.,
1919-   . II. Bunshah, Rointan Framroze.
TP156.C57D46   667'.9   82-7862
ISBN 0-8155-0906-5   AACR2

Preface

Almost universally in high technology applications, a composite material is used where the properties of the surface are intentionally different from those of the core. Thus, materials with surface coatings are used in the entire cross-section of applications ranging from microelectronics, display devices, chemical corrosion, tribology including cutting tools, high temperature oxidation/corrosion, solar cells, thermal insulation and decorative coatings (including toys, automobile components, watch cases, etc.).

A large variety of materials is used to produce these coatings. They are metals, alloys, refractory compounds (e.g. oxides, nitrides, carbides), intermetallic compounds (e.g. GaAg) and polymers in single or multiple layers. The thickness of the coatings ranges from a few atom layers to millions of atom layers. The microstructure and hence the properties of the coatings can be varied widely and at will, thus permitting one to design new material systems with unique properties. (A material system is defined as the combination of the substrate and coating.)

Historically, coating technology evolved and developed in the last 30 years in several industries, i.e. decorative coatings, microelectronics and metallurgical coatings. They used similar techniques but only with the passage of time have the various approaches reached a common frontier resulting in much useful cross-fertilization. That very vital process is proceeding ever more strongly at this time.

With this background in mind, a short course on Deposition Technologies and their applications was developed and given on five consecutive occasions in the last three years. This volume is based on the material used in the course.

It comprises chapters dealing with the various coating techniques, the resulting microstructure, properties and applications. The specific techniques covered are evaporation, ion plating, sputtering, chemical vapor deposition, electrodeposition from aqueous solution, plasma and detonation gun coating techniques, and polymeric coatings. In addition several other chapters are added. Plasmas are used in many of the deposition processes and therefore a special chapter on this topic has been added. Cleaning of the substrate and the related topic of adhesion of the coating are common to many processes and a brief exposé of this topic is presented. Characterization of the films, i.e., composition, impurities, crystal structure and microstructure are essential to the understanding of the various processes. Two chapters dealing with this area are included. Finally, a chapter on application of deposition techniques in microelectronics is added to give one example of the use of several of these techniques in a specific area.

This volume represents a unique collection of our knowledge on Deposition Technologies and their applications up to and including the state-of-the-art. It is hoped that it will be very useful to students, practicing engineers and managerial personnel who have to learn about this essential area of modern technology.

R.F. Bunshah

University of California
Los Angeles, California
April 1982

Authors

John M. Blocher, Jr.
Consultant
Oxford, Ohio

Thomas D. Bonifield
Texas Instruments, Inc.
Dallas, Texas

Rointan F. Bunshah
Department of Materials Science
 and Engineering
University of California
Los Angeles, California

John G. Fish
Texas Instruments, Inc.
Dallas, Texas

P.B. Ghate
Semiconductor Research
 Engineering Laboratories
Texas Instruments, Inc.
Dallas, Texas

Birgit E. Jacobson
Department of Physics and
 Measurement Technology
Linköping University
Linköping, Sweden

Donald M. Mattox
Surface Metallurgy Division
Sandia Laboratories
Albuquerque, New Mexico

Gary E. McGuire
Tektronix, Inc.
Beaverton, Oregon

Morton Schwartz
Electrochemical/Metal
 Finishing Consultant
Los Angeles, California

John A. Thornton
Telic Corporation
Santa Monica, California

Robert C. Tucker, Jr.
Materials Development
Union Carbide Corporation
Indianapolis, Indiana

Contents

PREFACE v
AUTHORS. vii

1. DEPOSITION TECHNOLOGIES: AN OVERVIEW: *Rointan F.
Bunshah* 1
   Introduction. 1
   Aim and Scope 2
   Definitions and Concepts 3
   Physical Vapor Deposition (PVD) Process Terminology 4
   Classification of Coating Processes 5
   Microstructure and Properties 7
   Unique Features of Deposited Materials and Gaps in Understanding 9
   Current Applications 10
      Decorative/Functional Coating. 10
      High Temperature Corrosion 10
      Environmental Corrosion 11
      Friction and Wear 11
      Materials Conservation 11
      Cutting Tools 11
      Nuclear Fuels 11
      Biomedical Uses 11
   "Frontier Areas" for the Application of the Products of Deposition
     Technology 12
   Selection Criteria. 12
   Summary. 13
   Appendix: Deposition Process Definitions. 15
      Conduction and Diffusion Processes 15
      Chemical Processes 15
      Wetting Processes 16

x   Contents

    Spraying Processes.................................16
  References..18

2. PLASMAS IN DEPOSITION PROCESSES: *John A. Thornton* 19
  Introduction..19
  Particle Motion....................................20
    Mean Free Path and Collision Cross Sections............20
    The Free Electron Kinetic Energy in a Plasma...........22
    Electron Energy Distribution Function..................23
    Processes Involving the Collision Frequencies..........24
    Particles in Magnetic Fields...........................28
  Collective Phenomena...............................30
    Basic Parameter..30
    Plasma Sheaths...30
    Ambipolar Diffusion....................................35
    Plasma Oscillations....................................36
  Plasma Discharges..................................37
    Introduction...37
    Ionization Balance-Paschen Relation....................40
    Cold Cathode Discharges................................41
    Magnetron Discharges...................................43
    RF Discharges..44
  Plasma Volume Reactions............................46
    Introduction...46
    Electron Interaction with Atoms........................46
    Electron Interaction with Molecules....................48
    Metastable Species.....................................49
    Applications of Volume Reactions.......................50
  Surface Reactions..................................50
    Introduction...50
    Ion Bombardment..51
    Electron Bombardment...................................53
    Glow Discharge Surface Cleaning and Activation.........58
  References...61

3. ADHESION AND SURFACE PREPARATION: *Donald M. Mattox* 63
  Nucleation...64
  Interface Formation................................65
  Film Formation.....................................66
  Microstructure Development.........................66
  Impurity Incorporation.............................67
  Intrinsic Stress...................................68
  Mechanical Properties..............................70
  Interfacial Fracture...............................70
  Time-Dependent Interfacial Changes.................71
  Adhesion Testing...................................71
  Surface Preparation................................72
    Cleaning Oxide Surfaces................................74
    Cleaning Metal Surfaces................................77

    Cleaning Organic Polymer Surfaces............................79
    Monitoring Surface Cleaning..................................80
    Clean Surface Storage..81
    Non-Cleaning Surface Treatments..............................81
Summary..82
References...82

4. EVAPORATION: *Rointan F. Bunshah*..............................83
General Introduction...83
Scope..84
PVD Processes..85
    Preamble...85
    PVD Processes..85
    Advantages and Limitations...................................89
Theory and Mechanisms..90
    Vacuum Evaporation...90
Evaporation Process and Apparatus................................92
    The System...92
Evaporation Sources..94
    General Considerations.......................................94
    Resistance Heated Sources....................................95
    Sublimation Sources..98
    Evaporation Sources and Materials...........................100
    Induction Heated Sources....................................101
    Electron Beam Heated Sources................................102
    Laser Beam Evaporation......................................108
    Arc Evaporation...108
Deposition Rate Monitors and Process Control....................108
    Monitoring of the Vapor Stream..............................109
    Monitoring of Deposited Mass................................110
    Monitoring of Specific Film Properties......................110
    Evaporation Process Control.................................113
Deposition of Various Materials.................................113
    Deposition of Metals and Elemental Semiconductors...........113
    Deposition of Alloys..115
    Deposition of Intermetallic Compounds.......................119
    Deposition of Refractory Compounds..........................122
Classification of Evaporation Processes.........................127
Microstructure of PVD Condensates...............................129
    Microstructure Evolution....................................129
    Texture...139
    Residual Stresses...139
    Defects...140
Physical Properties of Thin Films...............................143
Mechanical and Related Properties...............................143
    Mechanical Properties.......................................143
    Wear Resistance...155
    Corrosion Resistance..156

xii   Contents

      Influence of Coatings on Thermal Fatigue Resistance 156
Purification of Metals by Evaporation 157
Applications 158
      Present Uses 158
      Future Uses 159
Economics and Perspective 161
References 162
Suggestions for Further Reading 167
      Books 167
      Journals 167
      Papers 167
**Appendix: On Progress in Scientific Investigations in the Field of
Vacuum Evaporation in the Soviet Union:** *A. V. Demchishin* 167

5. COATING DEPOSITION BY SPUTTERING: *John A. Thornton* 170
    Introduction 170
      Sputtering Apparatuses 172
      Sputtering Applications 174
      Implementation of Sputtering Process 176
      Chapter Organization 177
    Basic Sputtering Mechanisms 178
      The Sputtering Yield 178
      Basic Momentum Exchange 180
      Alloys and Compounds 183
      Sputtering with Reactive Species 186
      Sputtered Species 187
      Energy of Sputtered Species 187
    Implementation 189
      Planar Diode and the dc Glow Discharge 189
      Assist Discharge Devices—Triodes 193
      Magnetrons 193
      RF Sputtering 204
      Ion Beam Sputtering 210
    Coating Growth, Structure, and Properties 211
      Substrate Environment 211
      Coating Nucleation and Growth 212
      Structure Zone Models 213
      Substrate Temperature Influence on Composition 218
      Substrate Roughness 219
      Oblique Deposition 220
      Inert Gas Effects 220
      Ion Bombardment—Bias Sputtering 220
      Internal Stresses 221
      Polymorphic Phase Formation 225
      Coating Properties 225
    Special Topics 227
      Target Fabrication 228
      Adhesion 228

  Sputter Cleaning and Bias Sputtering. 229
  Formation of Graded Interfaces. 231
  Reactive Sputtering. 231
  Predeposition Pumping. 235
 References. 237

6. ION PLATING TECHNOLOGY: *Donald M. Mattox* 244
 Introduction. 244
 Methods of Ion Bombardment. 245
  Ion Gun (Vacuum Environment) 245
  Gas Discharges (0.5–10 μ Gas Environment). 246
 Ion Bombardment Effects: Prior to Deposition 251
  Sputtering. 251
  Defect Production. 252
  Crystallographic Disruption. 253
  Surface Morphology Changes. 253
  Gas Incorporation. 253
  Temperature Rise 253
  Surface Composition 253
  Physical Mixing. 254
 Ion Bombardment Effects: Interface Formation. 254
  Physical Mixing. 254
  Enhanced Diffusion. 254
  Nucleation Modes 254
  Preferential Removal 255
  Surface Coverage. 255
 Ion Bombardment Effects: Film Growth. 255
  Morphology. 256
  Crystallography. 257
  Composition. 257
  Physical Properties. 257
 Sources for Ion Plating. 258
  Vapor Sources 258
  Enhanced Ionization Sources. 259
 Apparatus for Ion Plating 259
  Vacuum System 260
  Gas Discharge Chamber 261
  Gas Handling System. 262
  High Voltage Substrate. 263
  High Voltage Power Supplies (RF, AC, or DC) 264
 Process Control. 264
  Rate and Deposition Monitors 266
  Pressure Monitoring. 266
  Gas Composition Monitoring. 266
 Preparation of Substrate Surfaces. 267
 Applications of Ion Plating 267
  Adhesion. 267
  Wear and Erosion 267

Corrosion Inhibition 267
Joining 267
Incompatible Coating-Substrate Systems 267
Summary.................................... 268
References.................................. 268

7. MICROSTRUCTURES OF PVD-DEPOSITED FILMS CHARACTERIZED BY TRANSMISSION ELECTRON MICROSCOPY: *Birgit E. Jacobson* 288
Introduction................................. 288
Why Study the Microstructure? 288
What to Learn About the Microstructure? 289
Why TEM? 289
Content...................................... 290
The Possibilities in a TEM/STEM Instrument 290
Imaging..................................... 290
Electron Diffraction............................ 291
Microstructure and Properties of Electron Beam Evaporated Films ... 293
Chemical Analysis 294
Heating, Cooling and Deformation 294
Limitations and Complementary Techniques 294
Specimen Preparation for TEM 295
Growth Morphology and Cavity Formation in Single Phase
Structures 295
Influence of Fine Dispersed Particles................ 302
Variations in Growth Morphology with Composition in Two-Phase
Structures 311
Microstructure and Properties of Films Prepared by ARE 319
Microstructure and Hardness of Single Phase Carbide Films, TiC... 320
Effect of Carbide Alloying, VC-TiC.................. 320
Effect of in Situ TEM Annealing of Fine Grained Carbide Films... 327
Microstructure and Hardness of Poly-Phase Nitride Films,
Ti-Ti$_2$N-TiN 327
Microstructure and Properties of Sputter Deposited Films 330
Crystalline-to-Amorphous Transition in a Magnetron Sputter
Deposited Film 330
References.................................. 333

8. CHEMICAL VAPOR DEPOSITION: *John M. Blocher, Jr.* 335
Introduction................................. 335
Coating Procedures 338
Reactant Supply System......................... 339
Deposition System............................. 340
Gas Dynamics................................. 341
Substrate Heating 343
By-Product Recycle or Disposal 345
Surface Preparation 345
Diagnostic Tools 346
Cost of CVD.................................. 346

Fundamentals of CVD 348
Thermodynamics. 348
    Mass Transport 351
    Chemical Kinetics 356
Structure of CVD Coatings 357
Current CVD Products 359
Future Applications of CVD 361
Conclusion. 362
References. 362

9. PLASMA ASSISTED CHEMICAL VAPOR DEPOSITION: *Thomas D. Bonifield* 365
    Introduction. 365
        Scope 365
        Advantages and Limitations. 365
    Physical and Chemical Concepts of PACVD 366
        Glow Discharge Plasmas 366
        Collisional Dissociation by Electrons. 369
        Transport Kinetics. 369
        Surface Chemical Effects 370
    Practical Glow Discharge Reactors 371
        Reactor Designs 372
        Discharge Power Supplies 374
        Vacuum Requirements. 374
    Deposition of Various Materials 374
        Amorphous Silicon 374
        Amorphous Carbon 375
        Silicon Nitride 376
        Silicon Oxide and Oxynitride. 378
        Silicon Carbide 378
        Plasma Polymers 381
        Miscellaneous Materials. 381
    Summary. 382
    References. 383

10. DEPOSITION FROM AQUEOUS SOLUTIONS: AN OVERVIEW: *Morton Schwartz*. 385
    Introduction. 385
    General Principles 385
    Electrodeposition 392
        Mechanism of Deposition 392
        Parameters. 395
    Processing Techniques 402
    Selection of Deposits 404
        Individual Metals. 404
        Alloy Deposition. 404
    Selected Special Processes 412
        Electroless Deposition 412

Electroforming..415
Anodizing...419
Plating on Plastics..423
Plating Printed Circuit Boards...............................424
Structures and Properties of Deposits......................426
References..444
Appendix A: Preparation of Substrates for Electroplating.......449
ASTM Recommended Practices..................................449
Appendix B: Representative Electroless Plating Solution Formulations..450
Appendix C: Comparison of Aluminum Anodizing Processes (Types I, II and III)..452
Advantages of Type I Coatings................................452
Limitations of Type I Coatings...............................452
Advantages of Type II Coatings...............................452
Limitations of Type II Coatings..............................452
Characteristics of Hard Anodize Coatings.....................453
Effect of Alloying Elements on the Hard Coating..............453

11. PLASMA AND DETONATION GUN DEPOSITION TECHNIQUES AND COATING PROPERTIES: *Robert C. Tucker, Jr.*...........454
Introduction..454
Equipment..454
Plasma Coating Torches.......................................454
Detonation Gun...459
Auxiliary Equipment..460
Equipment Related Coating Limitations........................460
Coating Process..461
Powder...461
Substrate Preparation..461
Masking..462
Coating..462
Finishing..463
Coating Structure and Properties............................464
Macrostructure...464
Microstructure...464
Bond Strength..472
Residual Stress..473
Density..474
Mechanical Properties..475
Wear and Friction..479
Thermal..485
Corrosion Properties...486
Electrical Characteristics...................................487
References..488

12. ORGANIC POLYMER COATINGS: *John G. Fish*................490
Introduction..490

Contents   xvii

    History and Art of Organic Coatings 490
    Introduction to Polymer Chemistry and Physics 491
    Polymer Characteristics 493
        Thermoplastic and Thermosetting Polymers 493
        Molecular Weights of Polymers. 493
        Solubility of Polymers 494
        Solubility/Molecular Weight Considerations 497
    Typical Polymers Used for Coatings 498
        Natural Products. 498
    Mechanism of Final Film Formation. 498
    Coating Functions 502
        Decoration. 502
        Preservation of Substrate 503
        Self-Preservation of Paint Surfaces 503
        Lubrication and Wear Resistance 503
        Temporary Coatings. 504
        Antistatic Coatings 504
        Water Repellence. 504
        Hydrophilic Coatings 505
        Electrical Properties. 505
    Surface Preparation 506
        Cleanliness. 506
        Texture. 506
        Primer Technology 506
        Conversion Coatings. 507
    Deposition Methods. 508
        Mechanical Processes 508
        Dip Coating 509
        Spray Coating. 509
        Screen Printing 509
        Lamination 510
        Melt Extrusion 510
        Flow Coating 510
        Roll Coating. 510
        Graft Polymerization 510
        Fluid Bed Coating 511
        Electrophoretic Deposition 511
        Spin-On Coatings. 511

13. **DEPOSITION TECHNIQUES AND MICROELECTRONIC
    APPLICATIONS:** *P.B. Ghate* 514
    Introduction. 514
    Thin Film Resistors 514
        Resistor Characteristics. 514
        Conduction in Thin Films. 517
        Resistor Design. 519
        Resistor Materials 520
    Thin Film Capacitor. 523

Contents

      Capacitor...523
      Dielectric Films..524
      Tantalum Film Capacitors............................526
   Contacts...528
      The Schottky Model...................................529
      Ohmic Contacts...533
   Interconnections for Integrated Circuits...................538
      Single Level Interconnections........................538
      Two Level Interconnections..........................540
      Reliability and Electromigration......................543
   Bonding and Assembly...................................544
   Other Applications..545
   Future Trends...546
   References...546

14. CHARACTERIZATION OF THIN FILMS: *Gary E. McGuire* 548
   Introduction..548
   Principles..549
   Elemental Sensitivity and Resolution....................558
   Chemical Information....................................561
   Lateral Resolution..563
   Depth Resolution...564
   Sample Requirements....................................565
   Summary...567
   References...568

1

Deposition Technologies: An Overview

Rointan F. Bunshah

Materials Science and Engineering Department
University of California
Los Angeles, California

INTRODUCTION

Most materials used in high technology applications are composites, i.e., they have a near-surface region with properties differing from those of the bulk materials. This is caused by the requirement on the material to exhibit a combination of various and sometimes conflicting properties. For example, a particular engineering component may be required to have high hardness and toughness (i.e., resistance to brittle crack propagation). This combination of properties can be obtained by having a composite material with high surface hardness and a tough core. Alternately, the need may be for a high temperature, corrosion-resistant material with high elevated temperature strength as is the case with the hot stage blades and vanes in a gas turbine. The solution again is to provide the strength requirement from the bulk and the corrosion requirement from the surface.

In general, coatings are desirable or may even be necessary for a variety of reasons including economics, materials conservation, unique properties, or the engineering and design flexibility which can be obtained by separating the surface properties from the bulk properties.

This near-surface region is produced by depositing a coating onto it (i.e., *overlay* coating) by processes such as physical or chemical vapor deposition, electrodeposition, and thermal spraying or by altering the surface material by the in-diffusion of materials (i.e., *diffusion coating* or *chemical conversion coating*), or by ion implantation of new material so that the surface layer now consists of both the parent and added materials.

"Coatings" may also be formed by other processes such as melt/solidification, e.g., laser glazing technique, by mechanical bonding of a surface layer, e.g., roll bonding, by mechanical deformation (e.g., shot peening) or other processes which change the properties without changing the composition.

As stated above, the coating/substrate combination is a composite materials system. The behavior of this composite system depends not only on the properties of the two components, i.e., the coating material and the substrate material, but also on the interaction between the two, i.e., the structure and properties of the coating-substrate interface which is integral to the very important factor of adhesion of coatings. In some cases, e.g., for overlay coatings, this is a distinct region. For others, such as ion implantation or diffusion coatings, it is not a discrete region.

Historically, most solid metallic and some ceramic materials were produced by melting/solidification technology. Since the advent of deposition technologies, i.e., production of solid materials from the vapor, the diversity of materials that can be produced has more than doubled since the properties of solid materials produced from the vapor phase can be varied over a much wider range than the same material produced from the liquid phase. This is because melt techniques produce solid materials with properties close to "equilibrium" properties whereas the deposition conditions may be so chosen as to produce materials from the vapor phase with properties close to "equilibrium," i.e., similar to their melt-produced counterparts or properties far removed from "equilibrium" properties, i.e., "non-equilibrium" properties. Moreover, a much greater variation in microstructure is possible with vapor source materials. For example, a copper-nickel alloy produced by solidification from the melt will always consist of a single phase solid solution, whereas the same alloy produced by alternate deposition from two sources may consist of alternate layers of nickel and copper, i.e., a laminate composite or a solid solution depending on the deposition temperature.

There are a large number of materials used today with coatings. These may range from the naturally occurring oxide layer which protects the surfaces of many metals such as aluminum, titanium, and stainless steel, to those with very deliberate and controlled alloying additions to the surface to produce specific properties, as exemplified by techniques such as molecular beam epitaxy or ion implantation. Other examples with increasing degree of criticality range from paint coatings applied to wood and metals, electrostatically painted golf balls, the print in the daily newspaper, optical coatings on lenses and other elements, vapor deposited microcircuit elements such as resistors, diffusion or overlay coatings on superalloys used in gas turbines for high temperature corrosion protection, hard overlay coatings of engineering components and machine tools, etc. . . .

AIM AND SCOPE

The aim of this volume is to give the reader a perspective over several of the coating techniques with emphasis on these techniques which are used in critical or demanding, i.e., high technology, applications. Consequently, some of the techniques such as painting, dip coatings, or printing will not be emphasized except as they pertain to some special application, e.g., in "thick film" electrical components. Nevertheless, a wide variety of techniques and their applications will be covered. The material is intended to present a broad spectrum of deposition technologies to those who may be familiar with only one or two techniques. Hopefully, this will help them to select and weigh various alternatives when the next technological problem involving coatings faces them.

The specific deposition technologies to be covered are:

(1) Physical Vapor Deposition including evaporation, ion plating and sputtering.

(2) Chemical Vapor Deposition.

(3) Electrodeposition and Electroless Deposition.

(4) Plasma Spraying as well as a very special variant called Detonation Gun Technology.

(5) Polymeric Coatings.

There are some generic areas common to several of the deposition technologies, the most prominent example being the use of plasmas in many of the deposition technologies. Therefore, a chapter on plasmas in deposition processes is included. Another common topic is cleaning of the substrate and adhesion of the coating. A chapter is included on that topic.

A further common topic is the characterization of the chemical composition and the microstructure of the coating at various levels of resolution. Two chapters are included to satisfy this need, one dealing with the structural and crystallographic analysis of films and the other with the microstructural features of the coating as revealed by Scanning Electron Microscopy (SEM) and Transmission Electron Microscopy (TEM).

Finally a chapter on fabrication of microelectronic devices using various coating techniques is included to illustrate an important area of application.

It is realized that all specific applications cannot be satisfied within this framework. For example, specific applications such as coatings for optical or magnetic applications are not addressed per se. At the other end of the spectrum, coatings for the "first wall" of thermo-nuclear reactors cannot be discussed since the development of the subject is in an embryonic stage.

In each of the chapters on deposition technologies, the theory, methodology, advantages, limitations and applications will be discussed.

DEFINITIONS AND CONCEPTS

In order to clear up potential problems, it is necessary to clarify certain distinctions which are common and pertinent to deposition technologies. These are as follows:

(1) Diffusion vs Overlay Coatings—Diffusion coatings are produced by the complete interdiffusion of material applied to the surface into the bulk of the substrate material. Examples of this are the diffusion of oxygen into metals to form various sub-oxide and oxide layers, the diffusion of aluminum into nickel base alloys to form various aluminides, etc. . . . A characteristic feature of diffusion coatings is a concentration gradient from the surface to the interior, as well as the presence of various layers as dictated by thermodynamic and kinetic considerations. Ion implantation may be considered to be a special case where the coating material is implanted at a relatively shallow depth (a few hundred Angstrom units) from the surface.

An overlay coating is an "add-on" to the surface of the part, e.g., gold-plating on an iron-nickel alloy, or titanium carbide onto a cutting tool, etc. Depending upon the process parameters, an interdiffusion layer between the substrate and the overlay coating may or may not be present.

(2) Thin Films vs Thick Films—Historically, the physical dimension of thickness was used to make the distinction between thick films and thin films. Unfortunately, the "critical" thickness value depended on the application and discipline. In recent years, a "Confucian" solution has been advanced. It states that if a coating is used for surface properties (such as electron emission, catalytic activity), it is a thin film; whereas, if it is used for bulk properties, corrosion resistance, etc., it is a thick film. Thus, the same coating material of identical thickness can be a thin film or a thick film depending upon the usage. This represents a reasonable way out of the semantic problem.

(3) Steps in the Formation of a Deposit—There are three steps in the formation of a deposit:

1. Synthesis or creation of the depositing species.
2. Transport from source to substrate.
3. Deposition onto the substrate and film growth.

These steps can be completely separated from each other or be super-imposed on each other depending upon the process under consideration. The important point to note is that if, in a given process, these steps can be individually varied and controlled, there is much greater flexibility for such a process as compared to one where they are not separately variable. This is analogous to the Degrees of Freedom in Gibbs' phase rule. For example, consider the deposition of tungsten by CVD process. It takes place by the reaction:

$$WF_6(vapor) + 3H_2(gas) \xrightarrow[\text{Substrate}]{\text{Heated}} W(deposit) + 6HF(gas)$$

The rate of deposition is controlled by the substrate temperature. At a high substrate temperature, the deposition rate is high and the structure consists of large columnar grains. This may not be a desirable structure. On the other hand, if the same deposit is produced by evaporation of tungsten, the deposition rate is essentially independent of the substrate temperature so that one can have a high deposition rate and a more desirable microstructure. On the other hand, CVD process may be chosen over evaporation because of considerations of "throwing power," i.e., the ability to coat irregularly shaped objects since high vacuum evaporation is basically a line-of-sight technique.

PHYSICAL VAPOR DEPOSITION (PVD) PROCESS TERMINOLOGY

The basic PVD processes are those currently known as evaporation, sputtering and ion plating. In recent years, a significant number of specialized PVD processes based on the above have been developed and extensively used, e.g., reactive ion plating, activated reactive evaporation, reactive sputtering, etc. There is now considerable confusion since a particular process can be legitimately covered by more than one name. As an example, if the Activated Reactive Evaporation (ARE) process is used with a negative bias on the substrate, it is very often called Reactive Ion Plating. Simple evaporation using an R.F. heated crucible has been called Gasless Ion Plating. An even worse example of the confusion that can arise is found in the Chapter on ion plating in this volume where the material is converted from the condensed phase to the vapor phase using thermal energy (i.e., evaporation) or momentum transfer (i.e., sputtering) or supplied as a vapor (very similar to CVD processes). Carrying this to the logical conclusion, one might say that all PVD processes are ion plating! On the other hand, the most important aspect of the ion plating process is the modification of the microstructure and composition of the deposit caused by the ion bombardment of the deposit resulting from the bias on the substrate and the presence of a gas, i.e., what is happening on the substrate.

To resolve this dilemma, it is proposed that we consider all of these basic processes and their variants as PVD processes and describe them in terms of the three steps in the formation of a deposit as described above. This will hopefully remove the confusion in terminology.

Step 1: Creation of Vapor Phase Specie — There are three ways to put a material into the vapor phase—evaporation, sputtering or chemical vapors and gases.

Step 2: Transport from Source to Substrate — The transport of the vapor species from the source to the substrate can occur under line-of-sight or molecular flow-conditions (i.e., without collisions between atoms and molecules); alternately, if the partial pressure of the metal vapor and/or gas species in the vapor state is high enough or some of these species are ionized (by creating a plasma), there are many collisions in the vapor phase during transport to the substrate.

Step 3: Film Growth on the Substrate — This involves the deposition of the film by nucleation and growth processes. The microstructure and composition of the film can be modified by bombardment of the growing film by ions from the vapor phase resulting in sputtering and recondensation of the film atoms and enhanced surface mobility of the atoms in the near-surface and surface of the film.

Every PVD process can be usefully described and understood in terms of these three steps. It is proposed and hoped that in due course the terminology of evaporation, sputtering and ion plating processes will fade away!

CLASSIFICATION OF COATING PROCESSES

Numerous schemes can be devised to classify or categorize coating processes, none of which are very satisfactory since several processes will overlap different categories. For example, the Appendix contains a list and definitions of various deposition processes based upon those provided by Chapman and Anderson[1] with some additions. These authors classify the processes under the general heading of Conduction and Diffusion Processes, Chemical Processes, Wetting Processes and Spraying Processes. Here, the Chemical Vapor Deposition process falls under the Chemical Processes and the Physical Vapor Deposition Process (Evaporation, Ion Plating and Sputtering) falls under the spraying processes. The situation can easily get confused as, for example, when Reactive and Activated Reactive Evaporation, and Reactive Ion Plating are all classified as Chemical Vapor Deposition processes by Yee[3] who considers them thusly because a chemical reaction is involved and it does not matter to him whether evaporated metal atoms or stable liquid or gaseous compounds are the reactants. Another classification of the methods of deposition of thin films is given by Campbell.[4] He considers the overlap between physical and chemical methods, e.g., evaporation and ion plating, sputtering and plasma reactions,

6 Deposition Technologies for Films and Coatings

reactive sputtering and gaseous anodization.[5] He classifies the Chemical Methods of Thin Film Preparation as follows:

Chemical Methods of Thin Film Preparation

Basic Class	Method
Formation from the Medium	Electroplating
	Ion Plating
	Chemical Reduction
	Vapor Phase
	Plasma Reaction
Formation from the Substrate	Gaseous Anodization
	Thermal
	Plasma Reduction

In addition, he considers the following as chemical methods of thick film preparation: Glazing, Electrophoretic, Flame Spraying and Painting.

In contrast to the chemists' approach given above, the physicists' approach to deposition processes is shown in the following classification of vacuum deposition techniques by Schiller, Heisig and Goedicke[6] and by Weissmantel.[7]

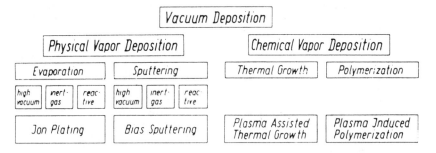

Figure 1.1: Survey of vacuum deposition techniques (Schiller[6]).

A different classification comes from a materials background where the concern is with structure and properties of the deposits as influenced by process parameters. Thus, Bunshah and Mattox[8] give a classification based on deposition methods as influenced by the dimensions of the depositing specie, e.g., whether it is atoms/molecules, liquid droplets or bulk quantities, as shown in Table 1.1.

In atomistic deposition processes, the atoms form a film by condensing on the substrate and migrating to sites, where nucleation and growth occurs. Further, adatoms do not achieve their lowest energy configurations and the resulting structure contains high concentrations of structural imperfections. Often the depositing atoms react with the substrate material to form a complex interfacial region.

Another aspect of coatings formed by atomistic deposition processes is as follows. The sources of atoms for these deposition processes can be by thermal vaporization (vacuum deposition) or sputtering (sputter deposition) in a vacuum, vaporized chemical species in a carrier gas (chemical vapor deposition), or ionic species in an electrolyte (electrodeposition). In low energy atomistic deposition processes, the depositing species impinge on the surface, migrate over the surface to a nucleation site where they condense and grow into a coating. The nucleation and growth modes of the condensing species determine the crystallography and microstructure of the coating. For high energy deposition processes, the depositing particles react with or penetrate into the substrate surface.

Particulate deposition processes involve molten or solid particles and the resulting microstructure of the deposit depends on the solidification or sintering of the particles. Bulk coatings involve the application of large amounts of coating material to the surface at one time such as in painting. Surface modification involves ion, thermal, mechanical, or chemical treatments, which alter the surface composition or properties. All of these techniques are widely used to form coatings for special applications.

Table 1.1: Methods of Fabricating Coatings

Atomistic Deposition	Particulate Deposition	Bulk Coatings	Surface Modification
Electrolytic Environment	*Thermal Spraying*	*Wetting Processes*	*Chemical Conversion*
Electroplating	Plasma Spraying	Painting	Electrolytic
Electroless Plating	D-Gun	Dip Coating	Anodization
Fused Salt Electrolysis	Flame Spraying	*Electrostatic Spraying*	(Oxide)
Chemical Displacement	*Fusion Coatings*	Printing	Fused Salts
Vacuum Environment	Thick Film Ink	Spin Coating	*Chemical-Liquid*
Vacuum Evaporation	Enameling	*Cladding*	*Chemical-Vapor*
Ion Beam Deposition	Electrophoretic	Explosive	Thermal
Molecular Beam Epitaxy	*Impact Plating*	Roll Bonding	Plasma
		Overlaying	Leaching
Plasma Environment		Weld Coating	*Mechanical*
Sputter Deposition			Shot Peening
Activated Reactive Evaporation			*Thermal*
Plasma Polymerization			*Surface Enrichment*
Ion Plating			Diffusion from Bulk
Chemical Vapor Environment			*Sputtering*
Chemical Vapor Deposition			*Ion Implanation*
Reduction			
Decomposition			
Plasma Enhanced			
Spray Pyrolysis			
Liquid Phase Epitaxy			

MICROSTRUCTURE AND PROPERTIES

In electrodeposition, typically, the growth process involves condensation of atoms at a kink site on the substrate surface, followed by layered growth of the deposit. Adatom mobility is increased by the hydrated nature of the ions and the adatom mobility may vary with crystal orientation. Field ion microscopy stripping studies of copper electrodeposited on tungsten has shown that there is surface rearrangement of the tungsten atoms during the electrodeposition process. Electrodeposited material does not grow in a uniform manner; rather it becomes faceted, develops dendrites and other surface discontinuities. Thus the microstructure of electrodeposited coatings may vary from relatively defect-free single crystals usually grown on single crystal substrates, to highly columnar and faceted structures. In the electroplating process, organic additives may be added to modify the nucleation process and to eliminate undesirable growth modes. This results in a microstructure more nearly that of bulk material formed by conventional metallurgical processes. Electrodeposition from a molten salt electrolyte allows the deposition of many materials not available from aqueous electrolytes.

In vacuum processes, the depositing species may have energies ranging from thermal (a few tenths of an electron volt) for evaporation to moderate energies (ten to hundreds of electron volts) for sputtered atoms to high energies for accelerated species such as those used in ion implantation. These energies have an important but poorly understood effect on interfacial interaction, nucleation and growth. Where there is chemical reaction betweeen the substrate atoms and the depositing atoms, and diffusion is possible, a diffusion or compound interfacial region is formed composed of compounds and/or alloys which modify the effective surface upon which the deposit grows. Low energy electron diffraction studies have shown that this interfacial reaction is very sensitive to surface condition and process parameters. If the coating and substrate materials are not chemically reactive and are insoluble, the interfacial region will be confined to an abrupt discontinuity in composition. This type of interface may be modified by bombardment with high energy particles to give high defect concentrations and implantation of ions resulting in a "pseudodiffusion" type of interface. The type of interface formed will influence the properties of the deposited coating. In many circumstances, these interfacial regions are of very limited thickness and pose a challenge to those interested in compositional, phase, microstructural and property analysis.

The microstructure of the depositing coating in the atomic deposition processes depends on how the adatoms are incorporated into the existing structure. Surface roughness and geometrical shadowing will lead to preferential growth of the elevated regions giving a columnar type microstructure to the deposits. This microstructure will be modified by substrate temperature, surface diffusion of the atoms, ion bombardment during deposition, impurity atom incorporation and angle of incidence of the depositing adatom flux. The structure zone model of Movchan and Demchishin[10] for vacuum deposited films is discussed in later chapters.

In chemical vapor deposition, the chemical species containing the film atoms is generally reduced or decomposed on the substrate surface often at high temperatures. Care must be taken to control the interface reaction between coating and substrate and between the substrate and the gaseous reaction products. The coating microstructure which develops is very similar to that developed by the vacuum deposition processes, i.e., small-grained columnar structures to large-grained equiaxed or oriented structures.

Each of the atomistic deposition processes has the potential of depositing materials which vary significantly from the conventional metallurgically processed material. The deposited materials may have high intrinsic stresses, high point defect concentration, extremely fine grain size, oriented microstructures, metastable phases, incorporated impurities, and macro and micro porosity. These properties may be reflected in the physical properties of the materials and by their response to applied stresses such as mechanical loads, chemical environments, thermal shock or fatigue loading. Metallurgical properties which may be affected include elastic constants, tensile strength, fracture toughness, fatigue strength, hardness, diffusion rates, friction/wear properties, and corrosion resistance. In addition, the unique microstructure of the deposited material may lead to such effects as anomalously low annealing and recrystallization temperatures where the internal stresses and high defect concentration aid in atomic rearrangement.

The high value of grain boundary area to volume ratio found in fine grained deposited material means that diffusion processes may be dominated by grain boundary rather than bulk diffusion. The fine grained nature of the materials also affects the deformation mechanisms such as slip and twinning. For thin films, the free-surface to volume ratio is high, and the pinning of dislocation by the free surface leads to the high tensile strengths often measured in thin films of materials.

In vapor deposition processes, impurity incorporation during deposition can give high intrinsic stresses or impurity stabilized phases which are not seen in the bulk forms of the materials. Reactive species allow the deposition of compounds such as nitrides, carbides, borides and oxides. Graded deposits can be formed.

Vapor deposition processes have the capability of producing unique and/or nonequilibrium microstructures. One example is the fine dispersion of oxides in metals, where the oxide particle size and spacing is very small (100-500 Å). Alternately, metals and alloys deposited at high substrate temperatures have properties similar to those of conventionally fabricated (cast, worked and heat-treated) metals and alloys.

UNIQUE FEATURES OF DEPOSITED MATERIALS AND GAPS IN UNDERSTANDING

It would be useful to state at this point some of the unique features of materials produced by deposition technologies. They are:

(1) Extreme versatility of range and variety of deposited materials.

(2) Overlay coatings with properties independent of the thermodynamic compositional constraints.

(3) Ability to vary defect concentration over wide limits, thus resulting in a range of properties comparable to, or far removed from conventionally fabricated materials.

(4) High quench rates available to deposit amorphous materials.

(5) Generation of microstructures different from conventionally processed materials, e.g., a wide range of microstructures—ultrafine (submicron grain or laminae size) to single crystal films.

(6) Fabrication of thin self standing shapes even from brittle materials.

Having listed the unique features, it is necessary to list some of the areas where our understanding of basic processes and phenomena is lacking and which obviously are the areas where research activities are essential. These are:

(1) Microstructure and properties in the range 500 to 10,000 Å—particularly important for submicron microelectronics, reflective surfaces and corrosion.

- (2) (a) Effect of the energy of the depositing species on interfacial interaction, nucleation and growth of deposit.
  - (b) Effect of "substrate surface condition," i.e., contamination (oxide) layers, adsorbed gases, surface topography.
- (3) Residual stresses—influence of process parameters.

CURRENT APPLICATIONS

The applications of coatings in current technology may be classed into the following generic areas:

> *Optically Functional*—Laser optics (reflective and transmitting), architectural glazing, home mirrors, automotive rear view mirrors, reflective and anti-reflective coatings, optically absorbing coatings, selective solar absorbers.
>
> *Electrically Functional*—Electrical conductors, electrical contacts, active solid state devices, electrical insulators, solar cells.
>
> *Mechanically Functional*—Lubrication films, wear and erosion resistant coatings, diffusion barriers, hard coatings for cutting tools.
>
> *Chemically Functional*—Corrosion resistant coatings, catalytic coatings, engine blades and vanes, battery strips, marine use equipment.
>
> *Decorative*—Watch bezels, bands, eyeglass frames, costume jewelry.

A few examples are chosen to illustrate them in greater detail.

Decorative/Functional Coating

Weight reduction is a high priority item to increase gas mileage in automobiles. Therefore, heavy metallic items such as grills are being replaced with lightweight plastic overcoated with chromium by sputtering for the appearance to which the consumer is accustomed.

Another extensive application is aluminum-coated polymer films for heat insulation, decorative and packaging applications.

A rapidly growing application is the use of a "gold-colored" wear resistant coating of titanium nitride on watch bezels, watch bands and similar items.

High Temperature Corrosion

Blades and vanes used in the turbine-end of a gas turbine engine are subject to high stresses in a highly corrosive environment of oxygen, sulfur and chlorine containing gases. A single or monolithic material such as a high temperature alloy is incapable of providing both functions. The solution is to design the bulk alloy for its mechanical properties and provide the corrosion resistance by means of an overlay coating of an M-Cr-Al-Y alloy where M stands for Ni, Co, Fe or Ni + Co. The coating is deposited in production by electron beam evaporation and in the laboratory by sputtering or plasma spraying. With the potential future use of synthetic fuels, considerable research will have to be undertaken to modify such coating compositions for the different corrosive environments as well as against erosion from the particulate matter in those fuels.

Environmental Corrosion

Thick ion plated aluminum coatings are used in various irregularly-shaped parts of aircraft and space-craft as well as on fasteners: (a) to replace electroplated cadmium coatings which sensitize the high-strength parts to hydrogen embrittlement or (b) to prevent galvanic corrosion which would occur when titanium or steel parts contact aluminum or (c) to provide good brazeability.

Friction and Wear

Dry-film lubricant coatings of materials such as gold, MoS_2, WSe_2 and other lamellar materials are deposited on bearings and other sliding parts by sputtering or ion plating to reduce wear. Such dry-film lubricants are especially important for critical parts used in long-lifetime applications since conventional organic fluid lubricants are highly susceptible to irreversible degradation and creep over a long time.

Materials Conservation

Aluminum is continuously coated on a steel strip, 2 feet wide and 0.006 inches thick to a 250 micro-inch thickness in an air-to-air electron-beam evaporator at the rate of 200 feet/minute. The aluminum replaces tin, which is becoming increasingly scarce and costly. The strip then goes to the lacquer line and is used for steel can production.

Cutting Tools

Cutting tools are made of high-speed steel or cemented carbides. They are subject to degradation by abrasive wear as well as by adhesive wear. In the latter mode, the high temperatures and forces at the tool tip promote microwelding between the steel chip from the workpiece and the steel in the high-speed steel tool or the cobalt binder phase in the cemented carbide. The subsequent chip breaks the microweld and causes tool surface cratering and wear. A thin layer of a refractory compound such as TiC, TiN, Al_2O_3 prevents the microwelding by introducing a diffusion barrier. Improvements in tool life by factors of 300 to 800% are possible as well as reductions in cutting forces. The coatings are deposited by chemical vapor deposition or physical vapor deposition. Some idea of the importance of such coatings can be assessed from the fact that the yearly value of cutting tools purchased in the U.S. is $1 billion and the cost of machining is approximately $60 billion.

Nuclear Fuels

Pyrolytic carbon is deposited on nuclear fuel particles used in gas-cooled reactors by chemical vapor deposition in fluidized beds. The coating retains the fission products and protects the fuel from corrosion.

Biomedical Uses

Parts for implants such as heart valves are made of pyrolytic carbon by CVD techniques. Metal parts are coated with carbon by ion plating in order to obtain biological compatability.

"FRONTIER AREAS" FOR THE APPLICATION OF THE PRODUCTS OF DEPOSITION TECHNOLOGY

(1) Reflective surfaces, e.g., for laser mirrors.

(2) Thermal barrier coatings for blades and vanes operating at high temperatures.

(3) Corrosion/erosion resistant coatings at high temperatures, e.g., valves and other critical compounds in coal gasification plants.

(4) Advanced cutting tools.

(5) Wear resistant surfaces without organic lubricants, particularly at high temperatures where lamellar solid state lubricants such as MoS_2 are ineffective.

(6) "First wall" of thermonuclear reactor vessels.

(7) High strength-high toughness ceramics for structural use.

(8) Ultrafine powders.

(9) Superconducting materials

High transition temperatures $>23.2°K$.

Fabricability of these brittle materials into wire or ribbons.

(10) Catalytic materials.

(11) Thin film photovoltaic devices.

(12) Transparent conductive coatings in opto-electronics devices, photodetectors, liquid crystal and electrochromic displays, solar photothermal absorption devices, heat mirrors.

(13) Biomedical devices, e.g., neurological electrodes, heart valves, artificial organs.

(14) Materials conservation.

(15) Sub-micron microelectronic devices. In this context, a good question is how far can dimensions be reduced without running into some limit imposed by physical phenomena.

SELECTION CRITERIA

The selection of a particular deposition process depends on several factors. They are:

(1) The material to be deposited;

(2) Rate of deposition;

(3) Limitations imposed by the substrate, e.g., maximum deposition temperature;

(4) Adhesion of deposit to substrate;

(5) Throwing power;

Deposition Technologies: An Overview   13

(6) Purity of target material since this will influence the impurity content in the film;

(7) Apparatus required and availability of same;

(8) Cost;

(9) Ecological considerations;

(10) Abundance of deposition material in the world.

In order to aid the reader in the task of selection, Table 1.2 lists several criteria for each of the processes. It is obvious that there are very few techniques which can deposit all types of materials. It is also impossible to detail the advantages and limitations of each of the techniques. However, in the evaluation of each application, the above factors will lead to a rational choice of the deposition technique to be used.

SUMMARY

In the above discussion, we have noted the following:

(1) There are a very large number of deposition techniques.

(2) There is no unique way to classify these techniques. Depending on the viewpoint, the same technique may fall into one or more classes.

(3) Each technique has its advantages and limitations.

(4) The choice of the technique to be used depends on various selection criteria which have been given above.

(5) More than one technique can be used to deposit a given film as shown in Figure 1.2 below from Campbell's article[4] on preparation methods in microelectronic fabrication.

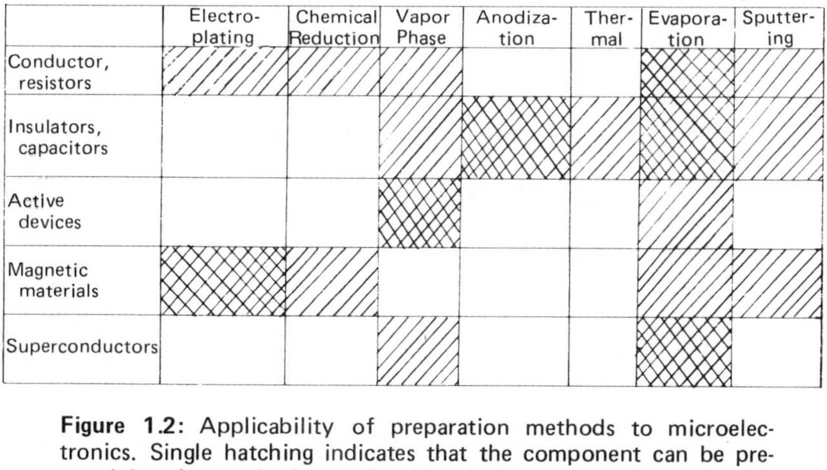

	Electro-plating	Chemical Reduction	Vapor Phase	Anodization	Thermal	Evaporation	Sputtering
Conductor, resistors	▨	▨	▨			▨	▨
Insulators, capacitors			▨	▨		▨	▨
Active devices			▨				▨
Magnetic materials	▨	▨				▨	▨
Superconductors			▨			▨	

Figure 1.2: Applicability of preparation methods to microelectronics. Single hatching indicates that the component can be prepared by the method; crosshatching indicates that the method is widely used.

Table 1.2: Some Characteristics of Deposition Processes

	Evaporation	Ion Plating	Sputtering	Chemical Vapor Deposition	Electro-deposition	Thermal Spraying
Mechanism of production of depositing species	Thermal energy	Thermal energy	Momentum transfer	Chemical reaction	Deposition from solution	From flames or plasmas
Deposition rate	Can be very high (up to 750,000 Å/min)	Can be very high (up to 250,000 Å/min)	Low except for pure metals (e.g., Cu—10,000 Å/min)	Moderate (200–2,500 Å/min)	Low to high	Very high
Depositing specie	Atoms and ions	Atoms and ions	Atoms and ions	Atoms	Ions	Droplets
Throwing power for:						
a. Complex shaped object	Poor line-of-sight coverage except by gas scattering	Good, but nonuniform thickness distributions	Good, but nonuniform thickness distribution	Good	Good	No
b. Into small blind holes	Poor	Poor	Poor	Limited	Limited	Very limited
Metal deposition	Yes	Yes	Yes	Yes	Yes, limited	Yes
Alloy deposition	Yes	Yes	Yes	Yes	Quite limited	Yes
Refractory compound deposition	Yes	Yes	Yes	Yes	Limited	Yes
Energy of deposit species	Low ≈0.1 to 0.5 eV	Can be high (1–100 eV)	Can be high (1–100 eV)	Can be high with plasma-aided CVD	Can be high	Can be high
Bombardment of substrate/deposit by inert gas ions	Not normally	Yes	Yes or no depending on geometry	Possible	No	Yes
Growth interface perturbation	Not normally	Yes	Yes	Yes (by rubbing)	No	No
Substrate heating (by external means)	Yes, normally	Yes or no	Not generally	Yes	No	Not normally

APPENDIX: DEPOSITION PROCESS DEFINITIONS

The definitions of various deposition processes are given below. They are grouped as proposed by Chapman and Anderson[1] and many of them are those proposed by these authors.

Conduction and Diffusion Processes

Electrostatic Deposition is the desposition of material in liquid form, the solvent used then being evaporated to form a solid coating. At the source, the liquid is atomized and charged, and then can be directed onto the substrate using an electrostatic field.

Electrophoretic Coating produces a coating on a conducting substrate from a dispersion of colloidal particles. The article to be coated is immersed in an aqueous dispersion which dissociates into negatively charged colloidal particles and positive cations. An electric field is applied with the article as anode (positive electrode); the colloidal particles are transported to the anode, where they are discharged and form a film. In the case of a paint coating, this requires subsequent curing, which further shows that electrophoresis itself is not a very effective transport process, so that electrodeposition may be a better term for the coating process.

Electrolytic Deposition is primarily concerned with the deposition of ions rather than of colloidal particles. Two electrodes are immersed in an electrolyte of an ionic salt which dissociates in aqueous solution into its constituent ions; positive ions are deposited onto the cathode (negative electrode).

Anodization is a process which occurs at the anode (hence its name) for a few specific metals. The anode reacts with negative ions from the electrolyte and becomes oxidized, i.e., it forms a surface coating.

Gaseous Anodization is a process in which the liquid electrolyte of the conventional wet process is replaced by a glow discharge in a low partial pressure of a reactive gas. Oxides, carbides and nitrides can be produced this way.

Ion Nitriding is a gaseous anodization to produce nitride diffusion coating on a metal surface usually steel.

Ion Carburizing is gaseous anodization to produce a carbide diffusion coating on a metal surface usually steel.

Plasma Oxidation is gaseous anodization to produce an oxide film on the surface of a metal, e.g., SiO_2 films on Si.

Diffusion Coating is produced by diffusion of a material from the surface into the bulk of the substrate.

Metalliding is a method using electrodeposition in molten fluorides.

Spark-hardening is a technique in which an arc is periodically struck between a vibrating anode and the conducting substrate (cathode); material is transferred from the anode and diffuses into the substrate.

Chemical Processes

Conversion and Conversion/Diffusion Coating is a process in which the substrate is reacted with other substances (which may be in the form of solids, liquids or gases) so that its surface is chemically converted into different compounds having different properties. (Anodization could probably be described as an electrochemical conversion process). Conversion coating usually takes place at elevated temperatures and diffusion is often an essential feature.

Chemical Vapor Deposition (CVD) is a chemical process which takes place in the vapor phase very near the substrate or on the substrate so that a reaction product is deposited onto the substrate. The deposition can be a metal, semiconductor, alloy or refractory compound.

Pyrolysis is a particular type of CVD which involves the thermal decomposition of volatile materials on the substrate.

Plasma Assisted CVD is a process where the reaction between the reactants is stimulated or activated by creating a plasma in the vapor phase using means such as RF excitation from a coil surrounding the reaction vessel.

Electroless Deposition is often described as a variety of electrolytic deposition which does not require a power source or electrodes, hence its name. It is really a chemical process catalyzed by the growing film, so that the electroless term is somewhat a misnomer.

Disproportionation is the deposition of a film or crystal in a closed system by reacting the metal with a carrier gas in the hotter part of the system to form the compound followed by dissociation of the compound in the colder part of the system to deposit the metal. Examples are epitaxial deposits of Si or Ge on a single crystal substrate and the Van-Arkel-deBoer process for metal purification and crystal growth.

Wetting Processes

Wetting Processes are coating processes in which material is applied in liquid form and then becomes solid by solvent evaporation or cooling.

Conventional Brush Painting and *Dip Coating* are wetting processes in which the part to be coated is literally dipped into a liquid, e.g., paint, under controlled conditions of, for example, withdrawal rate and temperature.

Hydrophilic Method is a surface chemical process known as the Langmeir Blodgett technique which is used to produce multimonolayers of long chain fatty acids. A film 25 Å thick can be deposited on a substrate immersed in water and pulled through a compressed layer of the fatty acid on the surface of the water. The process can be repeated to build up many layers.

Welding Processes are a range of coating techniques all of which rely on wetting.

Spraying Processes

Printing Process also relys on wetting and is a process in which the ink, conventionally pigment is a solvent, is transferred to and is deposited on a paper or other substrate, usually to form a pattern; the solvent evaporates to leave the required print.

Spraying Processes can be considered in two categories: (1) macroscopic in which the sprayed particle consists of many molecules and is usually greater than 10 μm in diameter; (2) microscopic in which the sprayed particles are predominantly single molecules or atoms.

Air and Airless Spraying are the first of the macroscopic processes. When a liquid exceeds a certain critical velocity, it breaks up into small droplets, i.e., it atomizes. The atomized droplets, by virtue of their velocity (acquired from a high pressure air or airless source) can then be sprayed onto a substrate).

Flame Spraying is a process in which fine powder (usually of a metal) is carried in a gas stream and is passed through an intense combustion flame, where it becomes molten. The gas stream, expanding rapidly because of the heating, then sprays the molten powder onto the substrate where it solidifies.

Detonation Coating is a process in which a measured amount of powder is injected into what is essentially a gun, along with a controlled mixture of oxygen and acetylene. The mixture is ignited, and the powder particles are heated and accelerated to high velocities with which they impinge on the substrate. The process is repeated several times a second.

Arc Plasma Spraying is a process in which the powder is passed through an electrical plasma produced by a low voltage, high current electrical discharge. By this means, even refractory materials can be deposited.

Electric-Arc Spraying is a process in which an electric arc is struck between two converging wires close to their intersection point. The high temperature arc melts the wire electrodes which are formed into high velocity molten particles by an atomizing gas flow; the wires are continuously fed to balance the loss. The molten particles are then deposited onto a substrate as with the other spray processes.

Harmonic Electrical Spraying is a process in which the material to be sprayed must be in liquid form, which will usually require heating. It is placed in a capillary tube and a large electrical field is applied to the capillary tip. It is found that by adding an a.c. perturbation to the d.c. field, a collimated beam of uniformly sized and uniformly charged particles is emitted from the tip. Since these particles are charged, they could be focused by an electrical field to produce patterned deposits.

Evaporation is a process in which the boiling is carried out in vacuum where there is almost no surrounding gas, the escaping vapor atom will travel in a straight line for some considerable distance before it collides with something, for example, the vacuum chamber walls or substrate.

Glow Discharge Evaporation and Sputtering are processes in soft vacuum (10^{-2} to 10^{-1} torr) operating in the range $10^{-1} < pd < 10^{-2}$ torr cm where p is the pressure and d is the cathode fall dimension).

Molecular Beam Epitaxy is an evaporation process for the deposition of compounds of extreme regularity of layer thickness and composition from well controlled deposition rates.

Reactive Evaporation is a process in which small traces of an active gas are added to the vacuum chamber; the evaporating material reacts chemically with the gas so that the compound is deposited on the substrate.

Activated Reactive Evaporation (ARE) is the Reactive Evaporation Process carried out in the presence of a plasma which converts some of the neutral atoms into ions or energetic neutrals thus enhancing reaction probabilities and rates to deposit refractory compounds.

Biased Activated Reactive Evaporation (BARE) is the same process as Activated Reactive Evaporation with substrate held at a negative bias voltage.

Sputter Deposition is a vacuum process which uses a different physical phenomenon to produce the microscopic spray effect. When a fast ion strikes the surface of a material, atoms of that material are ejected by a momentum transfer process. As with Evaporation, the ejected atoms or molecules can be condensed on a substrate to form a surface coating.

Ion Beam Deposition is a process in which a beam of ions generated from an ion beam gun impinge and deposit on the substrate.

Cluster Ion Beam Deposition is an ion beam deposition in which atomic clusters are formed in the vapor phase and deposit the substrate.

Ion Plating is a process in which a proportion of the depositing material from an evaporation, sputtering or chemical vapor source is deliberately ionized. Once charged this way, the ions can be accelerated with an electric field so that the impingement energy on the substrate is greatly increased, producing modifications of the microstructure and residual stresses of the deposit.

Reactive Ion Plating is ion plating with a reactive gas to deposit a compound.

Chemical Ion Plating is similar to Reactive Ion Plating but with a substitution of stable gaseous reactants instead of a mixture of evaporated atoms and reactive gases. In most cases, the reactants are activated before they enter the plasma zone.

Ion Implantation is very similar to ion plating, except that now all of the depositing material is ionized, and in addition the accelerating energies are much higher. The result is that the depositing ions are able to penetrate the surface barrier of the substrate and be implanted *in* the substrate rather than *on* it.

Plasma Polymerization is a process in which organic and inorganic polymers are deposited from monomer vapor by the use of electron beam, ultraviolet radiation or glow discharge. Excellent insulating films can be prepared in this way.

REFERENCES

(1) B.N. Champman and J.C. Anderson (eds.), *Science and Technology of Surface Coating*, Academic Press (1974).
(2) K.D. Mittal (ed.), *Adhesion Measurement of Thin Films, Thick Films and Bulk Coatings*, Am. Soc. for Testing Materials (1978).
(3) K.K. Yee, "Protective Coatings for Metals by Chemical Vapor Deposition," *International Metal Reviews*, No. 226, publishers, The Metals Society and American Society for Metals (1978).
(4) D.S. Campbell, "The Deposition of Thin Films by Chemical Methods," Chapter 5, from *Handbook of Thin Film Technology*, eds. L. Maissel and R. Glang, McGraw-Hill (1970).
(5) L. Maissel and R. Glang (eds.), *Handbook of Thin Film Technology*, McGraw-Hill (1970).
(6) S. Schiller, O. Heisig and K. Goedick, *Proc. 7th Int'l. Vacuum Congress*, R. Dobrozemsky (ed.), Vienna (1977), pg. 1545.
(7) C. Weissmantel, *ibid*, pg. 1533.
(8) R.F. Bunshah and D.M. Mattox, *Physics Today* (May 1980).
(9) J.A. Thornton, "Designing the Surface to the Job - Thin Film Application," *Proc. 19th National SAMPE Symposium* - Buena Park, CA. (April 23-25, 1974).
(10) B.A. Movchan and A.V. Demchishin, *Phys. Met. Metallogr. 28*, 83 (1969).

2

Plasmas in Deposition Processes

John A. Thornton

Telic Corporation
Santa Monica, California

1. INTRODUCTION

A glow discharge plasma can be defined as a region of relatively low pressure and low temperature gas in which a degree of ionization in a quasineutral state is sustained by the presence of energetic electrons. The character of such a plasma is a consequence of the mass difference between the electrons and the ions. When an electric field is applied to an ionized gas, energy is transferred more rapidly to the electrons than to the ions. Furthermore, the transfer of kinetic energy from an electron to a heavy particle (atom, molecule or ion) in an elastic collision is proportional to the electron-to-heavy particle mass ratio and therefore very small ($\sim 10^{-5}$). Consequently, at low pressures (low collision frequency) the free electrons can accumulate sufficient kinetic energy so that they have a high probability of producing excitation or ionization during the collisions that they do make with the heavy particles. The production of these excited species, and their interaction with surfaces and growing coatings, is the reason that low pressure glow discharge plasmas are assuming an ever-increasing role in materials processing. The following processes in which plasmas are of vital importance will be discussed.

- Sputtering
- Activated reactive evaporation
- Ion plating
- Plasma-assisted chemical vapor deposition
- Plasma-assisted etching
- Plasma polymerization

The purpose of this chapter is to review some of the features of glow discharge plasmas that are of importance in understanding the role of these plasmas in the above processes and to summarize some of the analytical relationships that are used in making engineering calculations.

2. PARTICLE MOTION

2.1 Mean Free Path and Collision Cross Sections

A glow discharge plasma can be viewed as a medium in which electrical energy is passed via an electric field to a gas. The energetic gas particles are then used to promote a chemical reaction or to interact with a surface in such a way as to produce a desirable effect such as sputtering. Thus the process of energy exchange during collisions involving the plasma particles is of vital importance.

The probability that a given type of collision will occur under given conditions is often expressed in terms of a collision cross section. A related parameter is the mean free path or average distance traversed by particles of given type between collisions of a specified type. The mean free path λ and collision cross section σ are generally defined by a simple relationship which treats the particles as impenetrable spheres. Thus the mean free path for electrons passing through a gas of particle density N, and producing a given type reaction (A) during collisions, is given by

$$(1) \qquad \lambda_A = \frac{1}{N\sigma_A} .$$

The atoms are imagined to be opaque spheres with projected cross sectional area σ_A. Thus if σ_A is the cross section for momentum exchange, then every time an electron trajectory is blocked out by one of these spheres, the electron loses all of its momentum.

The total collision cross section can be written as

$$(2) \qquad \sigma_t = \sigma_e + \sigma_{ex} + \sigma_{ion} + \sigma_a + \sigma_{oth} ,$$

where the subscripts e, ex, ion, a, and oth characterize the particular types of collisions, namely, elastic or momentum exchange, excitation, ionization, attachment, and other processes, respectively.

Figure 2.1 shows the cross sections for electrons interacting with argon gas. The cross sections are typically a strong function of the energy of the colliding species. For the case of electrons colliding with gas atoms, the kinetic energy of the gas atoms is generally much less than that of the electrons, so that it can be neglected. Consequently, only the electron energy is shown in Figure 2.1. Referring to the figure, it is seen that at low electron energies the primary collision process is momentum exchange ($\sigma_t \sim \sigma_e$), while at energies considerably larger than the ionization potential (15.75 eV for Ar), the primary process is ionization ($\sigma_t \sim \sigma_{ion}$).

Cross sections are most easily measured for reactions involving a species such as an electron or ion which can conveniently be formed as an energetic beam and passed through a stationary gas of the second species. Thus the cross section formulation is the one which is generally used for describing the interactions of electrons and ions in a plasma. Figure 2.2 shows the cross section for energetic O^+ ions passing through N_2 and producing the reaction $O^+ + N_2 \rightarrow NO^+ + N$. Note in comparing Figures 2.1 and 2.2 that the collision cross sections are typically a few x 10^{-16} cm^2 in magnitude (i.e., a few Angstroms in diameter).

For collision types that cannot be investigated in beam experiments, the cross sections are often deduced from measurements of macroscopic parameters such as viscosities, diffusion coefficients, and chemical reaction rates.[3] Thus one finds reference to viscosity cross sections, diffusion cross sections, etc. Cross sections are primarily of interest in making comparisons based on kinetic theory. In most plasma engineering calculations the macroscopic rate parameters are used directly if they are available.

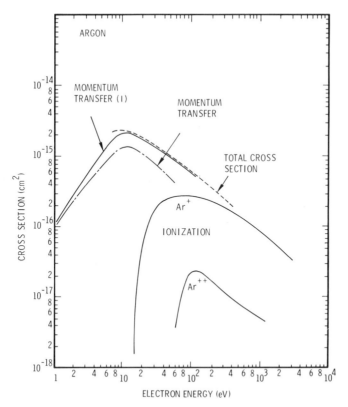

Figure 2.1: Collision cross sections for electrons in argon gas. From Reference 1. (Used courtesy Academic Press.)

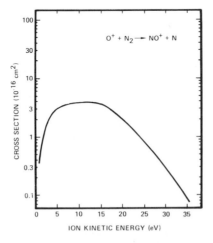

Figure 2.2: Example of cross section derived from molecular ion beam passing through molecular gas. Data from Reference 2.

2.2 The Free Electron Kinetic Energy in a Plasma

Consider a plasma electron in an electric field $\vec{E}$. Between collisions with the gas particles the electron will gain an energy from the electric field W_f that is equal to the force on the electron eE (where e is the electronic charge) times the distance that it moves in the electric field. This distance can be approximated by the mean free path so that $W_f = eE\lambda$ (we are now treating an average electron). In the steady state case this energy gain must be balanced against the energy loss in an average collision. We neglect inelastic collisions for the moment and consider collisions with heavy particles in which the electrons lose all of their momentum—i.e., are deflected by 90°. This permits us to use the momentum exchange cross section, as defined in the preceding section, in estimating λ. Application of the conservation of energy and momentum shows that loss of electron energy in such a collision is[4]

$$(3) \qquad \Delta W = \left(\frac{2 m_e}{m_H}\right)(W_e - W_H) ,$$

where m_e and m_H are the electron and heavy particle masses and W_e and W_H are the initial electron and heavy particle energies before the collision. Equating ΔW to the energy W_f gained from the electric field and using Eq 1 for λ yields

$$(4) \qquad W_e - W_H = \tfrac{1}{2}\left(\frac{m_H}{m_e}\right)\left(\frac{eE}{N\sigma_e}\right) .$$

In making calculations dealing with plasmas it is useful to note that:

- $m_e = 9.11 \times 10^{-31}$ Kg $= 9.11 \times 10^{-28}$ g.
- $m_H = 1.67 \times 10^{-24} \times$ (atomic mass number) g.
- $N = 3.2 \times 10^{24}$ particles/m³ $= 3.2 \times 10^{16}$ particles/cm³ at 1 Torr and 300°K (27°C).
- The electron volt (eV) is the unit of energy generally used in plasma calculations. One electron volt is the energy gained by a particle with unit charge which is accelerated in an electric field produced by a potential difference of one volt. (1 eV = 1.602 $\times 10^{-19}$ joules = 11,600°K).

Consider the case of electrons in an Ar plasma at 1 Torr and 300°K which is subjected to an electric field of 1 V/cm. Thus $N = 3.2 \times 10^{16}$ cm⁻³ and eE is 1 eV/cm. Using $\sigma_e \sim 10^{-15}$ cm² from Fig. 2.1 yields from Eq (1) that $W_e - W_H \sim 10^3$ eV. Thus the steady state condition implies that the electron energy will be much greater than that of the gas atoms (0.03 eV at 300°K). The actual electron energy will not reach 10^3 eV, because inelastic collisions will become important when the electron energy exceeds about 10 eV. However, the above analysis shows that even relatively weak electric fields can cause the electron kinetic energies in low pressure glow discharge plasmas to be elevated above the gas atom energies until they are finally "clamped" by losses due to inelastic collisions.

Figure 2.3 shows this elevation of electron energy at low pressures for the case of plasma arcs. (The energies here are expressed as temperatures; this will be discussed later.) At high pressures the electron-gas atom collisions are so frequent that the gas temperature becomes elevated. Such high pressure arcs are used for a variety of applications. However, our discussion will be limited to the low pressure case where $T_e \gg T_g$. The significant thing about this situation is that the energetic electrons can produce high temperature chemistry in a gas at low temperatures.[6]

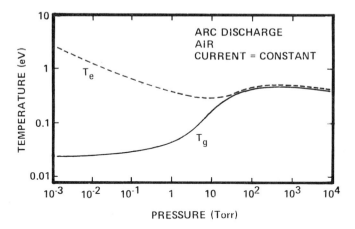

Figure 2.3: Dependence of electron temperature and gas temperature on pressure in electric arc. Data from Reference 5.

2.3 Electron Energy Distribution Function

For most purposes the state of a glow discharge plasma can be characterized by the densities of heavy particles ($[N_i]$, where i is species) the electron, density n_e, and the electron energy distribution function $F(E)$.[7] Under conditions of local thermodynamic equilibrium,[8] when the forward and reverse rates for all the electron energy exchange processes are equal (state of detailed balance[9]), the electrons will be in a Maxwellian velocity distribution and their state can be defined by an electron temperature T_e. Unfortunately, such a state of equilibrium seldom exists in a glow discharge plasma.

Figure 2.4 schematically illustrates the electron energy distribution function. The equilibrium energy distribution is also shown for comparison. The effect of an electric field is to shift electrons to higher energies and therefore to overpopulate the high energy region relative to the Maxwellian distribution. The cross section for a representative inelastic collision is shown superimposed (see Figure 2.1). Electrons undergoing inelastic collisions are transferred from the high energy to the low energy end of the distribution. Electron-electron collisions tend to smooth the distribution and drive it toward the Maxwellian form. If these collisions dominate so that a state of detailed balancing exists for one dominant process, then $F(E)$ can be approximated by a Maxwellian distribution and an electron temperature can be used to describe the state of the electrons. However, even this case seldom occurs in practice. In high pressure discharges the electric field perturbation is usually minimal, so that the distribution function approximates the Maxwellian form, although it may be somewhat depleted at high energies by inelastic collisions. In low pressure discharges, the electric field can generate relatively large numbers of energetic electrons and, in the extreme, produce a two-group or bi-modal distribution function. This is the case in low pressure negative glow discharges of the type that are important in sputtering.[7]

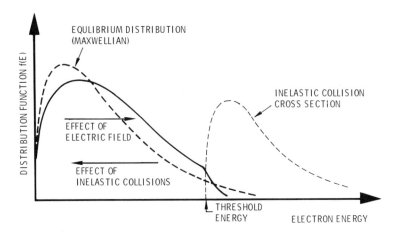

Figure 2.4: Schematic illustration of electron energy distribution function and factors that affect its form. (Used courtesy Telic Corporation.)

Electron energy distribution functions are usually measured by electrostatic analyzer and probe methods. However, these methods are difficult, particularly in material processing plasmas, because of the basic tendency of these plasmas to interact with probe surfaces. Therefore electron temperatures, although not strictly valid, are generally assumed in making engineering calculations. This approach will be used in these notes.

2.4 Processes Involving the Collision Frequencies

The collision frequency is one of the most important of the plasma parameters. It is defined as the rate at which an average particle of a particular type undergoes collisions of a specified type. Thus the total electron-atom collision frequency is the rate at which an average electron in a plasma under given conditions undergoes collisions of all types with gas atoms.

The general expression for the collision frequency is rather complex and involves the distribution functions of the colliding species.[10] For the electron-heavy particle case, the velocity of the heavy particles can be neglected, so that the frequency of collisions of some particular type (A) is given by

$$(5) \quad \nu_A = N \int_{E=0}^{E=\infty} (E/2m_e)^{1/2} \, \sigma_A(E) \, F(E) \, dE \ .$$

If the collision cross section $\sigma_A(E)$ is assumed to be independent of energy, and the electrons are assumed to have a Maxwellian velocity distribution at an electron temperature T_e, then Eq (5) reduces to

$$(6) \quad \nu_A = N \sigma_A \nu_e$$

where ν_e is the average electron speed.

$$\text{(7)} \qquad v_e = \left(\frac{8kT_e}{\pi m_e}\right)^{1/2}$$

In Eq (7) k is the Boltzmann constant. It is customary to write kT_e in units of eV.* Thus Eq (7) becomes

$$\text{(8)} \qquad v_e = (6.7 \times 10^7)\sqrt{kT_e(eV)} \text{ cm/sec}.$$

When Eq (6) is used in making approximate estimates, σ is generally approximated at the electron energy kT_e.

The electron-electron and electron-ion collision frequencies are of special interest. These are given by[11]

$$\text{(9)} \qquad v_{ee} = (3 \times 10^{-6})n_e \frac{\ln \Lambda}{[kT_e(eV)]^{3/2}} \text{ sec}^{-1},$$

and

$$\text{(10)} \qquad v_{ei} = (1.5 \times 10^{-6})n_e \frac{\ln \Lambda}{[kT_e(eV)]^{3/2}} \text{ sec}^{-1},$$

with

$$\text{(11)} \qquad \lambda_{ee} = (4.5 \times 10^{13}) \frac{[kT(eV)]^2}{n_e \ln \Lambda} \text{ cm},$$

where $\ln \Lambda$ is a weak function of kT_e and n_e. $\ln \Lambda$ is tabulated in most books on plasma physics, and has a value of approximately 10 for the glow discharge plasmas of interest here.[12] The $\ln \Lambda$ term arises because these collisions involve long range Coulomb forces, and the cross sections do not cut off as in the hard sphere case.

The primary use of Eqs (9) and (10) is in comparing v_{ee} and v_{ei} with the other collision frequencies that may be present. A plasma for which $v_{ee} \gg v_{eA}$, where v_{eA} is the electron atom collision frequency for elastic collisions, is said to be Coulomb-dominated. An approximate condition for Coulomb domination is easily derived from Eqs 6 and 9. (See Ref. 13.)

$$\text{(12)} \qquad \frac{n_e}{N} \gg \alpha_c = (2.23 \times 10^{13}) \frac{\sigma_{eA}[kT(eV)]^2}{\ln \Lambda}$$

The term α_c is known as the critical degree of ionization (see Section 3.1). Selecting $kT_e \sim 3eV$ and $\sigma_{eA} \sim 10^{-15}$ cm^{-3} (Figure 2.1) yields $n_e/N \sim 0.02$. Thus a moderate temperature glow discharge plasma with 20% ionization can be domi-

---

* From kinetic theory the average energy in one dimension is ½ kT. The average energy in three dimensions is 3/2 kT. Since T and E_{ave} are so closely related, it is customary in plasma physics to give temperatures in units of eV. To avoid confusion on the number of dimensions involved, it is not E_{ave} but the energy corresponding to kT that is used to denote the temperature.[11] Thus from kT = E one has 1eV = 11,600°K. By a 2eV plasma we mean that kT = 2eV, although it is realized that the actual average energy in three dimensions is 3/2 kT or 3eV.

nated by Coulomb collisions. Such high degrees of ionization are very uncommon in glow discharge plasmas.

A consequence of Coulomb domination is seen by examining Eqs 9 and 11. As the electron energy is increased, the electron collision frequency decreases and the mean free path increases. Thus electrons in an electric field will find that their energy gain is "unchecked" by collisions. Electron runaway is an important consideration in highly ionized plasmas[13] but seldom important in glow discharge plasmas because of inelastic collisions.

For the case of a heavy particle of mass M_1 in a gas with density N_2 of heavy particles of mass m_2, the collision frequency can be approximated by an equation very similar to Eq (6),[4,14] which can be written as

$$(13) \quad \nu_{12} \approx (2.5 \times 10^5) \frac{\sigma_{12} N_2}{(M^*)^{1/2}} \left(\frac{T}{300}\right)^{1/2},$$

where the cross section σ_{12} is assumed to be independent of the velocity of impact and all the heavy particles are at the common temperature T. In Eq (13)

$$(14) \quad M^* = \frac{M_1 M_2}{M_1 + M_2}$$

where the masses are molecular weights expressed in grams.

2.4.1 Reaction Rates: The gas-phase reaction rate is directly proportional to the collision frequency. For a process (A) involving electron collisions, one has

$$(15) \quad R_A = n_e \nu_A \quad \frac{\text{reactions}}{\text{cm}^3\text{-sec}}.$$

If the electrons are assumed to have a Maxwellian velocity distribution at a temperature T_e, and if the cross section for a given reaction is approximated by a step function of magnitude σ_0 and threshold energy E_0 as shown in Figure 2.5, then the reaction rate is given by

$$(16) \quad R_A = n_e N \sigma_0 v_e \left(1 + \frac{E_0}{kT_e}\right) \exp\left(-\frac{E_0}{kT_e}\right).$$

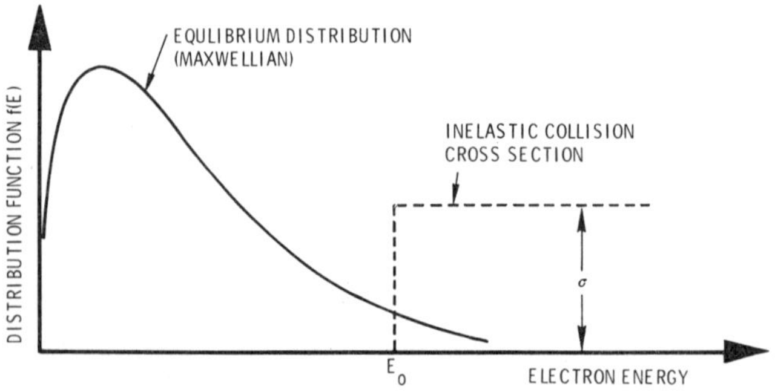

Figure 2.5: Reaction rate approximation for Maxwellian velocity distribution. (Used courtesy Telic Corporation.)

As a general rule, reaction rate constants rather than collision frequencies per se are measured and used for reactions involving heavy particle collisions. Thus, for a reaction occurring via a two body collision between species A and B in a gas at temperature T, with rate constant k(T), one has

(17) $$R = k(T) N_A N_B \ .$$

2.4.2 Mobilities: The plasma transport properties are dependent on the frequency ν_m of elastic (momentum exchange) collisions.

The mobility μ_j relates the electric field-driven drift velocity V_d of a given charged particle species (j) to the strength of the field E:

(18) $$V_{jd} = \mu_j E \ .$$

When the collision frequency is sufficiently large so that the drift velocity is small compared to the thermal velocity, one has

(19) $$\mu_j = \frac{1.6 \times 10^{-12}}{M_j \nu_m} , \ \frac{cm^2}{V\text{-sec}} \ ,$$

where M_j is the particle mass in grams.

The mobility description is generally used to describe the drift of ions through a plasma that is at a sufficiently high pressure to satisfy the collision frequency requirement. Mobilities for several gases of interest are given in Table 2.1.

Table 2.1: Mobilities of Ions in their Own Gas (From Reference 15)

.... 1 Torr and 0°C............

Ion-Gas	Mobility (cm^2/V-sec)
He$^+$-He	8,000
Ne$^+$-Ne	3,300
Ar$^+$-Ar	1,200
Kr$^+$-Kr	690
H$_2^+$-H$_2$	10,000
N$_2^+$-N$_2$	2,000
O$_2^+$-O$_2$	1,000
CO$_2^+$-CO$_2$	730

When a positive ion collides with a gas molecule or atom, two processes can occur. First, the ion and molecule can exchange momentum and energy in a "conventional" collision in which the particles preserve their "identity." Second, an exchange of charge can occur. For example, fast ions moving through a gas can engage in collisions where the ion extracts an electron from a gas atom, with the result that the fast ion becomes a fast neutral atom, while the "slow" atom becomes a slow positive ion. Charge exchange is particularly important for ions of low energy passing through their own gas (resonant charge exchange). Under these conditions the charge transfer cross section is about one half of the total cross section[16] and therefore contributes significantly to determining the value of the mobility.* Charge transfer is very important to high pressure sputtering and ion plating discharges.

---
* The charge exchange region surrounding an atom can be considered as a sphere inside of which the probability of charge transfer is ½ and outside of which it is zero. As the ion approaches the atom, it will simply be deflected by the dipole interaction if the distance of closest approach is greater than the sphere radius. If the ion enters the charge exchange sphere, half the time it emerges as a neutral and half the time as an ion; i.e., the electron is "up for grabs" between the two particles.[17]

28 Deposition Technologies for Films and Coatings

2.4.3 Electrical Conductivity and Diffusion Coefficients: The electrical conductivity σ is just $eN\mu$, so that

$$(20) \qquad \sigma_j = \frac{1}{\eta_j} = (2.6 \times 10^{-31}) \frac{N_j}{m_j \nu_m}, \quad (\Omega\text{-cm})^{-1},$$

where N_j is the particle density in cm^{-3} and m_j is the mass in grams of the current carrier. The resistivity η is often used to avoid confusion with σ, which is the common symbol used for both the electrical conductivity and the collision cross section.

The diffusion coefficient D_j relates the particle flux to the concentration gradient. Thus one has

$$(21) \qquad N_j V_{jd} = D_j \left(\frac{dN_j}{dx}\right),$$

where

$$(22) \qquad D_j = \frac{kT}{m_j \nu_m} = (1.6 \times 10^{-12}) \frac{kT(eV)}{m_j \nu_m}, \quad \frac{cm^2}{sec},$$

and m_j is again the particle mass in grams.

2.5 Particles in Magnetic Fields

Charged particle motion in a magnetic field is summarized in Figure 2.6. A charged particle in a uniform magnetic field $\vec{B}$ will orbit a field line as shown in Figure 2.6a and drift along the field with a velocity $V_\parallel$ that is unaffected by the field, as shown in Figure 2.6b. The orbiting frequency is called the gyro or cyclotron frequency and is given by

$$(23) \qquad \omega_c = \frac{eB}{m}.$$

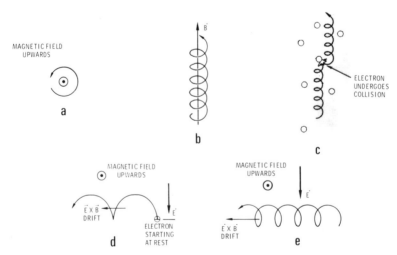

Figure 2.6: Electron motion in static magnetic and electric fields. (Used courtesy Telic Corporation.)

The orbiting radius is called the gyro, cyclotron or Larmor radius and is given by

(24) $$r_g = \frac{m}{e}\left(\frac{v_\perp}{B}\right).$$

Manipulation and confinement of plasma particles by a magnetic field requires that r_g be small compared to the apparatus size. Note in Eq (24) that r_g depends directly on the mass of the particle. Thus very large magnetic fields are required to influence the motions of the plasma ions. When magnetic fields are used with glow discharges, they are made strong enough to influence the energetic plasma electrons, but not the ions. However, magnetically confined electrons in a glow discharge also provide considerable confinement to the plasma ions, since electrostatic forces (see Section 3) prevent the ions from escaping from the electrons. For the electron case, Eqs (23) and (24) become[1]

(25) $$\omega_c = (1.76 \times 10^7)\, B(\text{gauss})\ \text{rad/sec},$$

and

(26) $$r_g = (3.37)\frac{[W_\perp(\text{eV})]^{1/2}}{B(\text{gauss})}\ \text{cm}.$$

Thus, for electrons with an average energy $W_\perp$ of 10 eV and a magnetic field strength B of 100 G, the gyro radius is ~0.1 cm. Magnetic field strengths of from 50 to 500 G are typically used with glow discharge devices.

An electron that is trapped on a given magnetic field line can advance to an adjacent field line by making a collision, as indicated schematically in Figure 2.6c. Collisional diffusion of electrons across magnetic field lines is an important consideration in many glow discharge devices.

When an electric field $\vec{E}$ is present and directed parallel to the magnetic field, the electrons are freely accelerated along the field lines. However, if the electric field has a component $E_\perp$ which is perpendicular to $\vec{B}$, the electrons undergo a drift in a direction perpendicular to both $E_\perp$ and $\vec{B}$, as shown in Figures 2.6d and 2.6e. This motion is known as the $\vec{E} \times \vec{B}$ drift. The $\vec{E} \times \vec{B}$ drift has the cycloidal form shown in Figure 2.6d if the initial electron energy is small compared to that gained on a half revolution from the electric field; it has the more circular form shown in Figure 2.6e if the initial electron energy is large compared to the electric-field-induced variations that occur during the course of the orbit. In both cases the electron drift speed is given by

(27) $$V_E = (10^8)\frac{E_\perp(\text{V/cm})}{B(\text{gauss})}\ \text{cm/sec}.$$

The drift of electrons along a magnetic field line can also be influenced by gradients in the magnetic field. An example of this behavior is shown in Figure 2.7. Electrons moving in such a field tend to conserve the magnetic moment, μ_M, defined by[11]

(28) $$\mu_M = \frac{W_\perp}{B}.$$

Therefore $W_\perp$ must increase as the electrons move in the direction of increasing field strength. Conservation of energy requires that $W_\parallel + W_\perp$ be constant. Therefore $W_\parallel$ must decrease, and the electron may be reflected as indicated in the figure. Pinched-field end confinement of this type is frequently used in glow discharge devices.

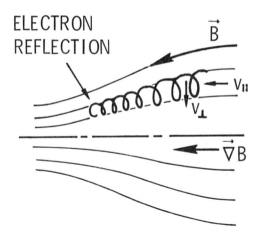

Figure 2.7: Electron reflection in magnetic field gradient. (Used courtesy Telic Corporation.)

3. COLLECTIVE PHENOMENA

3.1 Basic Parameter

Plasmas differ from nonionized gases by their propensity for undergoing collective behavior. Three parameters, derived from the basic plasma properties, N, n_e and kT_e, provide a useful measure of the tendency toward collective behavior.

The Debye length,

$$(29) \qquad \lambda_D = (743) \left[\frac{kT_e(eV)}{n_e(cm^3)} \right]^{1/2} cm \;,$$

provides a measure of the distance over which significant departures from charge neutrality can occur. A plasma cannot exist in a space having a characteristic size that is less than λ_D.

The plasma frequency, ω_p, written here as

$$(30) \qquad f_p = \frac{\omega_p}{2\pi} = (9000) [n_e(cm^{-3})]^{1/2} \; Hz \;,$$

provides a measure of the tendency for electrostatic waves to develop. Waves can form if $\omega_p \gg \nu_m$, where ν_m is the electron collision frequency for momentum exchange (see Section 2.4).

A critical degree of ionization α_c was defined by Eq (12) in Section 2.4. When the degree of ionization $\alpha = n_e/N \gg \alpha_c$, long range Coulomb collisions dominate, and the charged particles behave as though they were in a fully ionized gas. Coulomb domination can occur at degrees of ionization of a few percent for plasmas with low average electron energies (~1eV).

3.2 Plasma Sheaths

Given a gas of particle density $N(cm^{-3})$ and temperature T, the flux of particles passing to an adjacent wall is given by

$$(31) \quad J_w = \frac{N\bar{v}}{4} = \frac{N}{4}\left(\frac{8kT}{\pi m}\right)^{1/2},$$

where m is the particle mass.

For the case of electrons this becomes, see Eq (7) and (8),

$$(32) \quad J_{ew} = (1.67 \times 10^7) n_e \sqrt{kT_e(eV)} \; \frac{\text{particles}}{\text{cm}^2\text{-sec}},$$

which in terms of current is equal to

$$(33) \quad J_e = (2.7 \times 10^{-9}) n_e \sqrt{kT_e(eV)} \; \text{mA/cm}^2 .$$

Thus a typical glow discharge electron density of 10^9 cm^{-3} of 1 eV electrons tends to pass a current of ~3 mA/cm² to an adjacent wall.

For the case of heavy particles such as ions, Eq (31) can be written in the following useful form

$$(34) \quad J_{Hw} = (10^4) N \left(\frac{40}{M}\right)^{1/2} \left(\frac{T}{300}\right)^{1/2},$$

where T is the gas temperature (°K) and M is the species molecular weight. In terms of current Eq (34) becomes

$$(35) \quad J_i = (1.6 \times 10^{-9}) N_i \left(\frac{40}{M}\right)^{1/2} \left(\frac{T}{300}\right)^{1/2} \mu\text{A/cm}^2 .$$

Thus for an Ar sputtering plasma with an ion density of 10^9 cm^{-3} at 300°K, the wall current flux is 1.6 μA/cm².

It is clearly seen by comparing Eqs (33) and (35) that the electrons tend to pass from a plasma to an adjacent wall at a faster rate than the ions. Thus a space charge region in which one species is largely excluded forms adjacent to such surfaces. The potential variation between the surface and the plasma is largely confined to this layer, which is called a sheath. Sheaths are typically several Debye lengths in thickness.

The nature of the sheath will depend on the current density passing across it. Except for cases involving very high current densities to anodes, the space charge region will contain primarily the low mobility ion species. Such sheaths are known as positive space charge sheaths. The function of the sheath is to form a potential barrier, so that the more mobile species, which is the electrons except in the case of a strong magnetic field, are electrostatically reflected. Thus height of the potential barrier associated with a sheath adjusts itself so that the flux of electrons to the wall in question just equals the electron current that is drawn from the wall by the external circuit. If the wall is electrically isolated, the electron flux is reduced to the point at which it is equal to the ion flux.

Figure 2.8 shows a schematic illustration of a typical glow discharge plasma which is in contact with wall surfaces that are either cathodes, anodes, or electrically isolated (floating). The potential V_p is known as the plasma potential. The potential of a floating surface relative to the plasma potential is known as the floating potential V_f. For a Maxwellian velocity distribution, the floating potential is given by[18]

(36) $$V_f = \left(\frac{kT_e(eV)}{2e}\right) \ln\left(\frac{\pi}{2}\frac{m_e}{M}\right)$$

For argon V_f is about $5(kT_e)$. Typical values are -30 to -40V. Thus, when a floating surface is immersed into a plasma, the surface will be bombarded with ions having kinetic energies of up to eV_f.

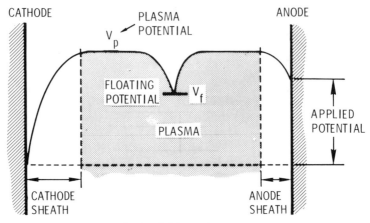

A. LARGE ANODE

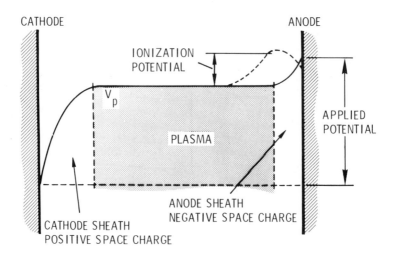

B. SMALL ANODE

Figure 2.8: Schematic illustration of the various types of sheaths that form between a plasma discharge and the surrounding apparatus walls. (Used courtesy Telic Corporation.)

In most cases the glow discharge anodes are large enough so that the current density is less than the thermal current given by Eq (33). In this case there is a positive space charge sheath at the anode, as shown in Figure 2.8A, and the sheath potential drop is between zero and V_f. The potential of a plasma locks onto the most positive surface, provided that the surface is large enough.[19] If the anode area is so small that the current density must exceed the thermal current, then the anode potential will be above the plasma potential, as shown in Figure 2.8B. The local electric field surrounding the anode will draw sufficient electrons to the anode to complete the external circuit.*

A large potential difference, approximately equal to the entire potential applied by the power supply, occurs in the cathode sheath. A cathode sheath is shown schematically in Figure 2.9. The sheath thickness d_s is taken to be the region where the electron density is negligible and where the potential drop V_s occurs. For the low pressure case where the ion mean free path is larger than d_s, the ion current density J_i is related to d_s and V_s by the Child-Langmuir law.[11] It is useful to write this relationship as

$$(37) \qquad J_i = 0.273 \left(\frac{40}{M_i}\right)^{1/2} \left(\frac{V_s^{3/2}}{d_s^2}\right) \text{ mA/cm}^2 ,$$

where V_s is in kV, d_s is in cm, and M_i is the ion molecular weight in grams. Thus, for an Ar sputtering plasma with $V_s = 1$ kV and $d_s = 1$ cm, $J_i = 0.27$ mA/cm². It is difficult to relate J_i to the density N_{io} of ions in the bulk plasma, because there is a quasi-neutral presheath region where a potential drop V_x of the order of ½ (kT_e/e) occurs. For estimates, the presheath density can be assumed to obey a Boltzmann distribution, so that[20]

$$\frac{N_{is}}{N_o} = \exp\left(-\frac{eV_x}{kT_e}\right) ,$$

and

$$(38) \qquad J_i \sim (0.6)eN_{io}\left(\frac{kT_e}{M_i}\right)^{1/2} ,$$

where M_i is in grams and kT_e is in ergs.

For the high pressure case, where collisions are so frequent that the ion drift velocity is of the order of the thermal velocity, a mobility description (Section 2.43) is used for the ion motion.[5] Under this condition

$$(39) \qquad J_i = (9.95 \times 10^{-5})\mu_i \left(\frac{V_s^2}{d_s^3}\right) \text{mA/cm}^2 ,$$

where μ is the ion mobility in cm²/V-sec, V_s is the sheath potential drop in kV, and d_s is the sheath thickness in cm. For an Ar plasma at 1 Torr, $\mu_i = 1{,}200$ cm²/V-sec from Table 2.1. Taking $V = 1$ kV and $d_s = 1$ cm yields $J_i = 0.11$ mA/cm².

---

* The potential rise surrounding a small anode cannot become much larger than the ionization potential of the gas atoms, since this potential causes the sheath electrons to be accelerated. If these electrons gain sufficient energy to produce ionization, then the electrons liberated from the ionization can provide the anode current flow requirement and no additional rise in potential is required.

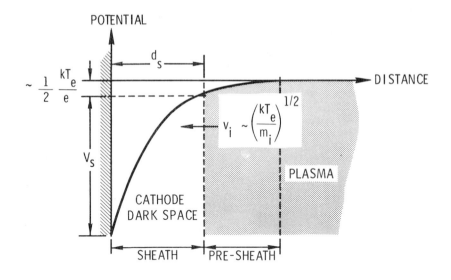

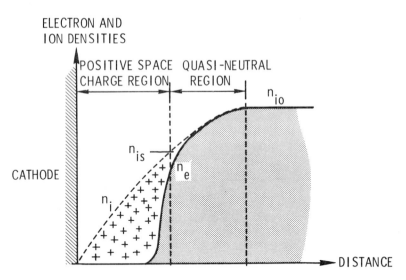

Figure 2.9: Schematic representation of the positive space charge sheath that develops over a cathode. From Reference 1. (Used courtesy Academic Press.)

In a low pressure plasma the ions will fall through the entire sheath potential and bombard the cathode with an energy about equal to V_s. At higher pressures, where charge exchange is important, the bombarding flux will consist of both ions and neutrals having energies considerably less than V_s, as indicated schematically in Figure 2.10. This is an important consideration in sputtering, ion plating, and reactive ion etching, as discussed in Section 6.2.

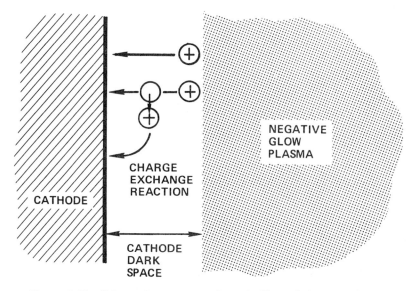

Figure 2.10: Schematic representation of effect of charge exchange reactions on ion flux passing across cathode dark space when ion-atom collisions are important. (Used courtesy Telic Corporation.)

3.3 Ambipolar Diffusion

Consider a plasma within a container having electrically isolated or floating walls. A sheath will develop on these walls such as to reduce the electron flux, so that it is equal to the ion flux as described in Section 3.2. Accordingly, an electric field in the sheath retards the loss of electrons and accelerates the loss of ions. This coupled diffusion is called ambipolar diffusion. The diffusion flux J of electrons or ions to a floating wall is given by

$$(40) \quad J_e = D_a \left(\frac{dn_e}{dx}\right) = J_i = D_a \left(\frac{dn_i}{dx}\right)$$

The term D_a is called the ambipolar diffusion coefficient. Noting that $\mu_e \gg \mu_i$ [see Eq (19)] permits D_a to be approximated as[11]

$$(41) \quad D_a \approx D_i \left(1 + \frac{T_e}{T_i}\right)$$

where D_i is given by Eq (22). Thus the effect of the ambipolar field is to enhance the diffusion of ions by a factor of more than two, but the diffusion rate of the two species together is primarily controlled by the slower species.

In the presence of a sufficiently strong magnetic field, directed perpendicular to the direction of diffusion, the electron mobility, and thus the electron diffusion coefficient, can be reduced to the point where it is lower than the ion diffusion coefficient and therefore rate controlling. Under this condition one can write

$$\text{(42)} \qquad D_a \approx \frac{D_e}{\left(1 + \frac{\omega_c^2}{\nu_m}\right)} \left(1 + \frac{T_i}{T_e}\right),$$

where D_e is the electron diffusion coefficient in the absence of a magnetic field. The effect of the magnetic field becomes strong when ω_c [given by Eq (25)] is much larger than the collision frequency ν_m; i.e., when the electrons are trapped on magnetic field lines as shown in Figure 2.6b, and the collisional hopping to adjacent field lines shown in Figure 2.6c is infrequent. It should be noted that Eq (42) is based on the assumption that electron losses along the field lines can be neglected. Attention to these losses should be given when analyzing the performance of an actual plasma device.[11,21]

3.4 Plasma Oscillations

The plasma state is rich in wave phenomena when the degree of ionization is large enough to make long range forces important, particularly when a magnetic field is present.[11] Departures from charge neutrality capable of generating waves can occur in the form of charge bunching and separation over distances of the order of the Debye length, Eq (29). A discussion of these behaviors is beyond the scope of these notes. However, one case will be mentioned because it is believed to be important in magnetron sputtering devices.

Consider the case of a plasma in a uniform magnetic and electric field, as shown at the left in Figure 2.11. There is an $\vec{E} \times \vec{B}$ drift perpendicular to both $\vec{E}$ and $\vec{B}$, but, in the absence of collisions, simple theory predicts no transport across the magnetic field in the direction of the applied electric field. Now suppose that a charge bunching occurs, as shown at the right in Figure 2.11. This perturbation produces an electric field $\vec{E}_p$ that can produce an $\vec{E} \times \vec{B}$ drift across the magnetic field in the direction of $\vec{E}$. This anomalous collisionless transport across a magnetic field is believed to be an important mechanism in Penning discharges and in some magnetron sputtering discharges.[22]

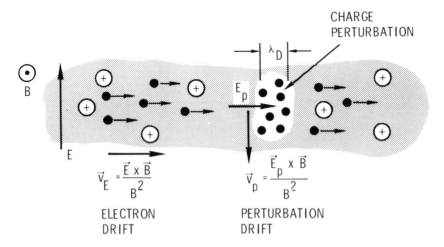

Figure 2.11: Schematic representation of a form of plasma instability that causes electron transport across a magnetic field. (Used courtesy Telic Corporation.)

4. PLASMA DISCHARGES

4.1 Introduction

A glow discharge plasma is a low temperature, relatively low pressure gas in which a degree of ionization is sustained by energetic electrons. The various glow discharge configurations that are used in materials processing differ in both their general geometry and in the configuration of electric field that is used to provide energy to the electrons.

In sputtering, simple planar diodes of the type shown schematically in Figure 2.12A are often used. They may be driven at radio frequencies (rf), as shown in the figure, or by a dc power supply. RF planar diode discharges are also used for sputter etching, plasma etching, and reactive ion etching, as indicated in Figures 2.13B, 2.13D and 2.13E respectively. Apparatuses with the configuration used in Figure 2.13D are also used for plasma-assisted chemical vapor deposition (CVD).

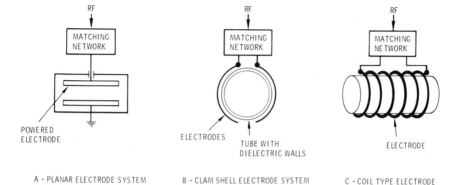

Figure 2.12: Schematic illustration of plasma discharge devices commonly used in plasma-assisted materials processing. (Used courtesy Telic Corporation.)

In activated reactive evaporation, a plasma discharge is sustained in a flux of evaporated material and reactive gas that is directed toward the substrates, as shown in Figure 2.14. The discharge may be driven by dc or rf means, using a variety of electrode configurations. The presence of the plasma has been shown to influence properties such as the stoichiometry of the resultant coatings.[51]

In ion plating the discharge is generally sustained in a mixture of the evaporated flux and an inert working gas, with the substrate holder biased negatively relative to the plasma potential. Usually this is done by simply making the substrate holder the cathode electrode for sustaining the plasma discharge, as shown in Figure 2.15. The ion bombardment of the growing coating has been shown to influence its structure.[42]

In plasma etching, plasma-assisted CVD, and glow discharge polymerization, discharges are often sustained in glass or quartz reactor tubes by surrounding electrodes which are driven at high frequencies (from 300 kHz to microwave frequencies).[23] Common electrode configurations are a pair of ring electrodes along the tube, clam shell electrodes as shown in Figure 2.12B, or a solenoidal coil electrode as shown in Figure 2.12C. It should be noted that all of these discharges are basically capacitive in nature. Although the coil electrode

38 Deposition Technologies for Films and Coatings

will introduce considerable inductance into the load seen by the matching network, the capacitive fields generated by the coil-to-coil potential drop dominate over those generated by the time rate-of-change of magnetic flux and therefore act as the primary source of ionization unless special precautions are taken to shield them. In the case of microwave discharges, the reactor tube is generally positioned within the waveguide at a location which places a strong electric field component within the tube.[6,23]

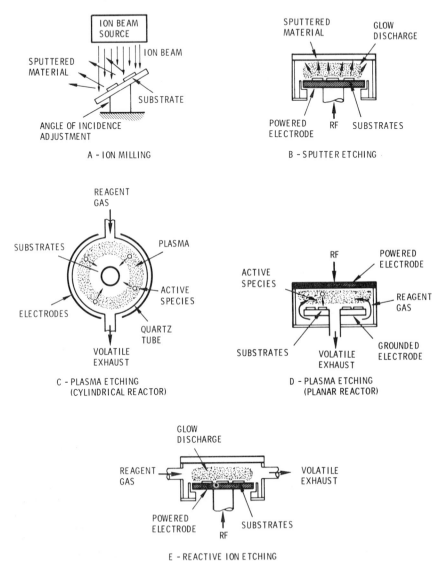

Figure 2.13: Apparatus configurations commonly used in plasma-assisted etching. (Used courtesy Telic Corporation.)

Plasmas in Deposition Processes   39

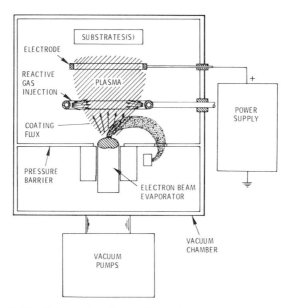

Figure 2.14: Schematic illustration showing activated reactive evaporation (ARE) process. See Reference 51. (Used courtesy Telic Corporation.)

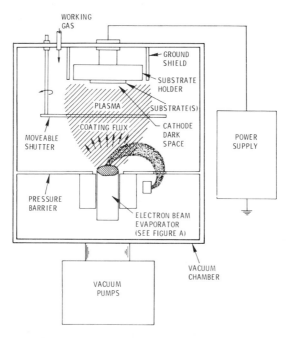

Figure 2.15: Schematic illustration showing a typical ion plating apparatus. (Used courtesy Telic Corporation.)

4.2 Ionization Balance-Paschen Relation

The degree of ionization in a glow discharge depends on a balance between the rate at which ionization is produced by the energetic electrons and the rate at which particles are lost by volume recombination and by passage to the walls of the apparatus. The rate of ionization depends on a relationship of the form (see Eqs. 6, 8 and 15)

$$R \propto N_A n_e \sigma_{ion} \sqrt{E}$$

Thus the rate of ionization depends on the type of gas (through the ionization cross section σ_{ion}), the gas pressure (through the particle density N_A), and the electric field strength (through the electron velocity). Wall losses generally dominate over volume recombination. Accordingly, the occurrence of a breakdown, and the resulting formation of a sustaining plasma discharge, in a given apparatus depend on the gas pressure and electric field strength and on the surface-to-volume ratio of the plasma. Figure 2.16 shows the inter-electrode breakdown voltage as a function of the gas pressure p and the electrode spacing d for plane parallel electrodes in air[5] and Ar.[24] Such curves are determined experimentally and are known as Paschen curves. Relationships of the same general form apply to the conditions under which a steady state discharge can be sustained. In such cases d may be replaced by Λ, where Λ is a characteristic diffusion length for the plasma vessel.[6,17,25]

The rise in voltage at the low pd side in Figure 2.16 happens because the apparatus is so small, or the gas density so low, that electrons are lost to the walls without colliding with gas atoms and producing ionization. The rise in the required voltage on the right side happens because the electron energy is becoming too low to produce ionization. This can occur at high pressures, because electron collisions with gas atoms become so frequent that the electrons cannot accumulate sufficient energy to overcome the ionization potential. It can occur at a given applied voltage in an apparatus that is so large that the local electric fields in the plasma are too weak to deliver sufficient energy to the electrons between collisions.

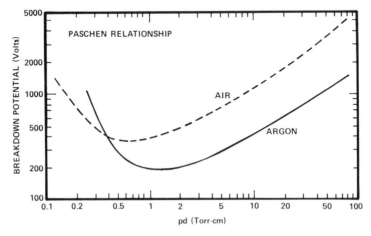

Figure 2.16: Paschen curves for breakdown between plane-parallel electrodes in air and argon at 20°C. (Used courtesy Telic Corporation.)

The functional form in Figure 2.16 provides a useful guide for adjusting the operating conditions within a given device in order to produce a plasma discharge. Conversely, the relation provides guidance for the prevention of discharges on surfaces such as the back of cathodes. One simply places a grounded shield over the surface to be protected. The spacing d between the shield and the cathode is made small enough so that the breakdown voltage is larger than the voltage required to form and sustain a plasma discharge at the operating pressure of interest.

The above considerations are also important in apparatus scaling. A discharge sustained in a small apparatus must have a high average electron energy to combat the wall losses. Such a discharge, with the same electron density but in a larger apparatus size, will be sustained at a lower average electron energy. This can in turn change the active species that are produced. Thus small bore discharge tubes are sometimes used in lasers to elevate the average electron energy to a desired value. Typical glow discharge electron densities are in the range 10^8 to 10^{12} cm^{-3}, with average electron energies of 1 to 30 eV. These conditions are shown, and compared with other forms of discharges, in Figure 2.17.

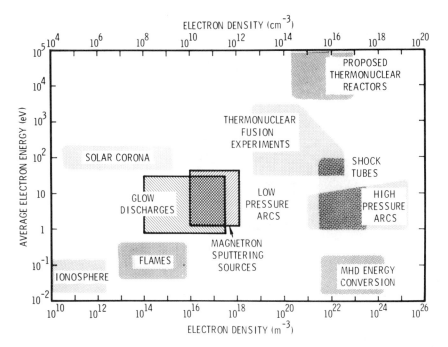

Figure 2.17: Typical regions of average electron density and energy that are representative of the plasmas found in various sources. From Reference 7. (Used courtesy American Institute of Physics.)

4.3 Cold Cathode Discharges

A low pressure cold cathode discharge is one which is maintained primarily by secondary electrons emitted from the cathode because of bombardment by ions from the plasma. These secondary electrons are accelerated in the cathode dark space and enter the negative glow as shown in Figure 2.18, where, they are known as primary electrons. Each primary electron must produce sufficient ions to release one further electron from the cathode.[15] The secondary emission coefficient is typically about 0.1 for low energy Ar^+ ions (such as are used in sputtering) incident onto clean metal surfaces.[26] The coefficient is larger, for example, for oxidized surfaces but still small enough so that each primary electron must produce or lead to the production of, a plurality of ions.[15]

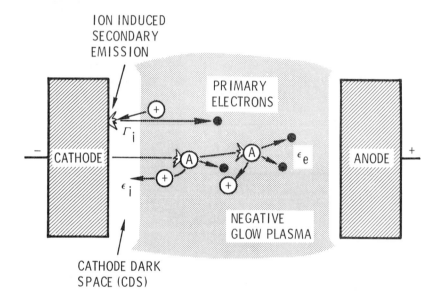

Figure 2.18: Schematic illustration of cold cathode discharge. (Used courtesy Telic Corporation.)

The negative glow (NG) is the region of the plasma where the primary electrons expend their energy, and its extent corresponds to the range of their travel from the cathode.[5,15] The electron energy distribution in the NG is multimodal. It consists of primary electrons, ultimate electrons (primaries that have transferred their energy), and much larger numbers of low energy ionization products. In the classical glow discharge described in most textbooks, a positive column (PC) extends from the NG to the anode.[5,15,17] The PC is a region where the electric field is just sufficient to transport the discharge current from the NG to the anode and to produce sufficient ionization to make up for wall losses.

In planar diode material processing sources of the type shown in Figures 2.12 and 2.13, the substrate mounting table or anode generally intercepts the NG, and there is no PC. A consequence of this small inter-electrode spacing is that the operating pressures are relatively high (see discussion of Paschen relationship in Section 4.2). For example, DC planar diode sputtering discharges in Ar typically operate at 75 mTorr with a substrate-to-cathode spacing of 4.5 cm, a current density of 1 mA/cm^2, and a discharge voltage of 3,000 V.

In order for a cold cathode discharge to operate effectively at low pressures, it is necessary that the primary electrons be preserved and not lost from the system until they have had a chance to expend their energy in making ionization. The hollow cathode geometry shown in Figure 2.19 is effective in this respect. Electrons which are accelerated in the cathode dark space and enter the NG cannot escape once they have lost an amount of energy about equal to their initial ejection energy (which is only a few eV[26]), since they encounter a sheath with repulsive forces whenever they approach the cathode. The only losses are out of the ends, and long hollow cathodes with minimized end losses can be operated effectively at low pressures and voltages. Accordingly, hollow cathodes are often used as ionization sources.[27]

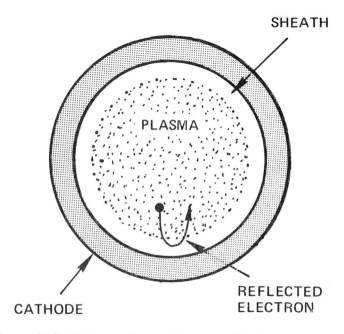

Figure 2.19: Hollow cathode discharge. (Used courtesy Telic Corporation.)

4.4 Magnetron Discharges

Magnetron discharge sources are assuming an increasing importance in sputtering. Therefore, these discharges are discussed in considerable detail in Chapter 5 (Ref. 28). It will simply be noted here that magnetrons are cold cathode discharge devices, in which magnetic fields are used in concert with cathode surfaces to form traps which are so configured that the $\vec{E} \times \vec{B}$ electron drift currents (see Section 2.5) can close on themselves.[1] The cylindrical post magnetron shown in Figure 2.20 provides one of the simplest examples of a magnetron. Primary electrons which leave the cathode barrel and enter the plasma find themselves trapped in an annular cavity which is closed on three sides by surfaces at cathode potential (hollow cathode effect) and on the fourth side by the magnetic field. The electrons can diffuse across the magnetic field and reach the anode only by making collisions (process shown in Figure 2.6c) and by plasma oscillations (see Section 3.4).[22] Because of the effectiveness of the collisions in producing ionization, these discharges are extremely efficient and operate at pressures of less than 1 mTorr with high current densities (10-200 mA/cm^2) and low voltages (700-1,000V). Planar magnetrons in which plasma rings are magnetically confined on planar cathodes are very important in sputtering technology.[29,30]

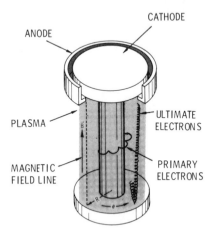

Figure 2.20: Cylindrical-post magnetron sputtering source with electrostatic end confinement. (Used courtesy Telic Corporation.)

4.5 RF Discharges

Rf-driven planar diode discharge devices of the type shown in Figures 2.12A, 2.13B, 2.13D, and 2.13E are used for sputter deposition, plasma-assisted etching and plasma-assisted CVD. Their application to sputtering is discussed in considerable detail in Chapter 5 (Ref. 28).

The operating frequency is generally 13.56 MHz, since this is the frequency in the 10 to 20 MHz range that has been allocated by the FCC for industrial applications. At this frequency only the electrons can follow the temporal variations in applied potential. Thus the plasma can be pictured as an electron gas that moves back and forth at the applied frequency in a sea of relatively stationary ions. As the electron cloud approaches one electrode, it uncovers ions at the other electrode to form a positive ion sheath. This sheath takes up nearly the entire voltage, the same as in the dc case. The ions are accelerated by this voltage and bombard the electrodes.

The rf discharge can be further understood by examining the electrode current flow. These discharges are capacitive in nature, because of external capacitance which is placed in the electrical circuits, and because one or both electrode surfaces are generally nonconducting. Consequently, the total ion and electron charge flow to a given electrode during an rf cycle must balance to zero. A consequence of this is that the plasma potential must become elevated relative to the average potential on the electrodes. Said another way, a self bias that is negative with respect to the plasma potential develops on any surface that is capacitively coupled to a glow discharge.[53] The basis for this behavior is illustrated in Figure 2.21, where the current voltage characteristic is shown for an electrode immersed in a glow discharge plasma. Because of the mobility difference between the electrons and the ions, much larger currents are drawn when the electrode is positive relative to the floating potential than when it is negative (upper figure). In order to achieve zero net current flow, it is necessary for the dc bias described above to develop on the electrode, so that the average potential is negative relative to the floating potential, as shown in this lower figure. Thus both electrodes exceed the floating potential (and become anodes) only for short portions of their rf cycle. Most of the time they are cathodes.

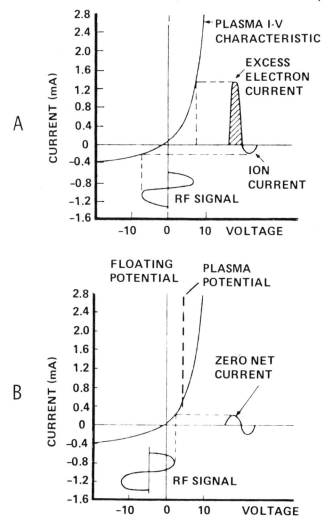

Figure 2.21: The formation of a negative bias on a capacitively coupled surface in an rf glow discharge. See Reference 53.

Because of their inertia, the motion of the ions can be approximated as if they follow the dc potential and pass to both electrodes throughout the cycle. This ion bombardment is the basis for using these discharges for sputtering and plasma-assisted etching.

RF discharges in planar diodes can be operated at considerably lower pressures than can dc discharges. Typical operating pressures are 5 to 15 mTorr. This ability to operate at low pressures is believed to be due to an increase in the volume ionization caused by the oscillating electrons and to a decrease in the loss of primary electrons, since both electrodes are negative relative to the plasma potential most of the time (hollow cathode effect discussed previously). As the pressure is increased, the volume ionization due to electrons accelerated by the oscillating electric field becomes increasingly important. Accordingly, when the planar and cylindrical plasma discharge devices shown in Figure 2.13 are used for plasma-assisted etching, CVD, and polymerization, the operating pressures are generally high enough so that volume-accelerated-electrons dominate in producing excitation and ionization. The same is true for the high frequency microwave type discharges.

5. PLASMA VOLUME REACTIONS

5.1 Introduction

Electron bombardment of atoms and molecules produces excitation, ionization, and dissociation, thereby leaving a variety of active species and radicals having much different chemical activities than those of the parent gas.[31,32] Thus, although He and Ar atoms are inert, He^+ ions with one valence electron are hydrogenic, and Ar^+ ions are similar to Cl and can react with H_2 molecules to form HA^+ ions.[31]

Electron ionization processes are obviously important in the sustaining of plasma discharges. The excitation and dissociation processes are important in plasma chemistry and form the basis for plasma-assisted etching, plasma-assisted CVD and plasma polymerization.

5.2 Electron Interaction with Atoms

An electron with a kinetic energy which exceeds the ionization energy of an atom has an approximately equal probability of producing either excitation or ionization as it passes in close proximity to the atom. A semi-classical picture of such a collision is shown in Figure 2.22. The Coulomb force from the electron produces an electric field at the atom. The component of this field which is perpendicular to the direction of electron motion ($E_\perp$) produces a time-varying "impulsive" electric field which can act on the components of the atom. The electric field pulse is equivalent to that which would be produced by a beam of photons having frequencies corresponding to the Fourier components of the pulse.[33] The point is that an electron passing closely by an atom does not simply knock an electron out of the atom, but produces a perturbation at the atom which may be approximated as a beam of white light that induces electronic excitation and ionization in proportion to the optical oscillator strengths.

In making plasma calculations, the average energy W spent by an electron in creating an electron-ion pair in a given gaseous medium is often used. Values of W for various atoms and molecules are shown in Table 2.2 along with values for the ionization potential X. It is seen that $W/X \sim 2$; i.e., there is an almost equal probability of producing either excitation or ionization, although excitation is a little more probable in molecules.

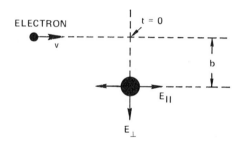

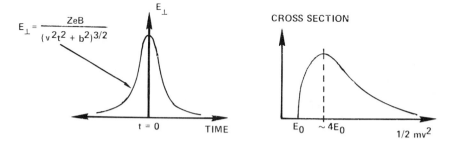

Figure 2.22: Electron-atom collision from the point of view of the virtual photon model. See Reference 33. (Used courtesy Telic Corporation.)

Table 2.2: Approximate Energy Spent to Create Electron-Ion Pairs*

Atom or Molecule	W (eV/ip)	X (eV)	$\frac{W}{X}$
He	46	24.58	1.87
Ne	37	21.56	1.71
Ar	26	15.76	1.65
Kr	24	14.0	1.71
Xe	22	12.13	1.81
H_2	36	15.43	2.33
N_2	36	15.59	2.31
NO	29	9.25	3.13
CO	35	14.04	2.49
O_2	32	12.15	2.63
CO_2	34	13.81	2.46
C_2H_2	28	11.40	2.45
CH_4	29	12.99	2.23
C_2H_4	28	10.54	2.65
C_2H_6	27	11.65	2.31
C_3H_6	27	9.73	2.77
C_3H_8	26	11.15	2.33
C_6H_6	27	9.23	2.92

*Data from Reference 33.

5.3 Electron Interaction with Molecules

Electron interactions with molecules produce excitation and ionization via mechanisms essentially identical to those for atoms as described above. Thus one sees the great similarity between the atoms and molecules and among the molecules themselves in Table 2.2. The thing that is different between the atoms and molecules is what ultimately happens to the excitation energy. In the atomic case, the excitation energy is lost by radiation unless the transitions are quantum-mechanically forbidden (see Section 5.4). In the molecular case, it may result in dissociation of the molecules.

Consider the case of CF_4, a gas which is commonly used in plasma etching. The threshold for producing excitation is 12.5 eV.[34] The excitation can be written as

$$e^- + CF_4 \rightarrow CF_4^* + e^-,$$

(where the symbol* refers to an excited atom or molecule). There is evidence that all electronic excitation processes in CF_4 produce dissociation.[34] Furthermore, because of the two-step nature of the excitation-dissociation process, one bond is broken, and the primary radicals produced are CF_3 and F rather than CF_2 and F_2.[35] The active F atoms produced in this way play a very important role in many plasma etching processes.

The ionization process can also result in dissociation. Thus one has dissociative ionization processes of the form

$$e^- + CF_4 \rightarrow CF_3^+ + F + 2e^-,$$

as well as simple molecular ionization

$$e^- + O_2 \rightarrow O_2^+ + 2e^-.$$

It has been noted that plasma discharges often contain relatively large numbers of low energy electrons which have spent their energy in making inelastic collisons (this is particularly true in regions of low electric field such as the negative glow). These electrons can attach to electronegative molecules to form negative ions.[23] Thus one has

$$e^- + O_2 \rightarrow O_2^-$$

The ion may then dissociate to give

$$O_2^- \rightarrow O^- + O$$

Atomic constituents of molecules, such as F atoms, cannot recombine in two-body gas phase collisions because the diatomic molecule formed cannot conserve both energy and momentum. Thus the gas phase recombination reaction requires a third body. Accordingly, the lifetime for such atoms in a plasma reactor can be long except when the working pressure is high.

However, when two molecular radicals associate, the energy of dissociation can be distributed within a large number of internal degrees of freedom. Accordingly, the association efficiency is close to unity for simple radicals.[23] Thus, for example, one has

$$CH_3 + CH_3 \rightarrow C_2H_6$$

Plasmas in Deposition Processes   49

The decay of initial reaction products in cascading reactions, with the development of high molecular weight species, is a well-known characteristic of the radiation chemistry of hydrocarbons and halocarbons in both the gas and solid phases.[32] The general hierarchy for the production of active species in a molecular gas plasma is shown schematically in Figure 2.23.

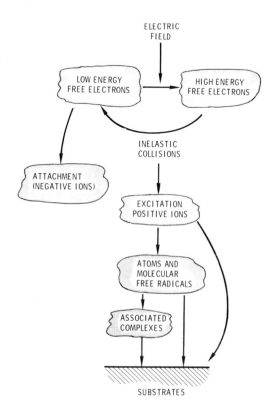

Figure 2.23: Schematic illustration of the hierarchy for the production of active species in a molecular plasma.

5.4 Metastable Species

An important consideration in using plasmas for materials processing is the ability of active species to flow or diffuse from the point of production to a point of reaction. Atoms or molecules that are excited into electronic states which can decay radiatively have very short lifetimes ($\sim 10^{-9}$ sec). However, some excited states are forbidden by quantum mechanical considerations from undergoing radiative transitions. Atoms and molecules in these metastable states can have sufficiently long lifetimes so that they can carry their stored electronic energy from the immediate vicinity of the discharge plasma to other points in a reactor.

Atoms or molecules can be excited directly into metastable states, or can arrive in these states by radiative decay after having been excited into states of higher energy. Consequently, a plasma may contain relatively large numbers of metastable species and they can have an important effect on the overall discharge chemistry. Metastable states are depopulated when the atoms undergo collisions. Thus, for example, a metastable atom may subsequently pass its excitation energy to another particle, thereby producing ionization or dissociative ionization in atoms or molecules of lower ionization potential, as indicated below.[36]

$$A^* + Y \rightarrow Y^+ + A + e^-,$$
$$A^* + XY \rightarrow XY^+ + A + e^-,$$
$$A^* + XY \rightarrow X^+ + Y + A + e^-.$$

These processes are known as Penning ionization processes.

5.5 Applications of Volume Reactions

Primary applications of interest here are plasma-assisted CVD,[37] plasma-assisted etching,[38] and plasma polymerization.[39] In all of these cases the advantage of the use of a plasma is that it can effectively produce reactions at low substrate temperatures. In some cases the reactions are unique.

An example is provided by the plasma-assisted deposition of Si_3N_4 using a SiH_4-NH_3 plasma. The plasma chemistry is not understood in detail at the present time. However, the overall reaction is

$$3SiH_4 + 4NH_3 \rightarrow Si_3N_4 + 12H_2$$

The important point is that the substrate temperature is typically 300°C or lower. When the same reaction is carried out by conventional chemical vapor deposition, the substrate temperatures are typically from 800-1200°C.[40] The lower substrate temperatures in plasma-assisted CVD are particularly important in electronic applications where coatings are deposited onto device structures.

Plasma-assisted etching is similar to plasma-assisted CVD, except that a volatile rather than an involatile compound is produced at the substrate. Thus, for example, Si etching is accomplished by using a glow discharge to generate active F atoms from an inert molecular gas such as CF_4 as discussed in Section 5.3. The F atoms cause etching of the Si by forming the volatile compound SiF_4 on the Si surface.

Plasma polymerization often proceeds in a series of steps.[39] Thus, for example, high molecular weight species can be formed in a glow discharge from low molecular weight starting material by the association processes discussed in the previous section. These high molecular weight species condense on the substrates, where they are cross-linked by plasma radiation and electron bombardment to form a polymer film.[41]

6. SURFACE REACTIONS

6.1 Introduction

Surfaces in contact with plasmas are bombarded by electrons, ions, and photons. The electron and ion bombardment is particularly important. Less is known about the influences of the plasma radiation. The relative number of ions and electrons which are incident on a surface depends on whether it is biased as a cathode or anode or is electrically isolated. In this section some of the effects of ion bombardment and electron bombardment, and of plasma bombardment of an electrically floating surface, are discussed.

6.2 Ion Bombardment

The momentum exchange associated with ion bombardment can cause rearrangement and ejection (sputtering) of surface atoms. The rearrangement can have dramatic effects on the structure and properties of a growing film and is of importance in the processes of ion plating and bias sputtering. The ejection is important in the processes of sputter cleaning and deposition. Accordingly, these mechanisms are discussed in considerable detail in Chapters 5 and 6 (Refs. 28 and 42).

At low working pressures (collisionless ion transport) the energy of ions bombarding a cathode surface will be about equal to the difference between the cathode potential and the plasma potential (see Section 3.2). This is usually about equal to the applied cathode-to-anode potential. The current density, bias voltage, sheath thickness, and plasma properties are related by Eqs (38) and (39).

At higher pressures, where ion collisions become important, the bombarding flux consists of both ions and energetic atoms because of charge exchange encounters. See Figure 2.10. Thus the average energies are considerably less than the potential drop across the cathode dark space. The effect is shown in Figure 2.24, where a histogram is given for the cathode arrival energies of 100 Ar^+ ions which cross a sheath having a voltage V_a in Ar gas at 3 Pa (22.5 mTorr). Approximately half (45%) of the ions arrive at the cathode with energies corresponding to less than 10% of the sheath voltage. The sheath parameters for the high pressure case are related by Eq (39).

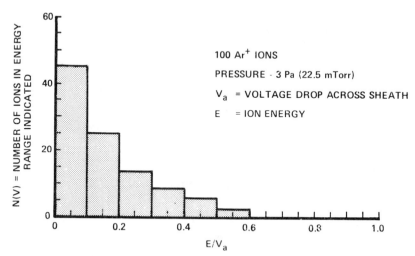

Figure 2.24: Histogram of ion energy distribution (calculated) showing effect of charge exchange. See Reference 52.

Ion bombardment can greatly influence the processes involved in the adsorption of molecules onto surfaces and their subsequent reactions. These processes are of obvious importance in plasma-assisted CVD and etching and in plasma polymerization. The process of molecular adsorption[43] and surface compound formation is illustrated in Figure 2.25 for the case of gas phase etching. The CVD case with the formation of a nonvolatile product is obviously very similar. Any of the steps shown in the figure can be rate-limiting. Physical adsorption is due to polarization (van der Waals) bonding. It is a nonactivated process and occurs with all gas-surface combinations under appropriate conditions of temperature and pressure. Adsorption energies are typically less than 0.5 eV. Chemisorption involves a rearrangement of the valence electrons of the adsorbed and surface atoms to form a chemical bond. It involves an activation energy and has a high degree of specificity between gas-surface combinations. Adsorption energies are typically 1 to 10 eV. Molecules may be chemisorbed in their molecular state or may dissociate into atoms. The latter case is known as dissociative chemisorption. Dissociative chemisorption is generally a precursor to compound formation, which is also an activated process. Various types of chemisorption bond sites can exist on a solid surface. Thus both molecular and dissociative chemisorption can occur simultaneously on the same surface. Ion bombardment can influence these processes in the following ways:

1) Ion bombardment can cause adsorbed molecules to dissociate, thereby overcoming the activation energy for this process. Ion bombardment dissociation is expected to be a sputter-type momentum transfer process.

2) Ion bombardment can create surface defect sites which have reduced activation energies for the occurrence of dissociative chemisorption or for the formation of a solid compound.

3) Ion bombardment can remove (by sputtering) foreign species from a surface. Such species may interfere with the dissociative chemisorption of a preferred species.

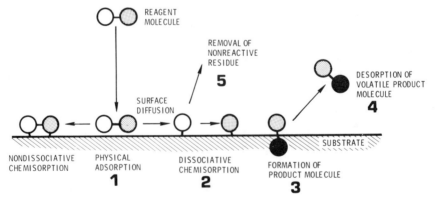

Figure 2.25: Schematic representation of surface chemisorption and volatile compound formation in gas phase etching. (Used courtesy Telic Corporation.)

Recent work[44] on radiation-induced ion etching illustrates the very significant and subtle effects which ion bombardment can have on the surface chemistry.

- F_2 *on Si.* F_2 has a very low probability for dissociative chemisorption on Si.[35] Consequently, the etch rate via the formation of volatile SiF_4 is low. The presence of Ar^+ bombardment greatly increases the etch rate, apparently by promoting the chemisorption step. This is shown in Figure 2.26. First, the Si was subjected only to Ar^+ ion bombardment. The sputter etch rate was about 2.5 Å/min. Then F_2 gas was introduced. The etch rate increased by a factor of about 3.5. When the Ar^+ bombardment was terminated, the Si etch rate due to the F_2 by itself was less than 0.1 Å/min.

- Cl_2 *on Si.* $SiCl_4$ is a volatile compound. Cl_2 chemisorbs on Si but exists as a stable structure. Apparently the activation energy for the formation of $SiCl_4$ is too large. However, in the presence of Ar^+ bombardment a significant etch rate occurs. This is shown in Figure 2.27. For the conditions of the experiment the sputter etch rate for Ar^+ bombardment of Si was 2.5 Å/min, as in the F_2 case above. Then Cl_2 was introduced. The etch rate increased to near 10 Å/min.

- XeF_2 *on Si.* XeF_2 chemisorbs on Si and forms SiF_4 to provide a significant etch rate, as shown in Figure 2.28. (Under the conditions of the experiment the XeF_2 etch rate was twice that for Ar^+ sputtering of Si.) When Ar^+ bombardment was superimposed on the XeF_2 etching, an order of magnitude increase in etch rate occurred. The increased etch rate is suggested to be due to an ion-bombardment-induced increase in the rate at which XeF_2 dissociatively chemisorbs on Si.[44]

- F_2 *on Al.* AlF_3 is nonvolatile at room temperature. Figure 2.29 shows the effect of introducing F_2 gas while sputtering Al with Ar^+. The sputtering rate drops by a factor of two. Apparently a compound with a lower sputtering yield is formed on the Al. This is the behavior commonly observed in reactive sputtering and discussed in Chapter 5 (Ref. 28).

6.3 Electron Bombardment

Electron bombardment of atoms and molecules adsorbed on surfaces is believed to produce excitation and ionization in processes which are similar to those which occur in the gas phase. Thus atoms are ionized and also excited into states from which there is a probability of dissociation or bond rearrangement. Electron bombardment can dissociate molecules and cause them to pass into a form that has a high probability of desorption (electron-stimulated desorption). Electron bombardment can also lead to the formation of dissociated species that have a high probability of diffusing into the bulk or interacting with other surface species (electron-stimulated adsorption). Finally, electron-induced bond rearrangement can cause polymerization of adsorbed surface species.

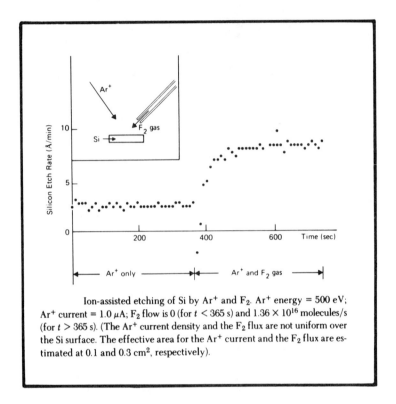

Ion-assisted etching of Si by Ar^+ and F_2. Ar^+ energy = 500 eV; Ar^+ current = 1.0 μA; F_2 flow is 0 (for $t < 365$ s) and 1.36×10^{16} molecules/s (for $t > 365$ s). (The Ar^+ current density and the F_2 flux are not uniform over the Si surface. The effective area for the Ar^+ current and the F_2 flux are estimated at 0.1 and 0.3 cm², respectively).

- F_2 HAS LOW PROBABILITY FOR DISSOCIATIVE CHEMISORPTION ON Si.
- F_2 ETCH RATE LESS THAN 0.1 Å/min.
- Ar^+ BOMBARDMENT PROMOTES CHEMISORPTION.

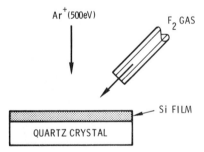

Figure 2.26: Ion-assisted gas surface chemistry experiments investigating the interaction between F_2 and Si. See Reference 44.

Plasmas in Deposition Processes 55

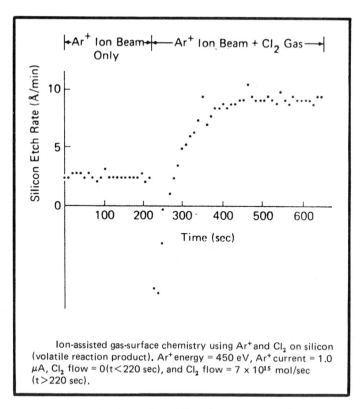

Ion-assisted gas-surface chemistry using Ar$^+$ and Cl$_2$ on silicon (volatile reaction product). Ar$^+$ energy = 450 eV, Ar$^+$ current = 1.0 µA, Cl$_2$ flow = 0 (t<220 sec), and Cl$_2$ flow = 7 × 10^{15} mol/sec (t>220 sec).

- Cl CHEMISORBS ON Si.
- SiCl$_4$ VOLATILE.
- HIGH ACTIVATION ENERGY FOR SiCl$_4$ FORMATION.
- ION BOMBARDMENT ASSISTS IN SiCl$_4$ FORMATION.

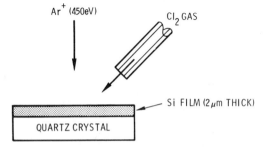

Figure 2.27: Ion-assisted gas surface chemistry experiments investigating the interaction between Cl$_2$ and Si. See Reference 44.

56 Deposition Technologies for Films and Coatings

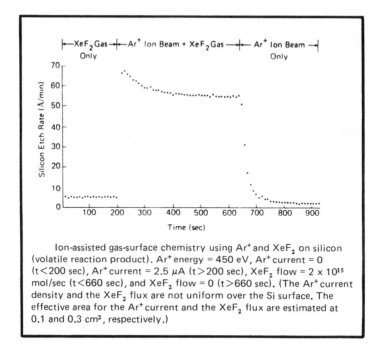

Ion-assisted gas-surface chemistry using Ar^+ and XeF_2 on silicon (volatile reaction product). Ar^+ energy = 450 eV, Ar^+ current = 0 (t<200 sec), Ar^+ current = 2.5 μA (t>200 sec), XeF_2 flow = 2 x 10^{15} mol/sec (t<660 sec), and XeF_2 flow = 0 (t>660 sec). (The Ar^+ current density and the XeF_2 flux are not uniform over the Si surface. The effective area for the Ar^+ current and the XeF_2 flux are estimated at 0.1 and 0.3 cm², respectively.)

- SiF_4 VOLATILE.
- XeF_2 ADSORPTION-DISSOCIATION PROCESS IS RATE LIMITING.
- ION BOMBARDMENT CREATES SURFACE DAMAGE.

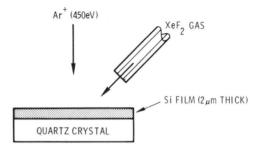

Figure 2.28: Ion-assisted gas surface chemistry experiments investigating the interaction between XeF_2 and Si. See Reference 44.

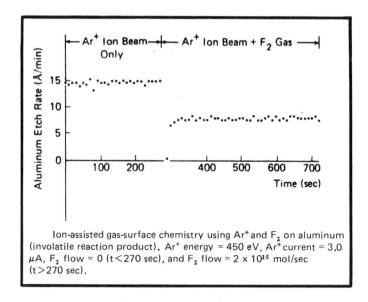

- AlF_3 NOT VOLATILE.
- SPUTTERING RATE OF MATERIALS WHICH CHEMISORB AN ACTIVE GAS TO FORM AN INVOLATILE COMPOUND IS DECREASED SIGNIFICANTLY WHEN INERT GAS SPUTTERING IS CARRIED OUT IN PRESENCE OF ACTIVE GAS.

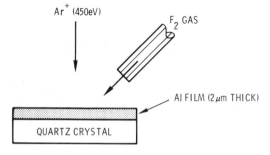

Figure 2.29: Ion-assisted gas surface chemistry experiments investigating the interaction between F_2 and Al. See Reference 44.

Two examples of electron-stimulated surface chemistry in the field of plasma etching are shown in Figures 2.30 and 2.31.

- XeF_2 *on* SiO_2. XeF_2 dissociatively chemisorbs on SiO_2 but etching does not occur, apparently because of a high activation energy for the formation of SiF_4. Electron bombardment has been observed to produce elemental Si on the surface of SiO_2,[45,46] but it does not cause etching. However, when SiO_2 is subjected to electron bombardment in the presence of XeF_2, etching at appreciable rates occurs in those regions of the substrate that are subjected to the electron bombardment. The effect is shown in Figure 2.30.

- XeF_2 *on* Si_3N_4. Auger analysis indicates that XeF_2 dissociatively chemisorbs on Si_3N_4 but it does not cause etching. Electron bombardment does not produce etching by itself, but in the presence of XeF_2 it produces a significant etch rate. The effect is shown in Figure 2.31.

6.4 Glow Discharge Surface Cleaning and Activation

Glow discharge cleaning, where electrically isolated work pieces are immersed into a low pressure plasma, has been used for many years,[47] particularly for glass and other nonconductive materials that cannot be cleaned effectively by simple dc sputtering. Working gases are typically air, oxygen, or argon. The process is empirical, but often very effective as a final cleaning step before vacuum depositing a coating. It probably involves the ion and electron stimulated desorption of physisorbed gas species via the processes described above.

Atmospheric corona discharges have been used for years to prepare plastic surfaces for processing. More recently, low pressure glow discharges are being used to promote the adhesion of vacuum deposited coatings. Some of the new analytical tools, such as X-ray photoelectron spectroscopy (ESCA), are providing insight into the mechanisms involved. Thus in ESCA studies of the effect of oxygen plasma treatment on ABS, polypropylene[48] and polystyrene surfaces,[49] the oxygen treatment was found to change the basic chemical nature of the polymer surfaces by increasing the number of single or double bonds between the carbon and oxygen atoms. The oxygen pretreatment was also found to increase the adhesion strength of evaporated copper films on the polymers. In the case of polystyrene, it was found that the copper atoms strongly altered the carbon-oxygen bond with the formation of a carbon-oxygen-copper complex.

A process called CASING (Crosslinking by Activated Species of Inert Gases) provides a rather dramatic final example of the influence of glow discharge plasmas on materials.[50] A polyethylene plate was treated by allowing excited He species produced in an electrodeless glow discharge to impinge on its surface. The treatment produced a highly crosslinked polymer which was insoluble in trichlorobenzene at 135°C. When the treated plate and an identical untreated plate were heated on a clean surface, the untreated polymer melted and flowed onto the surface, while the thin crosslinked shell formed during CASING prevented escape of molten polymer from inside the shell.

Plasmas in Deposition Processes 59

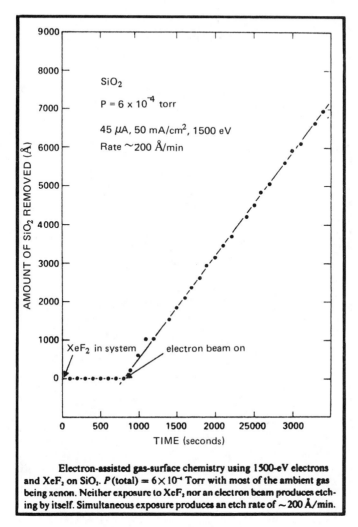

- XeF$_2$ DISSOCIATIVELY CHEMISORBS ON SiO$_2$.
- HIGH ACTIVATION ENERGY FOR SiF$_4$ FORMATION.
- ELECTRON BOMBARDMENT OF SiO$_2$ PRODUCES ELEMENTAL Si ON SURFACE.

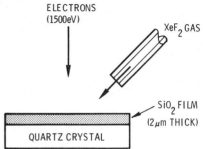

Figure 2.30: Electron-assisted gas surface chemistry experiments investigating the interaction between XeF$_2$ and SiO$_2$. See Reference 44.

60 Deposition Technologies for Films and Coatings

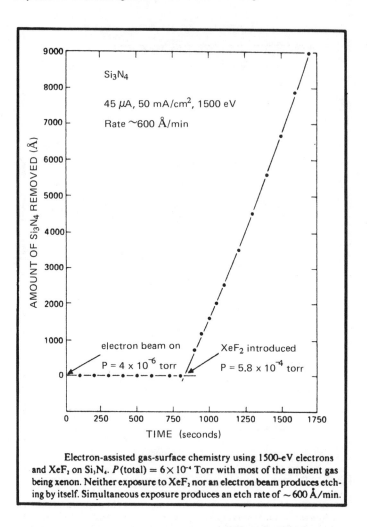

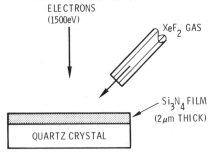

- AUGER SHOWS F ON SURFACE.
- RATE LIMITING STEP IS FORMATION OF SiF_4.
- ELECTRON BOMBARDMENT AND XeF_2 PRODUCES ETCHING.

Figure 2.31: Electron-assisted gas surface chemistry experiments investigating the interaction between XeF_2 and Si_3N_4. See Reference 44.

REFERENCES

(1) J.A. Thornton and A.S. Penfold, in *Thin Film Processes*, J.L. Vossen and W. Kern, eds., Academic Press, New York (1978), p. 75.
(2) E.W. McDaniel, V. Cermak, A. Dalgarno, E.E. Ferguson, and L. Friedman, *Ion-Molecule Reactions*, Wiley-Interscience, New York (1970), p. 345.
(3) J.O. Hirschfelder, C.F. Curtiss, and R.B. Bird, *Molecular Theory of Gases and Liquids*, Wiley, New York (1954), p. 523.
(4) G.W. Sutton and A. Sherman, *Engineering Magnetohydrodynamics*, McGraw-Hill, New York (1965).
(5) J.D. Cobine, *Gaseous Conductors*, Dover, New York (1958).
(6) R.F. Baddour and Robert S. Timmins, ed., *The Applications of Plasmas to Chemical Processing*, MIT Press, Cambridge, Mass. (1967).
(7) J.A. Thornton, *J. Vac. Sci. Technol., 15*, 188 (1978).
(8) H.R. Griem, *Plasma Spectroscopy*, McGraw-Hill, New York (1964), p. 129.
(9) D. ter Haar, *Elements of Statistical Mechanics*, Holt, Rinehart and Winston, New York (1960), p. 381.
(10) D.J. Rose and M. Clark, Jr., *Plasmas and Controlled Fusion*, MIT - Wiley, New York (1961), p. 80.
(11) F.F. Chen, *Introduction to Plasma Physics*, Plenum Press, New York (1974).
(12) L. Spitzer, Jr., *Physics of Fully Ionized Gases*, Interscience, New York (1956).
(13) J.L. Delcroix, *Introduction to the Theory of Ionized Gases*, Interscience, New York (1960), p. 128.
(14) S. Chapman and T.G. Cowling, *The Mathematical Theory of Non-Uniform Gases*, Cambridge Univ. Press, Cambridge, England (1960), p. 90.
(15) A. von Engel, *Ionized Gases*, Clarendon Press, Oxford, England (1965).
(16) E.W. McDaniel, *The Mobility and Diffusion of Ions in Gases*, Wiley, New York (1973), p. 132.
(17) S.C. Brown, *Basic Data of Plasma Physics*, MIT Press, Cambridge, Mass. (1959).
(18) F.F. Chen, in *Plasma Diagnostic Techniques*, R.H. Huddlestone and S.L. Leonard, eds., Academic Press, New York (1965), p. 113.
(19) M.H. Mittleman, in *Plasma Dynamics*, F.H. Clauser, ed., Addison-Wesley, New York (1960), p. 54.
(20) D. Bohm, E.H.S. Burhop and H.S.W. Massey, in *The Characteristics of Electrical Discharges in Magnetic Fields*, A. Guthrie and R.K. Wakerling, eds., McGraw-Hill, New York (1949), p. 13.
(21) S. Glasstone and R.H. Louberg, *Controlled Thermonuclear Reactions*, Van Nostrand, New York (1960), p. 459.
(22) J.A. Thornton, *J. Vac. Sci. Technol., 15*, 171 (1978).
(23) F.K. McTaggart, *Plasma Chemistry in Electrical Discharges*, Elsevier, New York (1967).
(24) B. Gänger, *Der Elecktrische Durchschlag*, Springer-Verlag, Berlin (1953).
(25) S.C. Brown and A.D. MacDonald, *Phys. Rev., 76*, 1629 (1949).
(26) E.S. McDaniel, *Collision Phenomena in Ionized Gases*, Wiley, New York (1964), Ch. 13.
(27) D.G. Williams, *J. Vac. Sci. Technol., 11*, 374 (1974).
(28) J.A. Thornton, "Coating Deposition by Sputtering," Notes prepared for use in course entitled "Deposition Technology and Applications," given at University of Calif., Los Angeles, CA, June 16-20, 1980. (See Chapter 5.)
(29) D.B. Fraser, in *Thin Film Processes*, J.L. Vossen and W. Kern, eds., Academic Press, New York (1978), p. 131.
(30) R.K. Waits, *Ibid*, p. 131.
(31) W.F. Libby, *J. Vac. Sci. Technol., 16*, 414 (1979).
(32) G.M. Burnett and A.M. North eds., *Transfer and Storage of Energy by Molecules*, Wiley-Interscience, New York (1969).
(33) L.G. Christophourou, *Atomic and Molecular Radiation Physics*, Wiley-Interscience, New York (1971), p. 6.
(34) H.F. Winters, J.W. Coburn, and E. Kay, *J. Appl. Phys. 48*, 4973 (1978).
(35) J.W. Coburn and H.F. Winters, *J. Vac. Sci. Technol., 16*, 392 (1979).

(36) E.E. Muschlitz, Jr., *Science, 159,* 599 (1968).
(37) J.R. Hollahan and R.S. Rosler, in *Thin Film Processes,* J.L. Vossen and W. Kern, eds., Academic Press, New York (1978), p. 335.
(38) C.M. Melliar-Smith and C.J. Mogab, ibid, p. 497.
(39) H. Yasuda, *Ibid*, p. 361.
(40) W. Kern and V.S. Ban, *Ibid*, p. 257.
(41) A.S. Penfold, J.A. Thornton and E.A. McLennan, *Bull. Am. Phys. Soc., 14,* 1026, (1969).
(42) D.M. Mattox, "Ion Plating," Notes prepared for use in course entitled "Deposition Technology and Applications," given at University of Calif., Los Angeles, CA, June 16-20, 1980. (See Chapter 6.)
(43) J.R. Anderson, ed., *Chemisorption and Reactions on Metallic Films*, Academic Press, New York (1971).
(44) J.W. Coburn and H.F. Winters, *J. Appl. Phys., 50,* 3189 (1979).
(45) S. Thomas, *J. Appl. Phys., 45,* 161 (1974).
(46) B. Carriere and B. Lang, *Surface Science, 64,* 209 (1977).
(47) L. Holland, *Vacuum Deposition of Thin Films*, Chapman and Hall Ltd., London (1966) Ch. 3.
(48) J.M. Burkstrand, *J. Vac. Sci. Technol., 15,* 223 (1978).
(49) J.M. Burkstrand, *Appl. Phys. Lett., 33,* 387 (1978).
(50) R.H. Hansen and H. Schonhom, *Polymer Lett., 4,* 203 (1966).
(51) R.F. Bunshah and A.C. Raghuram, *J. Vac. Sci. Technol., 9,* 1385 (1972).
(52) W.D. Davis and T.A. Vanderslice, *Phys. Rev., 131,* 219 (1963).
(53) H.S. Butler and G.S. Kino, *Phys. Fluids, 6,* 1346 (1963).

ns of the depositing material and the sur-
3
Adhesion and Surface Preparation

Donald M. Mattox
Sandia National Laboratories
Albuquerque, New Mexico

The interfacial region between a coating and a surface determines many physical and electrical properties of the couple. These include contact resistance, contact noise, acoustic coupling, electron trapping and recombination, thermal conductance, and film adhesion. Measurements of these properties may be used to define the state of an interfacial region.

A practical definition of "good adhesion" is that the interfacial region (or nearby material) does not fail under assembly or service conditions nor at unacceptably low stress levels under test conditions. Adhesion or adhesive strength is a macroscopic property that depends on the bonding across the interfacial region, local stresses, and the adhesive failure mode. The failure mode depends on the type of stress to which the interfacial region is subjected. Typical stresses encountered in assembly and service may result from mechanical loading (tensile, shear, fatigue), thermal (high and low temperatures, cycling), chemical environment (corrosion—both chemical and electrochemical), and electrical environment. Thus, the failure mode may be determined by the environment, the chemical and electrochemical properties of the film and substrate materials, film morphology, the mechanical properties, defect morphology of the interfacial region, and the manner in which external stresses are applied. In addition, adhesive failure may be time dependent.

Good adhesion is promoted by:

(1) Strong atom-atom bonding within the interfacial region.

(2) Low local stress levels.

(3) Absence of easy deformation or fracture modes.

(4) No long-term degradation modes.

These characteristics depend on the nature of the interfacial region, which in turn depends on the interactions between the depositing material and the surface.

In atomistic film disposition technologies, the nature and condition of the substrate surface determine many of the factors which control nucleation, interface formation, and film growth. These in turn control the interfacial properties such as adhesion. In order to understand the interrelationships involved, it is necessary to understand the various phases of film formation.

NUCLEATION

When atoms impinge on a surface, they lose energy to the surface and finally condense by forming stable nuclei. During condensation, the adatoms have a degree of mobility on the surface which is determined by their kinetic energy and the strength and type of interaction between the adatom and the surface. A strong surface-atom interaction will give a high density of nuclei and a weak interaction will result in widely spaced nuclei. If the depositing atom-substrate interaction strength is low the depositing atoms will nucleate on surface discontinuities such as steps or nucleate by collision with absorbed atoms or other atoms migrating on the surface.

The nuclei will grow to form a continuous film. It has been proposed that the nuclei density and the nuclei growth mode determine the effective interfacial contact area and the development of voids in the interfacial region. Nuclei density and orientation formed during deposition can be affected by ion bombardment, electric fields, gaseous environment, contaminant layers, surface impurities, surface defects, and deposition techniques. In addition to the effective contact area, the mode of growth of the nuclei will determine the defect morphology in the interfacial region and the amount of diffusion and reaction between the depositing atoms and the substrate material.

Bonding between unlike atoms and materials may be due to chemical, electrostatic, or polarization bonding. In chemical bonding, the interaction is due to the transfer or sharing of electrons. Electrostatic bonding is due to charge separation and the resulting electrostatic attraction. Polarization bonding is due to asymmetry in the electric field around atoms or molecules and the resulting attraction. Chemical bonds vary in strength depending on the degree of electron transfer or sharing. If the degree of charge transfer is high, compounds or ionic solids are formed which are usually strong but brittle. If electron sharing is the bonding mechanism, alloys or metallic-type materials are formed which are more ductile. Electrostatic bonding has been proposed as the principal adhesion mechanism for metal-polymer systems and is important in metal-atom nucleation on an insulator surface.

One would expect that a single atom on a surface would be bound with a different energy than if it were in a nucleus or a film, since the like and unlike atom coordination changes with coverage. The bonding across the interfacial region should be the summation of the individual adatom/substrate bonds less any bond energy lost due to strain or defect formation. Thermal desorption studies of an insoluble/noncompound-forming metal system (gold-tungsten) have shown that a single absorbed metal atom (gold) on a metal surface (tungsten) will be bound with an energy near the cohesive energy of the film material (3.5 eV/atom) but the film-substrate adhesion is usually low. Generally the adhesion between nonsoluble/noncompound-forming film substrate couples is poor unless energetic deposition processes are used. To improve adhesion, an intermediate layer of a metal which will react with both the film and substrate is used as a "glue" layer.

The problem of poor adhesion between insoluble materials must not lie in the intrinsic strength of the atom-surface bond but rather in the contact area, bond strain, or some other interfacial characteristics. When gold is deposited on an oxide surface, the adhesion is normally poor; and this is attributed to the low energy associated with the gold-oxygen bond. By sputter-depositing the gold in an oxygen discharge, very adherent gold films can be deposited on oxide surface. The only detectable difference is an increased nucleation density of the gold sputter deposited in the oxygen discharge. This may lead to an increased effective contact area.

INTERFACE FORMATION

Interfacial regions may be classed as mechanical, monolayer-to-monolayer (abrupt), compound, diffusion, or pseudo-diffusion and combinations of these types. The type of interfacial region formed during deposition depends on the substrate surface morphology, contamination, chemical interactions, the energy available during interface formation, and the nucleation behavior of the depositing atoms.

The mechanical interface is characterized by mechanical interlocking of the film material with a rough surface. The strength of this interface will depend on the mechanical properties of the materials. Surfaces may be deliberately roughened or made porous in order to increase the mechanical interlocking. It should be noted that often deposition of a film on a rough surface gives a porous film due to geometrical shadowing effects.

The monolayer-to-monolayer (abrupt) type of interface is characterized by an abrupt change from the film material to the substrate material in a distance on the order of the separation between atoms (2 to 5 Å). This type of interface may be formed when there is no diffusion and little chemical reaction between the depositing atoms and the substrate surface. This lack of interaction may be due to the lack of solubility between materials, little reaction energy available, or the presence of contaminant layers. In this type of interface, defects and stresses will be confined to a narrow planar region.

The compound interface is characterized by a constant composition layer, many lattice parameters thick, created by the chemical interaction of the film and substrate material. The compound formed may be either an intermetallic compound or some other chemical compound such as an oxide. Compounds are usually brittle materials. In this type of interface, there may be abrupt physical and chemical discontinuities associated with the abrupt-phase boundaries. Often during compound formation there is segregation of impurities at the phase boundaries, and stresses are generated due to lattice mismatching. Porosity may develop in the interfacial region.

In the diffusion type of interface, there is a gradual change in composition, intrinsic stress, and lattice parameters across the interfacial region (graded interface). If there is a difference of diffusion rates of the film atoms and the substrate atoms, Kirkendall porosity may be formed in the interfacial region. For very thin films, Kirkendall porosity may not develop because of rapid surface diffusion. Diffusion processes may be very important in the defect structure of the interfacial region. Diffusion will occur normal to and laterally along the substrate surface. If there are regions where depositing material has not nucleated, the diffusion process may allow the film material to cover (fill-in) the area.

Under energetic situations such as ion bombardment, codeposition, ion implantation, or melting/quenching, a "pseudo-diffusion" type of interface may be formed by materials which are normally insoluble. Ion bombardment prior to film deposition may increase the interfacial solubility by creating very high concentrations of point defects or stress gradients or both which will enhance diffusion.

Obviously, a combination of several types of interfacial regions is possible. Compound and graded types of interface may be formed by controlling the environment or film composition during the initial portion of the film deposition, as well as heating during and after film deposition.

FILM FORMATION

As the film nuclei join to become continuous and the interfacial region develops, the film begins to form. It is in general difficult to define where an interfacial region stops and the film begins, since processes such as grain boundary diffusion may allow substrate material to penetrate deeply into the film, possibly even to the film surface. The manner in which a film develops determines the properties of the coatings.

The factors which most strongly influence the mechanical properties of a coating are: (1) microstructure, (2) incorporated impurities, and (3) internal stresses. All of these are dependent on the deposition variables.

In considering the structure-property relationships for coatings, one should bear in mind that atomistically deposited materials often have unique mechanical properties which differ significantly from those of materials fabricated by melting-solidification or sintering processes. These properties may be different because of unique microstructures, unique phase compositions, impurity incorporation, and/or composition or phase gradients which may or may not be deliberate. Usually, it is difficult to obtain bulk properties in atomistically deposited materials because of the unique microstructures, graded properties and problems with compositional control.

MICROSTRUCTURE DEVELOPMENT

Atomistic deposition may occur from a liquid (electroplating, electroless plating), gaseous (chemical vapor deposition—CVD), or vacuum (vacuum evaporation, sputtering, ion plating) environment. In each case, the microstructure of the deposit is determined by nucleation and growth of films from the depositing atoms. Nucleation of the depositing atoms will depend on the atom-substrate interactions and the surface mobility of the depositing atoms. Growth will be determined by the surface mobility of the depositing atoms, atomic arrangement on condensation and subsequent rearrangement of the structure.

For coatings which have become continuous, when the mean free path between collisions for the depositing atoms is large and the surface mobility of the deposited atoms is low, the growth of the deposit will be controlled by the geometry of the surface. Deposition on the high spots and shadowing of the low areas will develop a columnar microstructure with the columns being composed of very fine grains of the material. This growth mode is shown in Figure 3.1 by the zone model proposed by Movchan and Demchishin (M-D diagram)[1]. In zone 1 the growth is controlled by the low surface mobility and geometrical shadowing effects and the columns are weakly bonded to each other with bonding decreasing with increasing degrees of texturing. In zone 2 (higher temperature), the mobility is greater and the column boundaries become more like grain boundaries. Zone 3 is characterized by recrystallization and grain growth giving a more equiaxed structure.

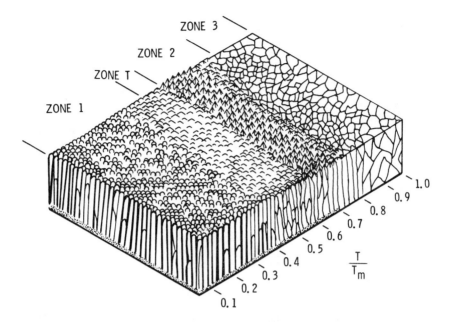

Figure 3.1: Movchan-Demchishin (M-D) diagram of structural zones in vacuum condensates as a function of the deposition temperature (T) and the melting point (T_m) of the condensate, zone T described by Thornton, Ref. 2.

As shown in the M-D diagram, the microstructure of the deposit may be altered by the temperature of deposition. Other methods of modifying this microstructure are by ion bombardment during deposition, gas pressure during sputter deposition[2], angle of incidence of the depositing atoms, mechanical disturbance during deposition, periodic reaction with other species, the formation of second phase material or the formation of a high mobility intermediate species (CVD)[3].

In electrolytic environments, a dendritic structure may be formed due to the high mobilities of hydrated ionic species and condensation on certain crystallographic planes. This type of growth is normally modified by the addition of brightening agents which are usually organics. These agents "poison" the surface causing the deposit to continuously renucleate.

IMPURITY INCORPORATION

The deliberate or unintentional incorporation of "impurity" atoms into the depositing material may be important in several ways. The atoms may react in such a way as to cause renucleation and the destruction of the columnar microstructure, stabilize phases of the material, stabilize a fine grain structure, form second phase material or stabilize the intrinsic stresses in the film. Since a small amount of impurity material can have a large effect on properties, it is not unexpected that there is appreciable variation in measured properties of deposited materials among various investigators.

68 Deposition Technologies for Films and Coatings

Impurities may arise from a number of sources. In vacuum processes oxygen, nitrogen, hydrogen, and carbon are common from residual gases, vacuum leaks, and outgassing. In sputter deposition, the implantation of several atomic percent of the sputtering gas is not at all unusual, particularly when a substrate bias is used. In chemical vapor deposition reaction products, carrier gases and unreacted or partially reacted source gases are commonly incorporated. Electrodeposited materials may contain hydrogen and co-deposited organic and inorganic additives in the deposited material. In electroless deposition, several atomic percent of the catalyzing agent (P,B) will be incorporated into the deposit.

Solute impurity atoms in a deposit may react with the host material to form compounds, stay in solution either in interstitial or substitutional positions (alloys) or segregate to surfaces or interfaces. Segregation and concentration of an impurity to grain boundaries may drastically affect the fracture mode of the material, generally resulting in embrittlement. Segregation to internal surfaces will also inhibit recrystallization and growth as does the formation of second phase material. Small quantities of foreign atoms may be deliberately incorporated into the deposit by codeposition to change the mechanical properties by dispersion strengthening mechanisms. Examples are oxides, carbides, and borides which have been codeposited with nickel and iron in electrodeposition processes to form dispersion strengthened material. Ion implantation may also be utilized to incorporate foreign atoms into a host.

Reactive deposition techniques may allow the bulk of the deposited material to be deposited as a compound. Reactive ion plating, reactive sputter deposition and activated reactive evaporation are some of the techniques used to deposit carbide, nitride, oxide, and boride coatings.

In addition to dispersed phases, multilayer (laminar) films may be used to obtain high strength composites of metals and compounds.

INTRINSIC STRESS

The intrinsic stress is the stress induced in a deposit during growth. It arises from sources such as impurity incorporation, incomplete structural ordering or structural reordering. The intrinsic stress plus the thermal stress which arises from differential thermal expansion between the deposit and substrate may be measured by substrate deflection techniques or x-ray lattice parameter measurements. Often the intrinsic stresses approach (or exceed) the yield stress of the material in bulk form. The total coating stress is a function of coating thickness. Thin stressed coatings may adhere while thick stressed coatings will not. Observations and theories of intrinsic stress have been reviewed, but no model seems to fit all the observations. The deposit microstructure is important to stress generation since a porous or distended structure will not sustain a high stress. In many cases use of a distended structure can resolve adhesion problems.

Any mechanism which impedes atomic rearrangement will allow the development of high stresses. Thus impurity incorporation probably plays a major role in stress retention, since without impurities low temperature recovery or recrystallization may occur. Stresses which exceed the normal yield strength of the bulk material imply that some strengthening mechanism is operational in the deposited material.

Figure 3.2 shows the flat beam stress (32.6 ksi-tensile) generated in a CVD deposit of TiB_2 on Poco graphite after cooling from the deposition temperature of $930°C$. On slow heating to the deposition temperature, an intrinsic stress of 10 ksi-tensile remains even though there is some stress relief at temperatures as low as 700°C. Heat treating at 1100°C gives a deposit which only shows thermal stresses on cycling between room temperature (15.6 ksi-tensile) and the deposition temperature. The initial intrinsic stress was on the order of 16 ksi-tensile. We attribute the stabilization of the intrinsic stress in the as-deposited material to chlorine incorporation in the deposit during deposition. The onset of stress relief at 700°C is far below the temperature at which one would expect appreciable atomic rearrangement in this refractory material (MP 3225°C). Sputter deposits of glass may also show atomic rearrangement at temperatures far below the strain point of the bulk glass.

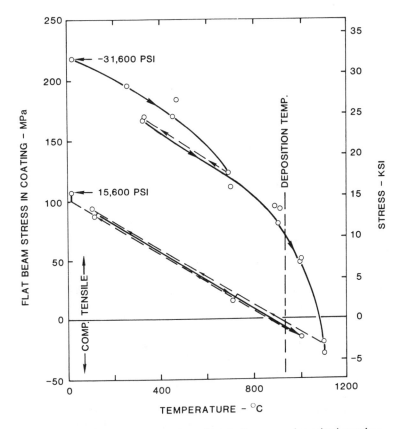

Figure 3.2: Flat beam stress in a chemically vapor deposited coating of TiB_2 on graphite as a function of heat treatment. After deposition the coating had a stress of 31,600 psi at room temperature. After annealing the stress at room temperature was 15,600 psi. Arrows indicate direction of temperature change.

Intrinsic stress may be affected by a number of processing and growth parameters. Film deposition rate, angle of incidence, presence of residual gas, deposition temperature, and gas incorporation affect the magnitude of stress in vacuum deposited films as well as whether the stress is tensile or compressive. Gas incorporation, deposition temperature, angle of incidence, sputtering pressure, reactive gases, and ion bombardment during deposition affect the stress in sputter deposited films. Ion bombardment during deposition seems to create high compressive stresses by "peening" the surface during deposition. By controlling the amount of ion bombardment during deposition, it is possible to sputter deposit coatings with zero intrinsic stress. Ion bombardment may be achieved in vacuum deposition by using a separate "ion gun" system or by applying a DC bias to the deposit during sputter deposition. In DC sputtering systems, high energy ions which are reflected from the cathode will bombard the growing film giving high compressive stresses in the deposits particularly at low sputtering pressures.

Ion bombardment during deposition can also alter the preferred crystallographic orientation of the deposited material, which may affect the properties of the deposited material.

MECHANICAL PROPERTIES

The mechanical properties of materials depend on their deformation and fracture mechanisms. These mechanisms in turn depend on purity and microstructure, density, solute content, second phase content, internal stresses, defect concentration, grain size, and other such factors which may be sensitive to the deposition process variables and growth modes. In addition, where the surface-to-volume ratio of a deposit is large, the free surfaces may play an important role in the properties of the deposit. Often these material properties are not controlled or defined so the measured mechanical properties may be unique to a particular sample. For that reason, specific values for mechanical properties of deposited materials must be used with caution. There have been a number of reviews of the mechanical properties of deposited materials.[3]

In some instances, the properties of the deposits may give directly observable effects without special measurements. The most obvious of these is stress. When the total stress (intrinsic and thermal) is high, the deposited film may deform the substrate. For thin substrates this is particularly obvious. If the stresses are high, the coating may crack (tensile stress) or buckle (compressive stress). For substrates of brittle materials, this stress may give delayed adhesive failure due to static fatigue.

The intrinsic and extrinsic stresses may be additive such than an interface with a high intrinsic stress may fail spontaneously or with little applied stress even though there is strong chemical bonding between film and substrate.

Adhesion testing utilizes externally applied stresses which may vary from tangential shear to tensile, though the stresses appear at the interfacial region in a very complex manner due to geometric effects, variations in the physical properties of the material and the usually nonhomogeneous nature of the film and interfacial material. The presence of pores and voids in the interfacial region will give stress concentration and alter the values of the tensile and shear components of the interfacial stress.

INTERFACIAL FRACTURE

Fracture is composed of crack initiation and propagation. Initiation will begin at a flaw which allows stress concentration or bond weakening. Propagation of the fracture will occur by repeated bond breaking. Bonds may be broken by physical straining, by chemical effects, or by a combination of both. In ductile material, the stress at the tip of a propagating crack is relieved by plastic deformation of the surrounding material. In a brittle material, there is a little plastic deformation. Fracture toughness is a measure of the energy which must be added to the system to give rapid fracture propagation, and less energy is used to propagate a fracture in a brittle material than in a ductile material. A fracture will preferentially propagate in a plane of weakness such as might be found in an abrupt interface, an interface with a large number of voids, or a region of high intrinsic stress.

In stress corrosion cracking or static fatigue, a foreign species at the crack tip will strain or weaken the bond, allowing the fracture to propagate. In metals, the corrosive agent may convert the material at the crack tip from a ductile to a brittle material, thereby reducing the amount of energy necessary to propagate the crack. Corrosion products may expand the crack by wedging.

Pores or cracks in the interfacial region act not only as crack initiators and stress concentrations but allow interfacial cracks to be propagated parallel to the direction of the applied force. Analysis of the fracture surfaces (fractography) may be used to determine the failure modes in some cases.

TIME-DEPENDENT INTERFACIAL CHANGES

It has been reported that in some metal film on oxide substrate systems the adhesion improves with time at low temperatures. This is attributed to the migration of active gases to the interface to form transition compounds or to stress relief. In the solid phase bonding of materials, a similar increase of bond strength with time has been noted and is attributed to stress relief in the interfacial region.

Treatment at high temperatures may improve adhesion by forming a desirable type of interface or may decrease adhesion by void formation or by diffusing a reactive species away from the interfacial region. Thermal and mechanical cycling likewise may lead to a time dependent failure by fatigue.

Corrosion may give a long-term loss of adhesion. It has been shown that only a small amount of halide and moisture can result in electrochemical corrosion of some common thin-film metallization systems such as titanium-gold. Extensive work has been done to identify metallization systems which are not susceptible to electrochemical corrosion.[4] Corrosion not only depends on the environment and the materials involved but also on the availability of the corrosive media at the interface. For this reason, film porosity and pinholes are important in that they not only serve to absorb and trap corrosive liquids and vapors but allow them access to the interfacial region. Usually film porosity is measured by corrosion or dissolution techniques.

ADHESION TESTING

There has been considerable effort devoted to developing various adhesion testing methods, but to be meaningful a testing program must be developed for each situation. This means that the testing methods must duplicate the stresses to which the components are subjected in assembly and service. In addition, aging of components in service environments must be performed in order to determine the long-term adhesion properties.

There have been many attempts to use adhesion tests to give quantitative information about the chemical bonding across an interface, though the analysis is highly questionable. In bulk materials, we find that the practical strength of a material never approaches the theoretical strength because failure is controlled by deformation and fracture processes and the same should be expected in film-substrate adhesion. The chemical environment can affect the mechanical properties of hard and brittle materials with hydrogen being an important chemical species. For bulk materials, a useful concept is that of fracture energy (toughness). If we consider that adhesive failure is a fracture process, then the energy necessary to fail an interface may be a more useful concept than that of chemical bond strength.

Loss of adhesion is the culmination of interfacial failure. Other techniques may be useful in detecting the degradation of interfacial adhesion. These include contact resistance, contact noise, acoustic coupling, shock wave coupling, and other processes which depend on the physical and electrical properties of the interface. Acoustic emission is used to detect fracture propagation in bulk materials and may be useful in interfacial studies. Other properties such as film porosity and stress may have an indirect effect on adhesion and, in particular, on the long-term stability of film-substrate specimens. Since these properties are very dependent on deposition conditions, they should be studied and controlled.

SURFACE PREPARATION

Surface preparation may differ from surface cleaning, but we will begin by discussing cleaning. As a practical matter, a "clean surface" is one that contains no *significant* amounts of *undesirable* material; thus what constitutes a clean surface depends on the requirements of the user. For someone studying surface crystallography, catalysts, or gas absorption, a clean surface is one that contains only a small fraction of a monolayer of foreign material (this is often called an atomically clean surface). For someone interested in joining surfaces using gross deformation in shear or high temperatures, quite a lot of contamination may be present without materially affecting the resultant bond. In some cases, such as the adhesion of deposited thin films, the presence of foreign material on the surface may be desirable or necessary. Adhesion is often equated with surface cleanliness*, so surface treatments or the deliberate addition of foreign materials which improve adhesion may be considered to be cleaning techniques.

In the utilization of surfaces, the reasons for cleaning vary. Cleaning may be used to get as-received material into a uniform condition before processing or it may be necessary for processing (e.g., in solid-solid bonding or to obtain adhesion of thin films) or, as in silicon technology, for uniform diffusion. Prior to packaging or sealing, surfaces may be cleaned to reduce the final level of contamination in the completed unit, or cleaning may be necessary for the material to function actively such as with an electron emitter, a catalytic surface or a gas getter. In some cases, industry has devoted a great deal of effort to providing a "clean" environment for processing as is the case with "clean" rooms where clean only refers to particulate contamination.

Surface contaminants may be classed as (1) reaction layers, (2) adsorbed layers, (3) variable composition contaminants, or (4) particulate contaminants. Reaction layers may be in the form of oxides, carbides, or other such layers. The

---

*Equating adhesion, which is a gross effect, with bonding or cleanliness may be very misleading. Failure of adhesion may be related more to fracture mechanisms than to bonding. In thin films the intrinsic stress may result in adhesive failure even though chemical bonding may be strong; also the interfacial morphology may lead to easy fracture even though the chemical bonding is strong.

material which forms these layers may come from the external environment or from the bulk material. Adsorbed layers form by the adsorption of material from the environment or by diffusion over the surface from the surroundings. Variable composition contaminants may occur as layers or regions near the surface, perhaps by the diffusion of a minor constituent from the bulk to the surface region giving a high surface concentration or they may be in the form of second-phase material strongly bound on the surface. Particulate contaminants, such as dust, are loosely bound contaminants whose origin is usually the external environment. In some cases, surface contamination may result from the "cleaning" techniques used. For example, vacuum heating may result in bulk diffusion, and the concentration of a minor constituent in the surface region or preferential etching may change the surface composition.

To design a cleaning process, we need to know the probable origins of contaminants. In a typical processing sequence the following factors may be important in cleaning and contamination control: (1) the as-received condition; (2) the time; (3) the component design; (4) processing; (5) handling; (6) storage.

Contaminants on as-received material may be deceptive since the prior history of the material is often not known. The supplier may have changed his procedures, so the type of contaminant may have changed, requiring a change in the cleaning procedure. Often, the cleaning of as-received material to put it into a known condition is a good first step in the design of a cleaning process.

Time may be an important factor in cleaning. Time, particularly in combination with an elevated temperature, will allow diffusion and reaction. The diffusion of contaminants may be from the bulk or across the surface. Reactions of the surface material may occur with these diffused species or with contaminants in the atmosphere. Time also allows the adsorption of contaminant vapors or the deposition of particulate matter from the surroundings. Some processes may be impossible to perform if the time between the processing steps is excessive. An example is the activation of a nickel surface, which is necessary prior to the electrodeposition of gold in order to obtain an adherent film. If the activated surface is exposed to air and allowed to dry, the gold will not adhere.

Component design is often a source of contamination. For example, a contaminant-sensitive electrical contact might be sealed in a component containing lubricants or volatile hydrocarbons such as are present in many adhesives and sealants. Over a period of time, or in subsequent processing such as sealing or vacuum baking, the contaminants may be transferred to the sensitive region; thus component designs must be reviewed with a view to possible sources of in-process or storage contamination.

Processing steps may lead to contamination. The contamination resulting from processing may be necessary or unnecessary. Necessary contamination may be residual polishing compounds, residuals from etching, residual from photoresist removal, etc. Unnecessary contaminants result from poor handling and storage procedures, an uncontrolled environment, contaminated cleaning solutions, etc. The necessary contaminants should be recognized and cleaning steps should be designed to remove them. Unnecessary contaminants should be avoided by proper design procedures and by control over the components and cleaning materials.

Contaminants which result from handling may also be necessary or unnecessary. Some handling is always necessary and procedures should be designed to minimize contaminants from this source. Unnecessary handling should be avoided.

In addition to designing a cleaning procedure to remove undesirable contaminants from the surface, we must be concerned with the effect of the cleaning procedure on the surface. The cleaning procedure may lead to an undesirable surface composition. For instance, a high temperature, high vacuum baking of stainless steel volatilizes the protective chromium oxide and allows the surface to rust on exposure to the atmosphere. Cleaning may also affect the surface morphology by selective etching, thermal faceting, etc. This change in morphology may affect subsequent processing. The cleaning process may leave undesirable residues on the surface. For instance, some electropolishing solutions leave inorganic compounds on the surface. Another concern is how to monitor surface cleaning both as a routine nondestructive in-process step and as a destructive lot sampling. This problem will be discussed in a later section.

Cleaning processes may be categorized as specific or general. Specific cleaning processes are designed to remove specific types of contaminants without changing the surface of interest. Examples are solvent cleaning techniques which dissolve or emulsify contaminants without attacking the surface. Other examples are volatilization techniques where the contaminant is vaporized by heating or is reacted with a gas to form a volatile compound (e.g., the oxidation of carbon contaminants). General cleaning techniques often involve the removal of some of the surface material as well as the contaminants. Etching procedures such as sputter cleaning, electropolishing, and chemical etches fall into this category, as does mechanical abrasion. Often a cleaning process will involve several procedures used in sequence. Unless the cleaning procedure is designed to take advantage of the strong points and weak points of each step, the cleaning steps may defeat each other. For example, an etching step may be used to remove the surface oxide on silver but, if it is followed by a vacuum bake, oxygen may diffuse from the bulk to the surface forming a new oxide layer. Generally, cleaning procedures should be kept as simple as possible.

Cleaning Oxide Surfaces

In the deposition of thin films on oxide substrates, it is generally considered that a chemical reaction between the film and the surface is necessary for adhesion. Thus metals which are strong oxide formers should adhere well to oxide surfaces by forming interfacial oxide regions, but metals such as gold, which do not form stable oxides, should have poor adhesion to oxide surfaces. There are notable exceptions to this rule which raise questions about the validity of the assumption. It has also been proposed that charge transfer between the surfaces in contact leads to an electrical double layer which provides a mechanism of adhesion without chemical reaction.

The nucleation of atoms deposited on a surface may be influenced by a variety of factors. These include deposition techniques, deposition conditions, deposition environment, substrate temperature, and substrate surface chemistry. Generally, it is found that the sputter deposition of films gives better adhesion than vacuum deposition. This may be attributed to the plasma cleaning and higher energy of the depositing atoms, as well as to the heating and electron/neutral atom bombardment of the substrate which normally accompanies sputter deposition. The gaseous environment during deposition can also affect the adhesion of metal films to oxide surfaces. Oxygen is sometimes used to generate an interfacial layer which is conductive to good adhesion.

Substrate surface chemistry may also have an important effect on the adhesion of metal films to oxide surfaces. It has long been known that polishing glass surfaces with chalk, calcium oxide, or calcium carbonate can give more adherent films. This is attributed to the embedding of the polishing compound in the substrate surface, which results in better nucleation of the depositing film material. It has also been shown that small amounts of calcium and silicon on Al_2O_3 surfaces are necessary for the good adhesion of Ta_2N films. Selective leaching of these materials from the surface may result in adhesion problems.

Other procedures may be used to modify the surface chemistry in order to improve adhesion. The most widely used technique is multilayer metallization,[4] in which a thin layer of reactive material such as titanium or chromium is deposited in contact with an oxide and then the desired metallization, usually gold, is deposited as the upper layer. The "adhesive layer" reacts chemically with the substrate and alloys with the gold to give good adhesion to both materials.

Other techniques may be used to modify the interfacial chemistry. By beginning the sputter deposition in a partial pressure of oxygen and by removing the oxygen as the deposition proceeds, a graded interfacial oxide composition can be obtained as the composite film metallizing system. Another technique to change the surface chemistry is to introduce desirable atoms into the substrate surface by diffusion or ion bombardment. An example of this is the improvement of the adhesion of tungsten to an SiO_2 surface by the introduction of aluminum into the surface. In this case aluminum, which has a lower valence (+3) than silicon (+4), is introduced into the surface, resulting in an unsatisfied oxygen bond, and a site becomes available for oxygen bonding to the tungsten.

Regardless of the mechanism of adhesion, it is necessary that the surface be free of contaminants which inhibit the interaction between the surface and the atoms of the depositing film. The most commonly encountered contaminants on oxide surfaces are hydrocarbons, hydroxyl ions, and alkali halides.

In order to remove water-soluble contaminants and some organic contaminants, the substrates should be cleaned in a good detergent or scrubbed with an abrasive cleaner. Mechanical scrubbing is more effective than ultrasonic agitation. For smooth glass surfaces, scrubbing with a lint-free de-sized cloth using a detergent water solution is adequate. If a surface is rough or porous, as are those of most sintered ceramics, the cleaning should be assisted by ultrasonic agitation in which jetting, which accompanies the collapse of bubbles generated by cavitation near a surface, increases the rate of solvation or emulsification. When solvents are used the surface should not be allowed to dry before the final rinse. The surface should be thoroughly rinsed in distilled/deionized water after detergent cleaning. In order to avoid water strains, the surface may be flushed with absolute alcohol which will displace the water. Ultrasonic rinsing helps to remove particulate contamination. Activated carbon black prepared from natural gas has been reported to be an excellent mild abrasive cleaner for organic contaminants. A slurry of carbon black is prepared using deionized water, and the surface is cleaned on a lint-free polishing pad. The slurry is easily removed by rinsing.

Solvent cleaning is widely used for cleaning insulator surfaces but with the advent of new industrial safety standards many of the organic solvents are difficult to use in production environments. Generally, solvent cleaning is no improvement on a good wash followed by air firing, oxygen plasma cleaning, glow discharge cleaning, or sputter etching of an oxide surface.

A number of cleaning procedures may be used to remove oxidizable contaminants such as hydrocarbons. These include boiling in hydrogen peroxide (H_2O_2), treatment with hot oxidizing solutions such as nitric acid or potassium dichromate, exposure to an oxygen plasma, high temperature air firing or exposure to short wavelength UV radiation in oxygen (UV/O_3 cleaning). If the oxidation product is volatile (e.g., CO or CO_2) the contaminant volatizes; if the product is soluble it may be dissolved in the cleaning solution. Air firing, oxygen plasma treatment, and UV/O$_3$ treatment have the advantage that organic conttaminants are reactively volatilized, and residues from the cleaning solutions are avoided.

For air fire cleaning the substrate should be heated to the highest temperature possible (about 1000°C for Al$_2$O$_3$ ceramics) in a furnace which contains no volatile species which might react with the substrate surface in an undesirable way. The substrates should be removed and placed in the deposition system while still warm or hot to minimize recontamination. Often furnaces have particulate contamination which originates from the insulation. Care must be taken to minimize this source of contamination.

While heating in air is a good way to remove carbon materials from oxide surfaces, vacuum heating may cause problems by carbonizing hydrocarbons and leaving a carbonaceous layer on the surface. Electron bombardment of surfaces in vacuum will also carbonize hydrocarbons on the surface.

Plasma oxidation is a very effective technique for removing oxidizable contaminants whose products are volatile. Hydrocarbons are removed very effectively and gross amounts of photoresist can be removed from a surface, although the process does leave non-volatile residues from the photoresist. Ozone, which is a strong oxidizing agent, is formed in the plasma, as is intense UV radiation. The UV radiation is very effective in causing bond scission which allows effective oxidation. It has been found that ozone by itself is not effective in removing gross contamination although UV radiation is effective. A disadvantage of oxygen plasma treatment is that it must be done in a vacuum system; this may be an advantage if in situ cleaning is desired and oxidation of other material in the system is not a problem.

In the UV/O$_3$ cleaning process the surface to be cleaned is exposed to short wavelength UV radiation (1849 Å) from a mercury vapor UV lamp in the presence of oxygen. As in oxygen plasma cleaning, bonds are broken and ozone is formed. This allows the efficient oxidation of surface contaminate. The UV/O$_3$ cleaning rates are lower than the oxygen plasma cleaning rates since the concentrations and intensities are lower. The UV/O$_3$ cleaning is as effective as sputter cleaning and much more effective as sputter cleaning and much more effective than an ultrahigh vacuum bakeout at 200°C. Generally, we clean oxide surfaces in a stagnant air chamber with UV lamps and aluminum reflectors inside the chamber.

Exposure of oxide surfaces to an inert gas discharge (glow discharge cleaning) has long been used to clean surfaces prior to film deposition. The cleaning mechanism is poorly defined since the surface is being bombarded with ions, electrons and high energy neutrals, as well as with radiation from the plasma. The exposure of an insulating surface to a d.c. plasma shielded from the cathode allows a small negative potential to build up on the surface (wall potential) due to the higher mobility of electrons compared with the ions in the plasma. This allows a low energy "ion scrubbing" of the surface to take place. If the surface is not shielded from the cathode, it will also be bombarded by high energy electrons and high energy neutrals from the cathode.

The exposure of an insulating surface to a plasma may give other effects which may contribute to thin film adhesion. The surface chemistry of the surface may be altered by gas sorption and desorption. It has been proposed that changes in the surface electronic states of insulator surfaces exposed to electron and ion bombardment and to UV radiation from the plasma contribute to improved adhesion. It has been reported that the adhesion of aluminum to silicon can be improved by the exposure of the silicon to UV illumination just prior to deposition. This effect is attributed to the increased chemical activity of the surface. It has also been shown that charge sites on the surface can originate from a variety of other sources which are dependent on deposition conditions. Electrostatic charging of the surface can affect nucleation density and may inhibit the coalescence of nuclei by electrostatic repulsion. An electrostatic field in the plane of the surface may enhance coalescence.

Sputter cleaning or sputter etching has the advantage that it may be done in situ in the deposition system just prior to the deposition process. In the sputter cleaning of an insulator surface an r.f. potential is impressed on the surface and the surface is then bombarded alternately with ions and electrons. The ion bombardment results in removal of material by a momentum transfer process called sputtering. Sputter cleaning is an effective cleaning technique, but problems may arise because of backscattering of sputtered material to the surface, particularly at high pressures; therefore, a low initial contaminant level should be achieved and a low pressure discharge and a high system throughput should be maintained in order to flush away sputtered contaminants. In some cases, gas incorporated into the sputtered surface may be released during subsequent processing, causing loss of adhesion of a film deposited on a sputter-cleaned surface. This can be avoided by heating the substrate during or after cleaning.

In the ion plating process, sputter cleaning is an integral step in the deposition procedure. In this technique the film deposition is begun while the surface is being sputtered. A film results only when the deposition rate is greater than the removal rate. For insulators, the substrate potential must be r.f. to prevent a build-up of charge which would halt d.c. sputter cleaning.

Various chemical etching techniques are also used to clean ceramic-type material. Hydrofluoric acid is a strong etchant for most of the oxides. Ultrasonic agitation or mechanical scrubbing in ammonium bifluoride may be used to etch or clean glass lightly. Other glass-cleaning formulations have been devised. Chemical etching has also been used to improve film adhesion by changing the surface morphology. If high alumina ceramics are etched in fused NaOH, it is found that the adhesion of copper to the surface is greatly improved. In this case the improved adhesion is attributed to microroughening of the surface, which improves the mechanical interlocking.

Particulate contamination may best be removed from surfaces by blow-dusting. In some cases an ionized gas may be used to avoid charging of the surface. Solvent sprays may also be used.

Cleaning Metal Surfaces

The adhesion between deposited metal films and metal surfaces depends on the interfacial morphology and the nucleation, chemical reaction, and diffusion in the interfacial region. It has often been assumed that to obtain good adhesion there must be mutual solubility in the metal-metal system. Solubility is a desirable characteristic, but it is not necessary, as has been demonstrated by the good adhesion which can be obtained between completely insoluble materials such as Cu-W and Au-W. Adhesion in the metal-metal system is probably more dependent on how the materials are in contact and on the fracture mechanisms than on the solubility. A lack of solubility will affect the nucleation of the depositing atoms on the surface and thus will affect the interfacial morphology. The extent of interactions in the interfacial region may also be important in adhesion. The growth of extensive, brittle intermetallic layers may give poor strength. Kirkendahl porosity, which weakens the interface, may occur if the diffusion rates are asymmetric.

The presence of surface contaminants may prevent diffusion and reaction between the depositing atoms and the surface. Very thin carbon or oxide layers

(10-100 Å) may provide effective surface barrier layers which prevent good bonding or adhesion. Since surface oxides and adsorbed hydrocarbons are common contaminants on metal surfaces, the elimination of these materials is often of prime concern in thin film deposition technology.

The surface contaminants on metals may be removed by a variety of techniques which may depend on the metals involved. As with oxide surfaces, the metal surfaces should be detergent or solvent-cleaned to rid the surface of soluble contaminants. Ultrasonic agitation will aid in this cleaning, although cavitation may affect the surface morphology of some metals, notably aluminum.

Gross contamination, e.g., thick oxides, may be removed by abrasive cleaning, etching or fluxing. Etching may be performed by chemical etching or electropolishing, depending on the metals involved. Various fluxes may be used to dissolve or float away thick oxides. These fluxes are often molten salts. After abrasive cleaning, etching or fluxing, care must be taken to remove residues from the surface by rinsing.

The most widely applicable cleaning technique for removing trace contamination is sputter cleaning followed by ultrahigh vacuum heating to desorb gases incorporated in the surface during cleaning. For conductive surfaces, the applied sputtering potential may be d.c. or r.f., although r.f. is preferred. Special sputtering configurations have been developed to sputter-clean metal surfaces in a magnetron sputtering system.

When sputter cleaning is used, care must be taken to flush away the sputtered contaminants by using a high system gas throughput. The discharge parameters may be used as an indication of cleaning and system condition. As the surface and system are cleaned and contaminants are removed from the discharge, the discharge current falls to an equilibrium value. The time for this value to be reached depends on the system throughput and the background contamination level. The total power input to the system is sometimes used as a measure of cleaning efficiency. Often it is not realized that sputter cleaning can do more harm than good if the proper conditions are not used. For sputter cleaning to be effective on reactive materials, the sputtering environment must be free of contaminating gases. If these gases are not eliminated, they will be ionized in the discharge and will react with the cleaned surface. Also, a sputter-cleaned surface is very reactive and will react rapidly with contaminants which are present after cleaning is complete. An advantage of sputter cleaning is that it may be done in situ just prior to film deposition in many systems. When an object is being sputter-cleaned, care must be taken that the surface is not contaminated by material sputtered from holders and shields. As was discussed earlier, the ion plating process has sputter cleaning and surface heating as integral parts of the deposition process so that recontamination is minimized.

Vacuum or hydrogen firing may also be used to clean some metal surfaces. Some metal oxides have a high volatility compared with the metal. For instance, molybdenum oxide is quite volatile at high temperatures in vacuum. Gold melted on a molybdenum surface in vacuum will not wet the surface initially because of the presence of the oxide. As the oxide volatilizes, the gold will wet the molybdenum surface very well even though gold is insoluble molybdenum. Hydrogen firing will clean the oxide from some metal surfaces, but care should be taken to remove the hydrogen from the system while the metal is hot if the hydrogen is soluble in the metal at lower temperatures. A number of chemical and electrochemical cleaning techniques exist for particular metals and contaminants.

Glow discharge cleaning in an inert gas or oxygen may be used to remove hydrocarbon contamination from metal surfaces. An oxygen discharge may cause excessive oxidation of some metals which do not form passive oxides. For instance, silver will oxidize heavily in an oxygen discharge. Direct absorption of ultraviolet radiation may cause the photodesorption of adsorbed gases. Pulsed laser bombardment has been used to clean metal surfaces in vacuum. Oxygen discharges, inert gas discharges, hydrogen discharges, and UV/O_3 cleaning have been used to clean the metal walls of ultrahigh vacuum systems.

In the semiconductor technology industry, reactive gases have been used to etch metal and dielectric surfaces to generate patterns. Recently, reactive plasma cleaning has been used to generate clean metal surfaces at a lower power input than is possible with sputter cleaning alone.[5]

The cleaning of gold presents a unique case. Since gold does not readily oxidize, it may be cleaned by the oxidizing treatments discussed under the cleaning of oxides. These include heating in air or oxygen, plasma cleaning, chemical oxidation treatments, and UV/O_3 cleaning.

Mechanical disruption may be used to clean a metal surface prior to film deposition. Bead or grit blasting is a favorite surface preparation process in plasma spraying. "Mechanical Activation" by wire brushing also seems to aid adhesion in some metallic coating systems.

Cleaning Organic Polymer Surfaces

The cleaning of organic polymer surfaces for thin film deposition poses some interesting problems since organic polymers are very complex materials. In addition, the mechanism of adhesion between metals and organics is not well defined. Some maintain that surface energies are the prime consideration, while others maintain that the type of bonding is important and that only dispersion forces are operational or important in metal-organic bonding. Charge transfer may also be relevant in metal-organic adhesion. The surface of organic materials is often composed of low molecular weight fractions which fracture easily. To obtain good adhesion, the surface must be modified in such a way that it does not represent a zone of weakness at the interface; thus, particularly in the case of organics, cleaning means modifying the surface region in a desirable way.

Some undesirable organics may be removed from the surface by a detergent wash or by solvent cleaning. Solvents may be absorbed into the polymer structure, causing it to expand and possibly to craze on drying, or they may leach some of the low molecular weight fractions from the surface.

Generally, surface treatments of organics that are conducive to the good adhesion of deposited thin films may be classed as (1) roughening, (2) activation, or (3) cross-linking. Surface roughness and mechanical interlocking have been shown to be important factors in adhesion. The roughness may be obtained mechanically by abrasion, grit blasting, etc., or it may be obtained by preferentially etching the organic surface. For instance, the copolymer ABS (acrylonitrile-butadiene-styrene) may be etched using a chromic-sulfuric acid solution to give microroughening of the surface.

The activation of some polymer surfaces may be accomplished by chemical treatments in which organometallic complexes are formed in the surface regions. Such activation treatments are the iodine treatment of nylon and the sodium treatment of Teflon. Activation may also be accomplished by the addition of a small amount of material which reacts with both the substrate surface and the deposited film (primer). These materials may be highly polar low molecular weight organics.

Cross-linking may be used to increase the strength of the surface material and to reduce the amount of undesirable low molecular weight components. Cross-linking can be encouraged by (1) a corona discharge, (2) exposure to a plasma, and (3) exposure to UV radiation. These treatments are probably not fundamentally different, since the UV radiation from a corona, gas discharge or UV lamp causes bond scission and adds carbonyl groups to the polymer surface. In thin film technology, exposure of the polymer surface to a gas discharge can often be done in situ before film deposition.

Surface modification may also occur during film deposition. If the film is deposited by sputter deposition, the depositing atoms have energies much higher than thermal energies and may be implanted into the surface to an appreciable depth. These physically incorporated atoms may then act as nucleating and bonding sites for further atom deposition. A problem with the sputter deposition of films on plastics is the heating by secondary electrons which may be present. This electron bombardment can be reduced by the use of a magnetron-type sputter deposition apparatus.

Monitoring Surface Cleaning

Since cleaning is "in the eye of the user," the best monitor of cleaning is subsequent processing or usage. This criterion is often not acceptable from either a practical or an aesthetic standpoint, so some type of in-process testing and acceptance specification is required. Unfortunately, most testing affects the surface and so testing must usually be performed on representative surfaces which then qualify the untested surfaces. This means that the surface to be tested must be shown to be representative of the surfaces to be utilized and that rigid cleaning specifications must be provided so that all surfaces are cleaned alike.

Monitoring techniques may be classed as direct or indirect. Direct monitoring involves identifying the surface species directly. This may be done by various surface analytical techniques such as Auger electron spectroscopy (AES), appearance potential spectroscopy (APS), ion scattering spectrometry (ISS), multi-reflectance IR analysis, and others. Of prime concern are the level of detection of the technique and the value judgment as to whether the amount and type of the detected species are important. AES, ISS, and APS usually only detect elemental species and do not determine their chemical state; therefore they do not readily differentiate between carbon as an element and carbon as a hydrocarbon.

Indirect monitoring utilizes the behavior of the surface to detect the presence of a contamination. Again, the technique may not indicate whether the type and amount of the contaminant is harmful or not. The most generally used indirect monitoring techniques for detecting low surface energy contaminants such as hydrocarbons may be classed as (1) wetting or contact angle, (2) coefficient of adhesion or coefficient of friction, (3) adsorption, (4) nucleation, and (5) adhesion.

In the wetting test, if a clean substrate is slowly withdrawn from a container filled with pure water, a continuous film of water remains on the surface. This test will not detect hydrophilic contaminants. One type of wetting test is contact angle measurement where a drop of pure liquid (usually water) is placed on the surface and the contact angle is measured. Spontaneous spreading or a low contact angle (less than $5°$) means wetting and a surface free of hydrophobic contaminants. Measurement of the resistance of a sliding surface on the substrate gives the coefficient of friction; the higher the coefficient of friction, the cleaner the surface. The coefficient of adhesion can be used in much the same way.

The absorption of dyes or radioactive tracer materials can be used to detect contaminants which will absorb these materials. The nucleation of water (black breath figure or atomizer test) may be used to detect hydrophobic contaminants. If the surface is clean, the water will form a continuous film; if it is contaminated, droplets will form. The nucleation of zinc can be used in much the same way. Thin film adhesion testing has been reviewed elsewhere, and these techniques may also be used for contaminant monitoring since they probably parallel most closely the conditions of use. Other tests include the measurement of contact potentials and electrolytic deposition techniques.

Ideally, we would like to generate an atomically clean surface and then contaminate the surface in a known way while monitoring the effect on subsequent processing and usage. This would allow the delineation of the values for significant amounts of undesirable materials. Unfortunately, in most cases this is impossible or impractical.

Clean Surface Storage

An often neglected aspect of cleaning is that of storage. After an elaborate cleaning procedure, it is not too unusual to find that the surface is stored in a plastic bag or tray in processing or in a contaminated atmosphere in use; therefore, storage and packaging must be considered as part of the cleaning process. In-process storage normally means keeping contaminants away. Storage in an ultra-clean controlled environment is usually unnecessary. The problem is to identify the undesirable contaminants and eliminate them from the storage environment.

Dust may be minimized by storing the cleaned surface in a closed container or in a clean bench. One of the most common contaminants is hydrocarbon vapor adsorbed from the air. Adsorptance of such contaminants may be minimized by storing in freshly oxidized aluminum containers which preferentially adsorb the hydrocarbons. A disadvantage of this technique is that extra steps are involved in periodically stripping and oxidizing the metal surfaces. The UV/O_3 cleaning technique may be used to keep oxide and gold surfaces clean indefinitely in an ambient environment. Cleaned surfaces should at least be stored in clean glass containers.

Often clean surfaces are sealed in a hermetic device. In some cases these surfaces are contaminated by the sealing operation which frequently involves heating of other non-clean surfaces. Before sealing, all interior surfaces should be as clean as possible. Even in the case of a vacuum bake, the conductance of the vacuum tubulation before sealing may be so small as to preclude the effective removal of the contaminants. In these cases, flushing with an inert atmosphere during the bakeout may help to eliminate recontamination. The design of hermetically sealed components should exclude sources of contamination such as adhesives and lubricants.

Since thin film systems normally have a high surface-to-volume ratio, they may be very sensitive to surface contaminants such as corrosive agents. For instance, the Ti-Au metallization system is very sensitive to chlorine contamination; thus care must be taken in process and in use to eliminate chlorine or halogenated organics from the cleaning process and the storage environment.

Non-Cleaning Surface Treatments

In most instances surface preparation (cleaning) does not give an atomically clean surface. "Cleaning" results in developing a surface with acceptable amounts of undesirable material. In some instances the "cleaning" procedure actually adds foreign material to the surface which aids adhesion. If foreign material is added deliberately, it may be called "activation" (iodine treatment of polymer, zincate or stanate treatment of metals), addition of nucleating agents (Bi_2O_3 on glass) or the formation of an adhesive interfacial layer (Ti in Ti-Au metallization of oxides). Nucleating agents and adhesion materials are usually ones that react with both the substrate and film material. For instance, titanium on an oxide surface will form a complex oxide reaction layer while the unreacted titanium will diffuse into the gold overlayer.

SUMMARY

Good adhesion (practical) is the result of many factors which have been discussed. Poor "cleaning," weak interfacial regions, high stresses or degradation can give rise to poor adhesion. In many instances, the origin of poor adhesion can be defined and remedied by consideration of the factors which lead to good adhesion. It has often been the case that the process as a whole is not considered and solutions are not found. For instance, improved cleaning may not solve the adhesion problem if the interface is porous or there is high intrinsic stress in the coating.

REFERENCES

(1) B.A. Movchan and A.V. Demchishin, *Phys. Metal. Metallogr. 28*, 83 (1969).
(2) John A. Thornton, *Ann. Rev. Mater. Sci. 7*, 239 (1977).
(3) D.M. Mattox, *Proc. of the Eighth International Vacuum Congress*, Cannes, Sept. 22-26, 1980, p. 297.
(4) D.M. Mattox, *Thin Solid Films 18*, 173 (1973).
(5) G.J. Kominiak and D.M. Mattox, *Thin Solid Films 40*, 141 (1977).

4

Evaporation

Rointan F. Bunshah

Materials Science and Engineering Department
University of California
Los Angeles, California

GENERAL INTRODUCTION

Physical Vapor Deposition (PVD) technology consists of the techniques of evaporation, ion plating and sputtering. It is used to deposit *films and coatings* or *self-supported shapes* such as sheet, foil, tubing, etc. The thickness of the deposits can vary from angstroms to millimeters. The application of these techniques ranges over a wide variety of applications from decorative to utilitarian over significant segments of the engineering, chemical, nuclear, microelectronics and related industries. Their use has been increasing at a very rapid rate since modern technology demands multiple and often conflicting sets of properties from engineering materials, e.g., combination of two or more of the following—high temperature strength, impact strength, specific optical, electrical or magnetic properties, wear resistance, fabricability into complex shapes, biocompatibility, cost, etc. A single or monolithic material cannot meet such demand in high technology applications. The resultant solution is therefore a composite material, i.e., a core material and a coating each having the requisite properties to fulfill the specifications.

PVD technology is a very versatile, enabling one to deposit virtually every type of inorganic materials—metals, alloys, compounds and mixtures thereof, as well as some organic materials. The deposition rates can be varied from 10 to 750,000 Å (10^{-3} to 75 μm) per minute, the higher rates having come about in the last 20 years with the advent of electron beam heated sources.

The thickness limits for thin and thick films are somewhat arbitrary. A thickness of 10,000 Å (1 μm) is often accepted as the boundary between thin and thick films. A recent viewpoint is that a film can be considered thin or thick depending on whether it exhibits surface-like or bulk-like properties.

Historically the first evaporated thin films were probably prepared by Faraday[1] in 1857 when he exploded metal wires in a vacuum. The deposition of thin metal films in vacuum by Joule heating was discovered in 1887 by Nahrwold[2] and was used by Kundt[3] in 1888 to measure refractive indices of such films. In the ensuing period, the work was primarily of academic interest being concerned with optical phenomena associated with thin layer of metals, researches into kinetics and diffusion of gases, and gas-metal reactions.[4,5] The application of these technologies on an industrial scale had to await the development of vacuum techniques and therefore dates to the post World War II era, i.e., 1946 and onwards. This proceeded at an exponential pace in thin films and is covered in an excellent review by Glang[6] on evaporated films and in other chapters of the Handbook of Thin Film Technology[7] as well as in the classic text by Holland.[8] A more recent reference on the Science and Technology of Surface Coatings[9] includes material on PVD techniques as well as the other techniques for surface coatings. The work on mechanical properties of thin films has been reported in several review articles.[10-15]

The work on the production of full-density coatings or self-supported shapes by high deposition rate PVD processes started around 1961 independently at two places in the U.S.A. Bunshah and Juntz at the Lawrence Livermore Laboratories of the University of California produced very high purity beryllium foil,[16-21] titanium sheets[22] and studied the variation of impurity content, microstructure and mechanical properties with deposition conditions, thus demonstrating that the microstructure and properties of PVD deposits can be varied and controlled. At about the same time, Smith and Hunt were working at Temescal Metallurgical Corporation in Berkeley on the deposition of a number of metals, alloys and compounds and reported their findings in 1964.[22,23] The development of evaporation processes in the U.S.S.R is described in the Appendix kindly supplied to the author by Dr. A.V. Demchishin of the Paton Electric Welding Institute, Kiev.

In the years between 1962 and 1969, there was considerable effort on the part of various steel companies to produce Al and Zn coatings on steel using HRPVD techniques on a production scale.[25,26] In 1969, Airco Temescal Corp. decided to manufacture Ti-6Al-4V alloy foil in pilot production quantities for use in homeycomb structures on the SST aircraft. The project was eminently successful but the patient, the supersonic transport aircraft "SST" died. The results of this work were published in 1970.[27] To give some idea of the production capability, 1,200 ft/run of Ti-6Al-4V foil, 12" wide, 0.002" thick was produced at the rate of 2 to 3 ft/min. The stated cost at that time was about one-fifth of the cost for similar material produced by rolling (i.e., $60/lb for HRPVD vs $300/lb for rolled material). It is very difficult to roll this alloy because it work-hardens very rapidly and therefore needs many many annealing cycles to be reduced to thin gauge (A.B. Sauvegot, TMCA Tech. Report AFML-TR-67-386, Dec. 1967).

The work on thick films and bulk deposits has matured later than the work on thin films and reviews on it have been given by Bunshah[61] and by Paton, Movchan and Demchishin[67] who summarized the work done at the Paton Electric Welding Institute up to 1973. In addition, the Soviet literature in the 1960s has numerous references to the extensive work on thin and thick films by Palatnick and coworkers of the Kharkov Polytechnic Institute (see Appendix). Note should also be made of a recent book in German on electron beam technology by Schiller, Heisig and Panzer in which many of the PVD aspects are treated.[27]

SCOPE

The scope of this chapter will be to review the evaporation technologies, theory and mechanisms, processes, deposition of various types of materials, the evolution of the microstructure and its relationship to the properties of the

Evaporation

deposits, preparation of high purity metals, current and future applications, and finally cost analysis as far as possible.

PVD PROCESSES

Preamble

In general, deposition processes may principally be divided into two types: (1) those involving droplet transfer such as plasma spraying, arc spraying, wire-explosion spraying, detonation gun coating, and (2) those involving an atom by atom transfer mode such as the physical vapor deposition processes of evaporation, ion plating and sputtering, chemical vapor deposition, and electrodeposition. The chief disadvantage of the droplet transfer process is the porosity in the final deposit which effects the properties.

There are three steps in the formation of any deposit:

(1) Synthesis of the material to be deposited —
   (a) Transition from a condensed phase (solid or liquid) to the vapor phase.
   (b) For deposition of compounds, a reaction between the components of the compound some of which may be introduced into the chamber as a gas or vapor.
(2) Transport of the vapors between the source and substrate.
(3) Condensation of vapors (and gases) followed by film nucleation and growth.

There are significant differences between the various atom transfer processes. In chemical vapor deposition and electrodeposition processes, all of the three steps mentioned above take place simultaneously at the substrate and cannot be independently controlled. Thus, if a choice is made for a process parameter such as substrate temperature (which governs deposition rate in CVD), one is stuck with the resultant microstructure and properties. On the other hand, in the PVD processes, these steps (particularly steps 1 and 3) can be independently controlled and one can therefore have a much greater degree of flexibility in controlling the structure and properties and deposition rate. This is a very important consideration.

PVD Processes

There are three physical vapor deposition processes, namely evaporation, ion plating, and sputtering.

In the evaporation process, vapors are produced from a material located in a source which is heated by direct resistance, radiation, eddy currents, electron beam, laser beam or an arc discharge. The process is usually carried out in vacuum (typically 10^{-5} to 10^{-6} torr) so that the evaporated atoms undergo an essentially collisionless line-of-sight transport prior to condensation on the substrate. The substrate is usually at ground potential (i.e., not biased).

Figure 4.1 is a schematic of a vacuum evaporation system illustrating electron beam heating. It may be noticed that the deposit thickness is greatest directly above the center-line of the source and decreases away from it.[28] This problem is overcome by imparting a complex motion to substrates (e.g., in a planetary or rotating substrate holder) so as to even out the vapor flux on all parts of the substrate; or by introducing a gas at a pressure of 5 to 200 μm into the chamber so that the vapor species undergo multiple collisions during transport from the source to substrate, thus producing a reasonably uniform (±10%) thickness of coating on the substrate. The latter technique is called gas-scattering evaporation or pressure plating.[29,30]

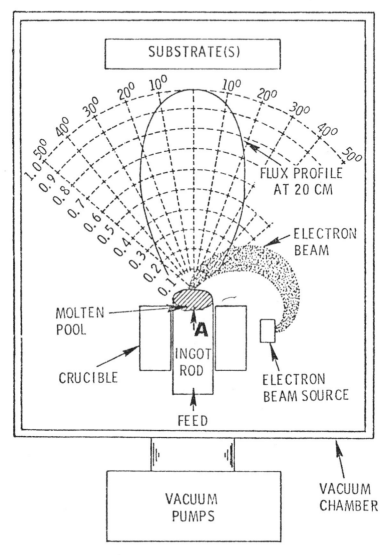

Figure 4.1: Vacuum-evaporation process using electron beam heating.

In the ion-plating process, the material is vaporized in a manner similar to that in the evaporation process but passes through a gaseous glow discharge on its way to the substrate, thus ionizing some of the vaporized atoms (see Figure 4.2). The glow discharge is produced by biasing the substrate to a high negative potential (-2 to 5 kV) and admitting a gas, usually argon, at a pressure of 5 to 200 μ into the chamber. In this simple mode, which is known as diode ion-plating, the substrate is bombarded by high-energy gas ions which sputter off the material present on the surface. This results in a constant cleaning of the substrate (i.e., a removal of surface impurities by sputtering) which is desirable for producing better adhesion and lower impurity content. This ion bombardment also causes a modification in the microstructure and residual stresses in the deposit. On the other hand, it produces the undesirable effects of decreasing the deposition rates since some of the deposit is sputtered off; as well as causing a considerable (and often undesired for micro electronic applications) heating of the substrate by the intense gas ion bombardment. The latter problem can be alleviated by using the supported discharge ion-plating process[31] where the substrate is no longer at the high negative potential, the electrons necessary for supporting the discharge coming from an auxiliary heating tungsten filament. The high gas pressure during deposition causes a reasonably uniform deposition of all surfaces due to gas-scattering as discussed above.

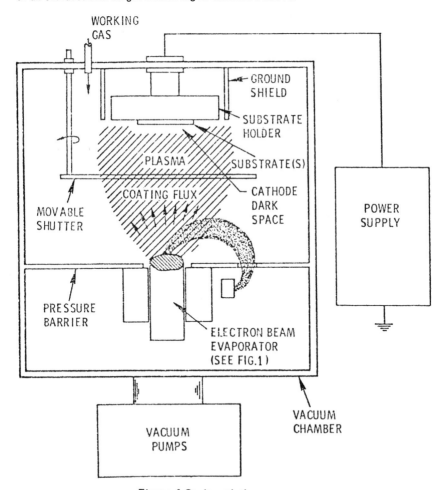

Figure 4.2: Ion-plating process.

In the sputtering process, illustrated schematically in Figure 4.3, positive gas ions (usually argon ions) produced in a glow discharge (gas pressure 20 to 150 μ) bombard the target material (also called the cathode) dislodging groups of atoms which then pass into the vapor phase and desposit onto the substrate. Alternate geometries of importance in various processing applications are shown in Figure 4.4. For example, hollow cathode sputtering would be the ideal geometry for coating the outer surface of a wire. Sputtering is an inefficient way to induce a solid-to-vapor transition. Typical yields (atoms sputtered per incident ion) for a 50 eV argon ion incident on a metal surface are unity. Thus the phase change energy cost is from 3 to 10 times larger than evaporation.[32] Thornton[32] has provided an excellent review on sputtering as it applied to deposition technology. The reader is also referred to the proceedings of a special conference on Sputtering and Ion-Plating.[33]

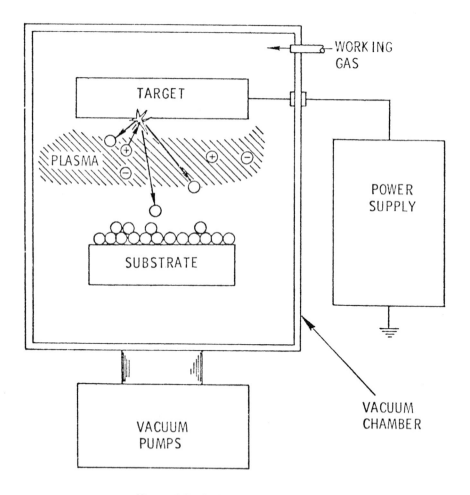

Figure 4.3: Basic sputtering process.

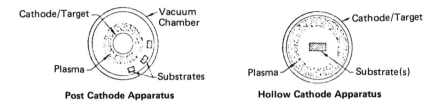

Figure 4.4: Cylindrically symmetric sputter-coating systems.

The deposition rates for the various processes are indicated in Table 4.1. The deposition rates of the evaporation and ion-plating processes are much higher than those of the sputtering process. Recently, Schiller and Jasch,[173] reported on large scale industrial applications of deposition of Al on strip steel continuously at a deposition rate of 20 μ/min thickness. It should be noted that sputtering rates at the high side (approximaely 10,000 Å/min.) can only be obtained for target materials of high thermal conductivity like copper, since heat extraction from the target is the limiting parameter. For most materials, it is much lower, i.e., 50 to 1,000 Å/min.

Table 4.1: Deposition Rates for Various PVD Processes

Evaporation, Å/min	100–250,000*
Ion Plating, Å/min	100–250,000
Sputtering, Å/min	25–10,000

*In special cases to 750,000 Å/min

Advantages and Limitations

There are several advantages of PVD processes over competitive processes such as electrodeposition, CVD, plasma spraying. They are:

(1) Extreme versatility in composition of deposit. Virtually any metal, alloy, refractory or intermetallic compound, some polymeric type materials and their mixtures can be easily deposited. In this regard, they are superior to any other deposition process.

(2) The ability to produce unusual microstructures and new crystallographic modifications, e.g., amorphous deposits.

(3) The substrate temperature can be varied within very wide limits from subzero to high temperatures.

(4) Ability to produce coatings or self-supported shapes at high deposition rates.

(5) Deposits can have very high purity.

(6) Excellent bonding to the substrate.

(7) Excellent surface finish which can be equal to that of the substrate.

(8) Elimination of pollutants and effluants from the process which is a very important ecological factor.

The present limitations of PVD processes are:

(1) Inability to deposit polymeric materials with certain exceptions.

(2) Higher degree of sophistication of the processing equipment and hence a higher initial cost.

(3) Relatively new technology and the "mystique" associated with vacuum processes, much of which is disappearing with education and familiarity.

THEORY AND MECHANISMS

Vacuum Evaporation

Reference to the various steps in the formation of a deposit ennumerated in the previous section shows that the theory of vacuum evaporation involves thermodynamic considerations, i.e., phase transitions from which the equilibrium vapor phase pressure of materials can be derived, as well as the kinetic aspects of nucleation and growth. Both of these are of obvious importance in the evolution of the microstructure of the deposit.

The transition of solids or liquids into the gaseous state can be considered to be a macroscopic or as an atomistic phenomenon. The former is based on thermodynamics and results in an understanding of evaporation rates, source-container reactions and the accompanying effect of impurity introduction into the vapor state, changes in composition during alloy evaporation and stability of compounds. An excellent detailed treatment of the thermodynamic and kinetic bases of evaporation processes is given by Glang.[6] He points out that the application of kinetic gas theory to interpret evaporation phenomena resulted in a specialized evaporation theory. Such well known scientists as Hertz, Knudsen and Langnuir were the early workers in evaporation theory. They observed deviations from ideal behavior which led to refinements in the theory to include concepts of reaction kinetics, thermodynamics and solid state theory. From the kinetic theory of gases, the relationship between the impingement rate of gas molecules and their partial pressure p is given by

$$(1) \quad \frac{dN_i}{A_w dt} = (2\pi mkT)^{-\frac{1}{2}} p$$

where N_i is the number of molecules striking a unit area of surface, and A_w is the area of the surface.

Hertz,[34] in 1882, first measured the evaporation rate of mercury in high vacuum and found that the evaporation rate was proportional to the difference between the equilibrium vapor pressure of mercury, p^*, at the evaporant surface and the hydrostatic pressure, p, acting on the surface, resulting from the evaporant atoms or molecules in the gas phase. Thus, the evaporation rate based on the concept of the equilibrium vapor pressure, (i.e., the number of atoms leaving the evaporant surface is equal to the number returning to the surface) is given by:

$$(1a) \quad \frac{dN_e}{A_e dt} = (2\pi mkT)^{-\frac{1}{2}} (p^* - p) \; cm^2 sec^{-1}$$

such that dN_e, the number of molecules evaporating from a surface area A_e in time dt, is equal to the impingement rate of gas molecules based on the kinetic theory of gases with the value of p^* inserted therein, minus the return flux corresponding to the hydrostatic pressure p of the evaporant in the gas phase. In the above equations, m is the molecular weight, k is Boltzmann's constant and T is the temperature in °K. The maximum possible evaporation rate corresponds to the condition p = 0. Hertz measured evaporation rates only about one-tenth as high as the theoretical maximum rates. The latter were subsequently measured by Knudsen[35] in 1915. Knudsen postulated that some of the molecules impinging on the surface were reflecting back into the gas phase rather than becoming incorporated into the liquid. As a result, there is a certain fraction $(1 - \alpha_v)$ of vapor molecules which contribute to the evaporant pressure but not to the net molecular flux from the condensed phase into the vapor phase. To this end, he postulated the evaporation coefficient α_v which is defined as the ratio of the real evaporation rate in vacuum to the theoretically possible value defined by Equation (1a). This then results in the well-known Hertz-Knudsen equation

$$(2) \qquad \frac{dN_e}{A_e dt} = \alpha_v (2\pi mkT)^{-\frac{1}{2}} (p^* - p)$$

α_v is very dependent on the cleanliness of the evaporant surface and can range from very low values for dirty surfaces to unity for clean surfaces. In very high rate evaporation with a clean evaporant surface, it has been found that the maximum evaporation given by Equation (2) has been exceeded by a factor of 2 to 3 for the evaporation of a light metal such as beryllium[21] using electron beam heating. The reason for this is that the high power input results in considerable agitation of the liquid evaporant pool resulting in a real surface area much larger than the apparent surface area.

The directionality of evaporating molecules from an evaporation source is given by the well-known cosine law. Figure 4.5 shows a small surface element dA_r receiving deposit from a small area source A_r. The mass deposited per unit area

$$(3) \qquad \frac{dM_r(\sigma,\theta)}{dA_r} = \frac{M_e}{\pi r^2} \cos\phi \cos\theta$$

where M_e is total mass evaporated.

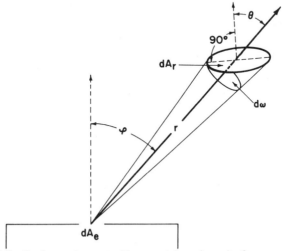

Figure 4.5: Surface element dA_r receiving deposit from a small-area source dA_e. (From *Handbook of Thin Film Technology*. Copyright © 1970, McGraw-Hill. Used with permission of McGraw-Hill Book Company.)

For a point source, Equation (4) reduces to:

(4)
$$\frac{dM_r}{dA_r} = \frac{M_e}{\pi r^2} \cos\phi$$

For a uniform deposit thickness, the point source must be located at the center of the spherical receiving surface such that r is a constant and $\cos\theta = 1$.

In high rate evaporation conditions, e.g., using a high power electron beam heated source, the thickness distribution is steeper than with a point or small area source discussed above. This has been attributed by some authors[28,36] to the existence of a virtual source of vapor located above the molten pool. On the other hand, at high power, the electron beam impact area on the surface of the molten pool is not flat but pushed down into an approximate concave spherical segment which as Riley shows[37] can equally well account for the steeper thickness distribution.

The above discussion points out one of the problems with evaporation technology, i.e., the variation in thickness of the deposit on a flat substrate. Numerous solutions are possible which involve either moving the substrate in a manner so as to randomly sample the vapor flux, the use of multiple sources or sources of special shapes. These have been discussed in some detail by Holland[8] as well as by Bunshah and Juntz.[38]

Models have also been presented for calculating the deposit temperature[39] and thickness distribution[40] during high rate evaporation and verified against experimental data. In a more recent paper, Szekely and Poveromo[41] have given a more general formulation describing the net rate of vapor desposition from a molten source onto an initial cold surface, making allowance for both molecular transport and diffusional effects.

EVAPORATION PROCESS AND APPARATUS

The System

A schematic of the evaporation apparatus has been illustrated in Figure 4.1. It consists of the following: chamber, vacuum pumps, vacuum gages, including total and partial pressure gages on sophisticated systems, evaporation sources, substrate holders, rate monitors, process controller, etc.

Vacuum Chamber: This is a simple bell jar or rectangular box for experimental or batch type production to more complex gear for production applications. The latter may consist of a deposition chamber with loading and unloading chambers attached to the deposition chambers by manifolds with isolation high vacuum valves. These are called fast cycle coaters. Alternate approaches are semi-continuous inline systems where a strip substrate stored in the vacuum chamber can be fed continuously over the source (Figure 4.6) or a continuous system where the strip or sheet substrate is inserted and removed from the deposition chamber through air-to-air seals[4,42] as shown in Figures 4.7 and 4.8.

Vacuum Pumping System: The gas loads in evaporation processes are fairly high due to outgassing from chamber walls promoted by the heat load from the evaporation source and substrate heaters, particularly for high deposition rate conditions. Therefore the pumping system is usually based on a diffusion pump with a liquid nitrogen cooled anti-creep type baffle-backed with a mechanical pump or a Roots blower/mechanical pump combination for large systems. For very high purity low deposition rate low heat flux conditions, ion pumped systems backed with cryosorption rough pumping are used, since a base pressure of 10^{-9} to 10^{-10} torr is needed. This is particularly true for molecular beam epitaxy where extreme control over composition and layer thickness are essential and deposition rates can be quite low. In such cases, the chamber and pumps are to be baked as with any other ultra-high vacuum operation.

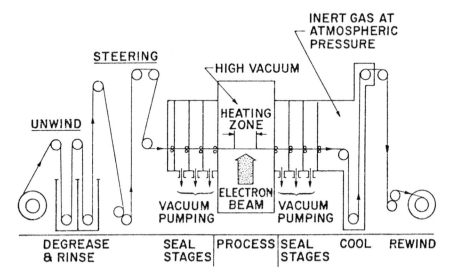

Figure 4.6: A schematic representation of a 24 inch continuous high vacuum strip processing line.

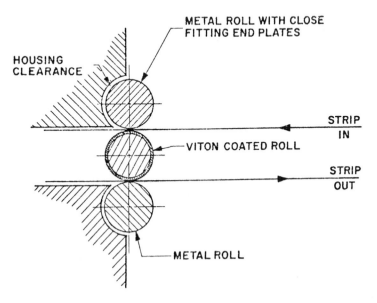

Figure 4.7: A three high roll seal arrangement for stripline.

94   Deposition Technologies for Films and Coatings

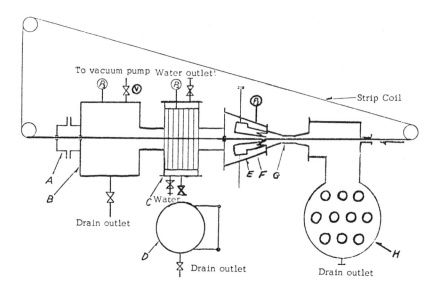

A : Seal Chamber
B : Vacuum Chamber
C : Condenser No. 2
D : Drain Tank

E : Suction Chamber
F : Nozzles
G : Diffuser
H : Condenser No. 1

Figure 4.8: Vacuum seal using steam jet or curtain.

Pressure Measurement: The vacuum gages used depend again on specific applications. A combination of high pressure gages (such as the Pirani or Thermocouple Gage) for monitoring the roughing of the system in combination with high vacuum gages (such as the hot cathode ionization gage and/or capacitance manometer). A partial pressure gage is highly desirable particularly for ultra-clean applications as well as for leak hunting.

Evaporation Sources: These are discussed separately below.

Substrate Holders and Heaters: Substrate holders may be very simple for stationary flat substrates or can incorporate quite complex motions as illustrated by planetary or nutating devices. The reason for this is to ensure deposition thickness uniformity and control over a large number of small parts such as lenses or silicon wafers.

Substrate heating can be accomplished by radiant heaters with refractory wires or quartz lamps acting as the heat source. Occasionally, substrates are directly heated by a scanning or diffuse electron beam.

Deposition Rate Monitors: These are discussed separately below.

EVAPORATION SOURCES

General Considerations

Evaporation sources are classified by the mode of heating used to convert the solid or liquid evaporant to the vapor phase. Thus one talks of resistance, arc, induction, electron beam, arc imaging, lasers and exploding wire types of sources. A very important fact to be noted is that we cannot evaporate every material from *any of the types of sources listed above for the following reasons.*

(1) Chemical interaction between the source material and the evaporant which would lead to impurities in the deposit. For example, evaporation of titanium from a MgO source would cause oxygen and magnesium contamination of the deposit; the titanium would reduce the MgO. Therefore, for the evaporation of reactive metals like titanium, zirconium, etc., we use water cooled copper crucibles.

(2) Reaction between metallic source (such as W or Ta boat) and evaporant (Ti) could occur. In many cases at high temperatures two metals can mutually dissolve in each other leading to a destruction of the source.

(3) The power density (i.e., Watts per sq. cm) varies greatly between the various heat sources.

Table 4.2 from Reference 6 from the article by Glang lists the temperature and support materials to be used in the evaporation of elements. Similar tables are found in many of the manufacturers' literature.

Evaporation of alloys and compounds pose additional problems and they are considered in a later section.

Resistance Heated Sources

The simplest vapor sources are resistance heated wires and metal foils of various types shown in Figure 4.9.

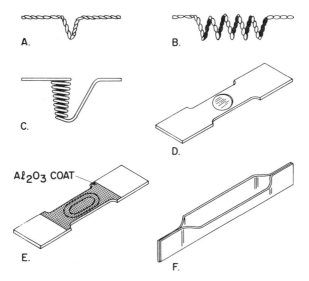

Figure 4.9: Wire and metal-foil sources. (A) Hairpin source. (B) Wire helix. (C) Wire basket. (D) Dimpled foil. (E) Dimpled foil with alumina coating. (F) Canoe type. (From *Handbook of Thin Film Technology.* Copyright © 1970, McGraw-Hill. Used with permission of McGraw-Hill Book Co.)

Table 4.2: Temperatures and Support Materials Used in the Evaporation of the Elements

Element and predominant vapor species	Temp, °C mp	Temp, °C $p^* = 10^{-2}$ Torr	Support materials Wire, foil	Support materials Crucible	Remarks
Aluminum (Al)	659	1220	W	C, BN, TiB$_2$-BN	Wets all materials readily and tends to creep out of containers. Alloys with W and reacts with carbon. Nitride crucibles preferred
Antimony (Sb$_4$, Sb$_2$)	630	530	Mo, Ta, Ni	Oxides, BN, metals, C	Polyatomic vapor, $\alpha_v = 0.2$. Requires temperatures above mp. Toxic
Arsenic (As$_4$, As$_2$)	820	~300	....	Oxides, C	Polyatomic vapor, $\alpha_v = 5.10^{-5}$–5.10^{-2}. Sublimates but requires temperatures above 300°C. Toxic
Barium (Ba)	710	610	W, Mo, Ta, Ni, Fe	Metals	Wets refractory metals without alloying. Reacts with most oxides at elevated temperatures
Beryllium (Be)	1283	1230	W, Mo, Ta	C, refractory oxides	Wets refractory metals. Toxic, particularly BeO dust
Bismuth (Bi, Bi$_2$)	271	670	W, Mo, Ta, Ni	Oxides, C, metals	Vapors are toxic
Boron (B)	2100 ± 100	2000	....	C	Deposits from carbon supports are probably not pure boron
Cadmium (Cd)	321	265	W, Mo, Ta, Fe, Ni	Oxides, metals	Film condensation requires high supersaturation. Sublimates. Wall deposits of Cd spoil vacuum system
Calcium (Ca)	850	600	W	Al$_2$O$_3$	
Carbon (C$_3$, C, C$_2$)	~3700	~2600	....	....	Carbon-arc or electron-bombardment evaporation. $\alpha_v < 1$
Chromium (Cr)	~1900	1400	W, Ta	....	High evaporation rates without melting. Sublimation from radiation-heated Cr rods preferred. Cr electrodeposits are likely to release hydrogen
Cobalt (Co)	1495	1520	W	Al$_2$O$_3$, BeO	Alloys with W, charge should not weigh more than 30 % of filament to limit destruction. Small sublimation rates possible
Copper (Cu)	1084	1260	W, Mo, Ta	Mo, C, Al$_2$O$_3$	Practically no interaction with refractory materials. Mo preferred for crucibles because it can be machined and conducts heat well
Gallium (Ga)	30	1130	....	BeO, Al$_2$O$_3$	Alloys with refractory metals. The oxides are attacked above 1000°C
Germanium (Ge)	940	1400	W, Mo, Ta	W, C, Al$_2$O$_3$	Wets refractory metals but low solubility in W. Purest films by electron-gun evaporation
Gold (Au)	1063	1400	W, Mo	Mo, C	Reacts with Ta, wets W and Mo. Mo crucibles last for several evaporations
Indium (In)	156	950	W, Mo	Mo, C	Mo boats preferred
Iron (Fe)	1536	1480	W	BeO, Al$_2$O$_3$, ZrO$_2$	Alloys with all refractory metals. Charge should not weigh more than 30 % of W filament to limit destruction. Small sublimation rates possible
Lead (Pb)	328	715	W, Mo, Ni, Fe	Metals	Does not wet refractory metals. Toxic
Magnesium (Mg)	650	440	W, Mo, Ta, Ni	Fe, C	Sublimates
Manganese (Mn)	1244	940	W, Mo, Ta	Al$_2$O$_3$	Wets refractory metals

(Continued)

Table 4.2: (continued)

Element and predominant vapor species	Temp, °C mp	Temp, °C $p^* = 10^{-2}$ Torr	Support materials Wire, foil	Support materials Crucible	Remarks
Molybdenum (Mo)	2620	2530	....	....	Small rates by sublimation from Mo foils. Electron-gun evaporation preferred
Nickel (Ni)....	1450	1530	W, W foil lined with Al$_2$O$_3$	Refractory oxides	Alloys with refractory metals; hence charge must be limited. Small rates by sublimation from Ni foil or wire. Electron-gun evaporation preferred
Palladium (Pd)	1550	1460	W, W foil lined with Al$_2$O$_3$	Al$_2$O$_3$	Alloys with refractory metals. Small sublimation rates possible
Platinum (Pt)	1770	2100	W	ThO$_2$, ZrO$_2$	Alloys with refractory metals. Multistrand W wire offers short evaporation times. Electron-gun evaporation preferred
Rhodium (Rh)	1966	2040	W	ThO$_2$, ZrO$_2$	Small rates by sublimation from Rh foils. Electron-gun evaporation preferred
Selenium (Se$_2$, Se$_n$; $n = 1-8$)[63]	217	240	Mo, Ta, stainless steel 304	Mo, Ta, C, Al$_2$O$_3$	Wets all support materials. Wall deposits spoil vacuum system. Toxic. $\alpha_v = 1$
Silicon (Si)....	1410	1350	....	BeO, ZrO$_2$, ThO$_2$, C	Refractory oxide crucibles are attacked by molten Si and films are contaminated by SiO. Small rates by sublimation from Si filaments. Electron-gun evaporation gives purest films
Silver (Ag)....	961	1030	Mo, Ta	Mo, C	Does not wet W. Mo crucibles are very durable sources
Strontium (Sr).	770	540	W, Mo, Ta	Mo, Ta, C	Wets all refractory metals without alloying
Tantalum (Ta)	3000	3060	....	....	Evaporation by resistance heating of touching Ta wires, or by drawing an arc between Ta rods. Electron-gun evaporation preferred
Tellurium (Te$_2$)	450	375	W, Mo, Ta	Mo, Ta, C, Al$_2$O$_3$	Wets all refractory metals without alloying. Contaminates vacuum system. Toxic. $\alpha_v = 0.4$
Tin (Sn)......	232	1250	W, Ta	C, Al$_2$O$_3$	Wets and attacks Mo
Titanium (Ti)	1700	1750	W, Ta	C, ThO$_2$	Reacts with refractory metals. Small sublimation rates from resistance-heated rods or wires. Electron-gun evaporation preferred
Tungsten (W)	3380	3230	....	....	Evaporation by resistance heating of touching W wires, or by drawing an arc between W rods. Electron-gun evaporation preferred
Vanadium (V).	1920	1850	Mo, W	Mo	Wets Mo without alloying. Alloys slightly with W. Small sublimation rates possible
Zinc (Zn).....	420	345	W, Ta, Ni	Fe, Al$_2$O$_3$, C, Mo	High sublimation rates. Wets refractory metals without alloying. Wall deposits spoil vacuum system
Zirconium (Zr)	1850	2400	W	....	Wets and slightly alloys with W. Electron-gun evaporation preferred

From *Handbook of Thin Film Technology*. Copyright © 1970, McGraw-Hill. Used with permission of McGraw-Hill Book Company.

They are available in a variety of sizes and shapes and at sufficiently low prices so that they can be discarded after one experiment if necessary. They are usually made from the refractory metals, tungsten, molybdenum and tantalum which have high melting points and low vapor pressure so as not to contaminate the deposit. Their properties are given in Table 4.3.

Table 4.3: Properties of Refractory Metals

Property	Tungsten	Molybdenum	Tantalum
Melting point, °C	3380	2610	3000
T, °C, for $p^* = 10^{-6}$ Torr	2410	1820	2240
Electrical resistivity, 10^{-6} ohm-cm			
At 20°C	5.5	5.7	13.5
At 1000°C	33	32	54
At 2000°C	66	62	87
Thermal expansion, %			
From 0–1000°C	0.5	0.5	0.7
From 0–2000°C	1.1	1.2	1.5

From *Handbook of Thin Film Technology.* Copyright © 1970, McGraw-Hill. Used with permission of McGraw-Hill Book Company.

Platinum, iron or nickel are sometimes used for materials which evaporate below 1000°C. *The capacity (total amount of evaporant) of such sources is small.* The hairpin and wire helix sources are used by attaching the evaporant to the source in the form of small wire segments. *Upon melting, the evaporant must wet the filament and be held there by surface tension.* This is desirable to increase the evaporation surface area and thermal contact. Multistrand filament wire is preferred because it increases the surface area. *Maximum amount held is about 1 gram.* Dimpled sources, basket boats may hold up to a few grams.

Since the electrical resistance of the source is small, low voltage power supplied, 1 to 3 KW are recommended. The current in the source may range from 20 to 500 Amps. In some cases, the evaporant is electroplated onto the wire source.

The principal use of wire baskets is for the evaporation of pellets or chips of dielectric materials which either sublime or do *not* wet the wire on melting. In such cases, if wetting occurs, the turns of the baskets are shorted and the temperature of the source drops.

The rate of evaporation from such sources may vary considerably due to localized conditions of temperature variation, wetting, hot spots etc. Therefore, for a given thickness of film, the procedure is to load the source with a fixed weight of evaporant and evaporate to completion or use a rate monitor and/or thickness monitor to obtain the desired evaporation rate and thickness.

Sublimation Sources

For materials evaporating above 1000°C, the problem of non-reactive supports may be circumvented for materials such as Cr, Mo, Pd, V, Fe and Si which reach a vapor pressure of 10^{-2} torr before melting. Hence, they can *sublime* and produce a sufficiently high vapor density. The contact area between the evaporant and the source crucible is held to a minimum. Figure 4.10 shows such a source designed by Roberts and Via.

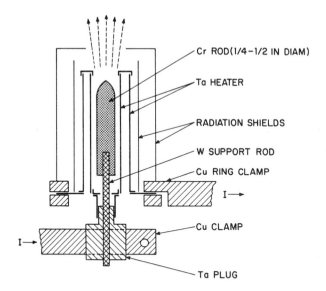

Figure 4.10: Chromium sublimation source after Roberts and Via. The electric current flows through the tantalum cylinder (heavy lines). (From *Handbook of Thin Film Technology*. Copyright © 1970, McGraw-Hill. Used with permission of McGraw-Hill Book Co.

A different type of sublimation source is used for the vaporization of thermally stable compounds such as SiO which are commonly obtained as powders or loose chunks. Such source material would release large quantities of gases upon heating thus causing ejection of particles of the evaporant which may get incorporated into the film. Figure 4.11 shows two sources which solve this problem by reflection of the vaporized material.

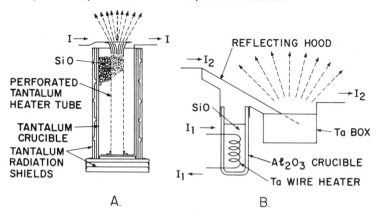

Figure 4.11: Optically dense SiO sources. (A) The Drumheller source. (B) Compartmentalized source. (After Vergara, Greenhouse and Nicholas.) (From *Handbook of Thin Film Technology*. Copyright © 1970, McGraw-Hill. Used with permission of McGraw-Hill Book Co.)

Evaporation Sources and Materials

We have already discussed the potential problems concerned with the reaction between metal sources and evaporants. Oxides and other compounds are more stable than metals. Table 4.4 gives the thermal stability of refractory oxides in contact with metals. There are many metals not listed in Table 4.4 which can be evaporated from refractory oxide sources. Note that there is no such thing as an absolutely stable oxide, nitride or other compound. Reaction is controlled by kinetics, i.e., temperature and time.

Table 4.4: Thermal Stability of Refractory Oxides in Contact with Metals*

Metal	Temp, °C for 10^{-2} Torr	Refractory oxides				
		ThO_2	BeO	ZrO_2	Al_2O_3	MgO
W....	3230	Slight reduction at 2200°C	Stable at 1700°C; reaction >1800°C	Stable <1600°C; little reaction up to 2000°C	Limited by the onset of Al_2O_3 sublimation at 1900°C	Limited by the onset of MgO sublimation at 1600–1900°C
Mo...	2530	Little reaction up to 2300°C	Stable at 1900°C but not above	Stable at 2000°C; ZrO_2 decomposes at 2300°C		
Ta....	3060	Stable up to 1900°C	Stable up to 1600°C	Stable up to 1600°C		
Zr....	2400	Interaction begins at 1800°C	Interaction begins at 1600°C	Slight interaction at 1800°C	Oxide attacked at 1600°C	
Be....	1230	Only slight interaction at 1600°C			Oxide discolored at 1400°C	Stable at **1400°C** but not at 1600°C
Si.....	1350	Little or no attack at **1400°C**; noticeable reaction at 1600°C			Slight reaction at 1400°C	Slight reaction at 1400°C, strong at 1600°C
Ti....	1750	Slight interaction at 1800°C	Little reaction at 1600°C but considerable at 1800°C		Little reaction at 1400°C but considerable at 1800°C	
Ni....	1530	Ni(l) is stable in contact with all oxides at **1800°C**				

Boldface type indicates metal-oxide pairs which can be used for thin film deposition.
* After Johnson, Economos and Kingery, and Kohl.

From *Handbook of Thin Film Technology.* Copyright © 1970, McGraw-Hill. Used with permission of McGraw-Hill Book Company.

Oxide crucibles have to be heated by radiation from metal filaments or their contents can be heated by induction heating. This is illustrated in Figure 4.12 and 4.13 for resistance heated sources.

Figure 4.12: Oxide crucible with wire-coil heater. (From *Handbook of Thin Film Technology.* Copyright © 1970, McGraw-Hill. Used with permission of McGraw-Hill Book Co.)

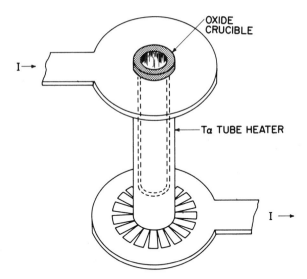

Figure 4.13: DaSilva crucible source. (From *Handbook of Thin Film Technology*. Copyright © 1970, McGraw-Hill. Used with permission of McGraw-Hill Book Co.)

Other source materials are nitrides such as boron nitride. A 50% BN-50% TiB_2 is also well established as a crucible material. This material (HDA composite Union Carbide) is a fairly good electrical conductor and hence can be directly heated to evaporate materials. It can be readily machined to shape.

Pyrolytic BN and carbon are also used.

Induction Heated Sources

Figure 4.14 shows the induction heated sources using a $BN-TiB_2$ crucible. Figure 4.15 shows an induction heated evaporation sublimation source using a water cooled copper crucible.[19] This is suited to the evaporation of reactive metals such as Ti, Be, etc... which will react with all the refractory oxides, nitrides, etc.

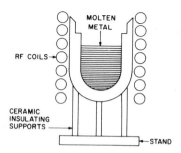

Figure 4.14: RF heated aluminum source with boron nitride-titanium diboride crucile. (After Ames, Kaplan and Roland). (From *Handbook of Thin Film Technology*. Copyright © 1970, McGraw-Hill. Used with permission of McGraw-Hill Book Co.)

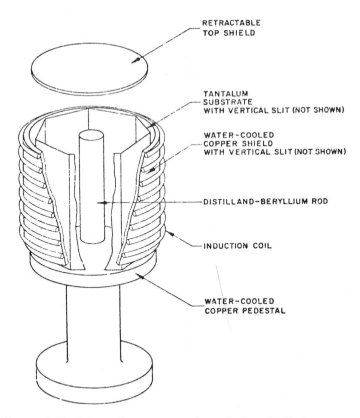

Figure 4.15: Schematic representation of the distillation setup.

Electron Beam Heated Sources

Electron beam heated sources have two major benefits. One, a very high power density and hence a wide range of control over evaporation rates from very low to very high. Two, the evaporant is contained in a water-cooled copper hearth thus eliminating the problem of crucible contamination.

The evaporation rate for pure metals like Al, Au, Ag, which are good thermal conductors from water-cooled copper crucibles decreases due to heat loss to the crucible walls. In such cases, crucible liners of carbon and other refractory materials are used.

Any gun system must consist of at least two elements — a cathode and anode. In addition, it is necessary to contain these in a vacuum chamber in order to produce and control the flow of electrons, since they are easily scattered by gas molecules. A potential difference is maintained between the cathode and the anode. This varies from as little as a few kilovolts to hundreds of kilovolts. In melting systems, a normal operational range is of the order of 10-40 kV. In the simple diode system, the cathode emits electrons, which are then accelerated to the anode across the potential drop. Where the anode is the workpiece to be heated, this is termed a work-accelerated gun. It is shown schematically in Figure 4.16a. In a self-accelerated gun structure and is located fairly close to the cathode, electrons leave the cathode surface, are accelerated by the potential difference between the cathode and anode, pass through the hole in the anode and continue onward to strike the workpiece. Self-accelerated guns have become the more common type in use and offer more flexibility than the work-accelerated gun.

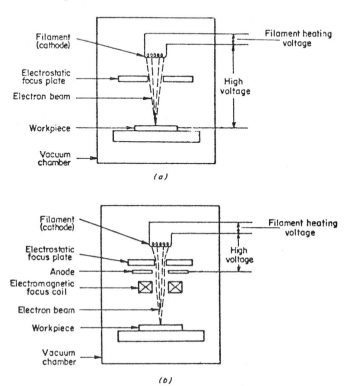

Figure 4.16: Simple electron beam guns (a) Work-accelerated gun. (b) Self-accelerated gun.

Electron beam guns may be further subdivided into two types depending on the source of electrons: (1) Thermionic gun and (2) Plasma gun.

Thermionic Gun: In thermionic guns, the source of electrons is a heated wire or disc of a high temperature metal or alloy usually tungsten or tantalum. Such guns have the limitation of a minimum operating gas pressure of about 1.10^{-3} torr. Higher pressures cause scattering of the electron beam as well as a pronounced shortening of the cathode life (if it is a wire or filament) due to erosion by ion bombardment. Figure 4.17 shows examples of thermionic electron beam heated work accelerated sources. The close cathode gun shown in Figure 4.17A is not a desirable configuration since molten droplet ejection from the pool impinging on the cathode will terminate the life of the cathode due to low melting alloy formation. Thus cathodes are hidden from direct line-of-sight of the molten pool and the electron beam is bent by electrostatic fields (Figures 4.17B and 4.17C) or magnetic field (Figures 4.18 and 4.19) generated by electromagnets. The latter is a preferred arrangement since variation of the X and Y components of the magnetic field can be used to scan the position of the beam on the molten pool surface.

104   Deposition Technologies for Films and Coatings

Figures 4.17, 4.18 and 4.19 show linear cathodes (i.e. wires or rods) and are referred to as transverse linear cathode guns. Figure 4.20 shows a disc cathode which is characteristic of a high power Pierce type electron beam gun. Low power Pierce type guns may have a hair pin filament or a wire loop as the cathode. In either case the beam geometry of the Pierce gun is different than that of the transverse linear cathode guns. In some instances, the electron emitter assembly is located at the distance from the crucible in a separately pumped chamber to keep the pressure below 1.10^{-3} torr, with a small orifice between the emitter chamber and the crucible chamber for the passage of electrons.

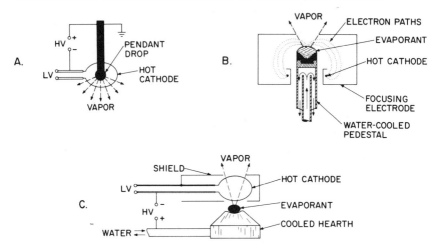

Figure 4.17: Work-accelerated electron-bombardment sources. (A) Pendant-drop method. (B) Shielded filament (Unvala). (C) Shielded filament (Chopra and Randlett). (From *Handbook of Thin Film Technology.* Copyright © 1970, McGraw-Hill. Used with permission of McGraw-Hill Book Co.)

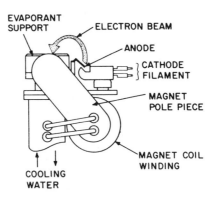

Figure 4.18: Bent-beam electron gun with water-cooled evaporant support. (With permission of Temescal Metallurgical Co., Berkeley, CA). (From *Handbook of Thin Film Technology.* Copyright © 1970, McGraw-Hill. Used with permission of McGraw-Hill Book Co.)

Evaporation 105

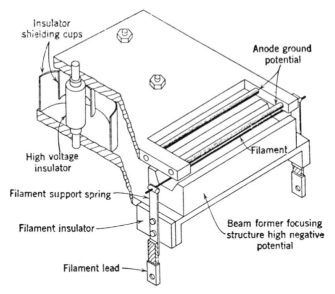

Figure 4.19: Transverse electron beam gun.

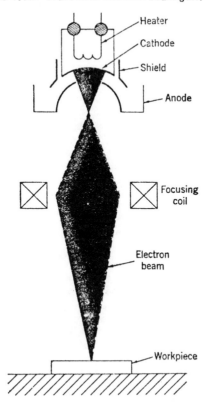

Figure 4.20: Schematic representation of a Pierce gun.

Plasma Electron Beam Gun: A plasma is defined as a region of high temperature gas containing large numbers of free electrons and ions. By a proper application of electrical potential, electrons can be extracted from the plasma to provide a useful energy beam similar to that obtained from thermoionic guns.

There ar two types of plasma e.b. guns:

(a) The Cold Cathode Plasma Electron Beam—The plasma electron beam gun has a cylindrical cathode cavity made from a metal mesh or sheet (Figure 4.21) containing the ionized plasma from which electrons are extracted through a small aperture in one end. The cathode is maintained at a negative potential, e.g., -5 to -20 kV, relative to the workpiece and remainder of the system, which are at ground potential. After evacuation of the system a low pressure of ionizable gas in the range of 10^{-3} to 10^{-1} torr is introduced. Depending upon the high voltage level, a long path discharge between the cathode and other parts of the system will occur in the gas at a particular pressure. Ionizing collisions in the gas then produce positive ions which are accelerated to the cathode, causing electrons to be released from the cathode surface. Although the cathode may heat up slightly due to ion bombardment, no heating is required for electron emission. Upon proper adjustment of cathode voltage and gas pressure a a beam mode of operation is established, since interaction between the plasma inside and outside of the cathode and the electric fields between cathode surface and plasma boundary will largely confine electron emission to the end of the cathode and its interior. In argon, a beam mode is supported at about 10^{-2} torr with 5-10 kV. Beam currents range up to 3A for a 3 inch diameter cathode in argon at 20 kV. With lighter gases, e.g., helium, higher pressure to about 10^{-1} torr will yield a beam mode in this same voltage range. Beam current will vary with voltage and pressure control, also. More specific information is given by Cocca and Stauffer.[44] The beam is self-collimating because of the focusing effect of positive ions in the beam path and the electrostatic lensing action of the aperture since it separates regions of different potential gradient. The beam is well collimated, having a cross section equal to that of the cathode aperature. Adjustment of focus can be achieved to some extent by varying pressure and voltage, but external focusing may also be used, if desired, with magnetic or electrostatic lenses, as with conventional electron beams.

(b) The Hot Hollow Cathode Discharge Beam—The hollow cathode discharge beam applied to vacuum processing has been reported by Morley[45] and differs in a number of respects from the plasma beam. A schematic of the hollow cathode discharge beam is shown in Figure 4.22. Here the cathode must be constructed of a refractory metal since it operates at elevated temperature. An ionizable gas, usually argon, is introduced into the system through the tubular-shaped cathode. A pressure drop across the orifice in the cathode provides a sufficient amount of gas inside the cathode to sustain the plasma, which generates the beam. A low voltage, high amperage d.c. power source is utilized. When rf power from a commercial welding starter is coupled to

Evaporation 107

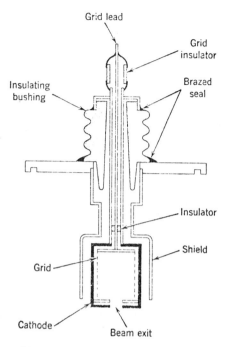

Figure 4.21: Cold cathode plasma electron beam gun.

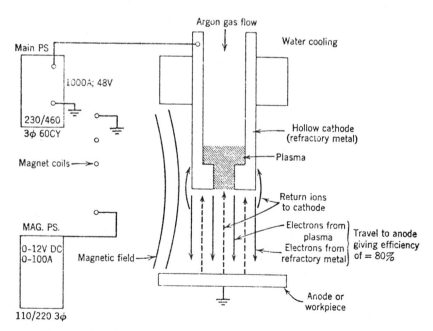

Figure 4.22: Schematic of the hot hollow cathode electron beam gun.

the gas, it becomes ionized and the plasma is formed. Continued ion bombardment of the cathode results in heating of the cathode and increased electron emission. Ultimately, a high current "glow discharge" will occur, analogous to that experienced in vacuum arc melting at higher pressures. At this point, the discharge appears as a low power density beam "flowing" from the cathode aperture and fanning out in conical shape into the chamber. However, a parallel axial magnetic field is imposed on the beam (as seen in Figure 4.22) which then forms a high power density, well-collimated beam. The hollow cathode discharge beam is operationally stable and efficient over the pressure range from 10^{-4} to 10^{-1} torr. A more detailed description of physical aspects, operational characteristics, and cathode design has been given by Morley.[45]

Comparisons: Thermionic as well as the plasma e.b. guns can be used equally well for evaporation. Focusing of the beam spot is easier for the thermionic guns. The plasma guns have the advantage of being able to operate at higher pressures which can be important for gas scattering evaporation, reactive evaporation and ion plating.

Laser Beam Evaporation

Laser beams have also been used to melt and evaporate materials. There are however practical limitations to its extended use at this time. They are:

(1) The laser would have to be placed inside the vacuum chamber or the beam transmitted through a window which is transparent for the laser wavelength for lasers located outside the vacuum chamber. The former is not a practical probability and the latter suffers from the limited availability of suitable window material and the blockage of the window by an evaporated layer since the laser has to sight the molten pool.

(2) The laser radiation wave-length must lie in the absorption band of the target material which is not always possible.

(3) Very low energy conversion efficiency.

All these problems preclude any serious consideration of the laser as a heat source for evaporation at this time.

Arc Evaporation

Arcs are high current electrical discharges which are quite appropriate heat sources for evaporation. Berghaus[178] describes the use of arcs to form refractory compounds by reactive evaporation. Since 1940, consumable and nonconsumable vacuum arc melting processes have been developed to melt and refine various reactive metals such as Ti, Hf, Zr, etc. More recently, arc techniques have been used to deposit metals[179,180] and refractory compounds and even for extraction of ions from the vacuum arc plasma for the deposition of metal films.[181]

DEPOSITION RATE MONITORS AND PROCESS CONTROL

The properties of deposits are dependent on the control exercised during

Evaporation   109

the process. The thinner the deposit, the more critical is the control of the operation.

Monitoring of the Vapor Stream

Ionization Gauge Rate Monitor: This device is very similar to hot cathode ionization gauge and monitors the atom density in the vapor phase by ionizing the vapors, collecting and measuring the ion current. Several arrangements are shown in Figure 4.23.

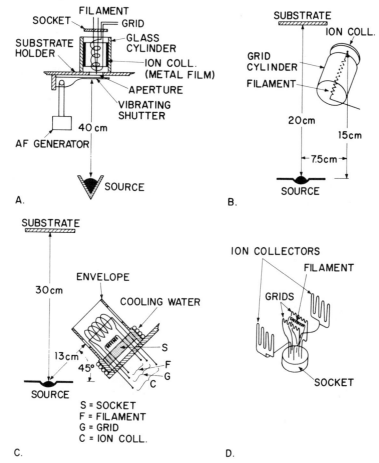

Figure 4.23: Ionization rate monitor designs and arrangements. (A) After Schwarz. (B) After Giedd and Perkins. (C) After Perkins. (D) After Dufour and Zega. (From *Handbook of Thin Film Technology*. Copyright © 1970, McGraw-Hill, Used with permission of McGraw-Hill Book Co.)

Particle Impingement Rate Monitors: The gauge which is a cylinder suspended by a wire or riding on a bearing is imparted a momentum by the impinging particles which can be measured by the torsional forces. They are illustrated in Figure 4.24.

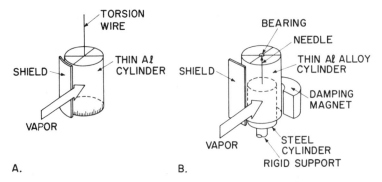

Figure 4.24: Particle-impingement-rate monitors. (A) Torsion-wire device. (After Neugebaur.) (B) Pivot-supported device. (After Beavitt.) (From *Handbook of Thin Film Technology.* Copyright © 1970, McGraw-Hill. Used with permission of McGraw-Hill Book Co.)

Ion Current Monitor for Electron Beam Heated Source: An electron beam heated molten pool has a plasma sheath above it. Positive ions from the plasma follow a very similar trajectory as the electrons with a slightly larger radius of curvature due to their higher mass and are beamed away from the molten pool by the same magnetic field which bends the electrons towards the pool. Therefore an ion collector can be placed so as to intercept this ion beam and the resultant ion current can be used in a feed-back loop to control the evaporation rate. Two manufacturers of electron beam guns have offered this option.

Spectroscopic Methods: Monitoring and control of the deposition rate can be done on the basis of mass spectrometry, atomic absorption spectrometry and electron emission impact spectrometry. Each of them involves the choice of an appropriate materials selective sensor. The principles, advantages and limitations of each of these are presented in a good review paper by Lu—Thin Solid Films *45,* 487 (1977). The reader is referred to this paper and the references cited therein.

Monitoring of Deposited Mass

Microbalances: There are various types of devices which measure a change in mass due to condensed atoms based on elongation of a thin quartz-fiber helix, the tension of a wire or the deflection of a pivot mounted beam. Examples are shown in Figures 4.25 and 4.26.

Crystal Oscillators: The crystal oscillator monitor utilizes the piezoelectric properties of quartz. *The resonance frequency induced by an a.c. field is inversely proportional to crystal thickness.* In practice, the change in frequency of a crystal exposed to the vapor beam is compared to that of reference crystal. An example is shown in Figure 4.27.

Monitoring of Specific Film Properties

In preparing thin films, often only one property is of interest, e.g. optical or electrical properties.

Optical Monitors: They measure phenomena such as light absorbance, transmittance, reflectance or related interference effects during film deposition. An example is shown in Figure 4.28.

Evaporation   111

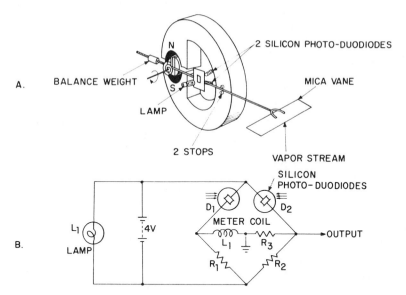

Figure 4.25: (A) Schematic drawing and (B) circuit diagram of a microbalance constructed from a microammeter movement. (Hayes and Roberts). (From *Handbook of Thin Film Technology.* Copyright © 1970, McGraw-Hill. Used with permission of McGraw-Hill Book Co.)

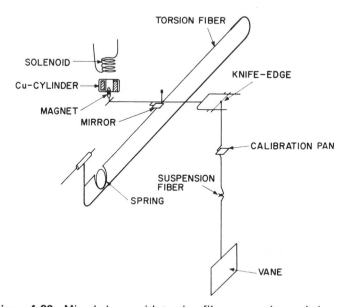

Figure 4.26: Microbalance with torsion-fiber suspension and electromagnetic force compensation at beam end (Mayer et al.) (From *Handbook of Thin Film Technology.* Copyright © 1970, McGraw-Hill. Used with permission of McGraw-Hill Book Co.)

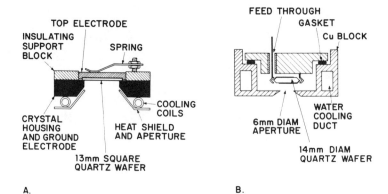

Figure 4.27: Oscillator crystal holders for deposition monitoring. (A) After Behrndt and Love. (B) After Pulker. (From *Handbook of Thin Film Technology.* Copyright © 1970, McGraw-Hill. Used with permission of McGraw-Hill Book Co.)

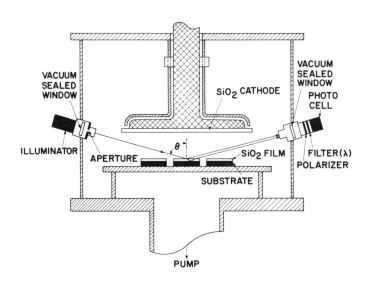

Figure 4.28: Schematic of an RF sputtering system (after Davidse and Maissel) with optical-thickness monitor. (From *Handbook of Thin Film Technology.* Copyright © 1970, McGraw-Hill. Used with permission of McGraw-Hill Book Co.)

Resistance Monitors: The film thickness can be continuously monitored using in situ resistance measurements as shown in Figure 4.29.

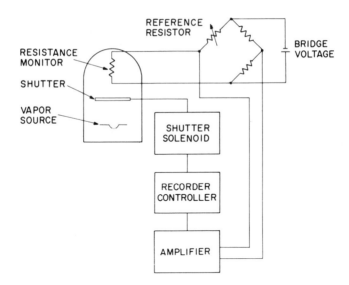

Figure 4.29: Wheatstone-bridge circuit for resistance monitoring. (From *Handbook of Thin Film Technology.* Copyright © 1970, McGraw-Hill. Used with permission of McGraw-Hill Book Co.).

Evaporation Process Control

Thickness Control: Usually monitoring of an evaporation process is combined with means to control film deposition. Frequently, the only requirement is to terminate the process when the thickness or a thickness related property has reached a certain value. The simplest way is to evaporate a weighed amount of source material to completion. Knowing the emission characteristics of the source will allow the film thickness to be calculated. Alternately, monitoring devices discussed earlier can be calibrated to measure thickness directly.

Rate Control: Rate Control is a more complex task and involves measuring the signal from a rate monitor and using it in a feedback loop to control the power to the source and hence its temperature and evaporation rate. Table 4.5 illustrates the pros and cons of Evaporation Process Control.

DEPOSITION OF VARIOUS MATERIALS

The family of materials which are deposited by evaporation include metals, semiconductors, alloys, intermetallic compounds, refractory compounds (i.e., oxides, carbides, nitrides, borides, etc...) and mixtures thereof. An important point is that the source material should be pure and free of gases and/or inclusions to forestall the problems of molten droplet ejection from the pool commonly called spitting.

Let us consider each of the materials.

Deposition of Metals and Elemental Semiconductors

Evaporation of single elements can be carried out from a variety of evapora-

Table 4.5: Evaporation Process Control

Control method	Electrical signal		Prerequisites for	
	Available	Related to	Thickness control	Rate control
Evaporation to completion	No	....	Weighed evaporant charge	Not applicable
Control of source temperature, power, or current	Yes	Evaporation rate	Rates must be integrated	Error-signal feedback
Ionization-rate monitor	Yes	Evaporation rate	Rates must be integrated	Error-signal feedback
Particle impingement devices	No	....	Direct observation on pivoted models, manual stop	Direct observation on torsion-wire models, manual power adjustment
Microbalances	Electromagnetic or electrostatic models: yes	Deposit thickness	Strip-chart recorder, preset microswitch for automatic termination	Programmed weight increase, or second signal by electronic differentiation
Crystal oscillator	Yes	Deposit thickness	Counter, meter, or recorder; preset microswitch for automatic termination	Second signal by differentiation; servoloop
Optical monitors	With photocell: yes	Deposit thickness	Strip-chart recorder; fairly complex circuitry for automatic termination	Has not been implemented yet
Resistance monitors	Yes	Deposit thickness	Strip-chart recorder and preset microswitch, or digital techniques	Second signal by electronic differentiation; servoloop
Capacitance monitors	Yes	Deposit thickness or rate	Planar capacitor substrate; oscillator circuit with reference capacity and comparator	Direct rate indication from Riddle's vapor capacitor

From *Handbook of Thin Film Technology*. Copyright © 1970, McGraw-Hill. Used with permission of McGraw-Hill Book Company.

tion sources subject to the restrictions discussed above dealing with melting point, reactions with container, deposition rate, etc. A typical arrangement is shown in Figure 4.1 for electron beam heating. As discussed above, either heating methods can also be used. These are the simplest materials to evaporate. Fortunately, at this time, it is estimated that 90% of all the materials evaporated is aluminum!

Deposition of Alloys

Alloys consist of two or more components, which have different vapor pressures and hence different evaporation rates. As a result, the composition of the vapor phase and therefore the deposit has a constantly varying composition. There are two solutions to this problem—multiple sources and single rod-fed or wire-fed electron beam sources.

Multiple Sources: This is the more versatile system. The number of sources evaporating simultaneously is equal to or less than the number of constituents in the alloy. The material evaporated from each source can be a metal, alloy or compound. Thus, it is possible to synthesize a dispersion strengthened alloy, e.g., Ni-ThO$_2$. On the other hand, the process is complex because the evaporation rate from each source has to be monitored and controlled separately. The source to substrate distance would have to be sufficiently large (15 inches for 2 inch diameter sources) to have complete blending of the vapor streams prior to deposition, which decreases the deposition rate (See Figure 4.30). Moreover, with gross difference in density of two vapors, it may be difficult to obtain a uniform composition across the width of the substrate due to scattering of the lighter vapor atoms. Some examples are given in Table 4.6.

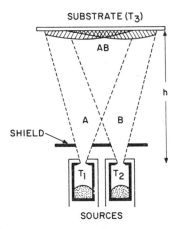

Figure 4.30: Two-source evaporation arrangement yielding variable film composition. (From *Handbook of Thin Film Technology*. Copyright © 1970, McGraw-Hill. Used with permission of McGraw-Hill Book Co.)

It is possible to evaporate each component sequentially thus producing a multi-layered deposit, which is then homogenized by annealing after deposition. This procedure makes it even more difficult to get high deposition rates. A multiple source arrangement for production of alloy deposits at high rates is not known.

Table 4.6: Two-Source Evaporation, Experimental Conditions, and Types of Films Obtained

Evaporated constituents	Evaporation conditions and method of control	Substrate temp, °C	Films obtained
Alloy and Multiphase Films			
Cu + Ni	Sequential evaporation	Low	Stratified films. Annealing at 200°C yields two-phase alloy films
Cu, Ag, Au, Mg, Sn, Fe, Co	Simultaneous evaporation from two sources. Ionization-rate monitor control, ±1%	−193	Binary alloy films of metastable structures
Cu, Ag, Au, Al, Ni, and others	Simultaneous evaporation from two sources. Rates adjusted by varying source temperatures	25–600	Binary alloy films of varying composition and structure
Ni + Fe	Two wire-ring sources, evaporation rates controlled by quartz-crystal oscillator	300	Permalloy films. $d' \approx 10$ Å s^{-1}
Nb + Sn	Two sources, rates monitored by particle impingement-rate monitor. Impingement ratio $N_{Nb}:N_{Sn} = 3$	25–700	Superconducting Nb$_3$Sn films. $d' \approx 2$ Å s^{-1}, $\alpha_c \approx 1$
V, Nb + Si, Sn	Two electron-gun sources, rates monitored by measuring ionization current	....	Superconducting films of approx composition Nb$_3$Sn and V$_3$Si
ZnS + LiF	Two sources, rates monitored by a microbalance. Variable impingement ratios	30–40	Mixed dielectric films of different composition. $d' = 10$–30 Å s^{-1}
Au, Cr + SiO, MgF$_2$	Two-source evaporation with ionization-rate monitor control (±1–2%)	25–300	Au-SiO, Au-MgF$_2$, Cr-SiO, and Cr-MgF$_2$ resistor films of different compositions
Cr + SiO	SiO source at 1100°C, Cr source at 1500°C. Impingement ratio varied with location on substrate	400	High-resistivity Cr-SiO films of variable composition

(Continued)

Table 4.6: (continued)

Evaporated constituents	Evaporation conditions and method of control	Substrate temp, °C	Films obtained
		Compound Films	
Cd + S	Two effusion ovens, Cd at 400–450°C, S at 120–150°C. Cd excess	400–650	Stoichiometric CdS crystals
Cd + Se	Impingement fluxes controlled by source temperature. $N_{Cd} = 2 \times 10^{16}$, $N_{Se} = 10^{16}$–10^{17} cm^{-2} s^{-1}	200	Stoichiometric CdSe films
PbSe + PbTe	Source temperatures varied around 700°C	300	Epitaxial films of PbSe$_{1-x}$Te$_x$ on NaCl crystals
Bi + Te	Bi source at 750°C, Te source temperature variable. $N_{Te}:N_{Bi} = 10$–40	400–500	Stoichiometric films of Bi$_2$Te$_3$. n-type, 2×10^{19} electrons cm^{-3}
Bi + Se	Rate control by quartz-crystal oscillator. Se source at 250°C; Bi source temperature variable	52	Vitreous, semiconducting films of nonstoichiometric composition
Al + Sb	Source temperatures adjusted by quartz-crystal oscillator to yield $N_{Sb}:N_{Al}$ ratios of 1.6–16	550	Stoichiometric AlSb films. $d' \approx 10$ Å s^{-1}
Ga + As	Ga source at 940–970°C, As source at 300°C. $N_{As}:N_{Ga} \approx 10$; Ga impingement flux: 10^{15} cm^{-2} s^{-1}	550	Stoichiometric GaAs films. Epitaxial on (100) NaCl, polycrystalline on quartz
Ga + As	Ga source at 910°C, As source at 295°C. Deposition rate: <2 Å s^{-1}	375–450	Stoichiometric GaAs films on GaAs, Ge, and Al$_2$O$_3$ single-crystal substrates. Fiber texture to single-crystalline
In + As	Incident fluxes: $N_{In} = 5 \times 10^{16}$ cm^{-2} s^{-1}, $N_{As} = 5 \times 10^{16}$–$5 \times 10^{17}$ cm^{-2} s^{-1}	230–680	Stoichiometric InAs films; n-type
In + Sb	Incident fluxes: $N_{In} = 5 \times 10^{16}$ cm^{-2} s^{-1} $N_{Sb} = 5 \times 10^{16}$–$5 \times 10^{17}$ cm^{-2} s^{-1}	400–520	Stoichiometric InSb films; n-type
	Source temperatures adjusted by microbalance to yield $N_{Sb}:N_{In} = 1.1$	250	Stoichiometric InSb films; α_c of Sb ≈ 0.6

From *Handbook of Thin Film Technology*. Copyright © 1970, McGraw-Hill. Used with permission of McGraw-Hill Book Company.

Single Rod-Fed Electron Beam Source: The disadvantages of multiple sources for alloy deposition can be avoided by using a single source.[46,46a] They can be wire-fed or rod-fed sources, the latter being shown in Figure 4.31. There is a molten pool of limited depth above the solid rod. If the components of an alloy, A_1B_1, have different equilibrium vapor pressures, then the steady state composition of the molten pool will differ from the feed rod, e.g., A_1B_{10}. Under steady state conditions, the composition of the vapor is the same as that of the solid being fed into the molten pool. One has the choice of starting with a button of appropriate composition A_1B_{10} on top of a rod A_1B_1 to form the molten pool initially or one can start with a rod of alloy A_1B_1 and evaporate until the molten pool reaches compositions A_1B_{10}. Precautions to be observed are that the temperature and volume of the molten pool have to be constant to obtain a constant vapor composition. A theoretical model has been developed and confirmed by experiment. Ni-20Cr, Ti-6Al, Ag-5Cu, Ag-10Cu, Ag-20Cu, Ag-30Cu, Ni-xCr-yAl-xY alloy deposits have been successfully prepared. To date, experimental results indicate that this method can be used with vapor pressure differences of a factor of 5,000 between the components. This method cannot be used where one of the alloy constituents is a compound, e.g., $Ni-ThO_2$.

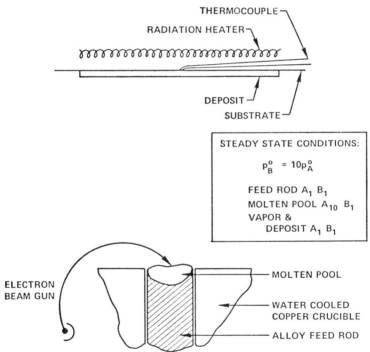

Figure 4.31: Schematic of direct evaporation of an alloy from a single rod-fed source.

In a recent paper, Shevakin et al.[174] investigated the relationship between the composition of the evaporant material and the condensates for alloy evaporation using electron beam evaporation techniques. They used this method to determine thermodynamic activities of the components of binary alloys at temperatures above the melting point of the alloy.

Evaporation   119

Deposition of Intermetallic Compounds

Intermetallic compounds which are generally deposited such as GaAs, PbTe, InSb, etc... have as their constituents elements with low melting points and high vapor pressures. These compound semiconductors need to have a carefully controlled stoichiometry, i.e. cation:anion ratio. Therefore, they can best be prepared by flash evaporation or sputtering.

In flash evaporation, powder or chips of the two components are sprinkled onto a superheated sheet to produce complete evaporation of both components. Various possible arrangements are shown in Figure 4.32. Table 4.7 gives examples of the use of this technique.

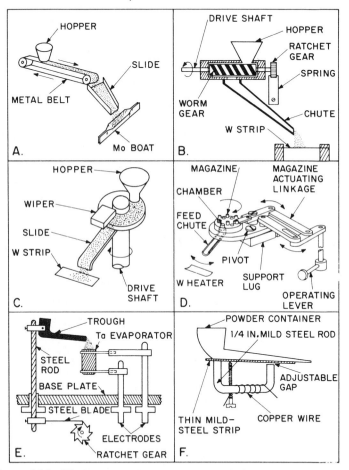

Figure 4.32: Flash-evaporation mechanisms. (A) Belt feeder. (Harris and Siegel) (B) Worm-drive feeder with mechanical vibrator. (Himes, Stout, and Thun, Braun and Lood) (C) Disk feeder (Beam and Takahashi) (D) Disk magazine feeder. (Marshall, Atlas and Putner) (E) Mechanically vibrated trough and cylinder source. (Richards) (F) Electromagnetically vibrated powder dispenser. (Campbell and Hendry) (From *Handbook of Thin Film Technology.* Copyright © 1970, McGraw-Hill. Used with permission of McGraw-Hill Book Co.)

Table 4.7: Flash Evaporation of Materials

Materials	Form of evaporant	Feeder mechanism	Filament temp, °C	Substrate temp, °C	Comments on films
Metals and Alloys					
Au(64)-Cd(36), Cu(52)-Zn(48)	Powdered alloys, 80/100 mesh	Moving belt	......	......	Au-Cd and β-brass films have composition of source
Ni + Fe	Mixed powders	Disk and wiper	~1930	......	Ni-Fe films with ±1% control of composition
Ni(86)-Fe(14)	Alloy wire	Spool and guide tube	2000	300	Ni-Fe film composition equal to that of source ±0.2%
Ni; Fe; constantan; chromel; alumel	Pellets	Disk magazine	2000	200-250	Thin film thermocouples. Rates: 5-300 Å s^{-1} across 12 cm distance
Ni(80)-Cr(20)	Alloy wire	Spool and guide tube	1620	......	Nichrome films, Cr content varies with filament temperature
Ni + Cr (20, 55, 70 % Cr)	Mixed powders, 100/300 mesh	Vibrating chute	1800	300	Alloy films within 1% of source composition. Rates: 1–10 Å s^{-1}
Sn; nylon	Wire; strands	Two spools and cutting knives	200 (nylon)	~0	Alternate layers of metal and insulator
Metal-Dielectric Mixtures					
Cu + SiO (1:5)	Mixed powders	Rotating tube	......	−269	Highly disordered films of high resistivity
Cr + 30 mol % SiO	Mixed powders, 325/400 mesh	Worm drive	~2000	200	250 ohms/sq resistor films with +20 to 50% deviations
Cr + SiO (62 and 74 mol % Cr)	Mixed powders, 125/325 mesh	Worm drive	2000	400	Resistor films; SiO content less than source. Rates: 4 Å s^{-1} across 23 cm distance
Cr + SiO (50-100 mol % Cr)	Sintered pellets, ~0.7 mm size	Disk and wiper	2050 ± 50	200	Resistor films, SiO content equals that of source ±1-3%. Rates: 20-30 Å s^{-1} across 70 cm distance
Cr(15)-Si(85)	Powdered alloy	Vibrating chute	2000	200-500	Resistor films
CrSi + TaSi$_2$ + Al$_2$O$_3$	Mixed powders	Worm drive	2500	200-400	Resistor films with ± 10% control

(Continued)

Table 4.7: (continued)

Materials	Form of evaporant	Feeder mechanism	Filament temp, °C	Substrate temp, °C	Comments on films
		Compounds			
AlSb	Powder, 100/150 mesh	Vibrating trough	1400–1600	700	Imperfect epitaxial films on Ge
GaP	Powder, 100/150 mesh	Vibrating trough	1500	540	Epitaxial films on Ge crystals
GaAs	Powder, 100/200 mesh	Vibrating trough	1450	300–670	Epitaxial films on Ge crystals above 600°C. Rates: 2–30 Å s^{-1}
	Powder, 100/200 mesh	Worm drive	1400–1800	530 ± 10	Epitaxial films on GaAs crystals. Rates: 2–5 Å s^{-1} across 21 cm distance
	Powder, 100/150 mesh	Vibrating trough	1300–1800	475–525	Epitaxial films on Ge crystals
	Powder, 40/60 mesh	Micrometer screw and piston	1325	525–575	Highly oriented films on Ge and GaAs crystals. Rates: 10–25 Å s^{-1} across 10 cm distance
GaSb	Powder, 100/150 mesh	Vibrating trough	1650	500	Epitaxial films on Ge crystals
InP	Powder, 100/150 mesh	Vibrating trough	1400–1650	300	Epitaxial films on Ge crystals
InAs	Powder, 100/150 mesh	Vibrating trough	1500	500	Epitaxial films on Ge crystals
InSb	Granules	Vibrating trough	1600	450–460	Epitaxial films on InSb crystals, n-type, 10^{15}–10^{17} donors per cm^3
	Powder, 100/150 mesh	Vibrating trough	1650	300–400	Epitaxial films on Ge crystals
Cu$_2$S; Cu$_2$Se	Powders of 250–300 μ	Vibrating chute	1400	25	Semitransparent, conductive films of Cu$_{1.8}$S and Cu$_{1.8}$Se
BaTiO$_3$	Sintered powder, 100/200 mesh	Vibrating trough	2300	500–700	Crystalline films, dielectric constants of 400–700. Rate: 3 Å s^{-1} across 8 cm distance
	?	?	2100	900–1000	Epitaxial films on sapphire and Si crystals, $p(O_2) = 10^{-4}$ Torr. Rates: 0.1–0.3 Å s^{-1}
Various perovskites	Sintered powder, 100/200 mesh	Vibrating trough	2050–2300	500–700	Perovskite films, epitaxial on LiF crystals. Rates: 1–3 Å s^{-1} across 8 cm distance

From *Handbook of Thin Film Technology*. Copyright © 1970, McGraw-Hill. Used with permission of McGraw-Hill Book Company.

Deposition of Refractory Compounds

Refractory compounds are substances like oxides, carbides, nitrides, borides, sulphides which characteristically have a very high melting point (with some exceptions). In some cases, they form extensive defect structure, i.e. exist over a wide stoichiometric range. For example, in TiC, the C/Ti ratio can vary from 0.5 to 1.0, demonstrating vacant carbon lattice sites. In other compounds, the stoichiometric range is not so wide.

Evaporation processes for the deposition of refractory compounds are further subdivided into two types: (1) Direction Evaporation[47] where the evaporant is the refractory compound itself; (2) Reactive Evaporation[48] or Activated Reactive Evaporation (ARE)[49] where the evaporant is a metal or a low-valency compound, e.g. where Ti is evaporated in the presence of N_2 to form TiN or where Si or SiO is evaporated in the presence of O_2 to form SiO_2.

Direct Evaporation: Table 4.8 gives the experimental conditions for the direct evaporation of refractory compounds. Evaporation can occur with or without dissociation of the compound into fragments. As seen from Table 4.8 the observed vapor specie show that very few compounds evaporate without dissociation. Examples are SiO, MgF_2, B_2O_3, CaF_2 and other Group IV divalent oxides (SiO homologs like GeO and SnO).

In the more general case, when a compound is evaporated or sputtered the material is not transformed to the vapor state as compound molecules but as fragments thereof. Subsequently, the fragments have to recombine most probably on the substrate to reconstitute the compound. Therefore, the stoichiometry (anion:cation ratio) of the deposit depends on several factors including the deposition rate and the ratios of the various molecular fragments, the impingement of other gases present in the environment, the surface mobility of the fragments (which in turn depends on their kinetic energy and substrate temperature), the mean residence time of the fragments of the substrate, the reaction rate of the fragments on the substrate to reconstitute the compound and the impurities present on the substrate. For example, it was found that direct evaporation of Al_2O_3 resulted in a deposit which was deficient in oxygen, i.e. which had the composition[50] Al_2O_{3-x}. This O_2 deficiency could be made up by introducing O_2 at a low partial pressure into the environment. In other cases, for example the direct evaporation of TiB_2 and ZrB_2 the deposit contains both the monoboride and diboride phases.[205]

Reactive Evaporation: In Reactive Evaporation,[48] metal or alloy vapors are produced in the presence of a partial pressure of reactive gas to form a compound either in the gas phase or on the substrate as a result of a reaction between the metal vapor and the gas atoms, e.g.

$$2Ti + C_2H_2 \rightarrow 2TiC + H_2$$

The reaction may be encouraged to go to completion by activating and/or ionizing both the metal and gas atoms in the vapor phase. It is called the Activated Reactive Evaporation (ARE) Process,[49] as illustrated in Figure 4.33. In this process the metal is heated and melted by a high acceleration voltage electron beam. The melt has a thin plasma sheath on top of the melt. The low energy secondary electrons from the plasma sheath are pulled upwards into the reaction zone by an interspace electrode placed above the pool biased to a low positive d.c. or a.c. potential (20 to 100V). The low energy electrons have a high ionization cross section, thus ionizing or activating the metal and gas atoms.

In addition, collisions between ions and neutral atoms results in charge exchange processes yielding energetic neutrals and positive ions. It is believed that these energetic neutrals condensing on the substrate along with the ions

Table 4.8: Direct Evaporation of Inorganic Compounds

Compound	Vapor species observed (in order of decreasing frequency)	mp, °C	T, °C, at which $p^* = 10^{-2}$ Torr	Comments on actual evaporation temperatures, support materials used, and related experience
		Oxides		
Al_2O_3	Al, O, AlO, Al_2O, O_2, $(AlO)_2$	2030	~1800	From W and Mo supports at 1850–2250°C. With telefocus gun at 2200°C, no decomposition. From W support: Al_2O_3 films have small oxygen deficits. O_2-dissociation pressure at 1780°C: 1.5×10^{-18} Torr
B_2O_3	B_2O_3	450	~1700	From Pt and Mo supports at 940–1370°C
BaO	Ba, BaO, Ba_2O, $(BaO)_2$, Ba_2O_3, O_2	1925	1540	From Al_2O_3 crucible at 1200–1500°C. From Pt crucible with only slight decomposition, p_{O_2} (1540°C) = 3.5×10^{-18} Torr
BeO	Be, O, $(BeO)_n$, $n = 1$–6, Be_2O	2530	2230	From W support at 2070–2230°C. With telefocus gun at 2400–2700°C, no decomposition
Bi_2O_3	. . . .	817	1840	From Pt support
CaO	Ca, CaO, O, O_2	~2600	~2050	Support materials: ZrO_2, Mo, W. The latter two form volatile oxides, molybdates, and wolframates at 1900–2150°C
CeO_2	CeO, CeO_2	1950	. . . .	From W support without decomposition
In_2O_3	In, In_2O, O_2	. . . .	. . . .	From Pt support with only little decomposition. Vapor species observed at 1100–1450°C. At 1000–1450°C from Al_2O_3 crucible, more In_2O than In
MgO	Mg, MgO, O, O_2	2800	~1560	Mo or W supports at 1840–2000° form volatile oxides, molybdates, and wolframates. With telefocus gun at 1925°C, no decomposition. From Al_2O_3 at 1670°C
MoO_3	$(MoO_3)_3$, $(MoO_3)_n$, $n = 4,5$	795	610	From Mo oven at 500–700°C, the trimer is the main species. Above 1000°C, there is some decomposition into $MoO_2(s) + O_2(g)$. At 730°C, the oxygen-decomposition pressure is 1.1×10^{-14} Torr. From Pt at 530–730°C
NiO	Ni, O_2, NiO, O	2090	1586	From Al_2O_3 crucible at 1300–1440°C. Heavy decom-

(Continued)

Table 4.8 (continued)

Compound	Vapor species observed (in order of decreasing frequency)	mp, °C	T, °C, at which $p^* = 10^{-2}$ Torr	Comments on actual evaporation temperatures, support materials used, and related experience
		Oxides		
Sb_2O_3	....	656	~450	position with $p_{O_2} = 4 \times 10^{-1}$ Torr at 1586°C Lower oxides result if evaporated from W supports. Pt heaters do not produce decomposition
SiO	SiO	....	1025	Usually evaporated from Ta or Mo heaters at residual gas pressures below 10^{-5} Torr and at temperatures between 1150 and 1250°C. Dissociation into Si and O_2 begins above 1250°C and may lead to oxygen-deficient films
SiO_2	SiO, O_2	1730	~1250	With telefocus gun at 1500–1600°C, no decomposition. Ta, Mo, W supports are attacked by SiO_2 and contribute volatile oxides. From Al_2O_3 at 1630°C, SiO_2 vapor species is present
SnO_2	SnO, O_2	....	....	From SiO_2 crucible at 975–1250°C. Films directly evaporated from W support are slightly oxygen-deficient
SrO	Sr, O_2, SrO	2460	~1760	From Al_2O_3 at 1830°C. Evaporation from Mo or W at 1700–2000°C produces volatile Mo and W oxides, molybdates, and wolframates.
TiO_2	TiO, Ti, TiO_2, O_2	1840	....	TiO_2 source material decomposes into lower oxides upon heating. p_{O_2} at 2000°C is 10^{-10} Torr. Nearly stoichiometric films by pulsed electron-beam heating
WO_3	$(WO_3)_3$, WO_3	1473	1140	From Pt oven at 1040–1300°C. From Pt support at 1220°C. From W heater with only slight decomposition; p_{O_2} at 1120°C is 3×10^{-10} Torr
ZrO_2	ZrO, O_2	2700	....	From Ta support at 1730°C, volatile TaO. From W support, oxygen-deficient films. ZrO_2 source material loses oxygen when heated by electron beams

(Continued)

Table 4.8: (continued)

Compound	Vapor species observed (in order of decreasing frequency)	mp, °C	T, °C, at which $p^* = 10^{-2}$ Torr	Comments on actual evaporation temperatures, support materials used, and related experience
Sulfides, Selenides, Tellurides				
ZnS	....	1830 ($p \approx 150$ atm)	1000	From Mo support. Minute deviations from stoichiometry if allowed to react with residual gases. From Ta at 1050°C
ZnSe	....	1520 ($p \approx 2$ atm)	820	
CdS	S_2, Cd, S, S_3, S_4	1750 ($p \approx 100$ atm)	670	From Pt oven at 740°C. Films tend to deviate from stoichiometry. Suitable support materials: graphite, Ta, Mo, W, SiO_2, Al_2O_3-coated W; evaporation at 600–700°C
CdSe	Se_2, Cd	1250	660	From Al_2O_3 crucible
CdTe	Te_2, Cd	1100	570	From Ta boat at 750–850°C; film stoichiometry depends on condensation temperature
PbS	PbS, Pb, S_2, $(PbS)_2$	1112	675	From quartz crucible at 625–925°C. From Mo support. Purest films from quartz furnace at 700°C; Fe or Mo boats react and form volatile sulfides
Sb_2S_3	....	546	550	From Mo support
Sb_2Se_3	Sb_4, $(SbSe)_2$, Sb_2, SbSe	611	....	From graphite at 725°C. From Ta oven at 500–600°C, fractionation and films of variable stoichiometry
Halides				
NaCl	NaCl, $(NaCl)_2$, $(NaCl)_3$	801	670	From Ta, Mo, or Cu ovens at 550–800°C
KCl	KCl, $(KCl)_2$	772	635	From Ni or Cu ovens at 500–740°C
AgCl	AgCl, $(AgCl)_3$	455	690	At 710–770°C. From Mo support, $p^* = 10^{-2}$ Torr at 790°C
MgF_2	MgF_2, $(MgF_2)_2$, $(MgF_2)_3$	1263	1130	From Pt oven at 950–1230°C. From Mo support. Very little dissociation into the elements
CaF_2	CaF_2, CaF	1418	~1300	From Ta oven at 980–1400°C. From Mo support
$PbCl_2$	....	678	~430	Direct evaporation possible

From *Handbook of Thin Film Technology.* Copyright © 1970, McGraw-Hill. Used with permission of McGraw-Hill Book Company.

and other neutral atoms "activate" the reaction, thus increasing the reaction probability between the reactants.

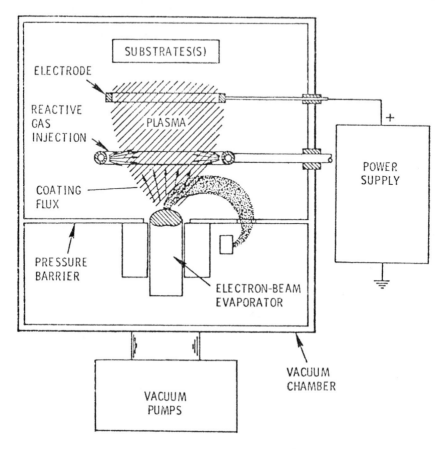

Figure 4.33: Schematic of the Activated Reactive Evaporation Process.

The synthesis of TiC by reaction of Ti metal vapor and C_2H_2 gas atoms with a carbon/metal ratio approaching unity was achieved with this process.[49,51] Moreover, by varying the partial pressure of either reactants, the carbon/metal ratio of carbides could be varied[52] at will. The ARE process has also been recently applied to the synthesis of all the five different Ti-O oxides.[49] These authors noted that in the ARE process (i.e., with a plasma) as compared to the RE process (i.e., without a plasma) a higher oxide formed for the same partial pressure of O_2 thus demonstrating a better utilization of the gas in the presence of a plasma. The same observation was noticed by Granier and Besson[52] for the deposition of nitrides.

A variation of the ARE process using a resistance heated source instead of the electron beam heated source has been developed by Nath and Bunshah[182] and is particularly useful for evaporation of low melting metals (such as indium and tin) where electron beam heating can cause splattering of the molten pool. The experimental apparatus is shown in Figure 4.33A. The plasma is generated by low energy electrons from a thermionically heated filament and pulled into the reaction zone by an electrical field perpendicular to the evaporation axis. The ionization probability is further enhanced by a superposed magnetic field which causes the electrons to go into a spiral path. This process has been used to deposit indium oxide and indium tin oxide transparent conducting films.

It should be pointed out that the two process of Direct Evaporation and Activated Reactive Evaporation are complementary to each other, having their own advantages and limitations.

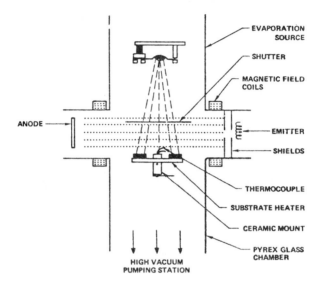

Figure 4.33A: The Activated Reactive Evaporation (ARE) Process[182] using resistance heated evaporation source.

CLASSIFICATION OF EVAPORATION PROCESSES

In the section above, the various evaporation processes for the deposition of various types of materials have been discussed. It was noted that the evaporation process for the deposition of refractory compounds included in some instances the use of a vapor plasma to enhance the reaction between the reactants. The plasma also causes the generation of ions and energetic neutrals.

A number of variations of the evaporation process have been reported in recent years and it would be useful to classify them.

Direct evaporation is the process where metals, alloys or refractory compounds are vaporized and deposited in the absence of any plasma intentionally generated. The substrate may be floating, grounded or electrically biased.

Activated evaporation is the evaporation of a metal, alloy or compound from a source in the presence of a plasma with the substrate floating, grounded or electrically biased.

Reactive evaporation is the evaporation of a metal in the presence of a reactive gas to deposit a compound without the presence of a plasma.

128   Deposition Technologies for Films and Coatings

Activated reactive evaporation (ARE) is the evaporation of a metal in the presence of a reactive gas and a plasma to deposit a refractory compound. The plasma is generated by an interspace positively biased electrode when an electron beam evaporation source is used and by low energy electrons injected into the vapor phase from a thermionically heated filament when a resistance heated source is used. The substrate can be floating, grounded or electrically biased.

The basic ARE process has several variations as shown in Figure 4.33B.

(1) If the substrate is biased in the ARE process, it is called biased activated reactive evaporation (BARE) process. This bias is usually negative to attract the positive ions in the plasma. The BARE process has been reinvented and called Reactive Ion Plating by Kobayashi and Doi.[175]

(2) Enhanced ARE process[200] is the conventional ARE process using electron beam heating with the addition of a thermionic electron emitter (e.g. a tungsten filament) for the deposition of refractory compounds at lower deposition rates as compared to the basic ARE process. The low energy electrons from the filament sustain the discharge.

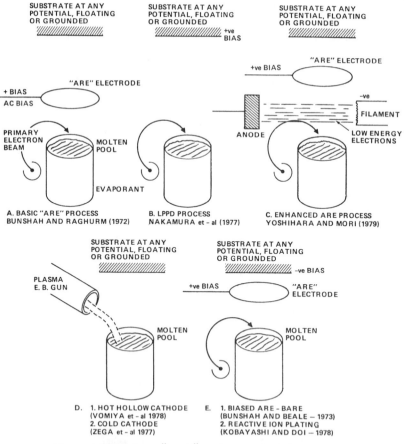

Figure 4.33B: Basic "ARE" Process and later variations.

(3) Using electron beam evaporation sources, the electric field may be generated by biasing the substrate positively instead of using a positively biased interspace electrode. In this case, it is called Low Pressure Plasma Deposition (LPPD) by Nakamura et al.[161] However, this version has a disadvantage over the basic ARE process since one does not have the freedom of choice to ground the substrate, let it float or bias it negatively (BARE process).

(4) The plasma electron beam gun instead of the thermionic electron beam gun can be used to carry out the ARE process. The hot hollow cathode gun has been used by Komiya et al.[176] to deposit TiC films whereas Zega et al.[177] used a cold cathode discharge electron beam gun to deposit titanium nitride films. A plasma assisted deposition process designated RF Reactive Ion Plating was developed and used by Murayama[201] to deposit thin films of In_2O_3, TiN and TaN. A resistance or electron beam heated evaporation source is used and the plasma is generated by inserting an r-f coil electrode of aluminum wire in the region between the evaporation source and substrate.

MICROSTRUCTURE OF PVD CONDENSATES

Microstructure Evolution

PVD condensates deposit as single crystal films on certain crystal planes of single crystal substrates, i.e., by epitaxial growth,[54] or in the more general case the deposits are polycrystalline. In the case of films deposited by evaporation techniques, the main variables are: (1) the nature of the substrate; (2) the temperature of the substrate during deposition; (3) the rate of deposition; (4) the deposit thickness; (5) the angle of incidence of the vapor stream; and (6) the pressure and nature of the ambient gas phase. Contrary to what might be intuitively expected, the deposit does not start out as a continuous film one monolayer thick and grow. Instead, three-dimensional nuclei are formed on favored sites on the substrates, e.g., cleavage, steps on a single crystal substrate, these nuclei grow laterally and in thickness (the so-called island growth stage) ultimately impinging on each other to form a continuous film. Figure 4.34 shows the growth of gold film on rock-salt. The average thickness at which a continuous film forms depends on the deposition temperature and the deposition rate (both of which influence the surface mobility of the adatom) and varies from 10 Å for Ni condensed at 15°K to 1000 Å for Au condensed at 600°K. This familiar model of island growth of a polycrystalline film during the initial stages of deposition illustrates the case where there is limited interaction between depositing atoms and the substrate. This is not always the case.

Important differences have been observed. Namba and Mori[183] found that by converting a significant fraction (~10%) of the vapor flux of Ag to positive ions, epitaxial growth of a single crystal Ag film on a single crystal Ag film on a single crystal NaCl substrate biased to -3,000 V was observed, whereas with vacuum evaporation the Ag film was polycrystalline. No clear explanation is possible except to note that the mobility of the deposited species is much greater when partially ionized than for neutral vapor specie. The effective surface temperature of the growing film is much higher due to ion bombardment thus permitting greater surface mobility and resulting in epitaxial growth. Taylor[184] used low energy electron diffraction (LEED) techniques to study the epitaxial deposition of Cu onto a single crystal[140] face of tungsten under ultra-

130 Deposition Technologies for Films and Coatings

high vacuum conditions. This represents the case where there is appreciable bonding between depositing atoms and the substrate. The deposit on a clean tungsten surface was a uniformly thin[111] Cu film, i.e. no island growth prior to the formation of a continuous film even at thicknesses as well as 1½ atomic layers. He further observed that chemisorption of even a half monolayer of oxygen severely inhibited epitaxial growth.

Sherman, Bunshah and Beale[64] studied the deposition of thick Mo films onto a rolled Mo sheet substrate as a function of deposition temperature. They observed polycrystalline deposits at all temperatures except in the range of 973° to 1188°K, where the surface oxide MoO_3 is unstable and evaporates rapidly, thereby leaving behind a "clean" Mo surface on which epitaxial growth can readily occur aided by the high surface mobility at the elevated deposition temperature.

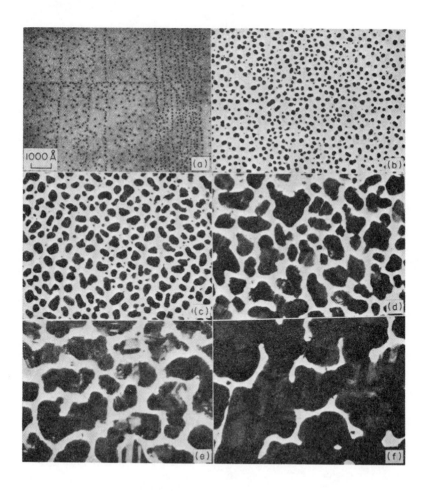

Figure 4.34: Sequence of micrographs illustrating the effect of increasing deposit thickness of gold on rock salt. X8000. (After Pashley-Ref. 101-with kind permission.)

Once a continuous film has formed, the subsequent evolution to the final structure of the thin film is poorly understood at present. It undoubtedly depends on the factors mentioned above which in turn influence the primary variables of nucleation rate, growth rate and surface mobility of the adatom. The problem has been tackled by Van der Drift[55] and is also the subject of a paper by Thornton.[56]

The microstructure and morphology of thick single phase films have been extensively studied for a wide variety of metals, alloys and refractory compounds. The structural model was first proposed by Movchan and Demchishin,[47] Figure 4.35 and was subsequently modified by Thornton as shown in Figure 4.36. Movchan and Demchishin's diagram was arrived at from their studies on deposits of pure metals and did not include the transition zone of Thornton's model, Zone T, which is not prominent in pure metals or single phase alloy deposits but becomes quite pronounced in deposits of refractory compounds or complex alloys produced by evaporation and in all types of deposits produced in the presence of a partial pressure of inert or reactive gas, as in sputtering or ion plating processes.

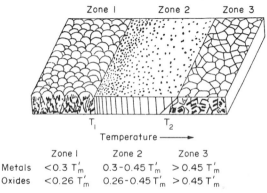

	Zone I	Zone 2	Zone 3
Metals	$<0.3\, T'_m$	$0.3 - 0.45\, T'_m$	$>0.45\, T'_m$
Oxides	$<0.26\, T'_m$	$0.26 - 0.45\, T'_m$	$>0.45\, T'_m$

Figure 4.35: Structural zones in condensates. (Movchan and Demchishan)

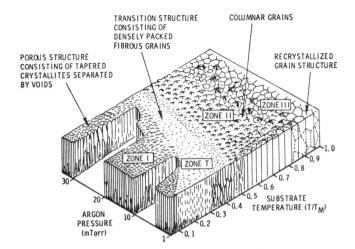

Figure 4.36: Structural zones in condensates. (Thornton)

The evolution of the structural morphology is as follows:

At low temperatures, the surface mobility of the adatoms is reduced and the structure grows as tapered crystallites from a limited number of nuclei. It is not a full density structure but contains longitudinal porosity of the order of a few hundred angstroms width between the tapered crystallites. It also contains a high dislocation density and has a high level of residual stress. Such a structure has also been called "Botryoidal" and corresponds to Zone 1 in Figures 4.35 and 4.36.

As the substrate temperature increases, the surface mobility increases and the structural morphology first transforms to that of Zone T, i.e. tightly packed fibrous grains with weak grain boundaries and then to a full density columnar morphology corresponding to Zone 2 (Figure 4.36).

The size of the columnar grains increases as the condensation temperature increases. Finally, at still higher temperatures, the structure shows an equiaxed grain morphology, Zone 3. For pure metals and single phase alloys, T_1 is the transition temperature between Zone 1 and Zone 2 and T_2 is the transition temperature between Zone 2 and Zone 3. According to Movchan and Demchishin's original model,[47] T_1 is 0.3 T_m for metals, and 0.22-0.26 T_m for oxides, whereas T_2 is 0.45-0.4 (T_m is the melting point in K).

Thornton's modification shows that the transition temperatures may vary significantly from those stated above and in general shift to higher temperatures as the gas pressure in the synthesis process increases.

It should be emphasized that:

(1) The transition from one zone to the next is not abrupt but smooth. Hence the transition temperatures should not be considered as absolute but as guidelines.

(2) All zones are not found in all deposits. For example, Zone T is not prominent in pure metals, but becomes more pronounced in complex alloys, compounds or in deposits produced at higher gas pressures. Zone 3 is not seen very often in materials with high melting points.

The reader is referred to a more extensive description given by Thornton in this book in the chapter on sputtering, which includes a discussion of the effects of substrate surface roughness and pressures.

Most thick deposits exhibit a strong preferred orientation (fiber texture) at low deposition temperatures and tend towards a more random orientation with increasing deposition temperature. Figure 4.37 shows the evolution of a large grained columnar morphology in a Be deposit from a much larger number of fine grains which were originally nucleated on the substrate. As growth proceeds, only those grains with a preferred growth direction survive, presumably due to considerations of the minimization of surface energy.

Elegant proof of the importance of surface mobility was also provided by Movchan and Demchishin.[47] Plots of the log of the grain diameter versus the inverse of deposition temperature in Zones 2 and 3 yield straight lines from which activation energies can be computed. It was found that the activation energy for Zone 2 growth corresponded to that for surface self-diffusion and for Zone 3 growth to volume self-diffusion.

The morphological results reported by Movchan and Demchishin for nickel titanium, tungsten, Al_2O_3 and ZrO_2 have been confirmed for several metals and compounds. The data are given in Table 4.9.[59,69,60,61]

Figure 4.37: Photomicrograph of a Be deposit showing the evolution of large columnar grains.

Table 4.9: Transition Temperatures Between Various Structural Zones

Material	Melting Temp., T_m (°K)	T_1 (°K)	$\frac{T_1 (°K)}{T_m (°K)}$	T_2 (°K)	$\frac{T_2 (°K)}{T_m (°K)}$	Reference
Ti	1945	653±10	0.3	923	0.5	47
Ti	1945	673±20	0.31	Phase transformation overlaps	—	62
Ni	1726	543±10	0.3	723±10	0.45–0.5	47
Ni	1726	—	—	777±20	0.45	63
W	3683	1133±50	0.3	1723±50	0.45–0.5	47
Mo	2883	923±20	0.3–0.34	1200	0.44	64
Fe	1810	—	—	Phase transformation overlaps	—	65
Be	1573	473±20	0.29	1023±50	0.63	66
Ni-20Cr	1673	500±50	0.3	870	0.52	69
ZrO_2	2973	648±10	0.22	1273	0.45–0.5	47
Al_2O_3	2323	623±10	0.26	1173	0.5	47
TiC	3340	1070±30	0.31	Not observed up to 1723°K or 0.51 T_m	—	62
NbC	....	Obeys the Movchan-Demchishin model				67
ZrO_2	....	Obeys the Movchan-Demchishin model				67

Note: From structures observed at a specific deposition temperature, Au-Cu[58] and V[37] appear to obey the Movchan-Demchishin model.

Bunshah and Juntz[62] studied the influence of condensation temperature on the deposition of titanium. Their microstructures, shown in Figure 4.38 agree substantially with those of Movchan and Demchishin for Zones 1 and 2 and T_1,

134 Deposition Technologies for Films and Coatings

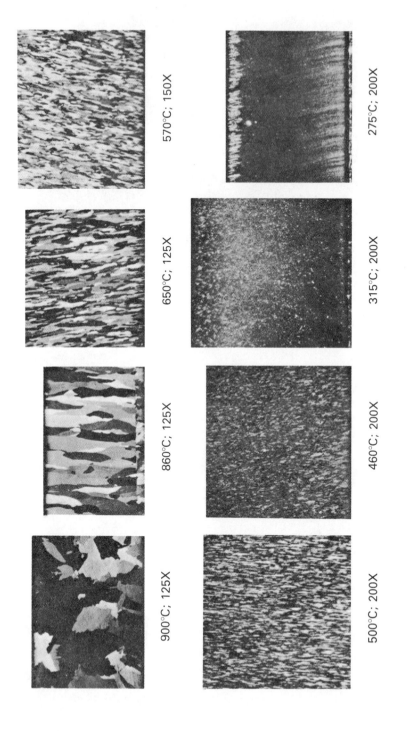

Figure 4.38: Structure of titanium deposits at various substrate temperatures (Bunshah and Juntz).

the transition temperature between Zones 1 and 2. However, they failed to observe Zone 3 at the temperatures above 700°C found by Movchan and Demchishin.[47] The structure was columnar up to 833°C, which is the $\alpha{:}\beta$ phase transformation temperature for titanium. At deposition temperatures above 833°C, the deposit crystallizes as the β phase and on cooling to room temperature, should transform to the α phase, resulting in the typical "transformed-beta" microstructure shown in Figure 4.38 (900°C deposit), which could be mistaken for an equiaxed microstructure. Hence, the claim of such a transition in structure from Zone 2 to 3 by Movchan and Demchishin for titanium deposits is confusing.

Kane and Bunshah[63] observed the change in morphology in deposited-nickel sheet. At 425°C deposition temperature, the deposit showed a Zone 2 morphology, whereas, at 554°C, the deposit showed a Zone 3 morphology.

Chambers and Bower[68] studied the deposition of magnesium, copper, gold, iridium, tungsten and stainless steel. Of the photomicrographs presented, gold and magnesium showed Zone 2 columnar morphology at the appropriate substrate temperatures.

Figure 4.39 shows surface and cross section photomicrographs of Ni-20Cr sheet deposited by Agarwal, Kane and Bunshah.[69] At 950°C, 760°C, 650°C and 427°C deposition temperatures, the surface and cross section showed an equiaxed Zone 3 morphology.

Mah and Nordin[66] found that the Movchan-Demchishin model was obeyed by beryllium also. They observed structures corresponding to all three zones with transition temperatures as predicted by the model.

Neirynck, Samaey and Van Poucke[70] studied the influence of deposition rate and substrate temperature on the microstructure, adhesion, texture, and condensation mechanism of aluminum and zirconium coatings on steel substrates and wires in batch and continuous-coating methods.

Kennedy[65] showed a change in morphology from columnar to equiaxed in Fe and Fe-10Ni alloy with higher deposition temperature. Deposits of Fe-1%Y which is a two phase alloy, showed columnar morphology only, the structure becoming coarser at higher deposition temperature. The second phase appears to nucleate new grains so that the grain size in Fe-1%Y alloys is much finer than that of iron.

The microstructure of copper-nickel alloys[67] produced by codeposition from two sources showed a single phase, as might be expected for this system, which shows a complete solid solubility. On the other hand, sequential deposition of Cu and Ni from two sources shielded from each other onto a rotating substrate produced a microlaminate structure in the deposit where the laminate size can be varied from 0.01 to 40 μm by adjusting the deposition parameters.[185] Similar structures were also developed in the Fe-Cu[185] and in the Ti-B$_4$C system.[185]

In alloy systems showing the presence of several phases, e.g., Ni-B and Cr-Si, the deposits showed the phases present corresponded to those expected from the diagram.[67]

Smith, Kennedy and Boericke[27] studied the deposition of the two phase ($\alpha+\beta$) type Ti-6Al-4V alloy deposited from a single rod-fed source. The microstructure was very similar to wrought material with the same characteristic $\alpha+\beta$ morphology present on a finer scale in the deposited material.

Dispersion-strengthened alloys produced by codeposition from multiple sources have also been produced. Paton et al[67] produced Ni-TiC, Ni-NbC and Ni-ZrO$_2$ alloys. The particle size increases from 100 to 1000 Å by changing the deposition temperature from 350° to 1000°C. The size of the dispersed carbide phase particles increased on annealing and 1000° to 1100°C due to their slight

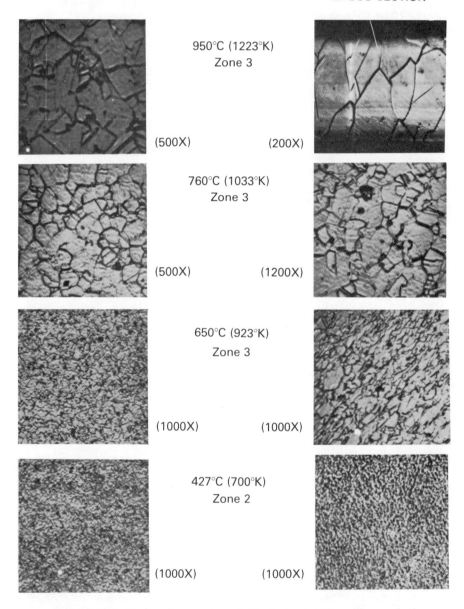

Figure 4.39: Photomicrographs of Typical Ni-20Cr deposits at various substrate temperatures. (Agarwal, Kane and Bunshah)

solubility in nickel. On the other hand, the size and distribution of ZrO_2 dispersion remained constant even after exposure at 1300°C for 5 hours as shown in Figure 4.40.

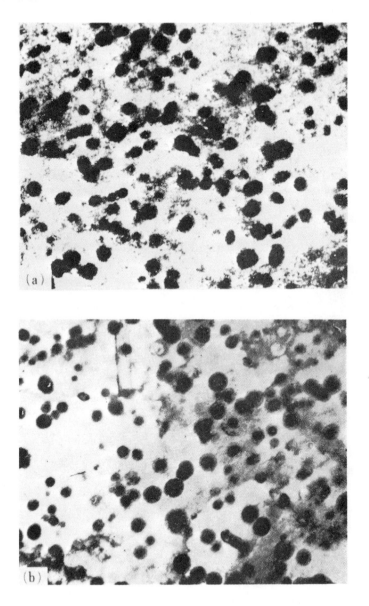

Figure 4.40: Microstructure of dispersion strengthened Ni-ZrO_2 alloy before and after exposure at 1300°C for 5 hours (Paton, Movchan, and Demchishin).

138 Deposition Technologies for Films and Coatings

Movchan, Demchishin and Kooluck[71] produced Fe-NbC and Fe-Ni-NbC dispersion strengthened alloys by coevaporation. The microstructure exhibited columnar morphology, with the inclusion of a fine dispersion of NbC particles.

Raghuram and Bunshah[72] studied the microstructure of TiC deposits from 500° to 1450°C shown in Figure 4.41. They observed the transition from the tapered crystallite (Zone 1) to columnar structure at 973°K, or 700°C (0.3 T_m). The highest deposition temperature (1450°C) used by these investigators was not sufficient to produce an equiaxed structure although this temperature corresponds to 0.51 T_m.

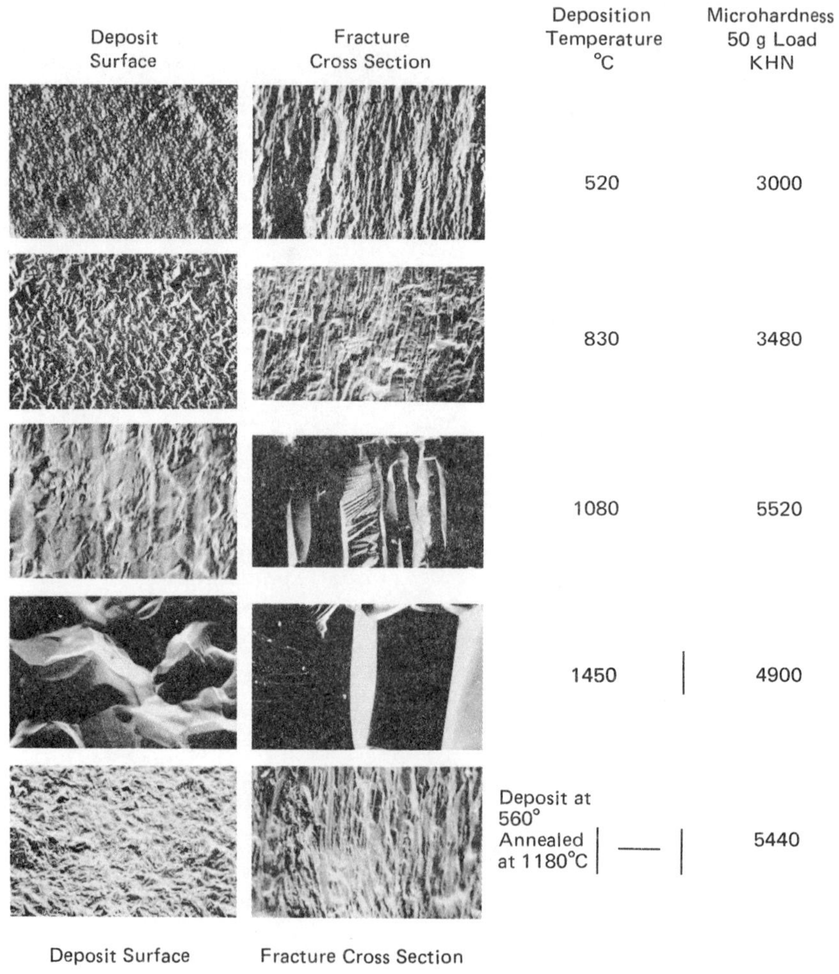

Deposit Surface	Fracture Cross Section	Deposition Temperature °C	Microhardness 50 g Load KHN
		520	3000
		830	3480
		1080	5520
		1450	4900
		Deposit at 560° Annealed at 1180°C	5440

Deposit Surface   Fracture Cross Section

Figure 4.41: Structure of TiC deposits at various substrate temperatures (Raghuram and Bunshah).

The energy of the depositing beam of atoms can be increased if some of them are ionized. It has been shown by Smith[23] that a small fraction of the vaporized species from an electron beam heated source is ionized due to collisions with electrons in the plasma sheath above the molten pool. Bunshah and Juntz[73] biased the substrate to -5,000 V during the deposition of beryllium at 570°C and found that the columnar grain size was markedly refined by the ion bombardment as compared to the grain size produced without biasing the substrate at the same deposition temperature. It may be postulated that the ion bombardment causes a localized increase in temperature at the surface where deposition is occurring, thus causing a higher nucleation rate and a finer grain size. Similar results have been reported for tantalum.[74] The use of hollow cathode gun intensifies the degree of ionization of the vapor species, resulting in a marked increase in kinetic energy of the vaporized atoms.[75] The effects of substrate bias are, therefore, easier to observe. Increasing the substrate bias results in a change in morphology from columnar to fine, equiaxed grains for silver deposited on beryllium and stainless steel,[76] and for silver and copper deposited on stainless steel.[77]

On the other hand, the presence of a gas at high pressures (5 to 20 μm) results in a net decrease in kinetic energy of the vaporized atoms due to multiple collisions during the transverse from source to substrate. This degrades the microstructure to loose columnar grains[77] and eventually to an agglomerate of particles. (This in fact is a way to produce fine powders by evaporation and subsequent gas-phase nucleation and condensation.) The negative effects of the presence of a high gas density on the kinetic energy and the mobility of adatoms on the deposit surface can be overcome by either biasing the substrate[77,78] and/or heating the substrate to a higher temperature.[79]

Texture

The texture of evaporated deposits is in general dependent on deposition temperature. At low deposition temperatures, a strong preferred orientation is generally observed: $\{211\}$ in iron,[65] $\{220\}$ in TiC,[72] $\{0002\}$ in Ti.[80] As deposition temperature increases, the texture tends to become more random. In the case of beryllium,[59] the texture changed to a $\{11\bar{2}0\}$ orientation at high deposition temperatures. The presence of a gas tends to shift the preferred orientation to higher index planes.[81] For silver, increasing the substrate bias changes the preferred orientation from $\{111\}$ to $\{200\}$ and back to $\{111\}$.[76]

Residual Stresses

Residual stresses in deposits are of two types. The first kind arises from the imperfections built in during growth (the so-called growth stresses). An increase in deposition temperature produces a marked decrease in the magntidue of this stress.[72,82] The other source of residual stress is due to the mismatch in the coefficient of thermal expansion between the substrate and the deposit. Its magnitude and size depend on the values of the thermal expansion coefficients as well as the thickness and size of the substrate and deposit. The influence of a negative bias on the substrate produces a compressive stress in the deposit, which reaches a maximum value at -200 to -300 V d.c. bias and then decreases.[78]

High residual stresses can cause plastic deformation (buckling or bending), cracking in the deposit or the substrate, or cracking at the substrate-deposit interface. The latter can be minimized by grading the interface, i.e., producing the change in material over a finite distance instead of producing it abruptly at a sharp interface. A graded interface can be produced by gradually changing the deposition conditions or by interdiffusion, which is enhanced by higher substrate temperature or increased kinetic energy of the vapor species.

Defects

Let us next consider the "defects" found in vapor-deposited materials. The first one is classified as a "spit", or small droplet ejected from the molten pool, which lands on the substrate and is incorporated into the coating.[83] An example is shown in Figure 4.42. The composition of the droplet is different from that of the coating in the case of an alloy and can therefore be the site of corrosion initiation. The bond between the droplet and the surrounding material is usually poor. Hence, corrosion attack can proceed down the boundary to the substrate or undermine the coating. The "spit" may also fall out, leaving a pinhole behind which can act as a stress concentrator and limit the ductility or the uniform elongation of a sheet material. Spits or pinholes do not affect the yield strength or reduction of area in a ductile material, but they can be stress raisers and sites for fatigue-crack initiation. Both spits and foreign particles on the substrate surface induce preferential growth of the deposit in that area because of higher exposure to the vapor flux than the general growing interface. This region of preferential growth is termed a flake; typical flakes are shown in Figure 4.43. There is marginal bonding between the flake and the deposit, which can lead to formation of a pit or crack, or to nucleation of corrosive attack.

Spits can be suppressed by eliminating porosity, oxide inclusions and compositional inhomogeneities in the evaporant source material, since spitting can be caused by included-gas release or by the release of bound gas through thermal decomposition. In electron-beam evaporation, the beam of electrons dissipates energy over a path extending as much as a mil (25 μm) or more into the melt. If this energy is delivered at a rate faster than the coating material can accommodate by evaporation, conduction or radiation, a pocket of vapor forms and spitting occurs. Spits are also caused by gas pockets included in the evaporant rod that suddenly expand when rapidly heated by the beam. Nonmetallic inclusions also can trap pockets of superheated vapor below them, which can erupt in a shower of molten droplets. Spits can be avoided by using a high purity vacuum melted rod as the evaporant. Flake formation can be avoided by avoiding the presence or impingement of foreign particles on the substrate (primarily by substrate surface cleaning and good housekeeping of the deposition apparatus). Deep grooves or ridges on the substrate can also produce flake-type defects by shadowing adjacent regions of the specimen surface.

Another type of defect occurs in complex alloys[83] such as M-Cr-Al-Y (where M can be nickel, cobalt or iron), where even at deposition temperatures of 955°C the deposit morphology corresponds to the fibrous transition zone between Zone 1 and Zone 2. The grain boundaries in this morphology are weak, causing intergranular corrosive attack (see Figure 4.44). The problem can be obviated by increasing the adatom mobility through the use of a higher substrate temperature or specimen bias of about -200 V, or by using a postcoating process that consists of a room temperature high intensity glass bead peening followed by a high-temperature anneal in hydrogen. Compound rotation of the specimen, which exposes higher surface irregularities to varying angles of impingement of incoming vapor atoms, produces a significant decrease in the number and size of open, columnar defects.

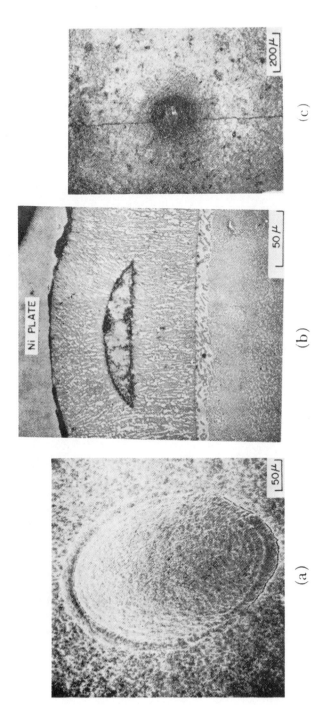

Figure 4.42: Vapor source droplet (spit) defect in M-Cr-Al-Y coatings. (a) and (b) show defects overcoated with additional material. (c) fatigue crack initiated at spit (Boone et al.). (Courtesy of Amer. Inst. of Physics.)

142 Deposition Technologies for Films and Coatings

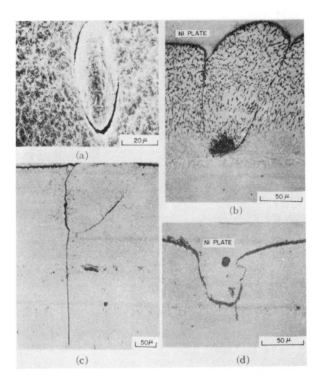

Figure 4.43: Flake defects in (a) and (b) produced by accelerated coating deposition on foreign particles. Glass bead peening incorporates flake into the coating (c) or knocks it out and forms a pit (d) (Boone et al.). (Courtesy of Amer. Inst. of Physics.)

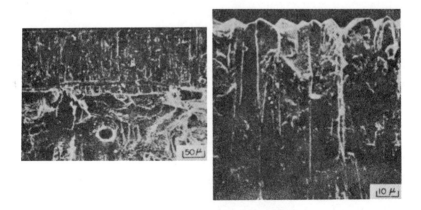

Figure 4.44: SEM photomicrograph of impact fracture surface of as-deposited overlay coating. Fracture is intercolumnar indicating weak boundaries (Boone et al.). (Courtesy of Amer. Inst. of Physics.)

Another problem in deposits of complex alloys is due to the variation in deposit chemistry attributable to segregation in the ingot and large pool temperature variations caused by the finite size of the electron beam.[83,84] Improved ingot quality, development of improved electron beam sources, and decrease in the temperature gradient at the crucible walls by using crucible liners or coolant of lower heat capacity, such as NaK, instead of water would minimize this problem.

In a more recent investigation on the origin of defects and continuing on the above,[85] it was found that spits in M-Cr-Al-Y type alloys consist of ejected pool material exhibiting enrichment in impurity elements of low vapor pressure as a result of superheating of non-metallic particles (carbides or oxides) in the melt initiating the ejection of pool material. Flakes, generally cone shaped, were found to originate at non-metallic particles loosely attached to the surface. Leader formation was found to be weakly dependent on the angle of incidence of the arriving vapor flux. Both flakes and leaders seem to be enhanced by preferential growth and shadowing phenomena.

PHYSICAL PROPERTIES OF THIN FILMS

The *Handbook of Thin Film Technology*[7] contains an extensive section on the electrical and electronic conduction, piezoelectrical and piezoresistive, dielectric and ferromagnetic properties of thin films. The reader is referred to it.

MECHANICAL AND RELATED PROPERTIES

Mechanical Properties

Mechanical Property Determination: A number of testing techniques have been used to determine the strength properties of thin films. They include the high speed rotor test,[86] the bulge test,[87-91] microtensile testing machines of the soft[92-95] and the "hard" categories[96-99] and even fixtures which can be operated in the electron microscope.[100,101] Hoffman[102,103] has reviewed the test techniques and the reader could do no better than to read Hoffman's article or the original references. The basic handling problem encountered with the preparation and mechanical property testing of thin film specimens is much less severe with thick films for which many of the standard test specimens, machines and techniques can be readily used. Therefore, the spectrum of mechanical properties measured on thick films is much broader than with thin films.

Tensile Properties of Thin Films: The tensile properties of thin films have been reviewed.[99,102,103,104,206] As Hoffman concludes,[101] the data reported are not very consistent even on the same material. The reader is advised to consult the references for details.

In general, the observed strength of vapor-deposited metal films consists of three parts

$$\sigma_{OBS} = \sigma_{Bulk} + \sigma_{Imperfections} + \sigma_{Thickness}$$

where σ_{Bulk} is the inherent strength level of bulk polycrystalline material in the annealed state, $\sigma_{Imperfections}$ is the contribution due to point defects in excess of those normally found in the bulk annealed state and $\sigma_{Thickness}$ is the contribution arising from the smallest dimension of the film and its limiting effect on grain size such that dislocation multiplication and migration are impeded.[94]

Table 4.10 gives the strength properties of thin films of some metals and compares them to bulk values.[103] In many cases, the strengths are about 200 times those of annealed bulk samples and 3 to 10 times those of hard drawn samples. The tensile strength values are given numerically as well as by fractions of the shear modulus. The ductility of the high strength films is very limited which is similar to the behavior of high strength fibers or whiskers. A principal point of contention is whether the ultimate tensile strength is a function of the film thickness or not. The discrepancy also appears to be dependent on the test method used, i.e. between the bulge test and tensile test. In many cases, it appears that the strength decreases as the film thickness increases from approximately the 200–300 Å range to about 2000–4000 Å range. At the greater thickness the strength is about the same as that of heavily worked bulk material. There are several papers relating the strength properties of thin films to the "crystallite size" and "block structure" as influenced by the deposition temperature, stress, recovery, and recrystallization process.[105-114] One manifestation of this is the phenomenon of creep or plasticity in room temperature tensile tests as exhibited by an irreversible initial loading curve but almost reversible unloading and reloading curves as long as the previous stress level is not exceeded. An example of this is shown in Figure 4.45 from Neugebauer[93] as the change in slope of the stress-strain curve. The possibility of this change in slope being related to an elastically soft measurement or to creep in the cementation of the grips cannot altogether be discarded.

Table 4.10: Strength of Properties of Thin Films

Material	Structure	Maximum Tensile Strength kg/mm^2	Shear Modulus*	Strain @ Fracture, %	Thickness-Dependent	Reference
Au	Bulk hard-drawn	28**	G/114	***	–	–
	(111) crystal film	81	G/36	1.2	No	36
	(100) crystal film	27	G/110	0.5	Yes	88
	(110) (111) Polycrystalline film	49	G/59	1	No	93
	Polycrystalline film	55	G/54	0.7	Yes	87
	(100) crystal film	26	G/115	3.5	No	95
	Polycrystalline film	32	G/92	2.3	No	95
Ag	Bulk hard-drawn	37**	G/75	–	–	–
	Polycrystalline film	59	G/47	0.7	Yes	78
	Polycrystalline film	42	G/68	0.3–0.4	–	116
Cu	Bulk hard-drawn	49**	G/98	–	–	–
	Polycrystalline film	93	G/51	1.8	Yes	117
	Polycrystalline film	88	G/54	–	No	91
	Rolled foil	19	G/256	10–15	No	118
Ni	Bulk cold-rolled	125**	G/67	–	–	–
	Polycrystalline film	210	G/40	1.8	Yes	94
Al	Bulk cold-rolled	16**	G/171	–	–	–
	Polycrystalline film	42	G/66	0.5–0.8	–	116

*Shear moduli from *AIP Handbook*, 1957.
**Bulk tensile strength from *Handbook of Chemistry and Physics*, 42nd ed., 1961.
***Fracture strain not quoted for bulk material.

Long term creep rates have been measured and for gold they vary from 10^{-7} to 10^{-4} min.$^{-1}$ depending on load, dimensions and the amount of prestrain.[93] The estimates of the relative elastic and plastic extension at fracture vary from completely elastic to an almost even mixture of elastic and plastic deformation.

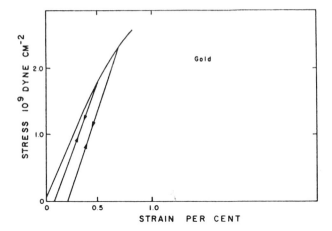

Figure 4.45: Typical stress-strain curve for thin film.

Fracture in ductile gold single crystal films[99] results from a localized plastic deformation with resultant thinning of the film and a rise in stress level. Eventually the smaller cracks formed in this manner join to cause fracture. The dislocations necessary for the deformation are not the grown-in dislocations but those which nucleate and multiply in discontinuous regions. Most observations show no necking prior to fracture. The maximum stress appears to correspond to that needed to propagate cracks from flaws existing in the specimen. In polycrystalline nickel, the fracture is the "clean-cleavage" type.[94]

Mechanical Properties of Thick Condensates and Bulk Deposits: Table 4.11 lists the mechanical properties of thick deposits of metals, alloys, refractory compounds and laminated structures. In many cases, the mechanical test data are quite extensive showing yield strength, ultimate tensile strength, hardness and ductility as a function of grain size, deposition temperature and test temperature. One of the features of the data is that the properties of thick deposits of metals and alloys are very similar to those of wrought materials which are produced by the conventional processes of melting, casting, mechanical working and heat treatment.

We will consider each type of material separately since the behavior of metals and alloys is vastly different from that of refractory compounds.

The early work in this area was that of Bunshah,[17,18] Bunshah and Juntz[22] and Smith[23] who deposited thick films of Be, Ti and Cu respectively and measured mechanical properties. In 1965, Palatnick and coworkers published a paper on mechanical properties of Al condensates.[106] It is impossible to review in detail all the papers. The pertinent data are shown in Table 4.11 and the discussion below will concentrate on the highlights.

Tensile Properties and Hardness of Metals and Alloys Deposits — Movchan and Demchishin studied the tensile properties and microhardness of Ni, Ti and W condensates produced at various deposition temperatures. No tensile tests were performed on specimens deposited in Zone 1 (Figure 4.45). Tests on specimens deposited in Zone 2 showed high strength and low ductility at low deposition temperature. The strength decreased and the ductility increased with deposition temperature. The strength and ductility values of specimens deposited in Zone 3 showed approximately the same values as for recrystallized specimens produced from wrought material. The microhardness variation

Table 4.11: Mechanical Properties of Thick Films or Bulk Condensates

Material	0.2% Yield Stress (kg/mm²)	Ultimate Tensile Strength	% Elong. (or % RA)	Microhardness (kg/mm²)	Deposition Temp. (°C)	Grain Size (μm)	Thickness (μm)	Test Temp. (°C)	Reference
Al	—	—	—	200	200	—	40	25	115
	—	—	—	160	300	—	—	—	—
	—	—	—	80	400	—	—	—	—
	—	—	—	8 Knoop @ 0.5 g load	250–350	—	1–2	25	119
Cu	15	—	—	—	—	1	4–10	25	109
	28	—	—	—	—	0.7	—	—	—
	40	—	—	—	—	0.1	—	—	—
	45	—	—	—	—	0.07	—	—	—
	—	18–22	35	—	400–800	—	1,000	25	67
	—	—	—	240	120	—	>15	25	120
	—	—	—	170	180	—	—	—	—
	—	50	2.38	—	—	—	3–20	25	121
	—	16 (annealed 500°C)	0.27	—	—	—	—	—	—
Ag	—	60	—	—	—	0.1	0.5	25	122
	—	40	—	—	—	0.6	3–5	—	—
	10	—	—	—	—	1	5–20	25	109
	35	—	—	—	—	0.1	—	—	—
Fe	—	66–72	—	250	—	—	20	25	123
	45	46	4	120	400	—	—	25	124
	20	35	20	90	550	—	1,000	—	—
	15	25	18	75	700	—	—	—	—
Be	—	140	0.1	240 (Knoop)	350–425	—	75–100	25	66
	—	281	0.3	215	480–550	—	—	—	—
	—	316	0.45	168	625–685	—	—	—	—
	—	295	1.0	170	760–790	—	—	—	—
	62 (bend test)	—	>4 (bend test)	—	—	100	1,000	25	18
	47 (bend test)	—	>5.2 (bend test)	—	—	—	—	—	—

(continued)

Evaporation 147

Table 4.11: (continued)

Material	0.2% Yield Stress (kg/mm²)	Ultimate Tensile Strength	% Elong. (or % RA)	Microhardness (kg/mm²)	Deposition Temp. (°C)	Grain Size (μm)	Thickness (μm)	Test Temp. (°C)	Reference
Ni	—	115	0.3	—	—	—	0.2–2	25	124
	—	66–74	—	—	—	—	20	25	125
	—	—	—	—	—	—	—	—	—
	—	—	—	—	—	—	—	25	63
	129	338	39 (20)	250	400–420	26	275	—	—
	135	336	32 (20)	450	260	19	375	—	—
	155	366	35 (21)	600	220	12.7	375	—	—
	247	403	21 (21)	96.6	554	3.7	375	—	—
	140 (melted and wrought)	407	44 (30)	109	425	18	357	—	—
				104	329				
				125	260				
				91					
Ti	—	138	0.6	—	250	620	2.6	25	90
	—	—	—	—	400	420	—	25	47
	—	—	—	—	600	380	—	—	—
	—	28	20	—	400	—	1,000	25	67
	189	38	12	—	800	32	—	25	62
	225	295	25 (66)	—	800	8	1,000	25	—
	436	302	20 (38)	—	600	1	250	25	—
	189 (melted and wrought)	471	3 (25)	—	450	50	250	—	—
		302	40 (61)				250		
							250		
Mo	—	—	—	560	200	—	1,000	25	47
	—	—	—	200	400	—	—	—	—
	—	—	—	130	500	—	—	25	80
Nb	280–420	—	20	—	—	—	100–250	200	64
	140–280	—	35	—	—	—	—	—	—
V	14	19	(5.3)	191	990	530	300	—	—
	24	28	(0.8)	169	704	85	—	—	64
	38	44	(1.0)	186	804	19	—	25	64
	44		7.3	112	745	10		25	64
	19		22	115	540	0.7		25	47
W	37	—	—	720	600	—	300	—	—
	—	—	—	460	1,000	—	1,000	—	—
	—	—	—	300	1,000	—	1,000	—	—
							1,000		

(continued)

148 Deposition Technologies for Films and Coatings

Table 4.11: (continued)

Material	0.2% Yield Stress (kg/mm²)	Ultimate Tensile Strength	% Elong. (or % RA)	Microhardness (kg/mm²)	Deposition Temp. (°C)	Grain Size (μm)	Thickness (μm)	Test Temp. (°C)	Reference
Ni-20Cr	337	540	26 (27)	80	950	13	375	25	69
	577	780	21 (26)	112	780	2.6	375	25	—
	752	871	18 (19)	131	680	1.8	375	25	—
	1,195	1,223	6 (8)	180	427	0.5	375	25	—
	393 (melted and wrought)	569	37 (31)	78	—	10.4	325	25	—
	25	49	25 (21)	—	950	13	375	1,000	—
	40.9	60	53 (41)	—	680	1.8	375	1,000	—
Ti-6Al-4V	970–1,195	1,068–1,195	1–12	—	—	—	79	25	27
80Ni-20Cu	—	35–45	8–25	—	400–800	—	375	1,000	67
50Ni-50Cu	—	40–55	5–20	—	400–800	—	375	1,000	—
20Ni-80Cu	—	27–42	10–30	—	400–800	—	375	1,000	—
80Ni-16Fe (permalloy)	—	180	—	650	200	—	—	25	120
	—	150	—	250	400	—	—	25	—
Al₂O₃	—	—	—	380	400	—	300	25	47
	—	—	—	80	600	—	300	25	—
	—	—	—	200	1,000	—	300	25	—
	—	—	—	1,000	1,400	—	300	25	—
	—	—	—	500–800	1,100	—	300	25	—
ZrO₂	—	—	—	450	400	—	300	25	126
	—	—	—	350	600	—	300	25	47
	—	—	—	400	1,000	—	300	25	—
	—	—	—	1,000	1,400	—	300	—	—
Y₂O₃	—	248	—	384	120	3.2	100	25	127
	—	226	—	256	210	5	100	25	—
	—	215	—	331	304	8	100	25	—
	—	208	—	320	416	14	100	25	—
	—	205	—	360	538	21	100	25	—
	—	198	—	502	721	58	100	25	—

(continued)

Evaporation 149

Table 4.11: (continued)

Material	0.2% Yield Stress (kg/mm^2)	Ultimate Tensile Strength	% Elong. (or % RA)	Microhardness (kg/mm^2)	Deposition Temp. (°C)	Grain Size (μm)	Thickness (μm)	Test Temp. (°C)	Reference
Ti$_2$O	—	—	—	960–1,960	750	—	—	25	52
TiO	—	—	—	700–2,015	1,100	—	—	25	—
SiO$_2$	—	—	—	300	—	—	10	25	127
Ta$_2$C	—	—	—	1,907–2,440	500	—	—	25	128
TaC	—	—	—	1,800–2,400	500	—	—	25	—
SiC (rf sputter)	—	—	—	3,500–4,000	—	—	2.6	25	129
TiC (ARE process)	—	—	—	2,710	520	—	50	25	72
	—	—	—	2,955	730	—	50	25	—
	—	—	—	2,955	830	—	50	25	—
	—	—	—	4,110	1,080	—	50	25	—
	—	—	—	4,160	1,120	—	50	25	—
	—	—	—	2,400	—	—	5	25	130
ZrC	—	—	—	2,500–2,800	550	—	50–75	—	51
(Hf-3Zr)C	—	—	—	2,700–3,030	550	—	50–75	—	—
VC	—	—	—	2,350	550	—	50–75	—	51
NbC	—	—	—	2,300	550	—	50–75	—	—
. . . . Laminates . . . .
Cu-Ni laminate	—	35–55	13	—	—	—	2	25	131
Cu-Fe laminate	35–70	110	1.8	400–800	240–360	—	15–25	25	132
Cu-Ni laminate	—	80	—	—	300	—	0.1	25	133
Cu-Fe laminate	—	125	0.88	—	150	—	0.04	25	—
Ni-Fe laminate	—	150	0.45	—	150	—	0.04	25	—
Ni-SiO	—	150	—	300	—	—	20	25	134
	—	105	—	300	—	—	20	200	—
	—	95	—	300	—	—	20	400	—
	—	80	—	200	—	—	20	600	—
Ti-TiC	—	251	1.8	—	500	—	250	—	135

with deposition temperature for Ni, Ti and W is shown in Figure 4.46. The tapered crystallite morphology in Zone 1 showed a high hardness much greater than that of annealed metal. The hardness decreased rapidly with increasing deposition temperature to a fairly constant value for Zone 3 morphology which corresponds to the hardness of recrystallized metals.

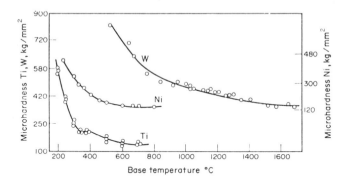

Figure 4.46: Variation of microhardness with deposition temperature of metals.

Bunshah and coworkers studied the effect of deposition temperature on the grain size, tensile properties and hardness of Ti,[22,62] Ni,[63] Nb, V, Mo[64] and Ni-20Cr[69] alloys for deposits made in Zones 2 and 3. They found that increasing deposition temperature produced larger grain size, lower strength, higher ductility, and lower hardness. Even at the lowest deposition temperature in Zone 2, the ductility was good (>20% RA for 1 μm grain diameter Ti at a yield strength of 56,000 psi). Moreover, they found that both the yield strength and hardness varied as the inverse square root of grain diameter, i.e. followed the Hall-Petch relationship[136,137] which is

$$\sigma_{ys} = \sigma_0 + kd^{-\frac{1}{2}}$$

where σ_{ys} is the yield strength; d is the grain diameter and σ_0, k are constants. Figure 4.47 shows an example of this relationship for Ni-20Cr alloy.

For all these metals and alloys, the yield strength, ductility and hardness values correspond to those of the same materials produced by casting, mechanical working and recrystallization. The variation of yield-strength and hardness with grain size, i.e. Hall-Petch type relationships were also very similar between the deposited and wrought materials, small variations being ascribable to differences in grain morphology and preferred orientations. The Ni-20Cr alloy showed good strength at 1000°C and also obeyed the Hall-Petch relationship.

The Hall-Petch relationship is also obeyed by thick films of Cu and Ag to grain-sizes as small as 0.05 μm as shown by Nenioto, Humbou and Suto.[138] Thus, these thick deposits behave as *true engineering materials*.

Chambers and Bower[139] studied the mechanical properties of 18-8 stainless steel, gold and magnesium and showed that their tensile properties were very similar to their wrought counterparts.

Smith, Kennedy and Boericke[27] studied the (α + β) type Ti-6Al-4V alloy. They showed that the tensile properties are very similar to the wrought material except for a much smaller value in percent elongation due to premature onset of plastic instability in a tensile test at pinholes in the deposited samples. The bend ductility was however superior to the wrought material.

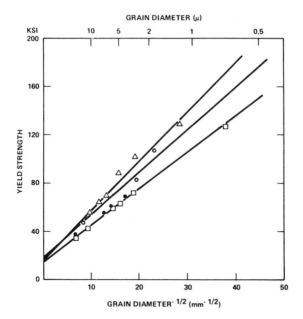

Figure 4.47: Yield stress vs. inverse square root of average diameter for Ni-20Cr alloy at 25°C. △ wrought; ○ deposited; □ Wilcox et al.; ● Webster. (*J. Vac. Sci. Technol.,* Vol. 12, No. 2, 662 (1975), References 12 to 13).

Shevakin et al.[174] studied the strength and hardness of aluminum and copper condensates as a function of the deposition parameters. They found that the mechanical properties varied widely with changes in process parameters. The deposited materials also showed higher strength and plasticity than the same materials conventionally fabricated, i.e. casting followed by the neo-mechanical treatments. They also found that the hardness values obeyed the Hall-Petch relationships.

Paton, Movchan and Demchishin[67] showed that it is possible to produce thick deposits of all the alloys across the Cu-Ni system and that the mechanical properties vary systematically with composition as would be expected.

Dispersion Strengthened Alloy Deposits — The first data on dispersion strengthened alloys produced by evaporation methods was reported by Paton, Movchan and Demchishin[67] who showed that Ni-ZrO_2 alloys produced by co-evaporation from two sources contained ZrO_2 particles in the size range of 150-3000 Å by changing the deposition temperature from 650° to 1100°C. They also showed that the creep strength at 1000°C increased with volume fraction of zirconia. These alloys showed remarkable stability in the microstructure and mechanical properties even after creep exposures of 5 hours at 1300°C. Subsequently, Movchan and coworkers studied the structure and properties of Ni-ZrO_2 alloys,[140] and Fe with Al_2O_3, ZrO_2, ZrB_2, TiB_2, NbC or TiC second phases.[141] The alloys were produced by coevaporation of the constituents from electron-beam heated evaporation sources.

One of the very striking effects of the incorporation of a dispersed phase in an evaporated metallic coating is a very pronounced refinement in grain size often by a factor of 10 to 100 or more and the inhibition of grain growth at elevated temperatures. This was first reported by Kennedy[65] for the incorporation of Y_2O_3 dispersions in Fe condensates. It was also observed by Majumder[148] for $Cu-Al_2O_3$ deposits and by Jacobson et al.[169] in $Ni-Al_2O_3$ deposits. In a very recent paper, Movchan et al,[47] show the grain size reduction in the $Ni-Al_2O_3$, $Fe-ZrO_2$, $Fe-ZrB_2$, and $Fe-NbC$ deposits. The most intense grain refining effect is observed at low volume fractions (0.5 V%) of the second phase

Of particular interest to this topic is a subsequent paper by Majumder[149] showing the strong effect of alumina content in increasing creep strength, which confirms the model proposed by Mott[170] who suggested that the ideal creep-resistant material is one with a fine grain size in which the grain boundaries are filled with some substance, say a refractory oxide, to inhibit the motion of grain boundaries.

Perhaps the most interesting result from Movchan's work[140,141,171] is that the dispersed phase alloys show a maximum in room temperature ductility in the $W-ZrO_2$ system at 1 vol % ZrO_2, in the $Fe-Al_2O_3$ system at 0.3 vol % Al_2O_3, and in the $Fe-NbC$ system at 0.1 vol % NbC. The yield strength and tensile strength do not show such a maximum but monotonically increase with volume fraction of the oxide phase. The significance of this observation lies in the possibility to increase the ductility of MCrAlY coatings which in turn would result in increased resistance to spalling, thermal shock and fracture, thus improving the performance of the coating. One might speculate on reasons for this effect including strain-relaxation sites at particle matrix interface, or at grain boundaries due to the greatly increased grain boundary area, favorable changes in residual stress distribution in the coating possibly due to changes in elastic modulus or strength, increased toughness or crack propagation resistance conferred by the dispersed phase particles, change in crystallographic texture, etc.

Movchan, Badilenko and Demchishin[172] have recently presented a very detailed treatment on the regulation of microstructure and mechanical properties of thick vacuum condensates with the help of dispersed phases. They give a detailed theoretical model of (1) the influence of dispersed phases on grain size; (2) the size and shape of dispersed particles as affected by deposition parameters; (3) strength and ductility of two phase condensates as influenced by the grain size, particle size, mean free path, nature of the particle (deformable vs. nondeformable) and particle-matrix adhesion energies; (4) steady-state creep behavior. The model is then confirmed by the experimental results. As a good illustration of one of these points, Figure 4.5 shows the difference in strength and ductility vs. volume fraction of second phase when the latter is deformable or nondeformable. For both types of particles, there is a ductility maximum at a particular D_g/λ ratio but the strength behavior is diametrically opposite showing a monotonic increase for a nondeformable particle and a minimum for the deformable particle. D_g is the grain size in the plane perpendicular to the vapor flux direction and λ is the interparticle spacing. This model forms an excellent basis for design of experiments to study the effect of dispersed phases on the structure and properties of MCrAlY alloys.

Another fascinating observation by Movchan et al.[172] applies to two phase alloys with deformable particles having a high adhesion to the matrix. The ductility of the alloys exceeds that of the pure matrix material at room temperature by a factor of 1.5 to 2 at a strain rate of $1:67.10^{-3}$ C^{-1} (0.1 min^{-1}). At high temperatures, the elongation at fracture exceeds 100%, i.e. superplasticity is developed.

Laminate Composites — Laminate composites are attractive and preferable over fibrous composites because of their uniform properties in the plane of the sheet. In comparison to mechanical methods of producing laminate composites, e.g. bonding of sheets or foil, physical vapor deposition techniques are very suited to the production of such composites, particularly if each lamellae is to be very thin (0.01 to 1 μm thickness) in order to improve the strength and toughness of the composite.

From theoretical considerations, it may be expected that the mechanical properties of microlaminate composites would follow an adaptation of the well known Hall-Petch relationship.[136,137] (Yield strength or hardness $\alpha d^{-\frac{1}{2}}$ where d is a characteristic microstructural parameter such as grain diameter, sub-grain diameter, laminae thickness, etc.). This correlation will be explored later.

In another approach, Koehler[187] proposed that a laminate structure which is formed of thin layers of two metals, A and B, where one metal (A) has a high dislocation-line energy and the other metal (B) has a low dislocation-line energy, should exhibit a resistance to plastic deformation and brittle fracture well in excess of that for homogeneous alloys. If the dislocation-line energies are so mismatched, the termination of the motion of dislocations in metal B is energetically favored over dislocation propagation across the layer interface into metal A. In the case of thick layers, the dislocations generated in either of the layers will pile-up in B at the A–B interface and thereby provide the stress concentrations needed for premature yield. Therefore, to suppress the generation of new dislocations in the layers, the thicknesses of A and B must be small. Thus, there is a critical minimum layer thickness required for the generation of dislocations.

This model does not take into account a high imperfection content in the laminate layers but assumes that their mechanical properties are similar to bulk annealed materials.

Most of the prior work on microlayer condensates was investigated in condensates produced at low deposition temperatures[186,188-195] (T <0.3 T_m) thus resulting in a high imperfection content. Moreover, the deposits were very thin (<25 μm in thickness), which makes it very difficult to measure the mechanical properties (particularly ductility) and draw good correlations with theory. The systems investigated were Ge/GaAg, Al/Mg, Be/Al, Al/Cu, Al/Ag, Ni/Cu, Mg/Cu, Al/Al$_2$O$_3$.

Recently Lehoczky[195] studied the layer thickness dependence of the yield strength of Al-Cu and Al-Ag laminates of thin specimens prepared by alternate vapor-deposition. Below the critical layer-thicknesses required for dislocation generation in the layers, the experimental results are in good agreement with Koehler's predictions. For layer thicknesses greater than those required for dislocation generation, he has extended the theoretical model to include dislocation pile-up groups.

A very recent investigation on the other hand by Bunshah et al.[185] used high deposition temperatures (T ≅ 0.4 - 0.45 T_m) where equilibrium structures are formed and thick specimens (200 to 1,000 μm thickness) containing a very large number of micro-layers were produced such that mechanical properties can be easily measured on standard test specimens. Fe-Cu and Ni-Cu microlaminate composites were prepared by sequential deposition from two evaporation sources. Very marked increase in strength were observed, by as much as a factor of 10 as compared to the pure metals and a factor of 5 as compared to the solid solution Cu-Ni alloy of the same composition. The ductility decreased somewhat but was still appreciable (5% elongation) for the highest strength alloys. The strength and hardness values followed the Hall-Petch relationship. Superplastic behavior was observed in Fe-Cu microlaminates when the average grain size of the metal equals the interlammellar spacing (approximately 0.45 - 0.50 μm) at a test temperature of 600°C at a strain rate of 0.005 min^{-1}.

High temperature creep properties of thick Fe/Cu and Ni/Cu microlaminate condensates were studied at 600°C as a function of layer thickness.[202] Steady state creep rate has been found to increase with a decrease in microlayer thickness. Microstructural study of the specimens after creep tests revealed the disintegration of iron and nickel layers in Fe/Cu and Ni/Cu condensates respectively with the formation of separate inclusions of an oval shape. The creep rate variation in the microlayer condensates is explained with the help of a structural model of high temperature creep.

Refractory Compounds — Deposits of refractory compounds, oxides, nitrides and carbides are very important for wear resistant applications in industry. Their structure and properties are strongly dependent on the deposition process. Their behavior is very different from metals and alloys. It is also very hard to measure the mechanical properties of ceramics by tensile tests similar to those used for metals and alloys because of their brittle nature. A very good test to measure the fracture stress of such brittle coatings is the Hertzian fracture test which measures the fracture stress and the surface energy at the fracture surface.[142] Colen and Bunshah[122] used this test to measure the fracture behavior of Y_2O_3 deposits of various grain sizes.

Figure 4.48 shows the variation in microhardness with deposition temperature for Al_2O_3 and ZrO_2 from the work of Movchan and Demchishin,[47] showing that the behavior of these oxides deposits is quite different in one respect from that of metals (Figure 4.46). The hardness falls when the structure changes from tapered crystallites (Zone 1) to columnar grains (Zone 2) as with metals. However, unlike metals the hardness increases markedly as the deposition temperature rises from 0.3 T_m to 0.5 T_m. The authors attribute this to a more "perfect" material produced at the higher deposition temperature due to "volume processes of sintering". A similar hardness curve was obtained for Y_2O_3 deposits.[127]

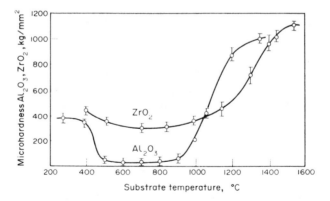

Figure 4.48: Variation of microhardness with deposition temperature for Al_2O_3 and ZrO_2.

Figure 4.49 from the work of Raghuram and Bunshah[72] also shows a very marked increase in microhardness of TiC deposits on going from 0.15 T_m (500°C) to 0.3 T_m (1000°C). The hardness increases for the oxides and TiC with increasing deposition temperature. Both these sets of results may be explained by the following concept. Since the strength of ceramics is very adversely effected by growth defects and at the higher deposition temperature, the occurrence of these defects is markedly reduced, the hardness (or strength) increased very significantly. However, it should be noted that the absolute value of the hardness of the oxides is much lower than that of the carbides. Thus the possibility of a different explanation for the "similar" behavior of these materials, i.e., the hardness increase with the deposition temperature needs to be investigated.

The hardness data on sputtered TiC and TiN coatings are quite similar to those produced by evaporation techniques.[130]

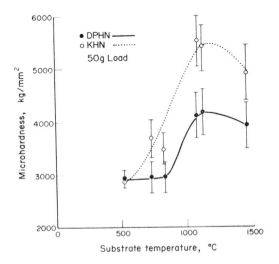

Figure 4.49: Variation of microhardness with deposition temperature for TiC.

Wear Resistance

Hard coatings of engineering surfaces and cutting tools are an important and growing application. In a recent investigation,[197] TiC and TiN films deposited onto stainless steel, titanium and aluminum substrates produced a very marked decrease in wear under adhesive wear conditions even with a conventional lubricant. This is in contrast to ion-implanted surfaces which show low wear rates only under lubricated conditions. Further studies on the adhesive wear of couples where both surfaces are hard,[198] showed that TiC/TiC and TiC/TiN couples show very low coefficients of friction (0.1–0.2) and correspondingly low wear rates. These hard coatings also show marked improvement in abrasive wear and impact erosion wear conditions.

Cutting tools of cemented carbides and high speed steel fail by adhesive and abrasive wear. Hard coatings of carbides, nitrides and oxides produce marked improvement in tool wear. The use of CVD techniques for such coatings on cemented carbide tools is well established. However, the CVD process is not suited to shaped high speed steel tools due to the high deposition temperature of the process resulting in dulling of the tool edge and distortion of the tool. The low deposition temperature PVD processes of evaporation, ion plating and sputtering are therefore particularly attractive. Laboratory data show improvements in tool life by a factor of 2 to 10 depending on the type of cutting tool.[161-164,175,196]

Type M-10 high speed steel drills, ¼" diameter, were coated with titanium carbide and titanium nitride using the Activated Reactive Evaporation Process. The coated drills were tested by drilling holes into 1" thick plates of AISI 4150 steel at a hardness level of 300 BHN. The speed used was 900 rpm and the feed rate was 0.007" per revolution. For comparison, uncoated and the standard black oxide coated drills were also tested using the same conditions. The average number of holes to failure for uncoated drills were 7, with black-oxide drills were 33 and the TiC and TiN coated drills were 141. These results show that the TiC and TiN coated drills have tool life of 20 times the uncoated and 4.2 times the black oxide drills under these test conditions.[203]

Corrosion Resistance

The microstructure of the coating influences the corrosion resistance afforded by coatings. In general, it is known that coatings with a columnar morphology are not as protective as those with an equiaxed morphology. Exact mechanisms for this behavior have not been determined. However, the real grain-boundary area is much smaller in the columnar morphology than in the equiaxed morphology. The concentration of impurities per unit grain-boundary area would be much higher in the columnar morphology, thus accounting for the lower corrosion resistance. The occurrence of coating defects such as weak grain boundaries in the transition fibrous morphology would also reduce the corrosion resistance, as reported by Boone, Strangman and Wilson.[83]

Meyers and Morgan[25] studied the influence of aluminum coatings on corrosion resistance of steel strip. The corrosion resistance was evaluated in terms of the number of spots per unit area produced on the steel strip exposed to a corrosive environment. The adhesion and corrosion resistance of the coating varied with porosity, which in turn was dependent on microstructure, substrate preheat temperature, deposition rate, surface finish and coating thickness. Unfortunately, they did not present microstructures of the coatings over the temperature range of deposition to enable any conclusions to be drawn on the influence of morphology on corrosion resistance.

Type M-50 high speed steel used in engineering components such as bearings coated with Ti, Cr, Mo, Ni-20 Cr, TiC and TiN by various evaporation processes eliminated the severe localized pitting corrosion of this steel which occurs when it is exposed to a hydrocarbon environment containing a few ppm of chloride ions.[204] The results are comparable to the corrosion prevention obtained with a much more expensive process, i.e., ion implantation. Rolling contact fatigue tests on M-50 steel specimens coated with a 5 μm TiC layer showed no decrease in fatigue life. This introduces the possibility of obtaining a coating for bearing steels to increase both the corrosion and wear resistance without degradation of fatigue properties.

Influence of Coatings on Thermal Fatigue Resistance

Boone and Sullivan[143] studied the influence of coatings on the thermal fatigue resistance of super alloys. On polycrystalline substrates, an inward type of diffusion aluminide coating improved fatigue resistance whereas an outward type was detrimental. "Overlay" coatings produced by vapor deposition with an equiaxed, fine grain morphology gave superior results. The authors point out that while the use of aluminide coatings can be optimized by thorough characterization, they are limited by the very nature of the process involved in their formation. On the other hand, "overlay" coatings produced by vapor deposition are much more versatile since their composition and structure can be controlled independently of the substrate and yet be compatible with it. Thus, the coatings can be tailored both for the required environmental protection and for improved fatigue properties.

PURIFICATION OF METALS BY EVAPORATION

Impurities in the deposit can be classified into two types, metallic and nonmetallic. Knowing the composition of the evaporant, the experimental conditions (temperature and time), certain thermodynamic data (vapor pressure and activities in solution), the composition of the vacuum environment during the experiment, and the types of melt-crucible reactions, if any, it is possible to estimate the impurity content of the distillate. The amount of impurity transfer to the vapor phase and hence in the deposit (assuming a sticking coefficient of unity) depends directly on the partial pressures of the impurity and the basis metal. For metallic impurities, one assumes that each impurity behaves independently of the other and, using Rayleigh's equation, the metallic impurity content of the distillate may be estimated. Experimental verification has been demonstrated by Bunshah for beryllium.[20]

The amount of nonmetallic impurity (C, O_2, N_2 and H_2) is estimated as follows: for example, for oxygen,

$$\text{ppm (atomic) } O_2 = \frac{\Sigma^{\nu}G}{\nu_M}$$

where $\Sigma^{\nu}G$ = sum of the impingement frequencies (number of atoms/cm^2/s) of the various gases and vapors present in the vacuum environment containing O_2, such as H_2), CO, CO_2 and Mo (metal suboxide), on the substrate ν_M = impingement frequency of metal atoms on the substrate, an experimentally determined parameter.

Implicit in this treatment is the assumption that the sticking coefficient for all the species is unity. This assumption is good for reactive gases such as CO, CO_2 and H_2O but poor for gases such as H_2, as has been shown by Bunshah and Juntz[31] for beryllium; they also demonstrated a satisfactory agreement between computed and experimentally observed values for the nonmetallic impurities. Table 4.12 shows the production of very high purity beryllium in sheet form by vacuum melting followed by vacuum distillagion. The oxygen content of the distillate is due to suboxide vaporization (Be_2O) from the melt and consequent contamination of the substrate, the suboxide having a higher vapor pressure than the evaporating species. The oxygen content of refractory metal deposits produced by vacuum evaporation can also be substantially increased by suboxide vaporization from the melt.[64] The suboxide can be that of the deposit itself, e.g., MoO in the case of molybdenum deposition; or that associated with an impurity in the evaporant, e.g., MoO in the evaporation of vanadium.

158   Deposition Technologies for Films and Coatings

Table 4.12: Purification of Beryllium by Vacuum Melting and Distillation
(in parts per million atomic)

Element Constituents	Starting Material Grade SR Flake	After 2 Vacuum Melts	Distillate	Residue after Distillation
C	135	19	5.8	81.2
N	15	1.0	0.1	3.3
O	108	29	17	23
Na	31	1.0	0.3	<0.1
Mg	<38	0.03	0.06	0.25
Al	17	5.0	0.07	0.8
Si	1.0	1.0	0.5	12.6
Cl	135	4.0	0.25	0.4
Ca	12	1.0	0.5	0.06
Fe	4	7.0	0.045	16
Ni	2	5.0	<0.005	32
Cu	1.4	1.2	0.15	14

APPLICATIONS

The products from high vacuum coating process are:

Decorative: Automotive trim (interior and exterior), toys, cosmetic packaging, pens and pencils, Christmas decorations, food and drink labels, costume jewelry, home hardware, eyeglass frames, packaging and wrapping materials, watch cases.

Optically Functional: Laser optics (reflective and transmitting), architectural glazing, home mirrors, automotive rear view mirrors, eyeglass lenses, projector reflectors, camera lenses and filters, instrument optics, auto headlight reflectors, TV camera optical elements, meter faces.

Electrically Functional: Semiconductor devices, integrated circuits, capacitors, resistors, magnetic tape, disc memories, superconductors, electrostatic shielding, switch contacts, solar cells.

Mechanically Functional: Aircraft engine parts, aircraft landing gear, solid film lubricants, tool bit hard coatings.

Chemically Functional: Corrosion resistant fasteners, gas turbine engine blades and vanes, battery strips, marine use equipment.

Present Uses

The present uses of evaporated coatings in industry are as follows:

(1) Deposition of Al and Al-Si alloys in microelectronics.

(2) Deposition of other metals, alloys and compounds (e.g., GaAs, SiO) in microelectronics.

(3) Deposition of Al polymer films for heat insulation and for decorative coatings.

Evaporation

(4) Deposition of metals such as Al and Ta on insulating film for capacitors.

(5) Deposition of refractory compounds such as MgF_2 in the optics industry.

(6) Deposition of metals and compounds on large panes of architectural glass (up to 10' x 12'). Sputtered coatings are also done on smaller sizes.

(7) Coating of nickel-base jet-engine blades used in the hot end of large aircraft-type jet engines, particularly for use in marine-environment and high performance applications.[83,143,144] The coating is a quaternary alloy of M-Cr-Al-Y where M is cobalt, nickel or iron. This is a unique application and it can be put by electron-beam-evaporation technology. Such coatings are deposited both in the U.S. and U.S.S.R.

(8) Coating of aluminum on Kovar for the manufacture of lead frames in the electronics industry. This is achieved by vacuum evaporation using electron beam, induction or resistance heating.

(9) Aluminum coating of steel strip. Steel strip, 2 feet wide and 0.006 inch thick is coated with 0.00025 inch (250 micro-inch) of aluminum on both sides in a continuous air-to-air electron beam evaporator using an S pass at 200 ft/min. This is to replace tin plate. Strip then goes directly to the lacquer line. Ten percent of the steel for can production in the German Democratic Republic is processed in such plants.[173]

(10) Coating of Ni on performated iron strip for Ni-Cd batteries in the U.S.S.R.

(11) Chromium on glass plates for masks to be used in photolithography *in Japan.*

(12) Coating of titanium and other alloys on steel wire for corrosion resistance in the U.S.S.R.

(13) The manufacture of ultrafine (<1 μm) powders of metals, alloys compounds by evaporation and condensation in high-pressure gas[145,146] on cold surfaces.[147] Ultrafine particles have been used in the manufacture of magnetic tapes, since they are expected to have a high memory density and high coercive force.

(14) Wear resistant coatings on high speed steel cutting tools.[175,196,203]

Future Uses

In this category are listed potential future applications in which research and/or development work has been carried out.

(1) General classes of Materials and Applications.

    i New alloys using (a) very fine grain size to obtain high strength with high toughness, (b) very fine particle sizes and spacings to produce dispersion-strengthened alloys,[32,140,141,148,149] and (c) very closely spaced and very thin single and multilaminate structures to produce high-strength, high-toughness composites.

ii Coatings for (a) corrosion resistance, e.g., a thin coating of titanium or a titanium alloy on a lower priced material as compared to fabrication of the entire item from bulk titanium, (b) abrasion resistance on common steel and alloys using deposits of oxides, nitrides, carbides, cermets, or graded deposits, (c) increasing tool performance of high speed steel and carbide tools, and (d) catalysis using the ability of the process to produce very high surface areas and nonequilibrium alloys.

iii Production of ultrafine powders of metals, alloys, ceramics mixtures thereof. These powders would have great utility in the production of tool steels, cermets, dispersion-strengthened alloys, etc.

iv **Production of** new superhard materials, e.g., alloy carbides, rare-earth borides, etc.

v New superconducting materials, particularly in the direction of high T_c and high field superconductors.[150,151]

vi Development of new optical and electrical materials, e.g., compounds of controlled stoichiometry.

vii Biomedical materials, e.g., controlled porosity surfaces for implants, fine coaxial cable,[152,153] etc...

viii Amorphous materials and finely dispersed alloys of immiscible components.[154,155]

The following are some specific examples of potential applications where the feasibility has been demonstrated in the laboratory and some industrial use has been initiated.

(1) Manufacture of Ti-6Al-4V alloy foil and other lightweight, high-strength materials for fabrication into honeycomb structures for aircraft.[27] A pilot plant for such free-standing foils 0.002 inch thick, 12 inches wide and up to 1,200 feet in length has been constructed and operated for Ti-6Al-4V, ferritic stainless steel (80% Fe, 20% Cr), Inconel 600, aluminum bronze, zirconium, hafnium, nicel and copper.[156]

(2) Hafnium foil for flashbulb applications, using electron-beam evaporation.[157]

(3) Single and multilayer laminates[158] with various material combinations, e.g., copper or aluminum on mild steel or stainless steel, stainless steel on mild steel, stainless steel on copper, aluminum bronze on mild steel, etc. The objective is to provide improved properties or similar properties, using cheaper material for the bulk of the structure. The technique is electron-beam evaporation.

(4) Coating of large plastic parts with metals and alloys to simulate the looks of a metal part but with significant weight savings.[158,159] This is of obvious importance in automobiles for weight reduction and energy conservation, electron-beam evaporation, ion plating and rf sputtering can be used. It is currently practiced on a limited scale.

Evaporation   161

(5) Hard coatings on engineering components for abrasion and wear resistance. Table 4.13 shows that TiC coating using the ARE process[160] on stainless steel can produce an excellent smooth wear surface comparable to that obtained with a nickel-diamond composite coating and 36 times better than hard chrome plate produced by electrodeposition. In Japan, there is extensive replacement of electroplated chromium deposits by vacuum deposition chromium.

(6) Hard coatings on cutting tools of cemented carbide[161,162] and high speed steel[162,164] showing improvements in tool life of 2 to 8 times in continuous cutting as well as reduction in cutting forces.

(7) Coatings of Si and other compounds such as CdS for solar energy applications.[165]

Table 4.13: Wear Tests on Various Coatings Using Al_2O_3 Abrasive Particles on Thread*

Coatings	Amount of Wear
Electroplated chromium	17.00 mils
Al_2O_3 particles in electroless Ni	3.5 mils
TiC-E-100	2.5 mils
TiC-E-104	0.5 mils
Diamond in electroless Ni	0.0 mils

*Test time was 15 minutes. Thread on test piece had 90° included angle.

ECONOMICS AND PERSPECTIVE

In many potential applications, the unique capabilities of PVD techniques make the economics a "go-no-go" situation, i.e, the market will have to bear the cost of making the desired component, which is otherwise impossible to fabricate. Good examples are coated turbine blades and aircraft parts. In other situations, PVD is but one step in the manufacture of a part, and the cost comparisons are hidden in systems costs. Nevertheless, some idea of economics can be obtained. For example, Thornton[82] states that sputter coating of a particular cylindrical machined part with 0.1 mil of refractory metal at a rate of 10,000 parts per day for a single machine would cost 10 to 25¢ per part. For difficult-to-roll materials, the cost of fabricating these foils (<10 mil) is expected to be significantly cheaper than producing the same foil by rolling, one factor being the high yield of the process.

In view of the current interest in solar energy, some estimates have been made for semicontinuous deposition of compounds such as CdS, CdTe, etc. Using vacuum evaporation, it is estimated that an apparatus costing $2.4 million could produce a continuous deposit of CdS 4 ft. wide and 15 μm thick at the rate of 1 ft/s at a cost of $1/sq. ft.[166] Another independent estimate comes up with a very similar figure.[167] Using high-rate sputtering technology, p-CdTe/n-CdS deposits can be manufactured from a system costing $300,000 at the rate of 288 panels, each 1 sq. ft. in area and a few micrometers thick per day at a cost of $4.40/sq. ft.[168]

A more recent estimate[199] is for CdS/Cu_2S solar cells with a 10% efficiency to be manufactured at a cost of less than $0.20/W ($1,979) as early as 1986 and a 15% efficiency solar cell at a cost of less than $0.15/C as early as 1990.

It is not difficult to see that materials requirements for energy-related applications (superconductors, high-temperature coal conversion, nuclear reactor materials), for higher productivity in machining and forming, for longer life of components (improved corrosion and wear resistance), etc., are spurring the development of coating technologies at a rapid tempo. Physical vapor deposition techniques are *not* just laboratory tools. They have also been demonstrated in large-scale applications. Therefore, the translation from laboratory to industrial practice can be rapidly achieved. The critical step is the product-development work in the labroatory. The versatility of PVD technology assures its rapid development and application in the next decade.

REFERENCES

(1) M. Faraday, *Phil. Trans. 147,* 145 (1857).
(2) R. Nahrwold, *Ann. Physik, 31,* 467 (1887).
(3) A. Kundt, *Ann. Physik, 34,* 473 (1888).
(4) F. Soddy, *Proc. Roy. Soc. London, 78,* 429 (1967).
(5) I. Langmeir, *J. Am. Chem. Soc., 35,* 931 (1913).
(6) R. Glang, in *Handbook of Thin Film Technology,* ed. L.I. Maissel and R. Glang, McGraw-Hill 1970, pg. 1-7.
(7) L.I. Maissel and R. Glang (eds.), *Handbook of Thin Film Technology,* McGraw-Hill, 1970.
(8) L. Holland, *Vacuum Deposition of Thin Films,* Chapman & Hall, 1956.
(9) B.N. Chapman and J.C. Anderson (eds.), *Science and Technology of Surface Coatings,* Academic Press, 1974.
(10) J.A. Allen, *Rev. Pure Appl. Chem. 4,* 1954, 133.
(11) G.A. Bassett and D.W. Pashley, *J. Inst. Metals, 87,* 1958, 449.
(12) R.W. Hoffman, *Thin Films,* p. 99, American Society for Metals, 1964.
(13) R.W. Hoffman, *Physics of Thin Films, 3,* 1966, 246, Academic Press, New York.
(14) W. Buckel, *J. Vac. Sci. Technol., 6,* 1969, 606.
(15) K. Kinosita, *Thin Solid Films, 12,* 1972, 17.
(16) R.F. Bunshah, *Physical Metallurgy of Beryllium,* April 1963, Conf. No. 170, Oak Ridge National Laboratory.
(17) R.F. Bunshah, in *Materials Science and Technology for Advanced Applications,* Vol. II, American Society for Metals, 1964, pg. 31.
(18) R.F. Bunshah, *Metals Engineering Quarterly,* Nov. 1964, pg. 8.
(19) R.F. Bunshah and R.S. Juntz, in *Beryllium Technology,* Vol. 1, Gordon and Breach Science Publishers, 1966, pg. 1.
(20) R.F. Bunshah, *Proceedings, Int. Conf. on Beryllium,* Grenoble, France, 1965, Press Universitaires de France, pg. 63.
(21) R.F. Bunshah and R.S. Juntz, *Trans. Vac. Met. Conf.,* 1966, American Vacuum Society, pg. 209.
(22) R.F. Bunshah and R.S. Juntz, *Trans. Vac. Met. Conf.,* 1965, American Vacuum Society, pg. 200.
(23) H.R. Smith, in *Materials Science and Technology for Advanced Applications,* Vol. II, American Society for Metals, 1964 pg. 569.
(24) H.F. Smith, Jr. and C.D'A. Hunt, *Trans. Vac. Met. Conf.,* 1964, American Vacuum Society, pg. 227.
(25) R.F. Meyers and R.P. Morgan, *Trans. Vac. Met. Conf.,* 1966, American Vacuum Society, pg. 271.
(26) J.F. Butler, *J. Vac. Sci. Tech. 7,* S-52 (1970).
(27) J. Schiller and U. Heisig, "Evaporation Techniques" (in German), *Veb Verlag Technik,* Berlin, 1975.
(28) E.P. Graper, *J. Vac. Sci. Tech. 8,* p. 333 (1971) and *J. Vac. Sci. Tech. 10,* p. 100 (1973).

(29) K.D. Kennedy, G.R. Schevermann, H.R. Smith, Jr., *Research Development Mag., 22*, p. 40 (1971).
(30) H.A. Beale, R.F. Bunshah, *Proc. 4th Int. Conf. on Vac. Met.*, Tokyo, Japan, (June 1973), published by the Iron and Steel Institute of Japan, p. 238.
(31a) C.T. Wan, D.L. Chambers, D.C. Carmichael, *ibid*, p. 231.
(31b) G.A. Baum, *Report No. RFP-686*, February 6, 1967, Dow Chemical Co., Golden, CO.
(32) J.A. Thornton, *SAE Transactions*, 1973.
(33) "Sputtering and Ion Plating", *NASA SP-511*, 1972.
(34) H. Hertz, *Ann. Physik 17*, 177 (1882).
(35) M. Knudsen, *Ann. Physik, 47*, 697 (1915).
(36) H.R. Smith, *Proc. 12th Annual Technical Conference*, Society of Vacuum Coaters, Detroit MI, 1969, pg. 50-54.
(37) T.C. Riley, "The Structure and Mechanical Properties of Physical Vapor Deposited Chromium", PhD. Thesis, Stanford University, November 1974.
(38) R.F. Bunshah and R.S. Juntz, *Trans. Vac. Met. Conf.* 1967, American Vacuum Soceity, pg. 799.
(39) R. Chow and R.F. Bunshah, *J. Vac. Sci. Tech. 8*, VM 73 (1971).
(40) R. Nimmagadda dn R.F. Bunshah, *J. Vac. Sci. Tech. 8*, VM 85 (1971).
(41) J. Szekely and J.J. Pvermo, *Met. Trans. 5*, 289 (1974).
(42) H.R. Smith and C. D'A. Hunt, *Trans. Vac. Met. Conf.*, 1965, published by American Vacuum Society, pg. 227.
(43) C. Hayashi and Y. Oku, *Trans. Vac. Met. Conf.*, 1966, published by American Vacuum Society, pg. 257.
(44) M.A. Cocca and L.H. Stauffer, *Trans. Vac. Met. Conf.*, 1963, published by Am. Vac. Society, pg. 203.
(45) J.R. Morley, *ibid*, pg. 186.
(46) R. Nimmagadda, A.C. Raghuram and R.F. Bunshah, *J. Vac. Sci. Tech. 9*, (1972).
(46a) T. Santala and M. Adams, *J. Vac. Sci. Tech. 7*, s22 (1970).
(47) B.A. Movchan and A.V. Demchishin, *Fizika Metall, 28*, 653 (1969).
(48) U.S. Patent 2,920,002 (1960) to M. Auwarter.
(49) R.F. Bunshah and A.C. Raghuram, *J. Vac. Sci, Techol. 9*, 1385 (1972).
(50) D. Hoffman and D. Liebowitz, *J. Vac. Sci. Technol. 9*, 326 (1972).
(51) A.C. Raghuram, R. Nimmagadda, R.F. Bunshah and C.N.J. Wagner, *Thin Solid Films, 20*, 187 (1974).
(52) W. Grossklaus and R.F. Bunshah, *J. Vac. Sci. Technol. 12*, 593 (1975).
(53) J. Granier and J. Besson, *Proc. 9th Plansee Seminar*, Reutte, Austria, May 23-26, 1977.
(54) D.W. Pashley, *Adv. Phsy. 5*, 1973 (1956).
(55) A. Van der Drift, *Phillips Res. Rep., 22*, 267 (1967).
(56) J.A. Thornton, *Ann Rev. Mater. Sci.*, p. 239 (1977).
(57) J.A. Thornton, *J. Vac. Sci. Technol. 11*, 666 (1974).
(58) R.F. Bunshah, *Proc. 4th Int. Conf. on Vacuum Metallurgy*, p. 17, Iron and Steel Inst. Japan (1973).
(59) R.F. Bunshah, *J. Vac. Sci. Technol. 11*, 633 (1974).
(60) R.F. Bunshah, *J. Vac. Sci. Technol. 11*, 814 (1974).
(61) R.F. Bunshah, *New Trends in Materials Processing*, p. 200, American Society for Metals (1976).
(62) R.F. Bunshah and R.S. Juntz, *Met. Trans. 4*, 21 (1973).
(63) N. Kane and R.F. Bunshah, in *Proceedings of the Fourth International Conference on Vacuum Metallurgy*, Tokyo, Japan, June 1973, Iron and Steel Inst. of Japan, pg. 242.
(64) M. Sherman, R.F. Bunshah and H.A. Beale, *J. Vac. Sci. Technol. 11*, 1128 (1974).
(65) K. Kennedy, in *Transactions of the Vacuum Metallurgy Conference*, 1968, American Vacuum Society, p. 195.
(66) G. Mah and C.W. Nordin, in *Proceedings of the Sixteenth Annual Conference*, Society of Vacuum Coaters, Chicago, 1973, p. 103.
(67) B.A. Paton, B.A. Movchan and A.V. Demchishin, *Proc. 4th Int. Conf. on Vacuum Metallurgy*, p. 251, Iron and Steel Inst. of Japan (1973).
(68) D.L. Chambers and W.K. Bower, *J. Vac. Sci. Technol. 7*, S63 (1970).

(69) N. Agarwal, N. Kane and R.F. Bunshah, *National Vacuum Symposium*, New York, 1973.
(70) M. Neirynck, W. Samaey and L. Van Poucke, *J. Vac. Sci. Technol. 11,* 647 (1974).
(71) B.A. Movchan, A.V. Demchishin and L. V. Kooluck, *J. Vac. Sci. Technol. 11,* 640 (1974).
(72) A.C. Raghuram and R.F. Bunshah, *J. Vac. Sci. Technol. 9,* 1389 (1972).
(73) R.F. Bunshah and R.S. Juntz, *J. Vac. Sci. Technol. 9,* 404 (1972).
(74) D.M. Mattox and G.J. Kominiak, *J. Vac. Sci. Technol. 9,* 528 (1972).
(75) J. Morley and H.R. Smith, *J. Vac. Sci. Technol. 9,* 1377 (1972).
(76) G. Mah, P.S. McLeod and D.G. Williams, *J. Vac. Sci. Technol. 11,* 663 (1974).
(77) W.R. Stowell and D. Chambers, *J. Vac. Sci. Technol. 11,* 653 (1974).
(78) R.D. Bland, G.J. Kominiak and D.M. Mattox, *J. Vac. Sci. Technol. 11,* 671 (1974).
(79) R.F. Bunshah unpublished research.
(80) C.F. Turk and H.L. Marcus, *Trans. AIME, 242,* 2251 (1968).
(81) H.A. Beale and R.F. Bunshah, unpublished research.
(82) J.A. Thornton, in *New Industries and Applications for Advanced Materials Technology,* Vol 19, SAMPE, 1974, p. 443.
(83) D.H. Boone, T.E. Strangman and L.W. Wilson, *J. Vac. Sci. Technol. 11,* 641 (1974).
(84) R.C. Krutenat, *J. Vac. Sci. Technol. 11,* 1123 (1974).
(85) W. Grossklaus, N.E. Ulion, and H.A. Beale, *Thin Solid Films, 40,* 271 (1977).
(86) J.W. Beams, J.B. Breazeale and W.L. Bart, *Phys. Rev. 100,* 1657 (1955).
(87) J.W. Beams, *Structure and Properties of Thin Films,* pg. 183, Wiley, NY (1959).
(88) A. Catlin and W.P. Walker, *J. Appl. Phys. 31,* 2135 (1960).
(89) S. Jovanovic and C.S. Smith, *J. Appl. Phys. 32,* 121 (1961).
(90) P.I. Krukover and V.A. Buravikhin, *Fizika Metall. 22,* 144 (1966).
(91) D.G. Brandon and Z. Bauer, *Israel J. Technol. 8,* 247 (1970).
(92) D.M. Marsh, *J. Sci. Instrum. 38,* 229 (1961).
(93) C.A. Neugebauer, *J. Appl. Phys. 31,* 1096 (1960).
(94) C. D'Antonio, J. Hirschorn and L. Tarshis, *Trans. AIME, 227,* 1346 (1964).
(95) J.M. Blakely, *J. Appl. Phys. 36,* 1756 (1964).
(96) D. Kuhlmann-Wilsdorf and K.S. Raghaven, *Rev. Sci. Instrum. 33,* 930 (1962).
(97) A. Lawley and S. Schuster, *Rev. Sci. Instrum. 33,* 1178 (1962).
(98) E. Orowan, *Z. Phys. 82,* 235 (1933).
(99) J. W. Menter and D.W. Pashley, in *Structures and Properties of Thin Films,* p.111, Wiley, NY (1959).
(100) H.G.F. Wilsdorf, *Rev. Sci. Instrum., 29,* 323 (1958).
(101) D.W. Pashley, *Proc. R. Soc. Lond., A225,* 218 (1960).
(102) R.W. Hoffman, *Thin Films,* p. 99, American Society for Metals (1964).
(103) R.W. Hoffman, *Physics of Thin Films, 3,* 246 (1966) Academic Press, NY.
(104) C.A. Neugebauer, *Physics of Thin Films,* Vol 2 (Edited by G. Hass and E. Thun) Academic Press, NY (1964).
(105) L.S. Palatnik, A.I. Ill'inski, G.V. Federov and V.S. D'yachenko, Izvestra VUZ, *Fizika (Soviet Physica Journal), 1,* 122 (1966).
(106) R.L. Grunes, C. D'Antonio and F.K. Kies, *J. Appl. Phys. 36,* 2735 (1965).
(107) M. Ya Fuks, V.V. Belozerov and Yu F. Boyko, *Fizika Metall. 33,* 571 (1972).
(108) K. Kinosita, K. Maki, K. Nakamizo and K. Takenchi, *Japan J. Appl. Phys. 6,* 42 (1967).
(109) M. Ya Fuks, V.V. Velozero and Yu F. Boyko, *Fizika Metall. 33,* 571 (1972).
(110) F.A. Doljack and R.W. Hoffman, *Thin Solid Films, 12,* 71 (1972).
(111) M. Ya Fuks, L.S. Palatnik, V.V. Belozerov, Yu V. Zolotnitsky and S.T. Roschchenko, *Fizika Metall. 36,* 316 (1973).
(112) B.Y. Pines and N.S. Tan, *Fizika Metall. 19,* 899 (1965).
(113) I.T. Aleksanyan, *Fizika Metall. 25,* 947 (1968).
(114) R.W. Hoffman, *Thin Solid Films, 34,* 185 (1976).
(115) L.S. Palatnik, G.V. Federov, A.I. Prokhavulov and A.I. Federenko, *Fizika Metall, 20,* 574 (1965).
(116) L.S. Palatnik, M. Ya Fuks, B.T. Boiko and A.T. Pugacheu, *Soviet Physics Dokl English Translation 8,* 713 (1964).
(117) A. Oding and I.T. Aleksanyass, *Soviet Phys. Dokl. 8,* 818 (1964).
(118) R.F. Bunshah, "Mechanical Properties of PVD Films", *Vacuum 27,* #4, 353 (1977).

(119) H. Yamamoto and G. Kamoshita, *Trans. Japan Inst. Metals, 12,* 12 (1971).
(120) L.S. Palatnik, M. Ya Fuks, A.I. Ill'inski and O.G. Alaverdova, *Fizika Metall. 22,* 744 (1966).
(121) C.A.O. Henning, F.W. Boswell and J.M. Corbett, *Acta Met. 23,* 177 (1975).
(122) K.K. Ziling, L.D. Pkrovskiy and V. Yu Pohelkin, *Fizika Metall. 29,* 1112 (1970).
(123) L.S. Palatnik, A.I. Ill'inski and A.G. Ravlik, *Fizika Metall. 19,* 310 (1965).
(124) C.A.O. Henning, F.W. Boswell and J.M. Corbett, *Acta Met. 23,* 187 (1975).
(125) L.S. Palatnik, A.I. Ill'inski, A.G. Ravhk, A.A. Nechitayls and G. Ye Lyakh, *Fizika Metall., 27,* 1114 (1969).
(126) R.F. Bunshah and R.J. Schramm, *Thin Solid Films, 41,* 1977 in press.
(127a) M. Colen and R.F. Bunshah, *J. Vac. Sci. Tech. 13,* 536 (1976).
(127b) S. Furuuchi, H. Sakata and K. Aiwaka, *Japan J. Appl. Phys. 13,* 1905 (1974).
(128) W. Grossklaus and R.F. Bunshah, *J. Vac. Sci. Technol. 12,* 811 (1975).
(129) K. Wasa, T. Nagai and S. Hayakowa, *Thin Solid Films, 31,* 235 (1976).
(130) G. Mah, C.W. Norden and J.F. Fuller, *J. Vac. Sci. Technol. 11,* 371 (1974).
(131) C.A.O. Henning, F.W. Boswell and J.M. Corbett, *Acta Met. 23,* 193 (1975).
(132) L.S. Palatnik, A.I. Ill'inski and N.P. Sapelkin, *Soviet Phys. Solid St. 8,* 2016 (1967).
(133) I.I. Solonovich and V.I. Startsev, *Problemy Prochn. 1,* 28 (1973).
(134) L.S. Palatnik, A.I. Ill'inski, N.M. Biletchanko and R.I. Sinel'nikova, *Fiz. Metall. 32,* 199 (1971).
(135) R.F. Bunshah, Y.D. Gupta and A.C. Raghuram, unpublished data.
(136) E.O. Hall, *Proc. Phys. Soc. Lond. B64,* 747 (1951).
(137) N.J. Petch, *J. Iron Steel Inst. 174,* 25 (1951).
(138) M. Nenioto, R. Jumbou, and H. Suto, *Trans. Japan Inst. Metals, 12,* 113 (1971).
(139) D.L. Chambers and W.K. Bower, *J. Vac. Sci. Technol. 7,* S62 (1970).
(140) B.A. Movchan, A.V. Demchishin and G.F. Badilienko, *Thin Solid Films, 40,* 237 (1977).
(141) B.A. Movchan, A.V. Demchishin and L.D. Kooluck, *Thin Solid Films, 44,* 285 (1977).
(142) F.C. Frank and B.R. Lawn, *Proc. R. Soc. Lond., 229A,* 291 (1967).
(143) D.H. Boone and C.P. Sullivan, in "Fatigue at Elevated Temperatures", *STP 520,* American Society for Testing and Materials, pg. 401 (1973).
(144) G.W. Goward, *J. Metals, 22,* 31 (1970)
(145) P.J. Clough, in *New Types of Metal Powders,* (H.H. Hausner, Ed.), Gordon and Breach, p. 9, (1964).
(146) C. Hayashi, *Jap. J. Appl. Phys. 12,* 1675 (1973).
(147) R.F. Bunshah, unpublished research.
(148) K.S. Majumder, *Thin Solid Films, 42,* 327 (1977).
(149) K.S. Majumder, *Thin Solid Films, 42,* 343 (1977).
(150) K.C. Chi, R.O. Dillon, R.F. Bunshah, S. Alterovitz, D.C. Martin and J.A. Wollam, *Thin Solid Films* (1978).
(151) R.F. Zubeck, C.N. King, D.F. Moore, T.W. Barbee, A.B. Hallak, J. Salem and R.H. Hammond, *Thin Solid Films, 40,* 249 (1977).
(152) P.L. Martin, R.F. Bunshah and A.M. Dymond, *J. Vac. Sci. Tech. 12,* 754 (1975).
(153) P.L. Agarwal, R.F. Bunshah and P.H. Crandall, unpublished research, UCLA, 1978.
(154) A.K. Sinha, B.C. Giessen and D.E. Polk, in *Treatise on Solid State Chemistry,* Vol. 3, N.V. Hannay ed., Plenum Press, NY, pg. 1, 1976.
(155) P.K. Keung and J.G. Wright, *Phil. Mag. 30,* 995 (1974).
(156) J.L. Hughes, Metals *Eng. Quart.* 14, No. 1, 1 (1974).
(157) R.F. Bunshah and R.T. Webster, *J. Vac. Sci. Technol. 8,* VM95 (1971).
(158) R.J. Hill, J.L. Hughes and H.R. Harker, in *Proceedings of the Fourth International Conference on Vacuum Metallurgy,* Tokyo Japan, June 1973, Iron and Steel Institute of Japan, p. 248.
(159) H.R. Harker and R.J. Hill, *J. Vac. Sci. Technol. 9,* 1395 (1972).
(160) R.F. Bunshah, U.S. Patent No. 3,971,582, Feb. 12, 1974.
(161) K. Nakamura, K. Inagawa, K. Tsuruoka and S. Komiya, *Thin Solid Films, 40,* 155 (1977).
(162) M. Kodama, R.F. Bunshah and A.H. Shabaik, *Thin Solids Films* (1978).
(163) R.F. Bunshah and A.H. Shabaik, *Res/Dev. 26,* 46 (1975).

(164) R.F. Bunshah, A.H. Shabaik, R. Nimmagadda and J. Covey, *Thin Solid Films, 45,* 1 (1977).
(165) G.H. Hewig and W.H. Bloss, *Thin Solid Films, 45,* 1 (1977).
(166) K.W. Boer, *Annual Progress Report, NSF/RANN/SE/G134872,* University of Delaware, January 1974.
(167) T.P. Brody and F.A. Shirland, in *Proceedings of NSF Workshop on Photovoltaic Conversion of Solar Energy for Terrestrial Applications,* Cherry Hill, NJ, October 1973, p. 168.
(168) N. Laegreid, in *ibid,* p. 63.
(169) B.E. Jacobson, J.R. Springarn and W.D. Nux, *Thin Solid Films, 45,* 517 (1977).
(170) N.F. Mott, *Phil. Mag., 44,* 742 (1953).
(171) B.A. Movchan, *Soviet Physics Doklady, 20,* #7, 575 (1975).
(172) B.A. Movchan, G.F. Badilenko and A.V. Demchishin, *Thin Solid Films* (1979).
(173) S. Schiller and G. Jäsch, *Thin Solid Films 54,* 9 (1978).
(174) Yu F. Shevakin, L.D. Kharitonova and L.M. Ostrovskaya, *Thin Solid Films 62,* 337 (1979).
(175) M. Kobayashi and Y. Doi, *Thin Solid Films 54,* 57 (1978).
(176) S. Komiya, N. Umezu and T. Narusawa, *Thin Solid Films 54,* 51 (1978).
(177) B. Zega, M. Kornmann and J. Amiguet, *Thin Solid Films 45,* 577 (1977).
(178) B. Berghaus, German Patent No. 683,414 (1939).
(179) L.P. Sabalev et al., U.S. Patent 3,783,231, January 1, 1974 and 3,793,179, February 19, 1974.
(180) A.M. Dorodnov, *Soviet Phys. Tech. Phys. 23,* 1058 (1978).
(181) V.A. Osipov et al., *Soviet Rev. Sci. Inst. 21,* 1651 (1978).
(182) P. Nath and R.F. Bunshah, *Thin Solid Films 69,* 63 (1980).
(183) Y. Namba and J. Mori, *J. Vac. Sci. Technol. 13,* 693 (1976).
(184) N.J. Taylor, *Surface Science 4,* 161 (1966).
(185) R.F. Bunshah, R. Nimmagadda, H.J. Doerr, B.A. Movchan, N.I. Grechanuk and E.V. Dabizha, "Structure Property Relationships in Microlaminate Ni-Cu and Fe-Cu Condensates", *Thin Solid Films 72,* p. 261 (1980).
(186) M.J. Hordon, "Development of Boron Carbide-Titanium Composites" in *Titanium Science and Technology,* Vol. 4, ed. R.I. Jaffee and H.M. Burte, Plenum Press A73, pg. 2347-57.
(187) J.S. Koehler, *Phys. Rev. B 2* (1970) 547.
(188) C.A.O. Henning, F.W. Boswell and J.M. Corbett, *Acta. Met. 23,* 193.
(189) L.S. Palatnik and A.I. Ill'inski, *Soviet Physics — Doklady 9,* #1 (1964) 93.
(190) L.S. Palatnik, A.I. Ill'inski and N.P. Sapelin, *Soviet Physics — Solid State 8,* #8 (1967) 2016.
(191) L.S. Palatnik, A.I. Ill'inski, N.M. Biletchenko and R.I. Sinel'nikova, *Fiz. Met. Metalloved. 32,* #6 (1971) 1312.
(192) I.I. Solonovich and V.I. Startsev. *Problemy Prochnosti,* No. 1 (1973) 28-30.
(193) M.J. Hordon and M.A. Wright in Metal-Matrix Composites, Symposium of the Metallurgical Society AIME, *DMIC Memorandum 243,* May (1969) 10-12.
(194) R.W. Springer and D.S. Catlett, *Thin Solid Films 54,* (1978) 197.
(195) S.L. Lehoczky, "Retardation of Dislocation Generation and Motion in Thin Layered Micro-Laminates", *J. App. Phys. 49,* 5479 (1978).
(196) A.P. Brodiansky, E.A. Anelchishina and I.A. Bunda, *Technology and Manufacturing Processes* (in Russian) 1977, #2, pg. 55.
(197) A.K. Suri, R. Nimanajadda and R.F. Bunshah, *Thin Solid Films 64,* 191 (1979).
(198) T. Jamal, R. Nimanajadda and R.F. Bunshah, "Function and Adhesive Wear of Titanium Carbide and Titanium Nitride Overlay Coatings", *Thin Solid Films.*
(199) A.M. Barnett, *IEEE Trans. on Electron Devices, ED-27,* 615 (1980).
(200) H. Yoshihara and H. Mori, *J. Vac. Sci. Technol. 16,* 1007 (1979).
(201) Y. Murayama, J. Vac. Sci. Tech. 12, 818, (1975).
(202) B.A. Movchan, E.V. Dabizha, R.F. Bunshah and R.R. Nimmagadda, *Thin Solid Films, 83,* p. 21 (1981).
(203) R.R. Nimmagadda, H.J. Doerr and R.F. Bunshah, *Thin Solid Films, 84* p. 303 (1981).
(204) P. Agarwal, P. Nath, H.J. Doerr, R.H. Bunshah, G. Kuhlman and A.J. Koury, *Thin Solid Films, 83,* p. 37 (1981).

(205) R.F. Bunshah, R. Nimmagadda, W. Dunford, B.A. Movchan, A.V. Demchishin and N.A. Chursanov, *Thin Solid Films, 54,* 85 (1978).
(206) R.W. Hoffman in *Physics of Non-Metallic Thin Films NATO Advanced Studies Institute Series, B14,* p. 273, Plenum Press (1976).

SUGGESTIONS FOR FURTHER READING

Books

(1) Vacuum Deposition of Thin Films by L. Holland, Chapman and Hall, 1968, *The Bible.*
(2) *Handbook of Thin Film Technology,* Edited by Leon I. Maissel and Reinhard Glang, McGraw Hill Book Co., 1970.
(3) *Thin Film Phenomena,* Kasturi L. Chopra, McGraw Hill Book Co., 1969.
(4) *Thin Film Technology,* Robert W. Berry, Peter M. Hall, Murray T. Harris, D. Van Nostrand Co., 1968.
(5) *The Use of Thin Films in Physical Investigation,* Edited by J.C. Andrews, Academic Press, 1966.
(6) *Physics of Thin Films,* published by Academic Press, Vol. 1-6 (1963-1971).
(7) *Thin Films,* American Society for Metals, 1964.
(8) *Techniques of Metals Research,* Vol. 1, part 3, edited by R.F. Bunshah, John Wiley & Sons, 1968.
(9) *Science and Technology of Surface Coatings,* ed. B.N. Chapman and J.C. Anderson, Academic Press, 1974.

Journals

(1) *Journal of Vacuum Science and Technology,* published by the American Physical Society, USA.
(2) *Thin Solid Films,* Elsevier S.A., Switzerland.
(3) *Journal of Materials Science,* England.
(4) *Vacuum,* England.
(5) *Journal of Electrochemical Society,* USA.
(6) *Journal of Applied Physics,* USA.
(7) *Japanese Journal of Applied Physics,* Japan.
(8) *Review of Scientific Instruments,* USA.
(9) *Surface Science.*

Papers

(1) Vacuum Evaporation by Reinhard & Glang in Book No. 1 above. A must for evaporation of thin films.
(2) Physical Vapor Deposition of Metals, Alloys & Compounds by R.F. Bunshah in *New Trends in Material Processing,* American Society for Metals, 1976. A comprehensive review of thick films, bulk coatings from the viewpoint of vacuum metallurgy.
(3) Instrumentation for monitoring and control of vacuum deposition, W. Von Steikelmacher in *Vakuum-Technik 20,* #5, 1971.

APPENDIX

"On progress in scientific investigations in the field of vacuum evaporation in the Soviet Union"

by A.V. Demchishin

E.O. Paton Electric Welding Institute
Kiev, Ukraine, U.S.S.R.

The first investigations dealing with the problems of evaporation and condensation were carried out by Soviet scientists as early as the twenties. Ya. I. Frenkel[1] found theoretically that there exists a critial temperature of reflection of metal atoms from a substrate.[1] Yu. B. Kharitonov and N.N. Semenov have shown experimentally that this phenomenon actually took place.[2] The problem of formation of chemical compound with a simultaneous condensation of molecular beams of cadmium and sulphur was studied by A.I. Shal'nikov and N.N. Semenov.[3] Structural studies of condensates of gold-copper alloys by electron and x-ray diffraction were carried out by M.M. Umanskii and V.A. Krylov.[4]

At the beginning of the forties, S.A. Vekshinskii and his colleagues performed a lot of work on a development of methods for production of specimens of condensates, on experimental verification of condensate distribution law, on studying physical and chemical properties of condensed metal films of pure metals and binary alloys.[5] S.A. Vekshinskii suggested the use of a method of co-condensation of vapor mixtures of several components for producing the films of variable composition thus enabling the structure and properties of an entire n-component system or its part to be studied at once without recourse to production of a great number of separate samples of constant composition alloys.

In the middle of the fifties, investigation of condensates was conducted by L.S. Palatnik and his collaborators at the Kharkov Polytechnic Institute towards the following trends:

- structure and substructure of thin and massive condensed films;
- mechanism of formation and kinetics of growth of continuous and island films;
- physical properties of films (mechanical, electrical, semiconductive, magnetic, thermal and other properties);
- studying the correlation between structure (substructure) and physical properties of films;
- the effect of physical and technological variables of evaporation processes and vacuum condensation on structure (substructure) and physical properties of continuous (thin and massive) and island films.

The main results of these investigations are published in the following papers, references 6, 7, and 8.

In addition to the said studies, in the sixties and seventies, the characteristics of macro-, micro- and submicroporosity of condensed films depending on substrate temperature, angle of incidence of molecular flow, condensation rate, film thickness, pressure and composition of residual gas atmosphere were investigated. Mechanisms of porosity formation processes were established and relationships between the porosity characteristics and physical-mechanical properties of films have been studied.[9,10]

In the middle of the sixties, B.A. Movchan and his collaborators developed an electron-beam technology for production of preparations of condensed systems and commenced the study of thick (up to 1 mm) condensates. In the sixties-seventies the effect of condensation conditions on structure and physical-mechanical properties of thick condensates of pure metals, refractory oxides, carbides, borides and their mixtures, ceramic-metallic materials and dispersion strengthened compositions were investigated. Their main results were published in references 11 to 15.

References

(1) J.I. Frenkel, *Zeitschr. f. Physik, 26,* 117, (1924).
(2) J.B. Chariton und N.N. Semenoff, *Zeitschr. f. Physik, 25,* 287 (1924).
(3) A.I. Shal'nikov, N.N. Semenov, *The Journal of Russian Physical and Chemical Society, 60,* 33, (1928).
(4) M.M. Umanskii, V.A. Krylov, *The Journal of Exp. and Theor. Physics, 6,* 691, (1936).
(5) S.A. Vekshinskii, *A New Method of Metallographic Studies of Alloys,* Bostechizdat, Moscow-Leningrad, (1944).
(6) L.S. Palatnik, I.I. Papirov, *Epitaxial Films,* Nauka, Moscow, (1971).
(7) L.S. Palatnik, M. Ya. Fux, V.M. Kosevich, *Mechanism of Formation and Substructure of Condensed Films,* Nauka, Moscow, (1972).
(8) L.S. Palatnik, V.K. Sorokin, *Fundamentals of Film Semiconductive Materials Technology,* Energia, Moscow, (1973).
(9) L.S. Palatnik, M. Ya. Fux, P.G. Cheremskoi, *Transactions of the Academy of Sciences of the U.S.S.R., 203,* 5, 1058, (1972).
(10) M. Ya. Fux, L.S. Palatnik, P.G. Cheremskoi, A.L. Toptygin, *Physics of Metals and Physical Metallurgy, 46,* 1, 114 (1978).
(11) B.A. Movchan, A.V. Demchishin, *Physics of Metals and Physical Metallurgy, 28,* No. 4, 653, (1969).
(12) B.E. Paton, B.A. Movchan, A.V. Demchishin, *Proceedings of the Fourth International Conference on Vacuum Metallurgy,* p. 251, Tokyo, June 4-8, (1973). Published by the Iron and Steel Institute of Japan, Tokyo (1974).
(13) B.A. Movchan, A.V. Demchishin, L.D. Kooluck, *Thin Solid Films, 44,* 285, (1977).
(14) B.A. Movchan, A.V. Demchishin, G.F. Badilenko, *Strength Problems,* No. 2, 61, (1978).
(15) B.A. Movchan, I.S. Malashenko, P.A. Pap, *Problems of Special Electro Metallurgy,* Naukova Dumka, Kiev, No. 8, 78, (1978).

5

Coating Deposition by Sputtering

John A. Thornton
Telic Corporation
Santa Monica, California

1. INTRODUCTION

Sputtering is a process whereby material is dislodged and ejected from the surface of a solid or a liquid due to the momentum exchange associated with surface bombardment by energetic particles. Sputter deposition is a vacuum coating process. A source of coating material called the target is placed into a vacuum chamber along with the substrates, and the chamber is evacuated to a pressure typically in the range 5×10^{-4} to 5×10^{-7} Torr.* The bombarding species are generally ions of a heavy inert gas. Argon is most commonly used. The sputtered material is ejected primarily in atomic form. The substrates are positioned in front of the target so that they intercept the flux of sputtered atoms.

The most common method of providing the ion bombardment is to backfill the evacuted chamber with the inert gas to a pressure of from 1 to 100 mTorr and ignite an electric discharge so that ionization of the working gas is produced in the region adjacent to the target. See Figure 5.1. Such a low pressure electric discharge is called a glow discharge, and the ionized gas is called plasma. The target is negatively biased so that its surface is bombarded by positive ions from the plasma. The most direct method for providing the plasma and the target ion bombardment is simply to make the target the cathode, or negative electrode, of the electric discharge. Applied potentials between the target and the anode are typically from 500 to 5,000 V. A sputtering apparatus with this design is called a diode. The glow discharge in such a device is of a form called an abnormal negative glow. Apparatuses in which the plasma is produced by independent means often have two electrodes in addition to the target, and are called triodes. The bombarding species may also be provided in the form of an ion beam by an ion source containing accelerating grids.

---

*The pressure unit of Torr (1 Torr = 1 mm Hg) is a carryover from the time when most pressures were measured with manometers. Starting about 1975, most technical publications began changing to the International System of Units (SI), where the unit of pressure is the pascal (Pa). The pascal is the MKS unit of pressure: 1 Pa = 1 N/m^2 = 7.5 mTorr (1 mTorr = 0.133 Pa). Most pressure gauges are still calibrated in Torr or microns (1 micron = 1 mTorr). Therefore Torr will be used in this text although both Torr and Pa are given on many of the figures.

Coating Deposition by Sputtering   171

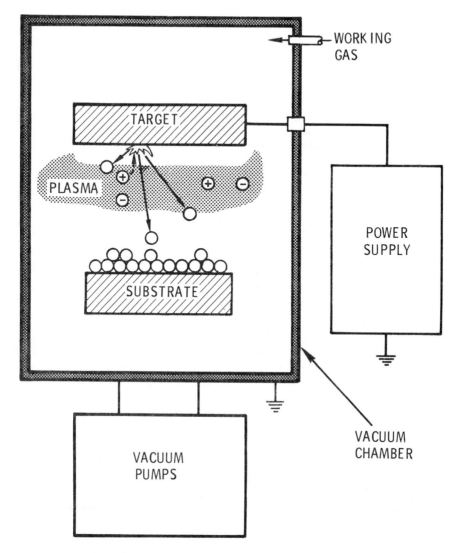

Figure 5.1: Schematic representation of sputter coating process. (Used courtesy Telic Corporation.)

The most striking characteristic of the sputtering process is its universality. Since the coating material is passed into the vapor phase by a mechanical (momentum exchange) rather than a chemical or thermal process, virtually any material is a candidate coating. Dc methods are generally used for sputtering metals. An rf potential must be applied to the target when sputtering nonconducting materials.

Sputter coating technology includes many variations of the basic process described above. For example, coatings may be formed by—

(1) Employing a target which is a mosaic of several materials.

(2) Employing several targets simultaneously, of identical or different materials.

(3) Employing several targets sequentially to create a layered coating.

(4) Biasing the substrate as an electrode prior to coating, so that contamination is removed by sputtering, and coating nucleation sites are generated on the surface. This is known as sputter cleaning.

(5) Biasing the substrate as an electrode to cause surface ion bombardment during deposition, in order to remove loosely bonded contamination or to modify the structure of the resulting coating. This is known as bias sputtering.

(6) Employing a gas to introduce one of the coating materials into the chamber. This process is known as reactive sputtering. It permits metal targets and dc power supplies to be used in preparing coatings of nonconducting compounds.

Particle energies in the sputtering process are generally referred to in units of the electron volt. One electron volt (eV) is the unit of energy that a particle with one unit of electronic charge accumulates while passing through a potential difference of 1 V. The bombarding ion energies are typically in the range from 100 to 1,000 eV. The atoms on the surface of a solid are typically bonded with energies (sublimation energies) of 2 to 10 eV. The average energies of sputtered atoms are in the range 10-40 eV. This is in comparison to evaporation, where the evaporated atoms typically have energies of 0.2 to 0.3 eV. By further comparison related to sputtering, electrons with energies in the 100 to 1,000 eV range are present in the negative glow and sustain the discharge by producing ionization in the working gas. The ionization energy for an argon atom is 15.76 eV. Ultraviolet radiation from an argon plasma has an energy of about 12 eV. Visible radiation has an energy of about 2 eV. Chemisorbed atoms have typical binding energies of 1 to 10 eV. Physisorbed atoms have binding energies of less than 0.5 eV. Thus the sputtering process is a relatively high energy process compared to many coating technologies.

Sputtering Apparatuses

Sputtering apparatuses can assume an almost unlimited variety of configurations, depending on the application. The simplest is the planar diode which consists of two planar electrodes, typically 10 to 30 cm in diameter, which are placed facing one another at a spacing of 5 to 10 cm, essentially as shown in Figure 5.1. One electrode contains the target. The substrates are placed on the other one. The substrate mounting electrode is made the anode when operation is with dc power. It is made larger than the target electrode to control unwanted sputtering when operation is with rf power. Planar diodes have played a major role in the development of sputtering technology over the past ten years and are probably still the most widely used form of apparatus. Figure 5.2 shows a planar diode sputtering installation of a type commonly used in research and for small production levels.

Coating Deposition by Sputtering   173

Figure 5.2: Planar diode sputtering system of type used for research and small production runs. System can be used for both dc and rf sputtering. (Photo courtesy of CVC Products, Inc., Rochester, NY.)

The substrates in a planar diode are in contact with the plasma. This makes it relatively easy to execute the processes of substrate sputter cleaning and bias sputtering. Sputtering technology has long enjoyed a reputation for providing coatings with superior adhesion. This distinction is probably due largely to sputter cleaning effects of this substrate plasma bombardment. However, the heating associated with plasma and electron bombardment prevents the use of planar diodes for coating thermally sensitive substrates.

It is difficult to sustain an intense plasma discharge in the planar diode electrode geometry. Thus working pressures are necessarily relatively high at 20 to 75 mTorr (3 to 10 Pa), and current densities are low ($\sim$1 mA/cm^2). The high pressure causes the target-to-substrate coating material transport to be largely by diffusion. This, coupled with the low current density, makes deposition rates generally less than 1,000 Å/min.

Triode devices can be used to produce intense sputtering discharges at low pressures. Therefore higher deposition rates can be achieved than with planar diodes. Thus, for example, high rate triode sputtering has been used to make a free standing 1.3 kg deposit of a Cu-alloy in the form of a cylinder 15 cm in diameter.[1] However, the complexity of triodes has, in general, limited their use to special applications.

The recent development of a class of sputtering sources with magnetic plasma confinement, called magnetrons, has greatly enhanced the capabilities of the sputtering process. There are many forms of magnetrons. They can vary from small ring sources—often referred to as Sputter-guns® (Sloan Technology, Santa Barbara, CA) and S-guns® (Varian Associates, Palo Alto, CA)—to long rectangular planar magnetrons, and cylindrical magnetrons with post or hollow cathodes. Magnetrons can be used for both dc and rf sputtering but are particularly effective for dc sputtering, where deposition rates can be more than an order of magnitude larger than those obtained with planar diodes. Planar and cylindrical magnetrons can be scaled to large sizes to provide uniform deposition over very large areas (many m^2). In addition, well-designed magnetrons virtually eliminate substrate heating due to electron and plasma bombardment.

Sputtering Applications

The range of sputtering applications reflects the enormous universality of the process. Films containing almost every element in the periodic table have been prepared by sputtering. Alloys and compounds can generally be sputter-deposited while preserving their compositions. For example, PTFE (Teflon) has been sputtered to produce lubricous films having many of the properties of the starting material.[2,3] The ability to control composition has caused sputtering to become widely used in the electronics industry. Typical applications are aluminum alloy[4] and refractory metal,[4,5] microcircuit metallization layers, oxide microcircuit insulation layers,[6] transparent conducting electrodes,[7,8] amorphous optical films for integrated optics devices,[9] piezoelectric transducers,[10] photoconductors[11] and luminescent films[12] for display devices, optically addressed memory devices,[13,14] amorphous bubble memory devices,[15] thin film resistors[16] and capacitors,[17] video-discs,[18] solid electrolytes,[19] thin film lasers,[20] and microcircuit photolithographic mask blanks.[21] Figure 5.3 shows a multisource sputtering system designed for wafer processing. In addition, one finds applications ranging from coating razor blades[22,23] to depositing wear-resistant coatings[24,25] for machine tools.

Planar diodes are still widely used, particularly for materials requiring rf power. However, recent trends find magnetrons replacing planar diodes for many dc and some rf applications. In addition, the magnetrons have opened up new applications because of their large area capability and the reduced substrate heating. For example, large in-line systems with vacuum interlocks use planar magnetron sources to coat 2 m x 3.5 m architectural glass plates at three-shift production volumes of about 10^6 m^2/yr.[25a] Sputtering is being investigated as a means for depositing selective absorber coatings for solar heating[26,27] and for manufacturing photovoltaic cells for direct solar-to-electrical energy conversion.[27a] Because of the reduced substrate heating,[28] magnetrons are being used on a production basis to deposit chromium decorative coatings on automobile grilles and other exterior trim.[29,29a] Figure 5.4 shows an automated load-lock sputtering system designed for metallizing plastic automotive parts.

The selection of a sputtering apparatus for a given application depends on the substrate size, shape, and sensitivity to heat and plasma bombardment. It also depends on the nature of the coating: i.e., single layer or multilayer, thickness, types of materials involved, and critical parameters such as hardness, porosity, resistivity, semiconductor charge carrier lifetimes, and magnetic anisotropy, as well as the production volume. Planar targets of an almost unlimited range of materials are available from a selection of suppliers. Thus planar diodes are attractive for depositing thin coatings of complex materials onto planar substrates, for research studies, or for small production volumes. The substrates must be capable of withstanding the plasma and particularly the electron bombardment. Triode devices are attractive if thicker coatings are required. For large production volumes, thick coatings, complex substrate shapes, or thermally sensitive substrates, magnetron type sources should be considered. The selection of a particular type of magnetron will depend on the nature of the coating and substrate and the availability of sputtering targets of the required material in the desired geometry. The procurement of targets is an important consideration for all sputtering apparatuses. A project can be doomed to failure before it starts if the targets are of unsuitable quality.

Figure 5.3: Multi-source sputtering system designed for wafer processing. Batches of wafers are passed into and out of coating chamber through interlocks. (Photo courtesy GCA Corporation, Vacuum Industries Division, Somerville, MA.)

Figure 5.4: Automated load-lock sputtering system designed for metallizing plastic automotive parts. System mounts 24 S-gun type magnetron sources (see Section 3). Throughput is 500 ft^2/hr of platen surface on which substrates may be mounted. (Photo courtesy of Varian Associates, Inc., Palo Alto, CA.)

Sputtering, like other vacuum coating processes, suffers from the disadvantage that the equipment is expensive. All substrates must be placed into a vacuum system and evacuated prior to coating. High rate sputtering equipment generally incorporates large nonstandard power supplies and automatic control systems. As a general rule, sputtering is most effective when production volumes are sufficient to permit the equipment cost to be written off over a large number of parts. It has the advantage that it is reliable and lends itself to automatic control. Thus labor costs can be minimized, and no particular technical training is required to operate well-designed coating equipment.

Implementation of Sputtering Process

Almost any vacuum chamber capable of evacuation into the 10^{-6} Torr range can be used for sputtering. Provisions are usually required for throttling the pumping system so that the desired working gas pressure can be sustained with the pumps in operation. Small planar magnetrons or the gun-type magnetrons are particularly easy to install. Some forms of cylindrical magnetrons require special chambers or more extensive chamber modification. Special chambers are generally used for applications involving large production volumes. They may, for example, include loading interlocks so that the target surface is not exposed to the atmosphere between depositions. Examples are shown in Figures 5.3 and 5.4.

Pre-deposition pumping and the criticalness of the vacuum will depend on the application. It is important to remember that sputtered coatings are deposited in an outgassing flux from the substrates and chamber walls, and that this flux can have a significant influence on the growth and properties of the coatings. Special problems may be encountered if the substrates undergo severe outgassing.[30] Generally pumping is continued until the total flux of outgassing from the chamber walls and substrates has decreased to a value that is significantly less than the total sputtering flux that will be used. Working gas is then injected into the chamber with the pumps throttled, and the sputtering is initiated. The sputtering is generally conducted for a period of time, with the substrates shielded in order to clean and condition the target and chamber surfaces.

The selection of deposition conditions is generally determined empirically. The primary control parameters are the deposition rate, target voltage, working gas species and pressure, and the substrate temperature and plasma bombardment. The available selection range for the deposition parameters is determined largely by the apparatus. In planar diodes many of the parameters are interrelated and unavailable for independent control. Much greater control is possible with magnetrons. However, other variables become important. For example, the many magnetron geometries, along with possible operation at low pressures where the sputtered atoms can pass to the substrates while making few collisions, make coating flux angle of incidence considerations important in determining coating properties.[30a] Thus in all applications where large production volumes are anticipated, it is wise to perform development work using an apparatus having the type and geometry that is as anticipated for the production facility. Scale-up jumps should generally not exceed a factor of three in apparatus size.

Chapter Organization

The following sections discuss many of the fundamental aspects of the sputtering process. The objective is to provide the reader with a background for making decisions such as the selection of targets, apparatuses, and deposition conditions and for following the sputtering literature. Section 2 discusses the first step in the sputtering process: i.e., the passage of coating material into the vapor phase and the associated target processes. Section 3 discusses use of glow discharges to provide the bombarding ions, with particular emphasis on planar diode and magnetron sources. Section 4 discusses the final step in the sputtering process; i.e., the formation of coatings on the substrate and their structure-property relationships. Section 5 briefly reviews several topics in processing technology such as reactive sputtering.

Several review papers written over the last twenty years permit the interested reader to follow the developments in sputtering technology. An extensive review of the basic process was written by G.K. Wehner, one of the most prominent of the early workers, in 1955.[31] Film properties were discussed by E. Kay in a 1962 review.[32] Execution of the process for depositing coatings was discussed in a 1966 review by L. Maissel[33] and a review by Kay.[34] The *Thin Film Handbook*, published in 1970, contains reviews by Wehner and Anderson[35] and by Maissel.[36] A review by Thornton in 1973 contains considerable discussion of sputtering equipment.[37] Process considerations in glow discharge sputtering were reviewed by Westwood in 1976.[38] A recent book edited by J.L. Vossen and W. Kern contains several chapters reviewing magnetron sputtering.[39] A new book by B. Chapman provides an introduction to glow discharges from the point of view of their application to sputtering and plasma etching. A chapter by J.E. Greene in a new book on semiconductors will review sputtered semiconductors.[39a]

Developments in sputtering technology are most commonly reported in the following journals: *Journal of Vacuum Science and Technology, Thin Solid Films, Journal of Applied Physics, Vacuum, Progress in Surface Science,* and the *Journal of the Electrochemical Society.*

2. BASIC SPUTTERING MECHANISMS

The Sputtering Yield

The sputtering process is quantified in terms of the sputtering yield, defined as the number of target atoms ejected per incident particle. The yield depends on the target species, and on the bombarding species and its energy and angle of incidence. It is insensitive to the target temperature.[35] It is also independent of whether the bombarding species is ionized or not, as ions have a high probability of being neutralized by a field-emitted electron prior to impact.[35,40,41,41a] Ions are generally used because of their ease of production and acceleration in a glow discharge. Molecular bombarding species behave as if the atoms of the molecule arrived separately with the same velocity as the molecule and initiated their own sputtering events.[35] The yield tends to be greatest when the mass of the bombarding particle is of the same order of magnitude or larger than that of the target atoms. The use of inert gas ions avoids chemical reactions at the target and substrate. Accordingly, argon is generally used because of its mass compatibility with materials of engineering interest and its low cost. However, krypton and xenon are sometimes used.

Sputtering yields are determined experimentally. Figure 5.5 shows yield versus ion energy data for several materials under normal ion incidence. Additional data are given in Table 5.1. The yield dependence on the bombarding ion energy is seen to exhibit a threshold of about 10–30 eV, followed by a near-linear range which may extend to several hundred eV. At higher energies the dependence is less than linear. The sputtering process is most efficient from the standpoint of energy consumption when the ion energies are within the linear range. The general dependence of the yield on the ion angle of incidence is indicated in Figure 5.6.[42] In glow discharge sputtering devices the ions generally approach the target in a direction normal to the target surface. The relationship shown in Figure 5.6 is of particular significance when the target surface is highly irregular. This will be discussed in the section on alloys and compounds.

Note that the yields of most metals are about unity and within an order of magnitude of one another. This is in contrast, for example, to evaporation, where the rates for different materials at a given temperature can differ by several orders of magnitude.[43] It is this universality that makes sputtering such an attractive process for many applications.

Sputtering apparatuses are generally calibrated to determine the deposition rate under given operating conditions. However, yield data of the type described above are often used in projecting rate changes when changing coating materials and in estimating the amount of material removed during sputter cleaning and bias sputtering. The erosion rate is given by

$$R = 62.3 \frac{JSM_A}{\rho} \text{ Å/min} , \qquad (1)$$

where J is the ion current density in mA/cm^2, S is the sputtering yield in atoms/ion, and M_A is the atomic weight in grams and ρ is the density in gm/cm^3 of the target material.

Coating Deposition by Sputtering

Table 5.1: Sputtering Yields for Various Materials Under Argon Ion Bombardment. Ion energy in eV. Data from Reference 36.

Target	200	600	1,000	2,000	5,000	10,000	Heat of Sublimation eV/atom***
Metal Sputtering Yields, Atoms/Ion							
Ag	1.6	3.4	...	...	...	8.8	2.94
Al	0.35	1.2	...	...	2.0	...	3.33
Au	1.1	2.8	3.6	5.6	7.9	...	3.92
C	0.05*	0.2*	...	...	...	...	7.39
Co	0.6	1.4	...	...	...	...	4.40
Cr	0.7	1.3	...	...	...	...	4.11
Cu	1.1	2.3	3.2	4.3	5.5	6.6	3.50
Fe	0.5	1.3	1.4	2.0**	2.5**	...	4.13
Ge	0.5	1.2	1.5	2.0	3.0	3.98	...
Mo	0.4	0.9	1.1	...	1.5	2.2	6.88
Nb	0.25	0.65	...	...	...	...	...
Ni	0.7	1.5	2.1	...	...	...	4.45
Os	0.4	0.95	...	...	...	...	8.19
Pd	1.0	2.4	...	...	...	...	3.90
Pt	0.6	1.6	...	...	...	...	5.95
Re	0.4	0.9	...	...	...	...	8.06
Rh	0.55	1.5	...	...	...	...	5.76
Si	0.2	0.5	0.6	0.9	1.4	...	4.68
Ta	0.3	0.6	...	...	1.05	...	8.10
Th	0.3	0.7	...	...	...	...	5.97
Ti	0.2	0.6	...	1.1	1.7	2.1	4.86
U	0.35	1.0	...	...	...	...	5.00
W	0.3	0.6	...	...	1.1	...	8.80
Zr	0.3	0.75	...	...	...	...	6.34
Compound Sputtering Yields, Molecules/Ion							
CdS(1010)	0.5	1.2	...	...	...	...	...
GaAs(110)	0.4	0.9	...	...	...	...	...
GaP(111)	0.4	1.0	...	...	...	...	...
GaSb(111)	0.4	0.9	1.2	...	...	...	...
InSb(110)	0.25	0.55	...	...	...	...	...
PbTe(110)	0.6	1.40	...	...	...	...	...
SiC(0001)	...	0.45	...	...	...	...	...
SiO$_2$	...	...	0.13	...	...	...	...
Al$_2$O$_3$	...	...	0.04	0.11	...	...	...

*Kr+ ions.   **Type 304 stainless steel.   ***From Ref. 240.

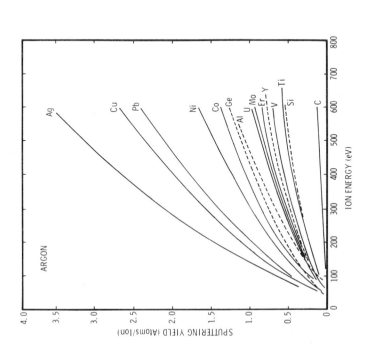

Figure 5.5: Variation of sputtering yield with ion energy at normal angle of incidence. Data from R.V. Stuart and G.K. Wehner, *J. Appl. Phys. 33*, 2351 (1962); D. Rosenberg and G.K. Wehner, *J. Appl. Phys. 33*, 1842 (1962); R. Behrisch, *Ergeb. Exakt. Naturw. 35*, 295 (1964).

180   Deposition Technologies for Films and Coatings

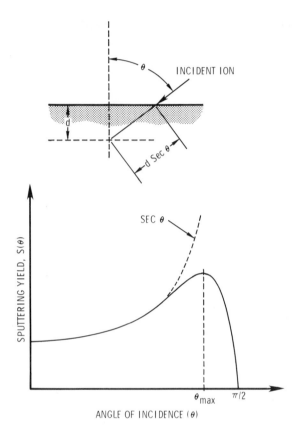

Figure 5.6: Schematic diagram showing variation of sputtering yield with ion angle of incidence. Ion energy constant. (Used courtesy Telic Corporation.)

Basic Momentum Exchange

Consider a particle of mass M_i and velocity v_i, which impacts on a line of centers with a target particle of mass M_t which is initially at rest, as shown in Figure 5.7a. Three simple observations can be made. First, the momentum passed to the target particle drives it into the target. Thus the ejection of a sputtered particle from a target requires a sequence of collisions* so that a component of the initial momentum vector can be changed by more than 90°. Computer modeling of the collision sequence under typical sputtering conditions has revealed that an incident ion does indeed produce a cascade of collisions and that its energy is partitioned over a region of the target material that may extend 50 to 100 Å or more below the target surface.[44] However, the sputtering momentum exchange occurs primarily within a region extending only about 10 Å below the surface.[44,45] The incident ion generally strikes

---

*The fact that a number of target particles are involved in the collision sequence is shown by the observation that for rare gas bombardment, the threshold energy for sputtered particle emission is about four times the sublimation energy of the target material.[43a]

two lattice atoms almost simultaneously. The low energy knock-on receives a side component of momentum and initiates sputtering of one or more of its neighbors, as shown in Figure 5.7b. The primary knock-on is driven into the crystal, where it may be reflected and on occasion return to the surface and produce sputtering by impacting on the rear of a surface atom, as indicated in the figure.

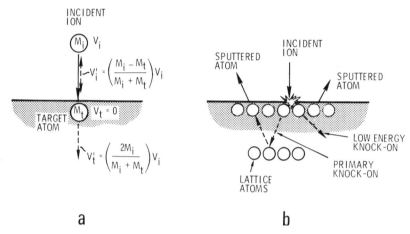

Figure 5.7: Schematic diagram showing some of the momentum exchange processes that occur during sputtering; v_i is ion velocity, v_t is target atom velocity, prime denotes velocities after collision. (Used courtesy Telic Corporation.)

The second observation that can be made from a simple line-of-centers atomic collision is that a fraction,

$$\epsilon = \frac{4 M_i M_t}{(M_i + M_t)^2} ,$$

of the kinetic energy of the incident particle is transferred to the target particle. An expression for the yield, which can be written in the form shown in Eq. (2), has been derived by assuming perpendicular ion incidence onto a target consisting of a random array of atoms (good approximation for polycrystalline material) with a planar surface.[46,46a,46b]

$$S = (\text{constant}) \, \epsilon \, \frac{E}{U} \, \alpha(M_t/M_i) . \tag{2}$$

The relationship is useful for illustrating the functional dependences of the important parameters and provides reasonably absolute agreement with measurements for medium mass (Ar, Kr) bombardment of many materials of engineering interest. The yield is seen to depend directly on the energy transfer function ϵ. The term $\alpha(M_t/M_i)$ is a near-linear function of M_t/M_i, E is the kinetic energy of the incident ion, and U is the heat of sublimation for the target material. For engineering materials the mass dependence of $\epsilon\alpha$ does not vary greatly from one material to another. The primary material-sensitive factor is the heat of sublimation, and this is only a first power dependence. This is in contrast to chemical and thermal processes that depend exponentially on an activation energy. It is this relative insensitivity to the properties of the target material that gives sputtering the universality that has been referred to previously.

Referring again to Figure 5.7a, we see thirdly that when the ion mass is lower than that of the target atom it may be reflected backward in a single collision, and that the energy of the reflected ion may be a significant fraction of its initial energy. For a 180° reflection this fraction is

$$\left(\frac{M_i - M_t}{M_i + M_t}\right)^2.$$

If $M_i > M_t$, the ion can be reflected backward only as the result of more than one collision.

Since the ions have a high probability of being neutralized prior to impact, they are reflected as energetic neutrals which are therefore not influenced by the electric field over the target surface.[47] The momentum flux represented by the reflected species contributes to substrate heating,[48] particularly in devices operating at low pressures or having cylindrical symmetry with central cathodes. In sputtering devices, which operate at low working gas pressures the reflected and neutralized ions may reach the substrates with little loss of kinetic energy due to gas phase collisions. Consequently, the reflected species bombard and can become entrapped in the growing coating.[49-52]

The energy flux which leaves the cathode via backscattering can be estimated using the sputtering efficiency.[53-55] The sputtering efficiency is defined as the fraction γ of the bombarding ion energy incident on the target surface, E_{in}, which leaves the surface in the backward direction, E_{out}, in the form of sputtered atoms or backscattered ions.

$$\gamma = \frac{E_{out}}{E_{in}} = \frac{E_{sputtered} + E_{backscattered}}{E_{in}}. \qquad (3)$$

The energy of the sputtered atoms is discussed in a subsequent section. Theoretical calculations for a target consisting of a random array of atoms in which the surface binding energy was neglected reveal a surprising aspect of the sputtering momentum exchange.[53] The sputtering efficiency is independent of the energy of the incident ion and is simply a monatomically increasing function of the target-ion mass ratio. This dependence, which has been confirmed for both low and high ion energies, is shown in Figure 5.8.

The surface nature of the sputtering momentum exchange provides an explanation for the angular dependence of the sputtering yield which is shown in Figure 5.6. An ion which is incident on the target surface at an angle θ will, to first order, have its path length increased by a factor Sec θ before it passes through the depth d where the primary sputtering momentum exchange occurs. At larger angles of incidence, ion reflection dominates and the yield decreases.

Another question of interest is the ultimate fate of the inert gas ions that bombard the target. The probability of their becoming trapped in the target surface increases with ion energy above a threshold of about 100 eV.[56] Thus an inert gas density will develop in the target surface, depending on a balance between the rates of implantation and release. The amount of gas entrapped in the target can be large enough to influence the sputtering yield.[56a] The mechanism of release is still unresolved.[57] Measured equilibrium argon densities in tungsten imply that if the release is by argon sputtering, then the inert gas yield is an order of magnitude larger than that of the host tungsten lattice.[56]

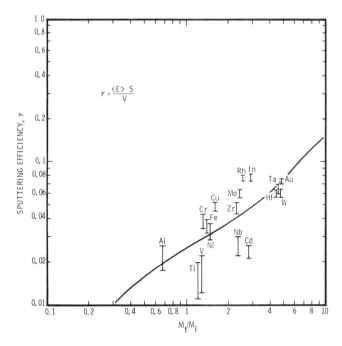

Figure 5.8: Sputtering efficiency versus target-to-ion mass ratio. Curve from theoretical work of Sigmund (Reference 53). Experimental data from substrate heating experiments with cylindrical-post magnetrons (Reference 48). (Used courtesy Telic Corporation.)

Alloys and Compounds

An important advantage of the sputtering process is that the composition of a sputtered film tends to be the same as that of the target, provided that (1) the target is maintained sufficiently cool to avoid diffusion of the constituents, (2) the target does not decompose, (3) reactive contaminants are not present, (4) the gas phase transport of the components is the same, (5) the sticking coefficients for the components on the substrate are the same.[58] Targets can be formed by casting or by hot pressing powders. In addition, composite targets can be formed by placing wires, strips or discs of one material over a target of another material.[38] However, special considerations are important in understanding the film composition produced by these different target forms.

Despite the simplicity implied by the fact that sputtering tends to produce a vapor flux having the chemical composition of the originating solid, the details of the sputtering interaction on multicomponent materials are complex and poorly understood.[59,59a] First consider the case of a homogeneous starting material composed of species having different individual sputtering yields or masses. When sputtering is first initiated from such a target, the sputtered flux will in general be rich in one of the constituents. The correct composition will not be achieved until after an adjustment period during which the compositions of the species in a surface layer adjust until the product of the effective sputtering yield times surface concentration for each species is proportional to its concentration in the target. The process is indicated schematically in Figure 5.9. Clearly, it is necessary that diffusion from the bulk not replenish the reduced concentrations of high yield materials in the altered layer. Thus the requirement on target cooling cited above. The thickness and composition of the altered layer will depend on the target material and sputtering conditions. Typical thicknesses are 30–100 Å.[59,60] A change in sputtering conditions will in general require an adjustment of the altered layer. It is important to note that the effective sputtering yield of a constituent in an alloy or compound will not be the same as that constituent by itself, because of the different binding energy and the different atomic masses involved in the collision sequence within the alloy or compound. Accordingly, if the species have similar binding energies, the low mass constituents can be expected to have higher effective sputtering yields. If the masses are similar, the weakly bound species can be expected to have higher sputtering yields.[46a,59a] Thus in the sputtering of most oxides the altered layer becomes deficient in the low mass oxygen component.[59a]

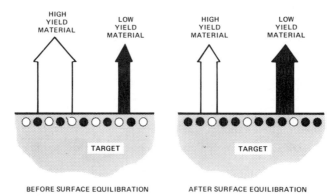

Figure 5.9: Schematic illustration of the modification in surface composition that occurs during the sputtering of a homogeneous multispecies material in which the species have similar atomic masses. (Used courtesy Telic Corporation.)

Next consider the case of a two-phase alloy in which the phases have significantly different sputtering yields. The inhomogeneous sputtering yield over the surface will cause an irregular surface topography to develop.[42,62-64] The sloping surfaces that survive tend to be those that make an angle with the sputtered flux such that the sputtering yield is maximized (see Figure 5.6). If the second phase, or any included impurity particles, have very low sputtering yields, the surface may develop into a forest of cones with side walls at the maximum sputtering angle, as shown in Figure 5.10.[61,65-68] The cones will of course sputter away; however, the receding target surface will expose new second phase regions and impurity particles (if they are distributed throughout the bulk) and new cones will form. Thus an equilibrium surface will develop. Surface diffusion on the target will, in general, make this situation more complex than the picture described above. The important point is that after an incubation period the composition of the sputtered flux leaving the target will become identical to that of the target. Nevertheless, the irregular surface topography may cause the overall yield to be considerably lower than what might be expected on the basis of the yields of the primary target constituents.

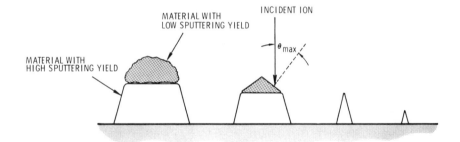

Figure 5.10: Schematic representation showing stages of cone formation. (Used courtesy Telic Corporation.)

Topographical evolutions such as cone formation can also influence the performance of composite sputtering targets.[69-70] When such targets are used in sputtering systems that operate at high pressures (greater than about 20 mTorr), some of the sputtering material will be backscattered by the working gas. Thus mixing of low and high yield, or low and high melting point, materials can occur on the target segments. Evidence has been seen which indicates that if atoms of a low yield material deposit on a high yield target surface, and if the low yield material can agglomerate into islands capable of protecting the material underneath, then cones will form.[68] Relative sputtering yields and melting points as well as the target temperature appear to be important in predicting this behavior.[68a] A typical combination would be copper and molybdenum, where the copper target surface can become covered with a cone forest. In the Cu-Mo case the resultant sputtering rate from the cone covered surface has been found to be very close to that for the low-yield material (Mo). The important point is that the film composition deposited from composite targets can be much different from that estimated from the individual sputtering yields and the relative areas of the target segments.

Special care should be exercised when using hot-pressed targets. Hot-pressed Au-Ni and Au-Co targets composed of powders in the 50 to 130 μm range were found to yield deposits with compositions that matched those of the targets after an equilibration period during which a layer only about 20 μm thick had been sputtered from their surfaces.[71] However, the overall yield dropped to a value equal to that of the low-yield constituents (Ni and Co), even when the volume fraction of that constituent was only about 30%.

Contamination can present a particular problem with hot-pressed targets because of the large surface area contained in the starting powder. Such contamination may be present throughout the target and in such case will not clean up as the target is used.[72,73]

Particular caution must also be exercised when using targets composed of compounds having poor electrical and thermal conductivity. Cracking often limits allowable current densities. The problem is particularly important for planar magnetrons where concentrated heating occurs under the plasma ring.[73a] Poor thermal conductivity leads to high surface temperatures and may result in the loss of volatile constituents by evaporation or sublimation. The high electric field in a poorly conducting target can act in concert with the high temperature and promote diffusion within the target. Thus the requirements listed at the beginning of this section are violated. It is not uncommon for films sputtered with such targets to be deficient in the more volatile constituents.[74,75]

A considerable amount of research is currently under way to investigate the composition that develops on the surfaces of multicomponent materials during sputtering. The impetus for this work is the increased use of sputtering as a depth profiling means for chemical analysis methods such as Auger electron spectroscopy. The results of this work should be very helpful in elucidating mechanisms that are also important to sputtering as a deposition process.

Sputtering with Reactive Species

The most complete data on the dependence of the sputtering yield on the ion species are those collected by Almen and Bruce, shown in Figure 5.11.[76] Although the ion energies were considerably above those generally used for sputter coating technology, they do illustrate the trends. It is seen that the yields increase with the mass of the ions and that for a given mass range the noble gas ions give the highest yields. Of particular interest is the fact that yields vary much more with ion species (factor of 100 or more) than they do with atom species (factor of 10).[35] This is believed to result because the bombarding ions form alloys or compounds with high binding energies on the surface of the target. Note that the yields for the three target materials examined in Figure 5.10 are particularly low for active species such as Be, C, Mg, Si, Ti, and Zr.

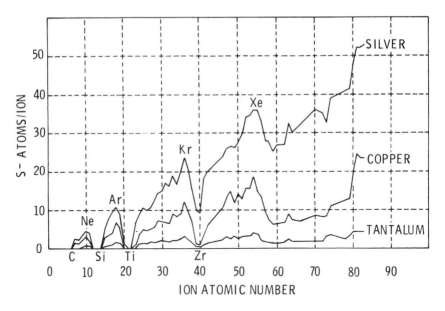

Figure 5.11: Sputtering yields for various ions impacting at normal incidence on silver, copper, and tantalum surfaces at high energies (45 keV). Data from Reference 76.

Reduced yields are commonly observed in reactive sputtering and attributed to compound formation on the target surface. Such surface interactions can also significantly influence the surface topography that develops on the target. Thus 20 keV O_2^+ bombardment of an Fe target yielded a considerably smoother surface than 20 keV Ar^+ bombardment.[77]

Another important consideration in reactive sputtering is the direct sputtering of reactive gas species that have become chemisorbed on the surface of the target. Calculations and measurements indicate that such species are sputtered as atoms with yields that are of the same order of magnitude or higher than those of elemental materials.[78]

Sputtered Species

The sputtered species are primarily atoms.[35,38,58] However, atom clusters and molecular fractions are also observed as well as positive and negative ions. Computer modeling has predicted the experimentally observed magnitudes for cluster formation (a few percent) and has provided insight into possible mechanisms for their formation.[79] The studies, based on Ar sputtering of Cu, indicate that typically one or more atoms are sputtered from the surface as the ion enters the target. Other (knock-on) atoms are driven into the target, where they may be reflected and may produce sputtering on their return to the surface, as shown in Figure 5.7b. The velocities of the knock-on atoms are considerably greater than those of the directly sputtered atoms. Thus all the atoms sputtered by a given incident ion can be considered to be leaving the surface at about the same time. Consequently, there is a significant probability that the sputtered atoms will come into sufficiently close proximity to one another, so that attachment to form atomic clusters can occur.

Molecular fractions have been observed in reactive sputtering.[80] The mechanism of formation may be similar to that described above for the clusters. For example, atoms sputtered from the targets may, on their exit from the surface, collide with reactive gas atoms chemisorbed on the surface, thereby forming molecular fractions. Consistent with this point of view, it was found, in an analysis of the ion fractions leaving target surfaces, that Ar^+ bombardment of oxidized W produced WO^+ and WO_2^+ in addition to W^+, and that O_2^+ bombardment of Mo yielded MoO^+ and MoO_2 in addition to Mo^+.[81]

The fraction of positive ions in the sputtered flux is generally less than one percent and relevant mainly to surface analysis methods such as SIMS.[38,58,82] In a glow discharge sputtering source, the electric field over the target surface prevents the escape of positive ions. However, negative ions can be accelerated to very high energies in this field and ultimately impact on the substrates. The yield of negative ions can be very high for targets composed of two materials, one of which has a low ionization potential I, and the other of which has a high electron affinity X.[83,84] For materials such as SmAu, where $I_{Sm} - X_{Au}$ <3.4 eV, it has been found that the flux of negative ions reaching the substrate can be large enough to provide a significant reduction in the deposition rate via back-sputtering.[83] The charge transfer probably occurs during target collision sequences in a manner similar to that described above with respect to the production of clusters and molecular fractions.

Energy of Sputtered Species

The sputtered atoms are ejected from the target surface with considerable kinetic energy—for example, 50–100 times higher than in vacuum evaporation.[35,38,58] The energy distribution is approximately Maxwellian, with a most probable energy of less than 10 eV and a slightly overpopulated high energy tail, so that the average energy is of the order of 10–40 eV. See Figure 5.12. Increasing the bombarding ion energy increases the population of the high energy tail. However, the average energy of the ejected particles ceases to increase for ion energies above about 1 keV.[35] The atoms ejected from most metals ($Z > 20$) under Ar bombardment have average velocities which lie in a relatively narrow range between 3 and 6 x 10^5 cm/sec.[35] The average energies therefore increase with the ion mass, as shown in Figure 5.13. (Ejection velocities under Kr bombardment shown in Figure 5.13 are about 20% higher than those for Ar bombardment.[35]) The target crystal orientation or structure does not significantly affect the ejection energies. However, the materials with high sputtering yields tend to have lower average ejection energies.[35]

Atoms sputtered from polycrystalline or amorphous targets under perpendicular ion incidence at typical working energies (1 to 3 keV) are ejected in nearly random directions as a consequence of the multiple collisions within the target, and therefore have near cosine distributions.[35] At low ion energies (~1 keV) the distribution may be slightly under cosine (more emission at large angles) and at higher energies (~3 keV) over cosine.[34] Under oblique incidence the target atoms are sputtered in the forward direction from smooth surfaces. However, the roughness of most practical targets causes the emission to be random. This is particularly true for polycrystalline targets, where the difference in yield for different crystallographic directions can lead to an increase in surface roughness as sputtering proceeds. Thus a cosine distribution is generally a good approximation for calculating deposition profiles.[35]

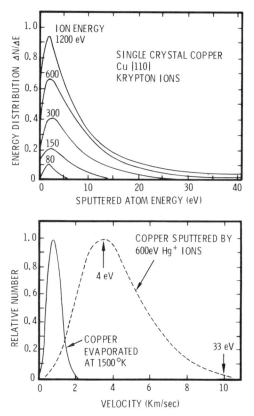

Figure 5.12: (Top) Energy distribution of sputtered copper atoms produced by krypton ions at various bombarding energies. (Bottom) Comparison of velocity distribution of sputtered and evaporated copper atoms. Data from Reference 58.

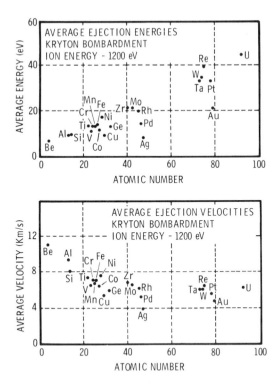

Figure 5.13: Average energies (top) and velocities (bottom) of sputtered atoms produced by 1.2 keV krypton ion bombardment. Data from Reference 35.

3. IMPLEMENTATION

The fundamental problem in implementing the sputtering process is to provide a uniform and copious supply of ions over the surface of the target. The low pressure glow discharge has proven to be the most cost-effective source of ions. A wide range of glow discharge apparatus geometries have been used in attempts to (1) increase the ion supply and thus the sputtering rate, (2) increase the target area and thus the available deposition area, (3) reduce the plasma heating of the substrates, (4) permit a lowering of the working gas pressures, and (5) facilitate the coating of particular substrate shapes. In the following discussion the essential features of the glow discharge and several of the more commonly used apparatus types are reviewed.

Planar Diode and the dc Glow Discharge

The planar diode shown in Figure 5.14 is the simplest and probably the most widely used sputtering configuration. The cathode diameter is typically 10 to 30 cm, and the cathode-to-anode spacing about 5 to 10 cm. Such systems are operated with both dc and rf power supplies. We first consider the dc case. In diodes the cathode serves a dual capacity. It is the target or source of coating

190   Deposition Technologies for Films and Coatings

material as well as the cathode electrode for sustaining the glow discharge. It is usually water-cooled. Often the target consists of a disc of the material, which is attached with solder or conducting epoxy to a backing plate which serves as part of the cathode cooling channel. A low pressure glow discharge of a type known as an abnormal negative glow[85] is maintained between the cathode and an adjacent anode which may also serve as the substrate mounting table, as shown in Figure 5.14.[86] A grounded shield, also shown in the figure, suppresses the occurrence of the glow discharge except over the front (target) surface. The current in such a discharge is carried in the vicinity of the negatively biased cathode, primarily by positive ions passing out of the plasma volume, and in the vicinity of the anode, by electrons passing from the plasma volume to the anode. Thus, a condition for sustaining the discharge is that the plasma volume be a suitable source of electrons and ions.

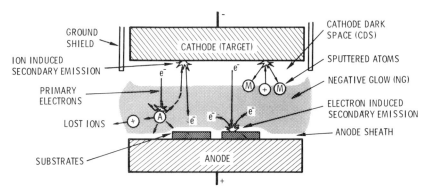

Figure 5.14: Schematic representation of the plasma in a planar diode sputtering source. (Used courtesy Telic Corporation.)

Because of the relatively low mobility of the ions compared to the electrons, most of the electrical potential that is applied between the anode and cathode by the power supply is consumed in a "cathode dark space," or sheath region.[85] Dark space thicknesses are typically 1 to 4 cm, depending on the pressure and current density.[87] Accordingly, strong electric fields are formed, and ions passing from the plasma volume to the cathode are accelerated by these fields and on impact at the cathode produce the desired sputtering. However, these ions also cause a small number of "secondary electrons" to be emitted from the surface (about one for every ten ions in the case of argon ions impacting on a metal cathode[88]). These electrons are accelerated in the cathode dark space to energies approaching the applied potential, and enter the plasma volume (negative glow) where, known as "primary electrons," they collide with gas atoms and produce the volume ionization required to sustain the discharge.[85,89].

The requirement for sustaining such a discharge is that each primary electron must produce sufficient ions to release one further electron from the cathode. Thus the interelectrode spacing must be large compared to the electron mean free path. The cross section, σ, for argon ionization by the impact of 500 eV electrons is about 10^{-16} cm^2.[90] Thus, for example, at an argon pressure of 1 mTorr (gas density N = 3.2 x 10^{13} atoms/cm^3) the electron mean free path ($\lambda = 1/N\sigma$) for the production of ionization is 300 cm; i.e., much larger than the cathode-to-anode spacing. Consequently, discharges of the form shown in Figure 5.14 can be sustained only at relatively high working pressures (50-100 mTorr), where a high density of argon collision targets is provided for the primary electrons. (The shields shown in Figure 5.14 successfully suppress the discharge from forming on the sides of the cathode because the spacing is too small to support the ionization mechanisms at the operating pressures of interest.)

Attempts to increase the discharge current in a planar diode by increasing the applied voltage are thwarted to a large degree by the fact that the ionization cross section decreases with increasing electron energy for energies greater than about 100 eV.[90] The current, and thus the sputtering rate, can be effectively increased at a given voltage by increasing the argon pressure. However, if the pressure is made too high, the deposition rate starts to decrease, since the motion of both the ions and the sputtered atoms is impeded by the working gas atmosphere, as discussed below. These conflicting requirements cause there to be an optimum operating pressure for producing the maximum deposition rate in a given apparatus. Typical operating conditions for metal sputtering in a dc planar diode sputtering source are listed below.

- Cathode current density—1 mA/cm^2.
- Discharge voltage—3,000 V.
- Argon pressure—75 mTorr (10 Pa).
- Cathode-to-substrate separation—4 cm.
- Deposition rate—400 Å/min.

At typical planar diode operating pressures the motion of the ions across the dark space is disrupted by collisitons with gas atoms. In such collisions there is a high probability of charge exchange, particularly when noble gas ions are passing through an atomic gas of their own species (resonance charge exchange).[85] A fast ion extracts an electron from a slow gas atom, as a result of which the fast ion becomes a "fast" neutral atom, while the "slow" atom becomes a slow positive ion, as indicated schematically in Figure 5.15. Thus, instead of being bombarded by a current of ions having an energy equal to the potential drop across the cathode dark space, the target is bombarded with a much larger number of ions and atoms having an average energy that is often less than 10% of the applied potential.[91] This can result in a reduction in the rate of sputtering (see Figure 5.5).

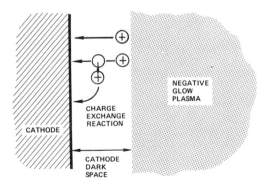

Figure 5.15: Schematic illustration of charge exchange process that affects ion transport across the cathode dark space. (Used courtesy Telic Corporation.)

The deposition rate in planar diodes is further reduced by gas scattering of the sputtered atoms. Optical emission measurements have confirmed that at typical sputtering pressures the sputtered atom transport within the negative glow region is largely by diffusion.[92] Thus surfaces adjacent to a substrate, that do not shield it optically from the target, can still rob it of coating flux. The charge transfer processes, and the diffusion transport, make it necessary to determine deposition rates experimentally for each set of operating conditions. Another consequence of the collision-dominated transport of the sputtered atoms is a reduction in their energy.[38] Figure 5.16 shows the results of an approximate calculation of the maximum transport distance required for sputtered atoms of various initial energies to have their energy reduced to the thermal energy of the gas atoms (~0.025 eV).[93,93a] At typical planar diode operating pressures the equilibration distances are short compared to target-to-substrate spacings.

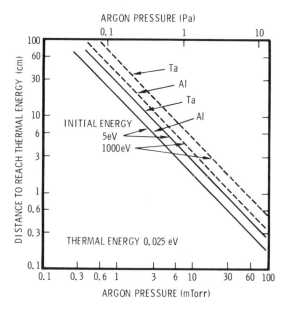

Figure 5.16: Maximum distance from the target at which sputtered Al and Ta atoms of different initial energies are thermalized in Ar at various pressures. Thermalized energy assumed to be 0.025 eV. Data from Reference 93.

Even under the relatively high pressure conditions that yield the maximum deposition rates, the planar diode discharge is inefficient. Many of the high-energy primary electrons fail to transfer their energy in the plasma volume, and they impact on the anode and substrates while still possessing considerable energy, as implied in Figure 5.14. Ions and electrons are also lost from the edges of the discharge. Note also that the substrates are in contact with the plasma and are therefore subjected to bombardment by the plasma electrons and ions as well as by the energetic primary electrons, as noted in Section 1. These bombardments preclude the coating of many heat-sensitive materials such as plastics.[28]

Coating Deposition by Sputtering

Planar diodes are widely used despite the low deposition rates, substrate heating, and relatively small deposition surface areas. The reason is their simplicity and the relative ease with which planar targets can be fabricated from a wide range of materials to take advantage of the versatility of the sputtering process, as discussed in Section 1. The plasma-substrate bombardment promotes adhesion. Sputter cleaning and bias sputtering are easily accomplished by adding an auxiliary anode and applying a negative bias to the substrate holder.

Assist Discharge Devices—Triodes

In assist discharge devices an electrode system that is independent of the target is provided for sustaining the glow discharge.[37] The most common such configuration is the hot cathode triode shown in Figure 5.17. Electrons are emitted at the cathode surface by thermionic emission rather than ion bombardment. This relaxes the volume ionization requirement for sustaining the discharge. Consequently, hot cathode triodes can be operated at low pressures (0.5 to 1 mTorr). The driving voltage is only 50–100 V, although the current may be several amperes. Radial plasma losses are often minimized through the confining effect (see next section) of an axial magnetic field, as shown in the figure. However, such a field produces a distortion of the current distribution over the target.

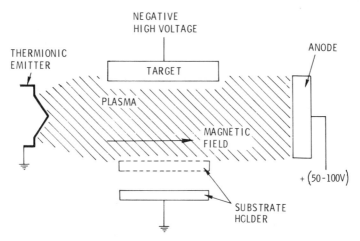

Figure 5.17: Schematic drawing of hot cathode assist discharge device (triode). From Reference 37.

Triodes permit high deposition rates (several thousand Å/min) to be achieved, even at low pressures (mTorr range).[21] Although impressive coatings have been deposited,[94] use of triodes has been limited by difficulties of scaling and the vulnerability of the thermionic emitter to reactive gases. Consequently, magnetron sources (next section) are assuming primary importance as high rate sputtering devices.

Magnetrons

The recent development of high performance magnetron sputtering sources that provide (1) relatively high deposition rates, (2) large deposition areas, and (3) low substrate heating, is revolutionizing the sputtering process by greatly expanding the range of feasible applications.[95]

194 Deposition Technologies for Films and Coatings

Magnetron sputtering sources can be defined as diode devices in which magnetic fields are used in concert with the cathode surface to form electron traps which are so configured that the $\vec{E} \times \vec{B}$ electron drift currents can close on themselves.[96,97] Magnetrons can be configured in various forms. Thus one has the planar magnetrons shown in Figure 5.18a, the S-gun type shown in Figure 5.18b, and the cylindrical type shown in Figure 5.19.

The magnetron configuration shown in Figure 5.19 has been termed the "cylindrical-post magnetron."[95-97] It provides the simplest geometry for explaining the principles of magnetron operation.[96-101] The cathode consists of a cylindrical barrel with end plates, all composed of the material to be sputtered. It is mounted in a chamber with a uniform magnetic field $\vec{B}$ directed parallel to the cathode axis. The magnetic field is of such strength (a few hundred gauss or less) that it affects the plasma electrons but not the ions. Figure 5.20 shows a chamber configuration that is often used for cylindrical magnetrons. A set of solenoidal field coils is positioned surrounding the cylindrical vacuum wall, which is constructed of a nonmagnetic material. A magnetic steel shell surrounds the coils and makes contact with the chamber top and bottom plates, which are also fabricated from a magnetic material such as low carbon steel. Thus a low reluctance return path is provided for the solenoidal flux, as indicated in the figure, with the consequence that the coil system of limited length efficiently provides a uniform magnetic field within the chamber.

Secondary electrons which are emitted from the cylindrical magnetron cathode barrel surface because of ion bombardment find themselves trapped in an annular cavity which is closed on three sides by surfaces at cathode potential and on the fourth side by the magnetic field. Anode rings are located adjacent to the end plates on one or both ends of the cathode. Therefore electrons emitted from the cathode must migrate radially across the cavity in order to reach the anode. Electron collisions of the type required to sustain a plasma discharge play an essential role in allowing this migration to occur. Thus, in contrast to the conventional discharge, the electrons are forced to make the required collisions, and an effective sputtering discharge is maintained in the cavity.

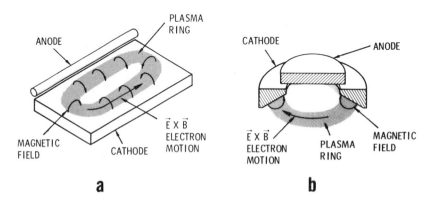

Figure 5.18: Magnetrons with magnetic end confinement: (a) planar magnetron, (b) gun type. From Reference 95.

Coating Deposition by Sputtering   195

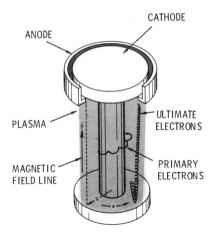

Figure 5.19: Cylindrical-post magnetron sputtering source with electrostatic end confinement. From Reference 95.

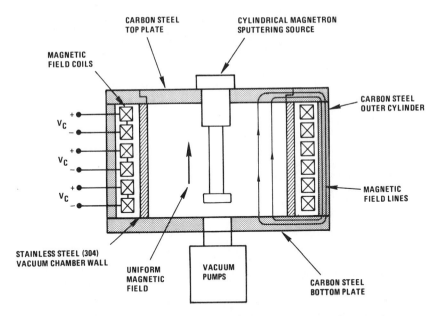

Figure 5.20: Chamber and magnetic field coil configuration that has been used for cylindrical-post magnetron sputtering sources (Courtesy Telic Corporation.)

The electron motion can be understood as follows.[102,103] When an electron is in a uniform magnetic field, its motion perpendicular to the field lines can be pictured as an orbit around a field line, as shown in Figure 5.21a. Its motion along the field is unimpeded, so that if it has a component of velocity along the field line, its net motion is a spiral along the field line, as shown in Figure 5.21b. Thus such electrons can be considered to be trapped on magnetic field lines. An electron trapped on a given field line can advance to an adjacent field line by making a collision, as indicated schematically in Figure 5.21c. An electron will also undergo a drift motion across the magnetic field if an electric field E is present. However, this motion, known as the $\vec{E} \times \vec{B}$ drift, is not in the direction of the electric field but in a direction perpendicular to both the electric and the magnetic fields. The $\vec{E} \times \vec{B}$ drift has the cycloidal form shown in Figure 5.21d if the initial electron energy is small compared to that gained from the electric field, and the more circular motion shown in Figure 5.21e if the initial electron energy is large compared to the electric-field-induced variations during the course of an orbit.

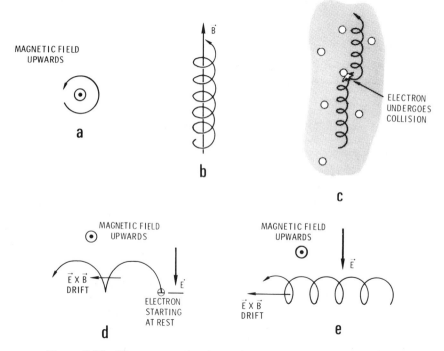

Figure 5.21: Electron motion in static magnetic and electric fields. Used courtesy Telic Corporation.)

Now, referring back to Figure 5.19, a radial electric field exists in the annular cavity. The field will be strong in the sheath region adjacent to the cathode but relatively weak at larger radii. Electrons emitted from the cathode will therefore undergo motions of the type shown in Figure 5.21d and will become trapped in orbits revolving around the cathode. They will be able to advance radially only by making collisions or by the action of plasma oscillations which produce azimuthal electric fields (in θ-direction shown in Figure 5.16) and radial drifts.[96,97] As noted previously, the collisions are particularly important. Many of these collisions will be with gas atoms, and since the electrons leave the cathode sheath with energies of several hundred eV, a large number of these collisions will produce ionization. The $\vec{E} \times \vec{B}$ drift motions of the primary electrons and the products of the ionization produce an intense azimuthal current sheet of trapped electrons adjacent to the cathode. Because of the free axial movement of the electrons along the field lines, the sheet tends to be uniform along the cathode length. Large numbers of ions are produced. These ions cause a uniform sputter-erosion and high sputtering rates along the cathode barrel.

As the electrons give up energy by collisions and become so-called "ultimate electrons,"[89] they move into regions of weak electric field at larger radii, and their motion becomes more like that shown in Figure 5.21e. The $\vec{E} \times \vec{B}$ drift velocity is relatively small, and the electrons move primarily up and down the field lines, reflecting off the end plates as indicated in Figure 5.19.[104] When they reach the anode radius R they immediately pass into the anode. Therefore in effect the high mobility of the electrons along the magnetic field lines causes the anode ring to be projected as a "virtual anode sheet" which surrounds and terminates the plasma discharge but is transparent to the sputtered flux.[96,97] Thus the electrons are trapped within the annular cavity throughout their lifetimes. The ions are constrained electrostatically to stay with the electrons and are therefore largely confined to this region as well. Consequently, there is virtually no plasma bombardment onto substrates located beyond the anode radius.[48]

Because of the efficiency of the ionization mechanisms in the magnetron cavity, intense plasma discharges capable of providing high sputtering rates can be maintained at moderate and near-constant voltages, even at low pressures. Deposition rates will depend on the radial position of the substrates. Typical operating conditions for cylindrical magnetrons of the type shown in Figure 5.19 are[97]

- Cathode current density—20 mA/cm^2;
- Discharge current—1 to 50 A;
- Discharge Voltage—800 V;
- Argon Pressure—1 mTorr (0.13 Pa);
- Cathode erosion rate—12,000 Å/min;
- Substrate position—radius equal to 6 cathode radii;
- Deposition rate—2,000 Å/min.

An important attribute of the cylindrical magnetrons is their capability for being scaled to a range of sizes while retaining common operating characteristics.[97] Cathodes have been operated which range in length from 0.1 m (4 in) to 2.1 m (7 ft). Figure 5.22 shows a 2.1 m cylindrical magnetron designed for depositing decorative coatings.[105] Such long cathodes provide a large substrate placement area located around the circumference. Substrates can also be passed on each side of a cylindrical post magnetron in systems that operate continuously or semicontinuously. However, the most common application is for batch processing, with the substrate arranged surrounding the source as shown in Figure 5.23. Post cathodes have been used to coat the insides of tubes up to six feet long.[106]

Cylindrical magnetrons can also be configured in the inverted or hollow cathode form shown in Figure 5.24.[97,99,107] Long hollow cathodes have the property that the coating flux at all points within the cathode is about equal to the erosion flux at the wall. This makes hollow cathodes particularly effective for coating objects of complex shape.[108]

198 Deposition Technologies for Films and Coatings

Figure 5.22: Large cylindrical-post magnetron sputtering source. Length 2.1 m (7 ft). See Reference 105. (Used courtesy Telic Corporation.)

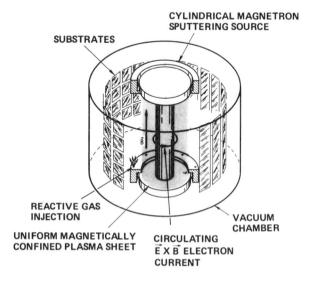

Figure 5.23: Typical arrangement of substrates for batch processing with a cylindrical-post magnetron sputtering source. (Used courtesy Telic Corporation.)

Coating Deposition by Sputtering   199

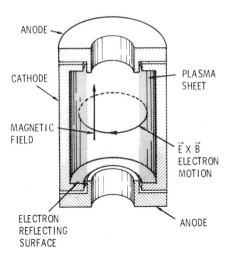

Figure 5.24: Cylindrical-hollow magnetron sputtering source with electrostatic end confinement. From Reference 95.

Cylindrical magnetrons can also be designed in an arrangement whereby the magnetic field lines are bent so that they intersect with the cathode barrel, as shown in Figure 5.25. The annular cross section of the electron trap is now closed on three sides by the magnetic field and on the fourth side by a surface at cathode potential. The plasma has the form of a ring rather than a sheet. Therefore such systems are generally configured with several such electron traps along the cathode cylinder.[99,100,109] Magnetrons of the type shown in Figure 5.25 are often referred to as having "magnetic end confinement," as opposed to those in Figures 5.19 and 5.24 which are referred to as having "electrostatic end confinement."[96,97] Magnetic end confinement devices have also been operated in the inverted or hollow cathode form.[109]

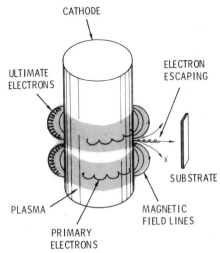

Figure 5.25: Cylindrical magnetron with magnetic end confinement. From Reference 95.

Plasma rings can also be confined over planar surfaces or within cylindrical surface cavities. This is the basis of the very important planar magnetron[110,111] and sputter or S-gun[112] configurations shown in Figure 5.18. At the present time these devices are the most widely used form of magnetrons. Like cylindrical magnetrons, planar magnetrons are attractive because of their ability to be scaled to large sizes. Elongated planar magnetrons of the type shown in Figure 5.18a are particularly useful for coating large substrate surfaces that are transported in a direction perpendicular to the long axis of the cathode. In this arrangement, close cathode-substrate spacing and deposition rates of 10,000 Å/min or more may be used. Proper cathode design, aided by minimal aperturing, can provide deposition uniformities of better than ±5%.[113] Figure 5.26 shows a schematic drawing of an in-line system with vacuum interlocks which is typical of the apparatuses that are used to achieve high production volumes with planar magnetrons. Large systems of this type are used to coat architectural glass panels several square meters in size at production volumes of 10^6 m²/yr. The planar magnetrons are typically 2 to 3 m long and are driven by currents in the 100 to 200 Å range. It is reported that cathodes as long as 6.5 m (20 ft) have been considered for architectural glass coating.[114] Large circular planar magnetrons with arrays of concentric plasma rings 0.6 m (24 in) in diameter have also been reported.[115] Patents have been granted for various configurations of the sputtering sources shown in Figure 5.18.[116,117-117c]

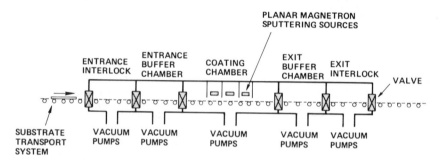

Figure 5.26: Schematic illustration of in-line vacuum system with vacuum interlocks which is typical of the type used to achieve high production volumes with planar magnetrons. (Used courtesy Telic Corporation.)

Gun type magnetrons do not have the scaling capabilities of the other forms of magnetrons. However, arrays can be used when large areas are to be coated. The system shown in Figure 5.4 uses an array of twenty-four S-gun sources in an in-line configuration of the type shown in Figure 5.26. Thus, in the case of sputter guns and planar magnetrons, the substrates are generally transported relative to the sputtering source. In the cylindrical magnetron case the substrates generally surround the source at radii where the deposition rates are 10x lower and the deposition areas are proportionately larger than in the planar cases.

Magnetron sputtering sources can be used to sputter magnetic as well as nonmagnetic materials. However, when a sputtering target composed of a magnetic material is used, it must be saturated magnetically so that its magnetic behavior is suppressed and a field of the desired shape can be maintained over its surface.[97]

dc magnetrons are generally operated at discharge currents in the 1 to 50 Å range. Current densities are generally higher in the plasma ring devices (Figures 5.18 and 5.25) than in the plasma sheet devices (Figures 5.19 and 5.24). Total coating material fluxes (which are dependent on the total discharge current) can be comparable at high currents to those often used with evaporation systems.

The current-voltage characteristic reveals a great deal about the ionization processes in a plasma discharge. The more efficient the discharge, the lower the voltage for a given cathode current density. Discharges operating in the magnetron mode obey an I-V relationship of the form $I \propto V^n$, where n is an index to the performance of the electron trap and is typically in the range 5 to 9. Typical I-V curves for various types of magnetron sputtering sources are shown in Figure 5.27 and compared with an I-V curve for a planar diode.

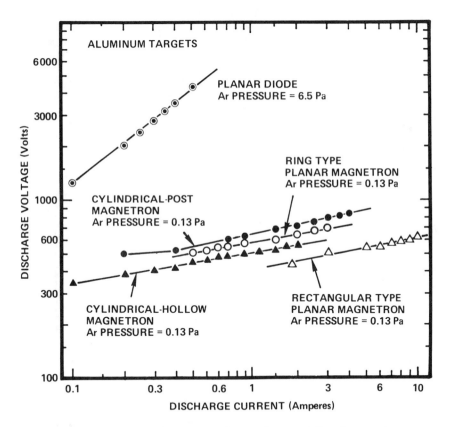

Figure 5.27: Typical current-voltage characteristics for a planar diode sputtering source and for various types of planar and cylindrical magnetrons. All sources operating with aluminum targets at the Ar working gas pressures indicated. (Used courtesy Telic Corporation.)

A basic disadvantage of the plasma ring devices of the types shown in Figures 5.18 and 5.25 is that sputtering occurs only under the plasma rings. The troughs are eroded in the cathode, and the source material is used relatively inefficiently. Sometimes relative motion between the cathode, and the magnetic field pattern is provided to improve the usage.[110] A further disadvantage of ring devices is that the complexity of the magnetic field shape makes effective anode placement more difficult than in the cylindrical cases shown in Figures 5.19 and 5.24.[97,110] Some field lines will intersect the substrates, as shown in Figure 5.25, thereby allowing electrons to escape and bombard the substrates. However, this bombardment is much less than in planar diode sputtering sources. All plasma ring magnetrons offer the advantage that the required magnetic field can be produced by permanent magnets located within the cathode rather than by magnetic field coils located at or beyond the chamber walls, as is required for the configuration shown in Figures 5.19 and 5.24. (See Figure 5.20.)

Cylindrical magnetrons with electrostatic end-confinement (Figures 5.19 and 5.24) have the advantage that they can be configured with large-diameter cathodes which provide a large inventory of coating material. Furthermore, the material is used very efficiently because of the uniform sputter erosion along the cathode length. However, a disadvantage with all cylindrical sources is that target fabrication from complex materials may be difficult.

At typical operating pressures (~1 mTorr) the sputtered flux from magnetron sources passes to the substrate with little gas scattering (see Figure 5.16). Thus the deposition flux can be predicted with considerable accuracy by assuming a cosine emission of sputtered material from the erosion area and collisionless passage to the substrates.[97,110] Figure 5.28 shows calculated and experimental profiles for the 2.1 m long cylindrical post magnetron shown in Figure 5.22.[105] Figures 5.29 and 5.30 show typical deposition flux profiles for planar magnetrons of the ring and rectangular types.

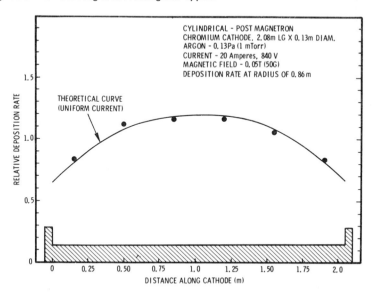

Figure 5.28: Comparison of experimental deposition profile with calculated profile for long (2.1 m) cylindrical-post magnetron with electrostatic end confinement. Profile measured parallel to cathode axis at radius of 0.86 m. Data from Reference 105. (Used courtesy Telic Corporation.)

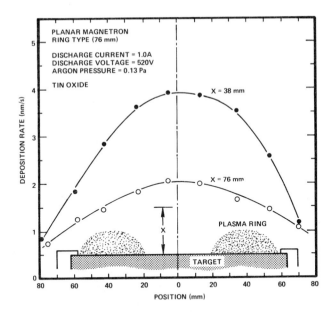

Figure 5.29: Deposition rate profile for ring type planar magnetron sputtering source at various distances from the cathode surface. (Used courtesy Telic Corporation.)

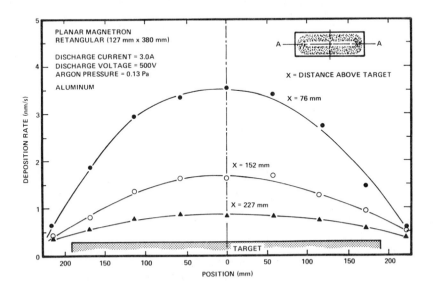

Figure 5.30: Deposition rate profile for rectangular type planar magnetron sputtering source on the long axis (A-A) at various distances from the cathode surface. (Used courtesy Telic Corporation.)

RF Sputtering

dc methods cannot be used to sputter nonconducting targets because of charge accumulation on the target surface. This difficulty can be overcome by using radio frequency (rf) sputtering.[39a,47,74,118-121] A single rf sputtering apparatus can be used to deposit conducting, semiconducting, and insulating coatings. Consequently, rf sputtering has found wide application in the electronics industry. Examples of nonconducting and semiconducting materials which have been deposited by rf sputtering include elemental semiconductors (Si^{122} and Ge^{123}), II-V compounds ($GaAs$,[124] $GaSb$,[125] GaN,[126] $InSb$,[125] InN^{126} and AlN^{127}), II-VI compounds ($CdSe^{128}$ and CdS^{75}), IV-VI compounds ($PbTe^{129}$), refractory semiconductors (SiC^{130}), ferroelectric compounds ($Bi_4Ti_3O_{12}^{131}$), oxides (In_2O_3,[132] SiO_2,[3,133] Al_2O_3,[134,135] Ta_2O_5,[136] Y_2O_3,[137] TiO_2,[138] ZrO_2,[139] SnO_2,[140] PtO,[141] HfO_2,[139] $La_2O_3^{139}$ Bi_2O_3,[142] ZnO,[143] CdO,[144]), pyrex glass,[145] and plastics.[2,146] Often several targets are placed within a common vacuum enclosure so that multilayer coatings can be deposited without breaking vacuum.

The usefulness of rf methods for sputtering nonconducting materials is based on the fact that a self-bias voltage that is negative relative to the plasma floating potential develops on any surface that is capacitively coupled to a glow discharge.[147] The basis for this potential, which forms as a consequence of the differences in mobility between the electrons and ions, is shown in Figure 5.31. The current-voltage characteristic for an electrode immersed in a plasma is shown in Figure 5.31a. The floating potential is negative relative to the plasma potential by an amount that depends on the gas species and plasma electron energy distribution function, but is typically -20 to -50 V and therefore too low to produce sputtering of most materials. When an alternating voltage is applied to such an electrode, more electron current flows when the electrode is positive relative to the floating potential than does ion current when the electrode is negative relative to the floating potential. This behavior is illustrated in Figure 5.31a. Capacitive coupling requires that there be no dc current flow; i.e., that the net current to the electrode in each rf cycle must be zero. Accordingly, a negative bias must form such that the electron current on the positive side of the cycle becomes equal to the ion current on the negative side, shown in Figure 5.31b. The negative bias is approximately equal to the zero-to-peak voltage of the rf signal and therefore can be made large enough to produce sputtering.

The behavior shown in Figure 5.31 applies strictly to the case where the electrode is passive; i.e., is not responsible for sustaining the plasma discharge. The planar diode shown in Figure 5.14 is the most commonly used apparatus for rf sputtering. The electrodes sustain the discharge and therefore have slightly different current-voltage characteristics than the one shown in Figure 5.31, particularly at negative voltages. However, the overall effect when an rf potential is superimposed on the IV characteristic is essentially identical to that shown in Figure 5.31.

Figure 5.32 shows a schematic drawing of a typical rf planar diode sputtering configuration in which a nonconducting target is placed over one electrode and substrates are placed on the other one. The electrodes reverse cathode-anode roles on each half cycle. The discharge is operated at a frequency that is sufficiently high so that significant ion charge accumulation does not occur during the cycle time when an electrode is serving as a cathode.[35] Frequencies in low MHz range are required. Most apparatuses are operated at a frequency of 13,560 MHz, since this is the frequency in the 10 to 20 MHz range that has been allocated by the Federal Communications Commission for industrial-scientific-medical purposes. Operation at another frequency will require careful shielding to assure compliance with FCC regulations on radio interference.

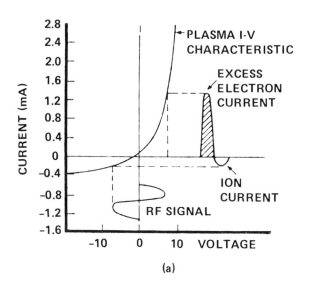

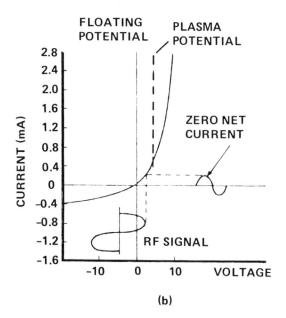

Figure 5.31: Schematic illustration of the development of a negative bias when an rf potential is capacitively coupled to a probe immersed in a plasma (after Butler and Kino, Reference 147).

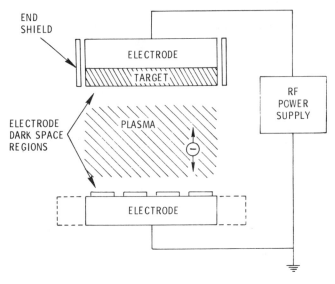

Figure 5.32: Schematic drawing of planar rf diode sputtering device. From Reference 37.

At MHz operating frequencies the massive ions cannot follow the temporal variations in the applied potential. However, the electrons can. Thus the cloud of electrons that constitute the electron component of the negative glow plasma can be pictured as moving back and forth at the applied frequency in a sea of relatively stationary ions. As the electron cloud approaches one electrode, it uncovers ions at the other electrode to form a positive ion sheath. This sheath takes up nearly the entire applied voltage, the same as in the dc case. The nonconducting target electrode constitutes a capacitor in the electrical circuit between the electrodes (an external capacitor would have the same effect). Thus there can be no dc component to the current flow. The total ion and electron charge flow to a given electrode during an rf cycle must balance to zero, as discussed previously. However, a large electron current flows to a given electrode as the electron cloud makes contact. Thus the electron cloud need approach a given electrode for only a small fraction of a half cycle for purposes of supplying sufficient electrons to fulfill the anode requirement; i.e., to balance the entire ion flux through the cycle. Accordingly, in the steady state both electrodes develop a negative dc bias relative to the plasma potential, such that the electrodes approach or exceed the plasma potential (and become anodes) only for very short portions of their rf cycle, as previously shown in Figure 5.31 and indicated in Figure 5.33. Because of their inertia, the motion of the ions can be approximated as if they follow the dc potential and pass to both electrodes throughout the cycle. The electron cloud spends most of its time near the center position between the electrodes. Visually, the discharge appears as a dc discharge with a cathode dark space over each electrode. Functionally, sputtering occurs continually off both electrodes, very much as in the dc case.

rf discharges in planar diode apparatuses can be operated at considerably lower pressures than can dc discharges. Typical operating pressures are 5 to 15 mTorr. This is believed to be caused by a reduction in the loss of primary electrons and, at high frequencies, by an increase in the volume ionization caused by the oscillating electrons. It was noted in the discussion of the dc discharge that many primary electrons are lost at the anode, as indicated in Figure 5.14, before they have used their full complement of energy in making ions. In the rf case both electrodes are at a negative potential relative to the plasma most of the time. Thus many of the primary electrons will be reflected back and forth between the electrodes (hollow cathode effect[85]) until they use their energy effectively in making ions.

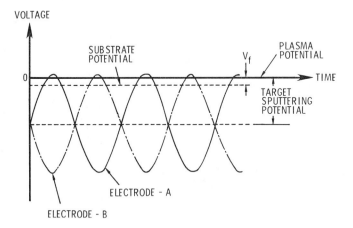

Figure 5.33: Approximate representation of target voltage waveforms (as function of time) relative to plasma potential for balanced rf system with two equal-area sputtering electrodes. V_f is the floating potential. See Figure 5.27. (Used courtesy Telic Corporation.)

The versatility of rf sputtering is not achieved without drawbacks. Implementation of the process is quite complicated.[74] It has all of the complexities of dc sputtering plus more, because of the complexity of the rf power source. A typical electrical circuit is shown schematically in Figure 5.34. It consists of an rf power supply, an inductive coupling to the load, and a matching network.

An equivalent circuit for the glow discharge is also shown in Figure 5.34. The equivalent circuit assumes that both electrodes and the chamber walls are in contact with the plasma, and that the impedance is dominated by that of the respective plasma sheaths. The sheath capacitances result from the charge separation across the dark space. These capacitors are shunted to the electrode surface by a resistor to account for the ion current, and by a diode to account for the high electron current that can flow from the plasma to an electrode that is biased positive relative to the plasma potential. The capacitor C_t accounts for capacitance of the target. C_b is a "blocking capacitor" that is added to make the system independent of variations in the target capacitance. The rf current through the plasma is principally an electron current caused by the relative motion of the electron cloud. To the extent that there is no volume power transfer from the oscillating electrons to the gas, this current is out of phase with the applied voltage. The primary power transfer occurs via the relatively small ion and electron current components that are in phase with the voltage. Thus in the equivalent circuit approximation the power transfer to produce the sputtering occurs as the ion currents pass through the sheath resistances. Efficient power transfer requires that the rf power supply operate into a resistive load.

Thus the purpose of the matching network is to introduce inductance, and often capacitance, into the circuit in such a way that, in combination with the load, they form a resonant circuit.[148] When the variable matching network components are tuned to resonance, high circulating currents flow within the resonant circuit. However, the power supply sees only the resistive component of the load, and the current passing from the power supply to the resonant circuit is in phase with the load and represents the power passing to the load. Many commercial sputtering sources monitor the "reflected power" from the load as an index of how effectively the matching network is adjusted. The reflected power should be minimized.[47,120,149]

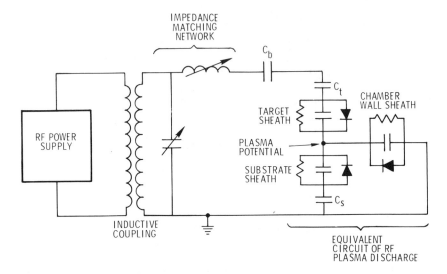

Figure 5.34: Schematic circuit for single-ended rf discharge system showing equivalent circuit for plasma discharge. See References 121 and 150. (Used courtesy Telic Corporation.)

The ion current, and thus the sputtering rate at a given electrode, is determined by the average difference in potential between the electrode and the plasma. Thus it is useful to consider the plasma potential as a zero-point reference voltage in examining the performance of various sputtering systems.

The electrical character of rf sputtering systems can be classified in general as being either balanced or single-ended. In the balanced system both electrodes are configured as identical sputtering targets. Their potentials are 180° out of phase and rise toward or exceed the plasma potential for a short portion of each cycle, during which the anode electron current flow occurs. The average sputtering voltage is about equal to zero-to-peak applied rf potential, as shown in Figure 5.33. The link center tap is placed at ground potential to stabilize the system. The chamber walls and substrates are connected to the center tap ground, as shown in Figure 5.35. Since this point is at zero potential relative to the rf voltage, no rf current will flow to these elements. Furthermore, because of the capacitance in series with each of the electrodes, there is no dc current path from the plasma to the wall and substrates and then back through the electrodes into the plasma.[150] Thus an asymmetrical charge will develop on the capacitors such that the substrates float at a potential slightly negative with respect to the plasma, just like a floating electrode in a dc plasma. See Figure 5.33.

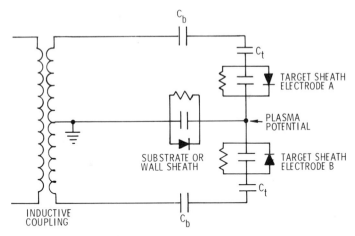

Figure 5.35: Schematic representation of equivalent circuit for balanced rf system with two equal-area sputtering electrodes and center-tap ground. Matching network not shown. (Used courtesy Telic Corporation.)

In the unbalanced system the substrates are placed on one electrode as shown in Figure 5.32. This electrode is made considerably larger than the target electrode.[121] This makes the sheath capacitance large, and the rf voltage drop across the substrate electrode small, as shown in Figure 5.36. The chamber and one side of the link are generally grounded (Figure 5.34). Again the capacitance in both electrode circuits prevents a dc current flow to the chamber, and a negative bias develops relative to the plasma potential. A slight rf potential relative to the plasma potential will exist on the substrates and chamber wall, as shown in Figure 5.36, unless the elective electrode area is large enough to reduce this potential to essentially zero and to move the rf balance point to the grounded end of the link. It is important that these voltage drops be small so that sputtering from uncontrolled surfaces does not introduce contamination into the coatings. An impedance may be added to the substrate electrode circuit so that the potential of this electrode relative to the plasma can be controlled for purposes of bias sputtering.[150]

The above discussion has been presented in the context of planar diode sputtering systems. Magnetron sputtering sources can also be used for rf sputtering. Cylindrical-post,[97,99] cylindrical-hollow,[99,151] and planar magnetrons,[110,151a] as well as gun type magnetrons,[112] have been successfully operated with rf power. However, some problems are encountered. Magnetron sputtering technology is basically a dc concept. The cathodes are shaped so that, in concert with the magnetic field, they form electron traps with specific symmetry. Anodes are placed to collect electrons which diffuse out of the trap. Effective double-ended rf magnetrons can be provided for some geometries. These configurations provide independent traps for both electrodes but allow magnetic coupling between them so that the electrons leaving one trap can diffuse freely to the vicinity of the other.[99] However, most magnetron configurations must be operated with single-ended arrangements. The magnetic confinement produces gradients in the plasma density, so that special care is required to minimize the voltage and therefore the sputtering at the counter electrode. Furthermore, in the planar magnetron case the current density concentration under the plasma ring requires that the power level be limited to avoid cracking when using targets with low thermal conductivities. When magnetron sources are driven single-ended they generally operate in hybrid modes with current-voltage characteristics which are not symptomatic of true magnetron behavior. They yield deposition rates that are typically a factor of three greater than are achieved with rf planar diodes. (This is to be compared to the factor of twenty-to-thirty improvement in deposition rate which dc magnetrons provide over dc diodes when sputtering metals.) Reduced substrate electron bombardment and heating are other advantages of magnetrons, as opposed to planar diodes, for rf sputtering.

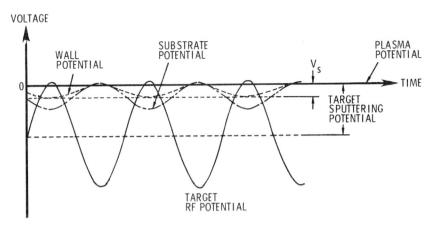

Figure 5.36: Approximate representation of voltages (as function of time) relative to plasma potential for single-ended rf sputtering system where wall area is much larger than target area. V_s is substrate ion bombardment potential. See Figure 5.26 and References 121 and 150. (Used courtesy Telic Corporation.)

Ion Beam Sputtering

Glow discharge sputtering technology is limited, in the sense that the target current density and voltage cannot be independently controlled except by varying the working gas pressure. An exception is the cylindrical magnetrons, where the voltage can be varied at a fixed current and pressure by varying the magnetic field strength.

Ion beam sputtering permits independent control over the energy and the current density of the bombarding ions.[152] A sputtering target is arranged to obliquely intersect an ion beam of given energy and flux density that is created by an independent ion source. Substrates are suitably placed to receive the coating flux, as shown in Figure 5.37. In addition to the independent control over the ion current and voltage, ion beam sources permit sputtered coatings to be deposited at very low inert working gas pressures (~0.1 mTorr) onto substrates that are not in contact with a plasma.

Early ion sources were of the duoplasmatron type, where an ion beam was extracted through an aperture from a low pressure arc.[153] Hollow cathode ion sources were also used.[154] These devices were limited for practical deposition because of the small ion beam sizes (~1 cm). The recent adaptation of space-type ion engine technology has provided distributed ion sources with ion beams of relatively large diameter (~10 cm).[152] Although these devices cannot compete as deposition sources with the very large substrate areas that are provided by magnetrons, they are attractive for ion beam etching and for special deposition applications and research studies. Sources designed for etching have provided 500 eV, 1 mA/cm^2 ion beams that are 30 cm in diameter with a uniformity of ±5% over 20 cm.[155] Ion beam deposition has led to unique film properties not obtained in conventional deposition methods. An excellent review of the fundamentals of ion beam deposition is provided in Reference 155a. Some of the structure property relationships for films deposited by ion beam techniques are reviewed in Reference 155b.

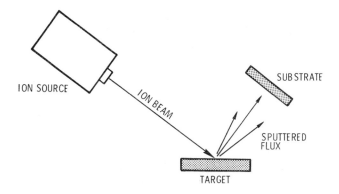

Figure 5.37: Schematic representation of ion beam sputtering source showing relative locations of target and substrate.

4. COATING GROWTH, STRUCTURE, AND PROPERTIES

Substrate Environment

In sputter deposition the substrate environment in which the coating nucleation and growth processes occur is very complex and depends greatly on the apparatus configuration.

In the case of planar diode devices the substrates are generally in contact with the plasma. Thus, even if the substrates are electrically isolated, they will be bombarded with plasma ions having kinetic energies of 5 to 30 eV and internal energies of about 10 eV as well as metastable atoms having similar internal energies. The substrates are also subjected to bombardment by energetic (100 to 1000 eV) primary electrons from the cathode.[156] These electrons are particularly important. They are the primary source of substrate heating in such devices and can cause radiation damage in sensitive substrates. Although the sputtered atoms are ejected with high energies, the high working gas pressure in planar diodes causes them to lose energy so that they are in near thermal equilibrium with the working gas by the time they reach the substrates.[38] See Figure 5.16.

It is useful to quantify the substrate heating in terms of the energy delivered to a surface per atom deposited. With this information it is possible to estimate the temperature that a given substrate will reach while depositing a particular coating. Typical heating rates for planar diodes are in the range 100 to 300 eV/atom.[157,157a] In most cases radiation is the primary cooling mechanism, even when the substrates are mechanically clamped to a cooled support. Thus even at moderate deposition rates (several hundred Å/min) substrate temperatures can reach 300 to 500°C.

In triode devices the substrates may or may not be in direct contact with the plasma. However, they will often be subjected to bombardment by energetic electrons originating from the target. Triodes operate at sufficiently low pressures so that the sputtered atoms may preserve much of their initial kinetic energy in passing to the substrates. Furthermore, energetic neutral working gas atoms, which are generated at the cathode by the neutralization and reflection of ions as discussed in Section 2, can also pass to the substrates with little loss of kinetic energy. These atoms can become trapped in the growing coating, particularly if the substrate temperature is low.[49-52] Thus the trapped argon content for coatings deposited with a triode has been found to be larger than that for similar coatings deposited with a planar diode.[51,52]

In the case of magnetrons the substrates are generally located beyond the magnetically confined plasma, although there may be some plasma bombardment in the hollow cathode magnetrons.[48,97] The substrates in cylindrical-post magnetrons are particularly free from plasma or electron influence.[97] Removal of energetic electron bombardment is one of the important advantages of these devices. Operating pressures are generally sufficiently low so that the sputtered atoms pass to the substrates with little loss of kinetic energy. The reflection of energetic neutral atoms is believed to be particularly important in cylindrical-post magnetrons because atoms undergoing reflections of considerably less than 180°, and therefore having considerable kinetic energy, pass in the direction of substrates.[97] The argon content of sputtered metal coatings has been found to increase with the atomic mass of the coating material and therefore with the probability of reflection at the cathode surface.[158]

Extensive substrate heating measurements for dc sputtering using cylindrical-post magnetrons have shown the heating rate to be largely independent of substrate radial position, deposition rate, or argon pressure (over the 1 to 10 mTorr range examined).[48] The heating rates per atom deposited could be consistently interpreted in terms of four heating processes that are fundamental to glow discharge sputtering: (1) heat of condensation, (2) sputtered atom kinetic energy, (3) plasma radiation, and (4) ion neutralization and reflection at the cathode. The ion reflection contribution became increasingly important for high mass materials. The energy per atom deposited varied from 15 to 25 eV/atom for light metals to over 50 eV/atom for heavy metals with moderate sputtering yields. Substrates in radiation equilibrium with the heating flux associated with a deposition rate of 1000 Å/min are projected to reach temperatures in the range of 100 to 200°C.[95]

Coating Nucleation and Growth

The coating atom condensation process can be pictured as occurring in three steps. First, incident atoms transfer kinetic energy to the substrate lattice and become loosely bonded "adatoms." This energy transfer is efficient, even for energetic sputtered atoms.[159] Second, the adatoms diffuse over the surface, exchanging energy with the lattice and other adsorbed species, until they either are desorbed, by evaporation or back-sputtering, or become trapped at low-energy sites. Once the coating becomes continuous, growth proceeds similarly, by adatoms arriving and diffusing over the coating surface until they are desorbed or become trapped in low energy sites. Finally, in the third step the incorporated atoms readjust their positions within the lattice by bulk diffusion processes.

In the case of vacuum evaporation, islands of coating material are well known to collect at the low-energy sites during the initial nucleation and finally to grow together to form a continuous coating. The same basic picture applies to the sputtering case, although much less is known about the details.[38] The complications arise because of the relatively high kinetic energies of the condensating atoms and the fact that the condensation surfaces may be subjected to a concurrent bombardment by energetic plasma species. Ion bombardment may be expected to increase the nucleation density by causing surface damage. It is also expected to enhance the mobility of adatoms by transferring kinetic energy to them. Considerable evidence is accumulating which indicates that the latter is particularly important and results in an acceleration of the nucleation, growth, and coalescence of the nuclei.[160a] Ion bombardment has been reported to have made the surface diffusion coefficients more than five orders of magnitude greater than the thermal values during the sputter-deposition of InSb/GaSb superlattice structures.[160b] Low energy (0-250 eV) ion bombardment during film growth has also been found to increase the growth temperature range over which stoichiometric films of compound semiconductors such as GaSb and InSb can be grown.[160c]

In addition to the above, ion bombardment has been found to have a large influence on the interface structure and thus on the adhesion of coatings.[161,162] The mechanism may be an increase in the density of nucleation sites, and therefore a reduction in the amount of agglomeration and void formation that occurs during the initial island growth.[162]

The nature of the subsequent growth process is such that the structure, and therefore the properties, of vacuum-deposited coatings in general and sputtered coatings in particular, are determined largely by the selection processes that are made evolutionary by the way the state of the coating surface changes as the coating grows.[163] The three process steps described above can be quantified in terms of the characteristic roughness of the substrate or coating surface, the activation energies for surface and bulk diffusion, and the adatom surface bonding energy.

Any process that causes a systematic nonuniformity in the arriving coating atom flux over the substrate surface can have drastic effect on the evolutionary growth process. Shadowing (a simple geometric interaction between the roughness of the growing surface and the line-of-sight directions of the arriving coating atoms) provides such an effect.[164] For many pure metals the adatom binding energies and the activation energies for both surface and bulk diffusion are related and proportional to the melting point.[165] Furthermore, the shadowing effect can be compensated by surface diffusion; i.e., if the surface diffusion is large enough, the point of arrival of a coating atom loses its significance. Thus various of the basic processes, including shadowing, can be expected to dominate over different ranges of T/T_m and to manifest themselves as differences in the resulting coating structures (where T is the substrate temperature and T_m is the coating material melting point). Such is the basis for the structure zone models described in the next section.

Structure Zone Models

Movchan and Demchishin studied the structures of thick evaporated coatings of Ti, Ni, W, ZrO_2 and Al_2O_3.[166] They concluded that the coatings could be represented as a function of T/T_m in terms of three zones, each with its own characteristic structure and physical properties. The model has been extended to sputtering in the absence of ion bombardment by adding an additional axis to account for the effect of the sputtering gas.[163,167] See Figure 5.38. The numbered zones correspond to those suggested by Movchan and Demchishin. In addition a transition zone (Zone T), not specially reported by Movchan and Demchishin, was identified between Zones 1 and 2.

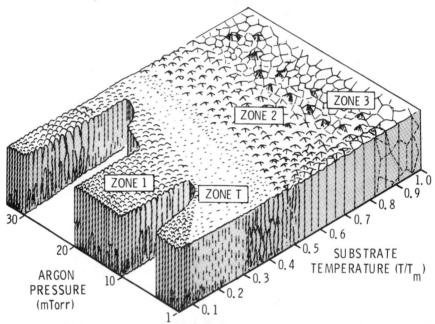

Figure 5.38: Schematic representation of the influence of substrate temperature and argon working pressure on the structure of metal coatings deposited by sputtering using cylindrical magnetron sources. T is the substrate temperature and T_m is the melting point of the coating material in absolute degrees. See References 163 and 167.

The Zone 1 structure results when adatom diffusion is insufficient to overcome the effects of shadowing. It therefore forms at low T/T_m and is promoted by an elevated working gas pressure. It usually consists of tapered crystals with domed tops which are separated by voided boundaries. The crystal diameter increases with T/T_m. Shadowing introduces open boundaries because high points on the growing surface receive more coating flux than valleys do, particularly when a significant oblique flux is present. The coating surface roughness can result from the shapes of initial growth nuclei, from preferential nucleation at substrate inhomogeneities, from substrate roughness (most common), and from preferential growth. The Zone 1 structure is promoted by hollow cathode sources and by substrate rotation because of the increased oblique component in the coating flux.[108,167] It can occur in amorphous as well as crystalline deposits.[164,168] The internal structure of the crystals is poorly defined, with a high dislocation density.[166]

The Zone T (transition) structure consists of a dense array of poorly defined fibrous grains without grossly voided boundaries. It has been defined as the limiting form of the Zone 1 structure at zero T/T_m on infinitely smooth substrates.[163] It is a columnar Zone 1 structure with crystal sizes that are small and difficult to resolve and appear fibrous, and with boundaries that are sufficiently dense to yield respectable mechanical properties. It forms the internal structure of the Zone 1 crystallites. Coatings with structures approaching the Zone 1 form grow under normal incidence on relatively smooth homogeneous substrates, at T/T_m values that permit the adatom diffusion to largely overcome the roughness introduced by the substrate and the initial nucleation.[163,169] Zone T is therefore viewed as a transition structure. Several examples of Zone 1 and Zone T structures are shown in Figures 5.39 and 5.40a. Recent experimental work, in which Zone T coatings of Ge and Si are being investigated by etching and electron microscopy, is providing an improved understanding of the Zone T microstructures and intergrain boundaries.[169a–169d]

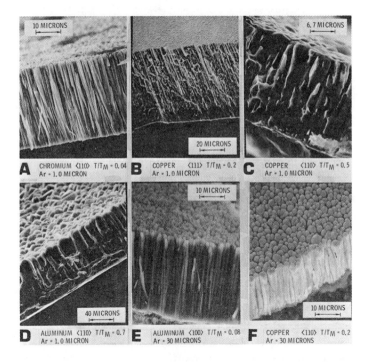

Figure 5.39: Fracture cross sections of metal coatings deposited using cylindrical-hollow magnetrons at various substrate temperatures and argon pressures: (A), (E), and (F) have Zone 1 structure; (B) has Zone T structure; (C) and (D) have Zone 2 structure. From Reference 167. (Used courtesy American Institute of Physics.)

It should be noted that a columnar structure is a fundamental consequence of forming a solid from a vapor flux which arrives from one direction and con-

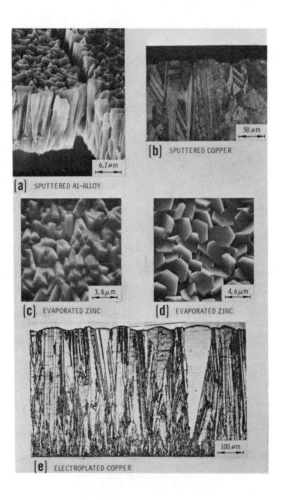

Figure 5.40: Examples of growth structures: (a) Aluminum sputtered onto glass at $T/T_m \sim 0.08$ and argon pressure of 30 mTorr, Zone 1 structure; (b) Copper sputtered onto stainless steel at $T/T_m \sim 0.7$ and argon pressure of 0.13 mTorr, Zone 2 structure; (c) Zinc evaporated onto steel at $T/T_m \sim 0.05$; (d) Zinc evaporated onto steel at $T/T_m \sim 0.6$; (e) Copper electroplated in sulfate-bath—note similarity to (b). From Reference 163; (c) and (d) courtesy M. Neirynck, (e) courtesy V.A. Lamb. Reproduced with permission from the *Annual Review of Materials Science*, Volume 7, copyright 1977, Annual Reviews, Inc.

denses with little adatom movement.[163,164] Thus computer simulations, in which hard spheres were deposited on smooth substrates at various angles of incidence and zero adatom mobility, yielded amorphous structures with columnar voids that could be interpreted as resulting from the "self-shadowing" of the spheres.[170] Larger voids formed as the angle of incidence was increased (similar to an increase in substrate roughness). At normal incidence, particularly when a slight degree of adatom mobility was programmed,[171] the coatings were sufficiently dense so that it was difficult to identify the columnar growth visually from computer displays which gave two-dimensional maps of the atom positions in the condensate. However, an examination of the distribution of directions for contacting pairs clearly showed that the columnar growth had persisted (Zone T structure). High temperature deposition, or deposition of materials with large mobility, leads to the elimination of the voided structure[164] (formation of Zone 2 structure).

The Zone 2 region is defined as that range of T/T_m where the growth process is dominated by adatom surface diffusion.[163] The structure consists of columnar grains separated by distinct dense intercrystalline boundaries. Dislocations are primarily in the boundary regions.[166] Grain sizes increase with T/T_m and may extend through the coating thickness at high T/T_m. Surfaces tend to be faceted. Figure 5.40b shows the cross section of a thick sputtered coating with a Zone 2 microstructure. Coatings with platelet structures and whisker growth can also grow under Zone 2 conditions of high surface diffusion (See Figures 5.40c and 5.40d.)

The Zone 3 region is defined as that range of T/T_m where bulk diffusion has a dominant influence on the final structure of the coating.[163] Movchan and Demchishin suggest a Zone 3 structure consisting of equiaxed grains at T/T_m greater than about 0.5 for pure metals.[166] However, the T/T_m value at which recrystallization occurs is dependent on the stored strain energy.[172] It occurs typically at T/T_m above about 0.33 for bulk materials, but has been observed at room temperature in sputtered Cu coatings that were deposited on a cold substrate with concurrent ion bombardment.[173] It is believed that the high temperature recrystallized structure in coatings need not be equiaxed.[163] Sputtered coatings will generally be deposited with a columnar morphology. Recrystallization to equiaxed grains may be expected if points of high lattice strain energy are generated throughout the coating during deposition. In contrast, large columnar grains may grow from columnar as-deposited grains by strain-induced boundary motion and grain growth.[172]

Figure 5.41 shows several examples of Zone 3 structures with equiaxed (Figures 5.41b and 5.41c) and clearly columnar structures (Figure 5.41e). Heterogeneous recrystallization[172] at the coating-substrate interface is seen in Figure 5.41a. Note the epitaxial relationship between the recrystallized copper coating and the copper substrate.

Epitaxial growth is of particular importance for semiconductor applications. Epitaxial growth can be achieved with sputtering at elevated temperatures,[38,58,174] although much less work has been done by sputtering than by evaporation. The deposition by sputtering of semiconductor coatings in general, and epitaxial coatings in particular, is reviewed in Reference 39b. In many cases essentially bulk values of carrier concentrations and mobilities have been obtained. For example, multitarget sputtering has been used to grow single crystal semiinsulating GaAs films [on (100) GaAs substrates] which yield room temperature electron mobilities of 5000 cm^2/V-sec.[39b] An important aspect of this work is the use of low energy ion bombardment during deposition.

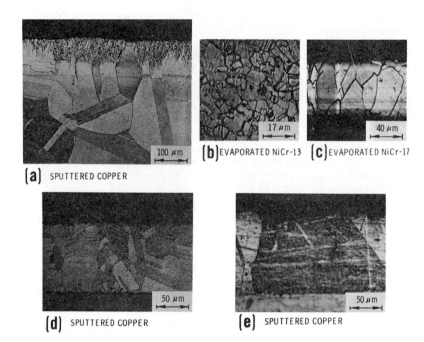

Figure 5.41: Cross sections of bulk diffusion structures: (a) Copper sputtered onto copper substrate at $T/T_m \sim 0.5$; (b) NiCr-13 evaporated at $T/T_m \sim 0.06$; (c) NiCr-17 evaporated at $T/T_m \sim 0.7$; (d) Copper sputtered onto tantalum substrate at $T/T_m \sim 0.8$; (e) Copper sputtered onto tantalum substrate at $T/T_m \sim 0.9$. From Reference 163; (a) courtesy E. McClanahan, (b) and (c) courtesy R.F. Bunshah. Reproduced with permission from the *Annual Review of Materials Science*, Volume 7, copyright 1977, Annual Reviews, Inc.

Substrate Temperature Influence on Composition

Even though all constituents of a given target material may be sputtered, the composition of the resulting deposit will depend on their relative sticking coefficients. Thus at elevated substrate temperatures sputtered coatings may be deficient in volatile constituents. The problem can be particularly troublesome in depositing compound semiconductors, where one species is often volatile and where an elevated substrate temperature is desired to promote a large grain size. This difficulty can be overcome in some cases by making the target rich in the volatile species and then adjusting the substrate temperature to control the composition; i.e., by a method analogous to the three-temperature method of evaporation.[175] Low energy ion bombardment can increase the temperature range over which stoichiometric films of compound semiconductors can be grown,[160c] as discussed previously.

It has been observed that an optimum occurs in the structure-sensitive properties (surface smoothness, crystallographic order, and charge mobility) for many compounds of electronic interest (ZnO, SnTe, and most of the sulfides, selenides, and tellurides of lead, cadmium, and zinc) at a substrate temperature that is within a few percent of $\frac{1}{3} T_b$ where T_b is the boiling point of the compound.[176] It has been estimated that $T/T_b \sim \frac{1}{3}$ is that point at which evaporation from amorphous regions of the film becomes significant.[176] Thus it has been proposed that film quality improves up to $\frac{1}{3} T_b$ because the existence of such amorphous regions decreases (for example, because of increasing grain size and reduced grain boundary area), and because slight evaporation from these regions creates vacancies and promotes crystallization and recrystallization.[176] At higher temperatures the more complete evaporation from the amorphous regions can make the films rough and voided. This, coupled with a loss of volatile species, leads to degraded properties. Thus an optimum structure occurs at $T/T_b \sim \frac{1}{3}$.

Substrate Roughness

The influence of substrate roughness on the Zone 1 structure has been discussed. The surface roughness of practical substrates often exists simultaneously on several size scales.[163] The resultant Zone 1 coating structure will then exhibit superimposed arrays of shadow growth boundaries, each associated with a size scale of substrate roughness. As T/T_m is increased, successively more severe shadow boundaries are overcome by diffusion.[169] Severe surface irregularities can lead to Zone 1 type boundaries that persist to the T/T_m ranges of Zones 2 and 3. Such boundaries have been referred to as columnar[177] and linear defects.[178]

Substrate surface irregularities such as inclusions or debris particles that fall on the substrate or growing coating can cause preferential nucleation and runaway growth.[179] These nodular or flake defects can form at both low and high T/T_m. Such a defect is shown in Figure 5.42.

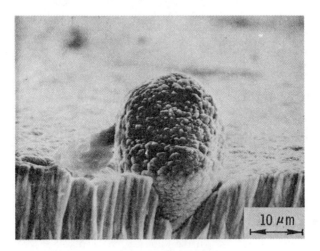

SPUTTERED CHROMIUM

Figure 5.42: Fracture cross section showing nodular growth flaw in Zone 1 sputtered chromium coating. From Reference 163. Reproduced by permission from the *Annual Review of Materials Science*, Volume 7, Copyright 1977, Annual Reviews, Inc.

As a general rule, if a critical coating is required for a high performance application, one of the single most effective steps that can be taken is to provide a smooth homogeneous substrate.

Oblique Deposition

Oblique deposition and substrate surface roughness are complementary in producing shadow-induced Zone 1 boundaries. Therefore it is difficult to achieve uniform dense microstructures in coatings deposited over complex shaped substrates by vacuum deposition at moderate T/T_m. Zone 1 crystals tend to point in the direction of the coating flux.[164,166,180] Machining roughness was found to cause open Zone 1 boundaries when a substrate was rotated.[177] Oblique deposition can be an important consideration when using a multiple array of sources to, for example, deposit a multilayer coating at a fixed substrate location. Similarly, hollow cathode sputtering is effective for coating all sides of a complex substrate without rotation. However, the oblique component within the hollow cathode flux tends to promote the Zone 1 structure.[108,167]

Inert Gas Effects

An elevated inert gas pressure at low T/T_m causes the Zone 1 boundaries to become more open, as indicated in Figure 5.38. This effect has been observed in coatings deposited by magnetron sputtering (no plasma bombardment) and by evaporation.[163] Planar diode sputtering sources must necessarily operate at high working gas pressures, as discussed in Section 2. However, the resulting coating structures are not so voided as would be implied by Figure 5.38, because the associated plasma bombardment tends to suppress the deleterious effects of the working gas (see next section).

The exact mechanism by which an inert gas promotes open boundaries is not known. The phenomenon may involve a reduction in adatom mobility.[163] Impurities such as oxygen are particularly effective in promoting open Zone 1 structures.[167] Gas scattering may also induce an increased oblique component in the coating flux by tending to randomize the coating atom directions at points close to the substrate surface.[158]

Ion Bombardment—Bias Sputtering

Intense substrate ion bombardment during deposition can suppress the development of open Zone 1 structures at low T/T_m. This has been demonstrated for both conducting[181,182] and nonconducting deposits.[183] The deposits have a structure similar to the Zone T type and appear to be free of nodular defects. Ion bombardment on uncooled substrates yields typical high T/T_m structures.[184]

The ion bombardment may suppress the Zone 1 structure by creating nucleation sites for arriving coating atoms, or by eroding surface roughness peaks and redistributing material into valleys. Ion bombardment has been found to be more effective at high than at low working gas pressures in suppressing the influence of a rough substrate.[184] Furthermore, it has been found that the ion flux must be of such magnitude and energy that it is capable of backsputtering a significant fraction of the arriving coating flux (i.e., 30-60%).[184] This is much greater than the normal plasma bombardment that occurs on an unbiased substrate in a planar diode source.

There are four disadvantages to the use of ion bombardment for suppressing Zone 1 structures. The first is the high ion fluxes required. This means that the current density passed through the substrates must be of the same order of magnitude as that which is passed through the sputtering target. Second, the substrates are often of an awkward shape, making it difficult to achieve a uniform density over the region being coated. Third, the ion bombardment may change the chemical content of the coating by preferential sputtering, as discussed in Section 2. Finally, inert gas ions with bombarding energies of more than 100 to 200 eV tend to become trapped in the growing coating.[49,181] At high ion energies the entrapped inert gas can have concentrations of a few atomic percent and may cause blistering in subsequent annealing.[184-186] Inert gas incorporation is generally greater in amorphous than in crystalline materials. Concentrations of Ne, Ar, and Kr of from a few to 30 atomic percent have been obtained in amorphous GdCo and GdCoMo films deposited with an rf substrate bias.[187]

Bias sputtering at low ion energies (<200 eV) is often used to improve the purity of growing coatings by removing loosely bonded impurity atoms.[189-191] The process is effective because such atoms have a high sputtering yield, as discussed in Section 2. Thus low energy ions preferentially sputter adsorbed gas.

The controlled use of low energy ion bombardment of a growing semiconductor film during sputter deposition has been found to affect all stages of crystal growth, ranging from adatom mobilities and nucleation kinetics to elemental incorporation probabilities,[39b] as discussed previously.

Internal Stresses

Virtually all sputtered coatings are in a state of stress. The total stress is composed of a thermal stress, due to the difference in the thermal expansion coefficients of the coating and substrate materials, and an intrinsic stress due to the accumulating effect of atomic forces generated throughout the coating volume by atoms which are out of position with respect to the minimums in the interatomic force fields.[192]

For low melting point materials, the deposition conditions will generally involve sufficiently high values of T/T_m so that the intrinsic stresses are significantly reduced by recovery during the coating growth. Thermal stresses are therefore of primary importance for such materials. Stress relief can occur as the coatings are brought to room temperature or annealed following deposition. The resultant material flow can lead to the formation of hillocks or holes, depending on whether the temperature change places the coating into compression or into tension. Thus extensive hillock growth has been reported in Al,[193] Au,[194] and Pb[195] films at elevated temperatures.

Higher melting point materials are generally deposited at sufficiently low T/T_m (<0.25) so that the intrinsic stresses dominate over the thermal stresses. For thin films (<5000 Å) the intrinsic stresses are generally constant throughout the coating thickness. They are typically tensile[58] for evaporated and often compressive for sputtered metal coatings.[196,197] Magnitudes can be near the yield strengths for the coating materials and can reach several hundred thousand psi (10^9 N/m^2) for refractory metals such as molybdenum and tungsten.

The interfacial bond must withstand the shear forces associated with the accumulated intrinsic stresses throughout the coating, as well as the thermal stresses. Since the intrinsic stress contribution increases with coating thickness, it can be the cause of premature interface cracking, and poor results in adhesion tests, for coatings with thicknesses exceeding critical values which may be as low as 1000 Å.

Only a relatively few quantitative determinations of stress in sputtered films have been reported.[38] However, an extensive series of measurements have been made of the internal stresses (primarily intrinsic) that develop in various metallic coatings deposited on glass substrates using cylindrical-post magnetron sputtering sources under various deposition conditions.[30a,158,198-201a] These measurements provide a poignant example of the way in which the deposition parameters can affect the properties of sputtered coatings in general and the internal stresses in particular.

The stresses were found to be particularly sensitive to the working gas species and pressure and to the apparatus geometry and angle of incidence of the coating flux relative to the substrate surface. Figure 5.43 shows the influence of the argon working gas pressure on the interface force per unit length (integrated stress) that developed in coatings, approximately 2000 Å thick, of several materials deposited at normal incidence on near-toom-temperature substrates. The general behavior has been observed for more than ten metals ranging in mass from Al to W, and for amorphous Si. Figure 5.44 shows the influence of the angle of incidence (measured from the normal) on the average stress in 2000 Å thick films of the same materials deposited at an argon pressure of 1 mTorr (0.13 Pa) and similar substrate temperatures. Figure 5.45 shows the influence of the working gas pressure on the internal stress for 1000 Å thick molybdenum films deposited using various inert working gases.

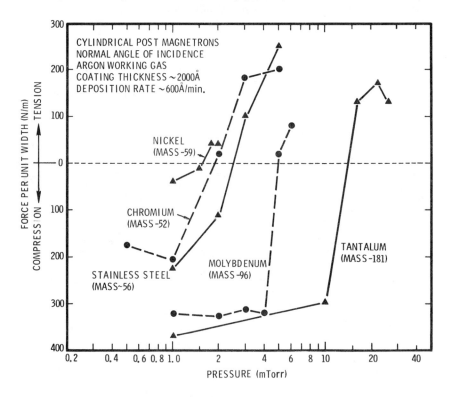

Figure 5.43: Force per unit width produced at substrate by interface by internal stress in coatings deposited at various argon pressures using cylindrical-post magnetron sputtering sources. (Used courtesy Telic Corporation.)

Coating Deposition by Sputtering 223

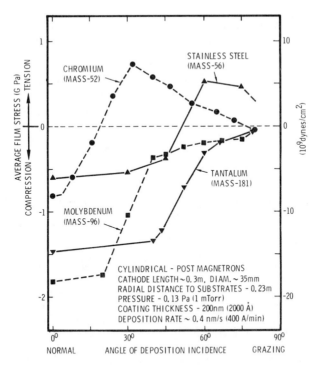

Figure 5.44: Dependence of internal stress on angle of incidence for several materials deposited using cylindrical-post magnetron sputtering sources. Data from Reference 201. (Used courtesy Telic Corporation.)

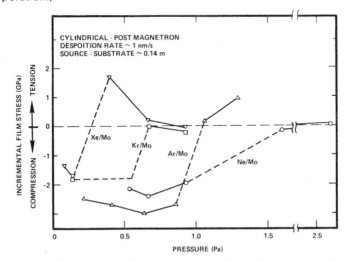

Figure 5.45: Influence of working gas species and pressure on the incremental (integrated) internal stress in Mo films sputter-deposited using a cylindrical-post magnetron source with Ne, Ar, Kr and Xe. From Reference 201a. (Used courtesy American Institute of Physics.)

The coatings deposited at low pressures and angles of incidence had smooth, highly reflective surfaces and, as seen in the figures, were in compression. Those deposited at higher pressures, or at larger angles of incidence, had rough surfaces of reduced reflectivity and tended to be in tension. The pressure at which the transition from compression to tension occurs is seen in Figures 5.43 and 5.44 to increase with the atomic mass of the coating material relative to that of the working gas. At large angles of incidence or high deposition pressures, the coating structure becomes so open that it cannot sustain a mechanical force, and the stress decreases to zero. The transition from compression to tension is believed to be related to the passage from Zone T to Zone 1 on the coating structure diagram shown in Figure 5.38.

The compressive stresses may result because of bombardment by energetic coating atoms and by ions which are neutralized and reflected at the cathode. The role of elevated working pressures, or oblique deposition angles, in promoting underdense Zone 1 structures has been discussed. The effects are similar, since an elevated working pressure enhances collisional scattering which introduces an oblique component in the deposition flux. It has also been found that the Zone 1 structure can be suppressed, and compressive stresses formed, if ion bombardment is present during deposition.[181] Thus it has been reasoned that the bombarding effect of the reflected and neutralized ions and the sputtered atoms in the low pressure magnetron sputtering case may densify the coating structure by an atomic peening mechanism and thereby delay the trend of the coatings to develop underdense Zone 1 structures at elevated pressures[198,199] (Figure 5.43) or oblique deposition angles (Figure 5.44). The dependence of the compression-to-tension transition pressure on the target-to-working gas mass ratio is the basis for concluding that the bombarding species may be composed of reflected and neutralized ions. In both of these cases the energy flux carried away from the cathode by the atoms increases with the mass ratio. Support for this point of view has been provided by experiments in which special shield configurations were used to isolate the particle flux which leaves the cathode at oblique angles from the flux which leaves at more normal angles.[30a,158,201a] The oblique flux may be expected to contain relatively large proportions of reflected and neutralized ions. Coatings that were formed from this flux exhibited larger compressive stresses, and contained more entrapped working gas, than coatings formed from the more normal flux. It should be noted, however, that large compressive stresses and entrapped gas concentrations were also seen for target-to-ion mass ratios of less than unity, where ion reflection would be expected to be less important.[158]

Although the compressive stress levels in the above experiments did correlate with the entrapped working gas content, the strains remained invariant over large changes in the amount of entrapped working gas. The entrapped gas has therefore been concluded to be a consequence of the atomic peening but not the cause of the compressive stresses.[30a]

The above experiments illustrate two important points: (1) the formation of coatings with compressive stresses and entrapped working gas is a common occurrence in sputtered coatings deposited at low pressures where the reflected ions do not lose their energy via collisions in their passage to the substrates; and (2) the angular dependence of the reflected ion flux emitted from the cathode makes the substrate bombardment dependent on the shape of the cathode.[30a] Thus preliminary measurements have indicated that chromium coatings deposited with a gun type magnetron did not possess the same high compressive stresses that are obtained with cylindrical-post magnetrons.[202]

A compressive state of stress is often desired, as many coating materials are stronger in compression than in tension. This consideration is especially important when the substrate has a greater thermal expansion coefficient than the coating, as is the case for decorative chromium coatings on plastics.[29a,95]

The dependence of the stress on the angle of incidence is of particular significance. It underscores the difficulty that is often encountered in obtaining a coating having uniform properties on a substrate of complex shape. It can also be the cause of variations in coating properties from one apparatus to another and can be an important consideration in apparatus scale-up.

Polymorphic Phase Formation

Many materials are capable of existing in more than one crystal structure. The stable phase at a given temperature will be the one with the minimum free energy F, given by $F = H - TS$ where H is the enthalpy, S is the entropy, and T is the temperature. At high temperatures the TS term becomes increasingly important. Therefore high temperature phases are characterized by large entropies, and more open and random structures which assume an amorphous liquid form in the high temperature limit.[203]

The influence of substrate temperature and the relative roles of surface and bulk diffusion on phase formation in coating growth are indicated schematically in Figure 5.46 for a hypothetical polymorphic material whose equilibrium phases are indicated along the vertical axis. Structural order is produced largely by the surface mobility of the adatoms. A highly disordered amorphous-like structure is expected if the mobility is negligible, so that the atoms come to rest at or near their points of impingement. Low substrate temperatures also favor the formation of open, more random crystallographic phases which are similar and in some cases identical to the high temperature phases.[204] At higher substrate temperatures the large adatom surface mobility is generally adequate to form the phases predicted by the phase diagram at a temperature equal to the substrate temperature.[151,163,205] In general, one expects a phase formed by surface diffusion to be the equilibrium phase if the difference between the surface free energy and the bulk free energy is the same for all phases and if the substrate surface does not promote one phase preferentially.[205] Equilibrium phases can also be generated by bulk diffusion within coating material that has already been deposited. Similarly, at high substrate temperatures a high temperature phase which has been formed in an initial deposit may be lost by diffusion during the postdeposition cool-down, so that a transformed structure characteristic of conventional metallurgical processing results.[206]

Thus high temperature phases can form in the initial deposits at both very low and very high temperatures, but the final phase will depend on the bulk diffusion properties of the material.[163] Considerable experimental support is available for the trends summarized in Figure 5.46.[163]

Coating Properties

Coatings with desired properties for a given application are generally optimized by a systematic variation of the deposition parameters applicable to the availabile sputtering apparatus. An advantage of the sputtering process is that once such an optimization is completed, coatings of the desired form can consistently be deposited. The large number of variables involved, and the success of the empirical approach, have limited the number of fundamental investigations of the process-property relationships.[38] However, a few generalizations can be made.

Coatings deposited at low substrate temperatures are characterized by a high density of structural defects, while those deposited at elevated temperatures approach bulk physical properties.[163,207] Simple metals with the fibrous

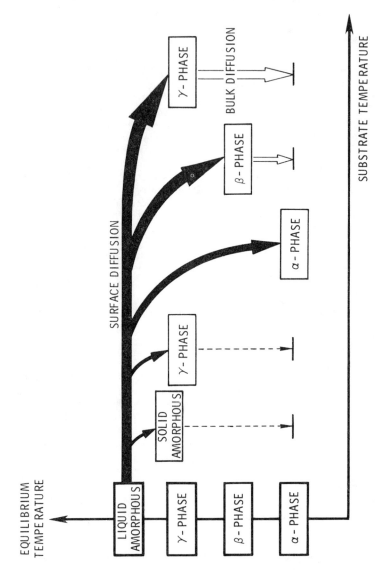

Figure 5.46: Schematic illustration of the influence of substrate temperature on polymorphic phase formation. From Reference 163. Reproduced with permission from the *Annual Review of Materials Science*, Volume 7, Copyright 1977, Annual Reviews, Inc.

Zone T structure have high hardness and strength but very little ductility.[167] Densely packed Zone 1 structures have the same hardness (two or three times that of annealed bulk material) but little lateral strength.[166] Within the Zones 2 and 3 regions the sizes of the columnar and equiaxed grains increase with T/T_m, and the hardness and strength of simple metals decrease in accordance with the Hall-Petch relationship, approaching the values for annealed bulk materials.[207] The ductility increases accordingly. In contrast to the simple metals, ceramic compounds tend to have lower than bulk hardness when deposited at low T/T_m but approach bulk values at high T/T_m.[166] Alloy deposits generally have smaller grain sizes and more highly dispersed phases than conventionally formed bulk materials.[94] In some cases this may lead to improved corrosion resistance.[208] Many opportunities exist for developing alloys and compounds with preferred combinations of properties such as mechanical strength, electrical conduction, and corrosion resistance.[94] Methods include co-deposition to produce dispersion-strengthened composites as well as deposition control to produce refinement of secondary phase morphology.[94,209-211]

Zone T coatings generally have electrical resistivities that are a few times above bulk values. At higher temperatures (Zones 2 and 3) bulk values are obtained. Passage into Zone 1 is signaled by a rapid increase in resistivity.[199] As one moves further into the Zone 1, order-of-magnitude increases in resistivity are produced by the open intergrain boundaries.

Zone T coatings generally have smooth, highly reflective surfaces. Transparent coatings deposited in this zone are clear. As one moves into Zone 1, coatings become increasingly rough and nonreflective, and transparent coatings become frosty from internal light scattering. Coatings deposited under extreme Zone 1 conditions may be used as light absorbers or as catalytic surfaces because of their high surface areas.[26] High temperature coatings (Zones 2 and 3) have reduced reflectivity because of the faceted surfaces of the grains.

Noncrystalline structures generally result when covalently bonded materials such as Al_2O_3, Si, and Ge are deposited at $T/T_m \leq 0.5$. Metal oxide deposits can have sufficiently high dielectric breakdown strengths (10^6 V/cm) for many applications. Amorphous silicon sputtered in hydrogen has been found to have favorable electrical properties, apparently because the hydrogen passivates the electrical activity of the structural imperfections.[212,212a]

Both epitaxial single crystal and polycrystalline semiconductor coatings can be grown by sputtering at elevated temperatures (Zone 2 and Zone 3 regions). As a general rule, the coatings are characterized by a relatively high density of structural defects as compared to "device grade" bulk materials.[175] The electrical character of these defects may determine the conductivity type of the coatings and limit the effectiveness of doping because of compensation.[212b] However, as discussed previously, epitaxial single crystal compound semiconductor films with essentially bulk values of carrier concentrations and mobilities have been obtained. Recent research indicates that low energy ion bombardment may be beneficial to forming semiconductor coatings with superior properties.[39b] However, as a general rule sputter-deposited semiconductors should be viewed as a generic class of materials with their own unique properties, rather than bulk-like materials in sheet form.

5. SPECIAL TOPICS

In this section several special topics of particular importance in processing technology are briefly discussed. The objective is to illustrate the types of factors that require consideration when establishing coating procedures for various applications.

Target Fabrication

Targets can be formed by an almost unlimited collection of methods ranging from vacuum casting of metals and hot pressing of powders to forming composite targets by placing wires, strips or discs of one material to cover a portion of the surface of another material. However, caution must be exercised. Poor coating quality or consistency can often be traced to an inadequate target.

Target cooling is important, particularly at the high current densities used with magnetrons. If the target material is a metal of moderate cost and good mechanical integrity, direct water cooling can usually be provided on the rear surface. Mechanically weak or nonconducting targets must be attached to a metal backing structure which serves as the vacuum seal and for nonconducting targets, as an electrode. The attachment is usually made with a low-temperature metal or alloy or a conducting epoxy. If target purity is critical, it is important to assure that the attachment medium does not diffuse into and contaminate the target material. In the case of nonconducting targets, the sputtering rate is limited by heat transfer within the target.[213]

A number of suppliers offer a wide selection of high purity metal and compound targets, particularly in the disk form used by planar diode sputtering sources. The compound targets are generally formed from powders by hot pressing. As noted previously, hot-pressed targets are particularly vulnerable to contamination. Therefore for critical application it is wise to measure the composition of new targets prior to placing them in service. It is also important to note that the purity levels quoted for most metal targets do not generally include gaseous impurities such as oxygen.

An increasing selection of target materials is being offered in shapes suitable for use with magnetrons. Ring type metal targets for sputter guns are designed to make contact with the support structure by undergoing thermal expansion. In the cylindrical magnetron case targets of common engineering materials can be easily fabricated from commercially available tube and rod stock. The tubular form possesses sufficient strength so that high purity targets have been cast using materials as soft as Cd. Very soft metals such as In, or compounds such as CdS, are generally attached to tubular sections that contain the vacuum seals.

Adhesion

Sputtering technology has gained a reputation for providing better adhesion that is generally achieved with simple evaporation.[38,161,162]

The adhesion strength of a coating bond depends on both the bonding strength between the coating and substrate materials and the microstructure in the interface region.[161,162] The bonding may be chemical, van der Waals, electrostatic, or a combination of these types.[162,214-216] Chemical bonds are the strongest (several eV) but require that atoms be in suitable positions to share electrons. Van der Waals bonds arise from polarization interactions. They do not require intimate atomic contact, but are weaker than chemical bonds (0.1 to 0.4 eV) and decrease rapidly with separation distance. Electrostatic bonds result from charge double layers between the coating and substrate and, for some metal insulator interfaces, may have strengths comparable to van der Waals bonds.[216,217] It has been shown theoretically that a typical van der Waals bond strength (0.2 eV) should be able to withstand a coating stress of about 5×10^8 N/m^2 (~72,000 psi).[218] Despite this seemingly high value, adhesion failures are often observed. There are two reasons for this. First, as noted in Section

4, the internal stresses in coatings are often greater than 5×10^8 N/m². Second, the strengths of interfaces, like bulk solids, are dominated by microstructure flaws and are much weaker than predicted by interatomic bonds. Thus in analyzing adhesion, attention must be directed to the overall nature of the interface region.

Interface regions may be classed as abrupt, compound, diffusion, pseudodiffusion, irregular, or a combination of these.[162] Abrupt interfaces form when there is little chemical reaction between the coating and substrate materials. Compound interface regions form when the materials react to form intermetallic or chemical compounds, and diffusion interfaces form when they have considerable solubility with one another. Pseudodiffusion interfaces may be formed, for example, by using codeposition to form graded interfaces. Irregular interfaces can be formed by bead blasting or etching the substrate.

When depositing metals onto metals, a diffusion type interface region with metallic bonds is ideal because of the gradual change in composition and stress across the interface. Such interfaces are promoted by elevated substrate temperatures, which enhance diffusion, but are prevented by oxide or other contamination layers that can act as diffusion barriers. Compound interface regions also involve chemical bonds and are strong. They are also promoted by elevated substrate temperatures but have the disadvantages that they are brittle. Thus it is desirable that the compound layer be kept thin. In the case of metals, a third metal which is soluble in both the substrate and coating materials can be used to form the interface. Often sputter deposition involves coating-substrate combinations which tend not to interdiffuse or form compounds because of their chemical nature or the substrate temperature. The resultant abrupt interfaces are vulnerable to contamination layers which preclude the formation of chemical bonds and greatly weaken van der Waals bonds. If the nucleation density is low, the lateral coating growth between the initial nuclei may form voids in the interface region.[161] The voids lead to poor adhesion because of the reduced contact area and ease of crack propagation.

Sputtering's reputation for superior adhesion has come largely from work with planar diode sources, where the substrates are subjected to plasma bombardment. The good adhesion is believed to be primarily due to this plasma bombardment.[38] It removes loose contamination and creates nucleation sites. In most magnetron configurations the substrates are not in contact with the plasma. The relatively high energies of the sputtered atoms in magnetron sputtering sources operating at low pressures and in ion beam systems may act to some degree to promote adhesion by mechanisms similar to those in plasma bombardment.[95] However, poor adhesion has been seen when using cylindrical magnetrons to deposit metals onto metals without sputter-cleaning. Poor adhesion has also been encountered when depositing Ta films using an ion beam system.[93]

The use of sputter cleaning to prepare substrates for deposition is discussed in the next section. Bead blasting can also improve coating adhesion. It is believed that the irregular bead-blasted surface makes interface crack propagation more difficult, because a fracture propagating along a plane of weakness must often change direction or pass through a stronger region.[162] Mechanical interlocking is not important as a rule in vacuum coating, because the deposition is primarily line-of-sight over the distance involved. Bead blasting is very effective for coating-chamber wall shields that accumulate thick coatings.

Sputter Cleaning and Bias Sputtering

Sputter cleaning is performed by biasing the substrate negatively relative to the plasma potential (which is about equal to the anode potential). Contamination is thereby removed from the substrate surface by sputtering. The ion bombardment flux and energy in the sputter cleaning process are therefore generally considerably larger than those for electrically isolated or grounded substrates in contact with a plasma (e.g., the substrates in a planar diode). Typical conditions are: current density, 1 to 5 mA/cm^2; voltage, 100 to 1000 V; and time of treatment, 5 to 20 min. Conditions are often selected to remove 200 to 5000 Å of material from a surface. An rf bias is required if the substrates are nonconducting: Sputter cleaning improves coating adhesion by removing that contamination which forms diffusion barriers and precludes bond formation. In addition, it promotes diffusion by creating a high density of surface defects.[162] These defects also increase the nucleation density and thereby minimize the formation of interface voids.[162]

Bias sputtering is the process of maintaining a negative bias on the substrate during deposition. It is particularly effective when the sputter cleaning process is simply continued, usually at reduced bias potential, after the formal coating process has been started. Bias sputtering can be used to reduce the amount of contamination which is incorporated into the growing coating and to cause favorable modifications to the microstructure that develops in the coating. See Section 4.

Although simple in principle, the implementation of sputter cleaning and bias sputtering, particularly for production applications, requires some precautions. For example, a suitable shield should be provided in sputter cleaning, so that the contamination sputtered from the substrate does not pass onto the target. Flat plate substrates on a planar holder, as shown in Figure 5.14, provide a relatively effective arrangement for sputter cleaning. Bias sputtering is also easily executed in such a device, because ions can be drawn from the plasma created by the target electrode. However, this is often not the case in magnetrons, particularly when the plasma containment is very effective. In such cases the plasma does not extend to the substrates, and an independent discharge must be sustained. This may be difficult, since the substrate geometry is frequently not an effective cathode shape for sustaining a plasma discharge, particularly at the low pressures where magnetrons are generally operated. Complex substrate shapes also make it difficult to achieve uniform ion bombardment and erosion over the surface. Sometimes this difficulty can be overcome by designing the substrate holder so that, in concert with the substrate, it forms an efficient magnetron electrode.[219] rf sputter cleaning and bias sputtering of nonconducting substrates of complex shape are particularly difficult to accomplish.

The substrate holders should be designed so that they do not sputter excessively and introduce contamination onto the substrates.[220] For example, substrate contamination, due to atoms sputtered from the holder and backscattered onto the substrate by collisions with the Ar working gas, was observed when sputter cleaning was done at typical dc planar diode operating pressures (25 to 80 mTorr).[221] The impurity level was much reduced at lower pressures.

Sputter cleaning (particularly at high incident ion energies) can drive contamination into the substrate because of the forward momentum exchange, as shown in Figure 5.7a, or implant working gas, as discussed in Section 4. It is important also to note that the degree of surface cleanness that can be obtained is generally limited by the purity of the working gas, which is influenced in turn by outgassing from the chamber walls and fixturing.[222] Sputter cleaning can modify the composition of the substrate surface because of the difference in the sputtering yield of the constituents. Note in Figure 5.5 that carbon, a common surface contaminant, has a very low sputtering yield. Oxygen is sometimes used as a working gas for sputter cleaning. Carbon is then removed by forming CO.

Despite the apparent limitations described above, when suitable precautions are taken, effective production processes can be developed which incorporate bias sputtering and/or sputter cleaning.

Formation of Graded Interfaces

By changing the operating conditions during deposition, it is possible to form graded interfaces at temperatures that are far below those that would be required to cause significant thermal diffusion. Such a method can be useful when depositing a refractory coating on a substrate with a low melting point or a much different thermal expansion coefficient from that of the coating material. Reactive sputtering (next section) is particularly useful in forming graded interfaces because of the ease with which the coating composition can be varied. For example, a thick Ti_xC coating was deposited on an aluminum alloy substrate, using a titanium cylindrical magnetron system and the following procedure.[37] The substrate was first sputter cleaned with the cathode shielded. Sputtering was then caused to occur simultaneously on the titanium target and on the substrate, so as to form a mixed Ti-Al layer on the substrate. The bias sputtering was then phased out so that a Ti layer was formed. This was continued until the Ti coating reached a thickness of about 10 μm. Methane was then gradually introduced into the argon working gas until the injection rate was sufficient that the deposited material assumed the form $TiC_{0.3}$. The deposition was then continued until the total coating thickness reached 25 μm. The graded bond was sufficiently effective that the rear surface of the ⅛-inch-thick aluminum alloy substrate could be melted (~660°C) without causing damage to the coating.

Oxygen-active metals often yield sufficiently good adhesion to glass and ceramics so that no sputter cleaning is required. Accordingly, metals such as titanium, chromium, niobium, tantalum, and tungsten are often used to form interfacial zones for depositing materials such as copper, silver, and gold. Solderable silver coatings with excellent adhesion have been deposited onto fused quartz using a sequence incorporating several of the techniques discussed above.[223] First the substrate was cleaned with a glow discharge. Then niobium oxide was deposited by reactive sputtering. The reactive sputtering was phased out, and a 1500 Å thick layer of niobium was deposited. Then a second source was used to simultaneously sputter silver along with the niobium. The niobium was finally phased out, and a 1 μm coating of silver was deposited.

Caution must be exercised in forming graded interfaces by codeposition, particularly with refractory materials, because of the large amount of atomic disorder which can be introduced into the interface region. Thus when simultaneous sputtering of a hafnium target and aluminum oxide substrate was used in a planar diode to form a graded interface by passing material back and forth between the target and substrate, high internal stresses and poor adhesion resulted.[224] Low stresses and very high bond strengths were obtained when an abrupt interface was formed.

Thus it is seen that, although guidelines are available, an empirical approach is generally required to obtain desired interface properties.

Reactive Sputtering

Reactive sputtering is that process where at least one of the coating species enters the system in the gas phase. Examples of reactive sputtering include sputtering Al in O_2 to form Al_2O_3,[225] Ti in O_2 to form TiO_2,[138] In-Sn in O_2 to form tin-doped In_2O_3,[5,226] Nb in N_2 to form NbN,[227] Cd in H_2S to form CdS,[228] In in PH_3 to form InP,[229] and Pb-Nb-Zr-Fe-Bi-La in O_2 to form a ferroelectric oxide.[230]

The advantages of reactive sputtering are that (1) many complex compounds can be formed using relatively easy-to-fabricate metallic targets; (2) insulating compounds can be deposited using dc power supplies; and (3) graded compositions can be formed as described in the preceding section. The difficulty in the reactive sputtering process is the complexity which accompanies its versatility.

Reactions can occur on the cathode surface, following which the reacted material is sputtered. They also occur at the substrate, and in cases of high working pressures, in the gas phase. When sputtering with a reactive-gas/argon mixture, the relationship between film properties and the reactive gas injection rate is generally very nonlinear. The condensing film can be considered as an additional pump for the reactive gas. The nonlinearity occurs because the sticking coefficient or getter pump speed of the condensing coating depends in a complex way on its growth rate, composition, film structure, and temperature. The composition dependence is shown in Figure 5.47 for N_2 incident on a growing Ti film.[231] Note that as the number of N_2 molecules sorbed per Ti atom deposited approaches 0.5 (i.e., a stoichiometric TiN film), the sticking coefficient drops by more than two orders of magnitude. The sticking coefficient drops with increased nitrogen concentration, because the number of unoccupied surface adsorption sites decreases, as one would expect from the concept of the Langmuir adsorption isotherms.[232] Thus, for example, when sputtering in an oxygen/argon mixture at low reactive gas injection rates, virtually all of the injected gas can react with the film. Consequently, the oxygen is largely removed from the working gas, and the cathode process becomes primarily one of simple argon sputtering of a metal. The coatings deposited under such conditions are generally metallic in nature. As the reactive gas injection rate approaches that required to produce a stoichiometric coating, there is an increase in the reactive gas partial pressure in the sputtering system because of the reduced getter pumping rate of the depositing coating. This increase in reactive gas partial pressure changes the composition of the gas within which the sputtering discharge is being maintained and greatly changes the processes which occur at the cathode surface. The result is that, for most metal/reactive-gas combinations, the sputtering discharge undergoes a transition into a mode in which the metal sputtering rate and therefore the reactive compound deposition rate is reduced. The cathode surface reactions in this mode (reflected atoms, sputtered chemisorbed atoms, sputtered surface compounds—see Section 2) produce an energetic flux of highly reactive gas atoms and molecular fractions which accompany the sputtered metal atoms to the substrate and produce a near-stoichiometric coating. It is this large flux of reactive species that makes the reactive sputtering process so effective for producing a wide range of compounds. The variation in discharge voltage and relative deposition rate during a typical transition is shown in Figure 5.48. The voltage decrease at high oxygen injection rates is the metal-to-compound transition. The voltage increase at low injection rates is the compound-to-metal transition.

The reduction in sputtering rate shown in Figure 5.48 is believed to result primarily from compound formation on the cathode surface and from the reduced sputtering yield of the reactive gas molecules. The compounds often have higher electron secondary emission coefficients, which result in a reduction in both the discharge voltage and the ion component in the cathode current, for discharges driven at constant currents (as in the case shown in Figure 5.48). The hysteresis effect, which is shown for discharge voltage but applies also the deposition rate, is believed to result because, once formed, the compound layer will remain until the working gas is made sufficiently lean in the reactive species so that a net sputter removal of the layer can occur. A cathode on which such a layer has formed is often referred to as being "poisoned." The effect of cathode poisoning on the reactive sputtering process depends on the metal-reactive gas combination and the properties of the cathode surface layer. Thus the very pronounced poisoning effect shown in Figure 5.48 occurs for the oxygen reactive sputtering of materials such as Al, Cr, Ti, and Ta that form tenacious oxides. The decrease in deposition rate is generally less for other reactive gases such as N_2 and C_2H_2. No poisoning occurs for Au, where the sputtering rate for pure O_2 is not much different from that for Ar.

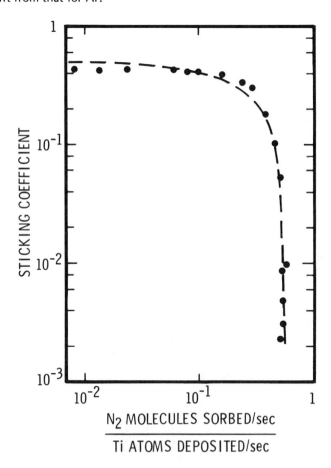

Figure 5.47: Sticking coefficient measured during the continuous deposition of titanium as function of ratio of nitrogen flux that is getter-pumped to the titanium deposition flux. Data from Reference 231.

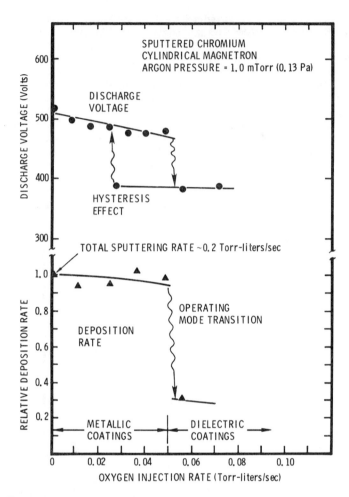

Figure 5.48: Transitions in steady state operating mode of chromium cylindrical-post magnetron sputtering source due to injection of oxygen. From Reference 30.

The poisoning effect introduces two practical problems. One is the loss in depositing rate. The second is that during the transition the material being deposited often passes abruptly from a metal to a nearly full-stoichiometric compound. Intermediate materials such as suboxides therefore become difficult to deposit. Consequently, considerable work has been directed toward trying to operate sputtering sources right at or very near the transition point, with, in most cases, not much success. Transition behavior has been seen in planar diodes,[232] planar magnetrons,[233] and cylindrical magnetrons.[97] Many papers have been written concerning the transition mechanism.[232,234-239] Most are incomplete, however, because they concentrate on the cathode processes and do not consider the total system. It is important to realize that the reactive sputtering process is dependent on the total system; i.e., its geometry, the accumulation of coating on walls and fixtures, and the positions of gas injection. All these

Coating Deposition by Sputtering   235

parameters must be carefully controlled in order for reactive sputtering to be effectively used on a production basis.

Recent reactive sputtering work with planar magnetrons, which takes advantage of the nonuniform cathode current densities in these devices, is encouraging with respect to the achievement of high deposition rates.[240,241] The behavior is illustrated schematically in Figure 5.49. The reactive gas flux, and therefore the tendency for reactive gas adsorption, is relatively uniform over the cathode surface (Figure 5.49A). However, the ion flux is nonuniform and causes sputter removal of adsorbed reactive species, thereby reducing absorbed specie surface coverage under the plasma ring (Figure 5.49B). At higher current densities (Figure 5.49C) the sputtering rate is adequate to maintain a fresh metal surface, which in turn yields a high rate of sputtering. By exerting control over the total system—i.e., by arranging the reactive gas injection adjacent to the substrate and the Ar injection adjacent to the target—it appears to be possible, with the assistance of suitable getter surfaces as shown in Figure 5.49D, to maintain a gradient in the composition of the reactive gas in the sputtering atmosphere. If this gradient is adequate, the target surface under the plasma rings can remain unpoisoned and yield a high flux, even when the reactive gas flux to the substrate is adequate to produce a stoichiometric compound.[240] Substrate ion bombardment is sometimes used to promote the substrate surface reaction. Deposition rates of 10,500 Å/min for Ta_2O_5 and 6800 Å/min for TiO_2 have been reported using this method. The coatings were found to have optical properties which were close to those of the bulk materials.[241]

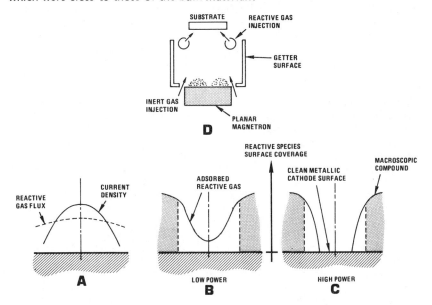

Figure 5.49: Schematic illustration of various elements of the planar magnetron reactive sputtering process. (Used courtesy Telic Corporation.)

Predeposition Pumping

When a vacuum chamber is exposed to the atmosphere, several monolayers of water vapor adsorb on the metal surfaces. Once the pumpdown starts, the volume of gas that is present in the chamber, and the gas which is physisorbed on the walls (bond energy <0.5 eV), are removed very quickly.[242] Thereafter the pressure p is determined by a balance between the net rate Q of desorption and diffusion of gas from the substrates and the walls of the chamber and the pumping speed S,[242] i.e., p = Q/S. Strongly chemisorbed species (bond energy 1.5 eV) release at such slow rates that they do not contribute significantly to the pressure. However, the water vapor, which is bonded with intermediate energies (~1 eV), releases slowly and dominates the chamber pressure during typical pumpdown times (<10 hr) in room temperature chambers.[242] The rate of gas release decreases slowly as the reservoirs of stored gas are depleted. It is often found that the net outgassing rate obeys a relationship of the form $Q = Q_0/t^n$, where t is the pumping time and n varies from unity for metal surfaces to $\sim 1/3$ for plastic substrates.[30] Thus the outgassing rate from the chamber walls and substrates depends, to first order, only on the time that the chamber is pumped. The pressure, on the other hand, obeys the relationship $p(t) = Q_0/(t^n S)$, and depends on the pumping speed.

The above considerations are important in determining the amount of pumping required before deposition is begun. If the sputtering source and deposition area occupy a small volume within the coating chamber, so that the getter pumping speed is small compared to the chamber pumping speed, then the flux of contamination that enters this volume and becomes incorporated into the coating depends on the chamber pressure. A large pumping system that can provide a low pressure before the start of deposition should improve the coating purity. However, if the sputtering source and deposition area occupy most of the sputtering chamber, as is often the case for cylindrical magnetrons, then the gettering rate of the sputtered flux will in general be large compared to the speed of the vacuum pumping system. Under these conditions essentially the entire outgassing flux Q(t) is incorporated into the coating, so that the contamination level depends on the time of pumping and, to first order, not on the size of the pump. Thus the achievement of a low starting pressure in a short pumping time through the use of a large pump can lead to a false sense of security if the sputtering system geometry causes most of the outgassing flux to be gettered by the coating flux.

Target-cleanup provides a specific example of a case in which the outgassing flux is of critical importance. The change in deposition conditions that occur with increased reactive gas injection rate was shown in Figure 5.48. As discussed previously, the reduction in sputtering rate is believed to occur largely because a surface layer forms on the target. When a clean metal target in a batch processing apparatus is exposed to the atmosphere for purposes of loading the chamber (exposure time 1 to 10 minutes), a thin oxide layer will form on the cathode. Experiments with a Cr cathode indicate that this layer is very similar in its sputter-removal characteristics to the one that forms during reactive sputtering in the compound mode.[30] Thus a target that has been exposed to the atmosphere must undergo a compound-to-metal transition similar to that shown in Figure 5.48 before metal coatings can be deposited. This requires that the chamber be pumped for sufficient time so that the outgassing flux entering into the plasma and deposition region is less than the critical flux for permitting the compound-to-metal transition at the target current density being used.[30] If the outgassing flux exceeds the critical value, the target will not "clean-up," and a dielectric coating will be deposited. For materials such as aluminum, the critical flux is relatively low. This has led inexperienced workers, using devices operating at low cathode current densities, to conclude that aluminum cannot be sputtered.

Finally, we note that the species on the walls of an evacuated chamber and in the outgassing flux is primarily water vapor. Its dissociation in the plasma discharge produces hydrogen, which has a low sputtering yield and thereby causes a reduction in the sputtering rate. Significant variations in deposition rate from run to run and within a run have been attributed to this effect.[243] Variations within a run can be particularly severe when the plasma reaches significant portions of the chamber and substrate surfaces, thereby causing the release of a large water vapor flux during the initial period of sputtering. Accordingly, the hydrogen partial pressure is often monitored with a mass spectrometer in critical production applications. There is evidence that the effect of hydrogen on the deposition rate is less pronounced in the case of magnetron sources operated under typical condtions.[244]

REFERENCES

(1) S.D. Dahlgren, "Vapor-Quenching Techniques," Proceedings, 3rd Intl. Conf. on Rapidly Quenched Metals, Univ. Sussex, Brighton, England (1976).
(2) T. Robertson and D.T. Morrison, *Thin Solid Films 27*, 19 (1975).
(3) D.T. Morrison and T. Robertson, *Thin Solid Films 15*, 87 (1973).
(4) J.L. Vossen, G.L. Schnable, and W. Kern, *J. Vac. Sci. Technol. 11*, 60 (1974).
(5) R.S. Wagner, A.K. Sinha, T.T. Sheng, H.J. Levinstein, and F.B. Alexander, *J. Vac. Sci. Technol. 11*, 3 (1974).
(6) H.U. Schreiber and E. Froschle, *J. Electrochem. Soc. 123*, 30 (1976).
(7) G. Haacke, Ann. *Rev. Mater. Sci. 7*, 73 (1977).
(8) J.A. Thornton and V.L. Hedgcoth, *J. Vac. Sci. Technol. 13*, 117 (1976).
(9) P.K. Tien and A.A. Ballman, *J. Vac. Sci. Technol. 12*, 892 (1975).
(10) F.S. Hickernell, *J. Vac. Sci. Technol. 12*, 879 (1975).
(11) D.B. Fraser, *Thin Solid Films 13*, 407 (1972).
(12) T.G. Maple and R.A. Buchanen, *J. Vac. Sci. Technol. 10*, 616 (1973).
(13) E. Sawatsky and G.B. Street, *J. Appl. Phys. 44*, 1789 (1973).
(14) A.H. Meitzler and J.R. Maldonado, *Electronics* (Feb. 1, 1971), p. 34.
(15) P. Chaudhari, J.J. Cuomo, and R.J. Gambino, *Appl. Phys. Lett. 22*, 337 (1973).
(16) L. Maissel, in *Handbook of Thin Film Technology*, L. Maissel and R. Glang, ed., McGraw Hill, New York (1970), p. 18-3.
(17) D. Gerstenberg, in *Handbook of Thin Film Technology*, L. Maissel and R. Glang, ed., McGraw-Hill, New York (1970), p. 19-1.
(18) *Business Week* (Sept. 15, 1975), p. 58.
(19) M. Croset, J.P. Schnell, G. Velasco, and J. Siejka, *J. Appl. Phys. 48*, 775 (1977).
(20) D.J. Quinn, J.M. Berak, and D.E. Cullen, *J. Appl. Phys. 46*, 3866 (1975).
(21) L. Mei and J.E. Greene, *J. Vac. Sci. Technol. 11*, 145 (1975).
(22) G.C. Lane, "Razor Blade Sputtering," Proceedings 21st Technical Conference, Society of Vacuum Coaters, Detroit, MI (May 1978), p. 44.
(23) I.W. Fischbein, B.H. Alexander, and A. Sastri, U.S. Patent 3,682,795 (1972).
(24) W.D. Sproul and M. Richman, *J. Vac. Sci. Technol. 12*, 842 (1975).
(25) E. Eser and R.E. Ogilvie, *J. Vac. Sci. Technol. 15*, 401 (1978).
(25a) A.D. Grubb, Airco Temescal, Carleton, Michigan (private communication).
(26) J.A. Thornton, in *Plating and Surface Finishing 67*, 46 (1980).
(27) J.C.C. Fan, F.J. Bachner, G.H. Foley, and P.M. Zavracky, *Appl. Phys. Lett. 25*, 695 (1974).
(27a) W.W. Anderson, A.D. Jonath, and J.A. Thornton, "Magnetron Reactive Sputtering Deposition of Cu_2S/CdS Solar Cells," Proceedings 1976 Photovoltaic Solar Energy Conference, Commission of European Communities, Berlin (April, 1979).
(28) J.A. Thornton, *Metal Finishing 74*, 46 (1976).
(29) L. Hughes, R. Lucariello, and P. Blum, "Production Sputter Metallization of Exterior Plastic Automotive Parts," Proceedings 20th Technical Conf., Society of Vacuum Coaters, Atlanta, GA (April 1977), p. 15.

(29a) J.A. Thornton and A.S. Penfold, "The Application of Sputtering to Plastic Substrates," Proceedings Pactec V, Society of Plastics Engineers, Los Angeles, CA (Feb. 1980), pp. 67-101.
(30) J.A. Thornton, "Considerations Relating to Pumping Requirements for Sputtering Metallic Coatings on Plastic Substrates," Proceedings 20th Tech. Conf., Society of Vacuum Coaters, Atlanta, GA (April 1977), p. 5.
(30a) J.A. Thornton, J. Tabock, and D.W. Hoffman, *Thin Solid Films 64*, 111 (1979).
(31) G.K. Wehner, *Advances Elec. and Elec. Phys. 7*, 239 (1955).
(32) E. Kay, *Advances Elec. and Elec. Phys. 17*, 245 (1962).
(33) L.I. Maissel, *Physics of Thin Films 3*, 61 (1966).
(34) E. Kay, in *Techniques of Metals Research*, R.F. Bunshah, ed., Vol. 1, Part 3, Interscience, New York (1968), p. 1269.
(35) G.K. Wehner and G.S. Anderson, in *Handbook of Thin Film Technology*, L. Maissel and R. Glang, ed., McGraw Hill, New York (1970), p. 3-1.
(36) L. Maissel, in *Handbook of Thin Film Technology*, L. Maissel and R. Glang, ed., McGraw Hill, New York (1970), p. 4-1.
(37) J.A. Thornton, *SAE Transactions 82*, 1787 (1974).
(38) W.D. Westwood, *Progress in Surface Sci. 7*, 71 (1976).
(39) J.L. Vossen and W. Kern, ed., *Thin Film Processes*, Academic Press, New York (1978).
(39a) B. Chapman, *Glow Discharge Processes; Sputtering and Plasma Etching*, Wiley, New York (1980).
(39b) J.E. Greene, in *Handbook of Semiconductors*, Vol. 3, ed., S. Keller, North Holland Pub. Co. (to be published).
(40) H.D. Hagstrum, *Phys. Rev. 104*, 317 (1956).
(41) H.D. Hagstrum, *Phys. Rev. 123*, 758 (1961).
(41a) G.M. McCracken, *Rep. Prog. Phys. 38*, 241 (1975).
(42) B. Navinsek, *Progress in Surface Sci. 7*, 49 (1976).
(43) J.A. Thornton, Proceedings 19th SAMPE Symposium, Buena Park, CA (April 1974).
(43a) R.V. Stuart and G.K. Wehner, *J. Appl. Phys. 33*, 2345 (1962).
(44) T. Ishitani and R. Shimizu, *Phys. Lett. 46A*, 487 (1974).
(45) D.E. Harrison, Jr., N.S. Levy, J.P. Johnson III, and H.M. Effron, *J. Appl. Phys. 39*, 3742 (1968).
(46) P. Sigmund, *Phys. Rev. 184*, 383 (1969).
(46a) P. Sigmund, *J. Vac. Sci. Techol. 17*, 396 (1980).
(46b) P.D. Townsens, J.C. Kelly, and N.E.W. Hartley, *Ion Implantation, Sputtering and Their Applications*, Academic Press, New York (1976), p. 111.
(47) J.L. Vossen, *J. Vac. Sci. Technol. 8*, S12 (1971).
(48) J.A. Thornton, *Thin Solid Films 54*, 23 (1978).
(49) H.F. Winters and E. Kay, *J. Appl. Phys. 38*, 3928 (1967).
(50) I. Brodie, L.T. Lamont, Jr., and R.L. Jepson, *Phys. Rev. Lett. 21*, 1224 (1968).
(51) W.W. Lee and D. Oblas, *J. Vac. Sci. Technol. 7*, 129 (1970).
(52) W.W.Y. Lee and D. Oblas, *J. Appl. Phys. 46*, 1728 (1975).
(53) P. Sigmund, *Can. J. Phys. 46*, 731 (1968).
(54) H.H. Anderson, *Rad. Effects 3*, 51 (1970).
(55) E. Oechsner and W.R. Gesang, *Phys. Lett. 37A*, 235 (1971).
(56) E.V. Kornelsen, *Can. J. Phys. 42*, 364 (1964).
(56a) P. Blank and K. Wittmaack, *J. Appl. Phys. 50*, 1519 (1979).
(57) W.A. Grant and G. Carter, *Vacuum 15*, 477 (1965).
(58) K.L. Chopra, *Thin Film Phenomena*, McGraw Hill, New York (1969).
(59) J.W. Coburn, *J. Vac. Sci. Technol. 13*, 1037 (1976).
(59a) J.W. Coburn, *Thin Solid Films 64*, 371 (1979).
(60) E. Gillam, *J. Phys. Chem. Solids 11*, 55 (1959).
(61) A.D.G. Stewart and M.W. Thompson, *J. Matl. Sci. 4*, 56 (1969).
(62) M.J. Nobes, J.S. Colligon, and G. Carter, *J. Matl. Sci. 4*, 730 (1969).
(63) G. Carter, J.S. Colligon, and M.J. Nobes, *J. Matl. Sci. 6*, 115 (1971).
(64) P. Sigmund, *J. Matl. Sci. 8*, 1545 (1973).
(65) M.H. Witcomb, *J. Matl. Sci. 9*, 551 (1974).

(66) J.E. Green, B.R. Natarjan, and F. Sequeda-Osorio, *J. Appl. Phys. 49*, 417 (1978).
(67) G.K. Wehner and D.J. Hajicek, *J. Appl. Phys. 42*, 1145 (1971).
(68) M.L. Tarng and G.K. Wehner, *J. Appl. Phys. 43*, 2268 (1972).
(68a) G.K. Wehner, private communication (1980).
(69) T. Oohashi and S. Yamanaka, *Jpn. J. Appl. Phys. 11*, 1581 (1972).
(70) R. Shimizu, Japan. *J. Appl. Phys. 13*, 228 (1974).
(71) S.D. Dahlgren and S.D. McClanahan, *J. Appl. Phys. 43*, 1514 (1972).
(72) J.L. Vossen, *J. Vac. Sci. Technol. 8*, 751 (1971).
(73) D.R. Wheeler and W.A. Brainard, *J. Vac. Sci. Technol. 15*, 24 (1978).
(73a) A.B. Cistola, *J. Vac. Sci. Technol. 17*, 420 (1980).
(74) J.L. Vossen and J.J. O'Neill Jr., *RCA Rev. 29*, 149 (1968).
(75) D.B. Fraser and H. Melchior, *J. Appl. Phys. 43*, 3120 (1972).
(76) O. Almen and G. Bruce, Transac. 8th National Vac. Sym., New York, Pergamon Press (1962), p. 245.
(77) K. Tsunoyama, T. Suzuki, and Y. Ohashi, *Jpn. J. Appl. Phys. 15*, 349 (1976).
(78) H.F. Winters and P. Sigmund, *J. Appl. Phys. 45*, 4760 (1974).
(79) D.E. Harrison, Jr., and C.B. Delaplain, *J. Appl. Phys. 47*, 2252 (1976).
(80) J.W. Coburn, E.W. Eckstein, and E. Kay, *J. Vac. Sci. Technol. 12*, 151 (1975).
(81) A.R. Krauss and D.M. Gruen, *Appl. Phys. 14*, 89 (1977).
(82) J.M. Schroeer, R.N. Rhodin and R.C. Bradley, *Surface Sci. 34*, 571 (1973).
(83) J.J. Cuomo, R.J. Gambino, J.M.E. Harper, J.D. Kuptsis, and J.C. Webber, *J. Vac. Sci. Technol. 15*, 281 (1978).
(84) J.J. Hanak and J.P. Pellicane, *J. Vac. Sci. Technol. 13*, 406 (1976).
(85) A. von Engle, *Ionized Gases*, Oxford University Press, London (1965).
(86) J.L. Vossen and J.J. Cuomo in *Thin Film Processes*, J.L. Vossen and W. Kern, ed., Academic Press, New York (1978), pp. 1-73.
(87) W.D. Westwood and R. Boynton, *J. Appl Phys. 43*, 2691 (1972).
(88) E.S. McDaniel, *Collision Phenomena in Ionized Gases*, Wiley, New York (1964), Chapter 13.
(89) J.A. Thornton, *J. Vac. Sci. Technol. 15*, 188 (1978).
(90) L.G. Christophorou, *Atomic and Molecular Physics*, Wiley, New York (1971), p. 379.
(91) W.D. Davis and T.A. Vanderslice, *Phys. Rev. 131*, 219 (1963).
(92) A.J. Stirling and W.D. Westwood, *J. Appl. Phys. 41*, 742 (1970).
(93) W.D. Westwood, *J. Vac. Sci. Technol. 15*, 1 (1978).
(93a) W.D. Westwood, "Sputtering—Notes from AVS Short Course." Available from American Vacuum Society, 335 East 45th Street, New York, NY 10017.
(94) R. Busch and E.D. McClanahan, *Thin Solid Films 47*, 291 (1977).
(95) J.A. Thornton, *Metal Finishing 77*, 45 (1979).
(96) J.A. Thornton, *J. Vac. Sci. Technol. 15*, 171 (1978).
(97) J.A. Thornton and A.S. Penfold, in *Thin Film Processes*, J.L. Vossen and W. Kern, ed., Academic Press, New York (1978), p.75.
(98) F.M. Penning and J.H.A. Moubis, *Proc. Ned. Akad. Wet. 43*, 41 (1940).
(99) A.S. Penfold and J.A. Thornton, U.S. Patents 3,884,793 (1975); 3,919,678 (1975); 3,995,187 (1976); 4,030,996 (1977); 4,031,424 (1977); 4,041,053 (1977); 4,111,782 (1978); 4,116,793 (1978); 4,116,794 (1978); 4,132,612 (1979); 4,132,613 (1979); and A.S. Penfold, 3,919,678 (1975).
(100) N. Hosokawa, T. Tsukada, and T. Misumi, *J. Vac. Sci. Technol. 14*, 143 (1977).
(101) K.I. Korov, N.A. Ivanov, E.D. Atanasova, and G.M. Minchev, *Vacuum 26*, 237 (1976).
(102) L. Spitzer, Jr., *Physics of Fully Ionized Gases*, Interscience, New York (1956).
(103) F.F. Chen, *Introduction to Plasma Physics*, Plenum Press, New York (1974).
(104) J.A. Thornton, *J. Vac. Sci. Technol. 16*, 79 (1979).
(105) A.S. Penfold, *Metal Finishing 77*, 33 (1979).
(106) A.S. Penfold, Telic Corporation (unpublished).
(107) F.M. Penning, *Physica 3*, 873 (1936).
(108) J.A. Thornton and V.L. Hedgcoth, *J. Vac. Sci. Technol. 12*, 93 (1975).

(109) U. Heisig, K. Goedicke, and S. Schiller, "High Rate Sputtering with Torous Plasmatron," Proceedings 7th Intl. Sym. Electron and Ion Beam Science and Technology, Washington, D.C. 1976 (Electrochemical Society, Princeton, NJ, 1976), p. 129.
(110) R.K. Waits, *J. Vac. Sci. Technol. 15*, 179 (1978).
(111) R.K. Waits, in *Thin Film Processes*, J.L. Vossen and W. Kern, ed., Academic Press, New York (1978), p. 131.
(112) D.B. Fraser, in *Thin Film Processes*, J.L. Vossen and W. Kern, ed., Academic Press, New York (1978), p. 115.
(113) A. Aronson and S. Weinig, *Vacuum 27*, 151 (1977).
(114) T. Van Vorous, *Optical Spectra*, (November, 1977), p. 30.
(115) H.R. Smith, Jr., "Current Industrial Applications of High Rate Sputtering," Proceedings 20th Annual Tech. Conf., Society of Vacuum Coaters, Atlanta, GA (1977), p. 1.
(116) J.F. Corbani, U.S. Patent 3,878,085 (1975).
(117) P.J. Clarke, U.S. Patent 3,616,450 (1971); U.S. Patent 3,711,398 (1973).
(117a) P.S. McLeod, U.S. Patent 3,956,093 (1976).
(117b) R.M. Rainey, U.S. Patent 4,100,055 (1978).
(117c) J.S. Chapin, U.S. Patent 4,166,018 (1979).
(118) P.D. Davidse, *Vacuum 17*, 139 (1967).
(119) B.S. Probyn, *Vacuum 18*, 253 (1968).
(120) G.N. Jackson, *Thin Solid Films 5*, 209 (1970).
(121) H.R. Koenig and L.I. Maissel, *IBM J. Res. Develop. 14*, 168 (1970).
(122) M.H. Brodsky, R.S. Title, K. Weiser, and G.D. Pettit, *Phys. Rev. B 1*, 2632 (1970).
(123) R. Messier, T. Takamori, and R. Roy, *J. Vac. Sci. Technol. 13*, 1060 (1976).
(124) S.B. Hyder, *J. Vac. Sci. Technol. 8*, 228 (1971).
(125) G. Moulton, *Nature 195*, 793 (1962).
(126) H.J. Hovel and J.J. Cuomo, *Appl. Phys. Lett. 20*, 71 (1972).
(127) A.J. Shuskus, T.M. Reeder, and E.L. Paradis, *Appl. Phys. Lett. 24*, 151 (1974).
(128) R.W. Glew, *Thin Solid Films 46*, 59 (1977).
(129) C. Corsi, *J. Appl. Phys. 45*, 3467 (1974).
(130) K. Wasa, T. Nagai, and S. Hayakawa, *Thin Solid Films 31*, 235 (1976).
(131) W.J. Takei, N.P. Formigoni, and M.H. Francombe, *J. Vac. Sci. Technol. 7*, 442 (1970).
(132) J.L. Vossen, *RCA Rev. 32*, 289 (1971).
(133) T.W. Hickmott, *J. Appl. Phys. 45*, 1050 (1974).
(134) C.A.T. Salma, *J. Electrochem. Soc. 117*, 913 (1970).
(135) I.H. Pratt, *Solid State Technol. 12*, 49 (1969).
(136) P.L. Young, F.P. Fehler, and A.J. Whitman, *J. Vac. Sci. Technol. 14*, 176 (1977).
(137) R.M. Goldstein and S.C. Wigginton, *Thin Solid Films 3*, R41 (1969).
(138) K. Wasa and S. Hayakawa, *Microelectron. & Reliab. 6*, 213 (1967).
(139) R.M. Goldstein and F.W. Leonhard, "Thin Film Dielectric Capacitors Formed by Reactive Sputtering," Proceedings Electronic Components Conf., AIEE (1967), p. 312.
(140) T. Takao, K. Wasa, and S. Hayakawa, *J. Electrochem. Soc. 123*, 1719 (1976).
(141) W.D. Westwood and C.D. Bennewitz, *J. Appl. Phys. 45*, 2313 (1974).
(142) J.G. Titchmarsh and P.A.B. Toombs, *J. Vac. Sci. Technol. 7*, 103 (1970).
(143) E.L. Paradis and A.J. Shuskus, *Thin Solid Films 38*, 131 (1976).
(144) T.K. Lakshmanan, *J. Electrochem. Soc. 110*, 548 (1963).
(145) Y. Shimomoto, H. Matsumaru, and T. Nishimura, *Jpn. J. Appl. Phys.*, Suppl. 2, Pt. 1, 701 (1974).
(146) H. Biederman, S.M. Ojha, and L. Holland, *Thin Solid Films 41*, 329 (1977).
(147) H.S. Butler and G.S. Kino, *Phys. of Fluids 6*, 1346 (1963).
(148) F.E. Terman, *Electronics and Radio Engineering*, McGraw Hill, New York (1955), p. 458.
(149) J.S. Logan, N.M. Mazza, and P.D. Davidse, *J. Vac. Sci. Technol. 6*, 120 (1969).
(150) J.W. Coburn and E. Kay, *J. Appl. Phys. 43*, 4965 (1972).
(151) J.A. Thornton and J. Chin, *Ceramic Bulletin 56*, 504 (1977).
(151a) R.S. Nowicki, *J. Vac. Sci. Technol. 14*, 127 (1977).

(152) H.R. Kaufman, *J. Vac. Sci. Technol. 15*, 272 (1978).
(153) K.L. Chopra and M.R. Randlett, *Rev. Sci. Instr. 38*, 1147 (1967).
(154) P.D. Reader, D.P. White, and G.C. Isaacson, 14th Symp. Elect. Ion Photon Beam Tech., Palo Alto, CA (May 1977).
(155) R.S. Robinson, *J. Vac. Sci. Technol. 15*, 277 (1978).
(155a) J.M.E. Harper, in *Thin Film Processes*, J.L. Vossen and W. Kern, ed., Academic Press, New York (1978), pp. 175-206.
(155b) C. Weissmantel, *Thin Solid Films 72* (1980). (In publication.)
(156) D.J. Ball, *J. Appl. Phys. 43*, 3047 (1972).
(157) L.T. Lamont, Jr., and A. Lang, *J. Vac. Sci. Technol. 7*, 198 (1970).
(157a) S.S. Lau, R.H. Mills, and D.G. Muth, *J. Vac. Sci. Technol. 9*, 1196 (1972).
(158) J.A. Thornton and D.W. Hoffman, *J. Vac. Sci. Technol. 18*, 203 (1981).
(159) K.L. Chopra, *Thin Film Phenomena*, McGraw Hill, New York (1969), p. 138.
(160) C.A. Neugebauer, in *Handbook of Thin Film Technology*, L. Maissel and R. Glang, ed., McGraw Hill, New York (1970), p. 8-1.
(160a) M. Marinov, *Thin Solid Films 46*, 267 (1977).
(160b) A.H. Eltoukhy and J.W. Greene, *Appl. Phys. Lett. 33*, 343 (1978).
(160c) A.H. Eltoukhy, S.A. Barnett, and J.E. Greene, *J. Vac. Sci. Technol. 16*, 321 (1979).
(161) D.M. Mattox, *Thin Solid Films 18*, 173 (1973).
(162) D.M. Mattox, "Thin-Film Adhesion and Adhesive Failure—A Perspective," in *Adhesion Measurements of Thin Films, Thick Films and Bulk Coatings*, ASTM STP 640, K.L. Mittal, ed., American Society for Testing Materials (1978), p. 54.
(163) J.A. Thornton, *Ann. Rev. Mater. Sci. 7*, 239 (1977).
(164) A.G. Dirks and H.J. Leamy, *Thin Solid Films 47*, 219 (1977).
(165) J.H. Brophy, R.M. Rose, and J. Wulff, *The Structure and Properties of Materials*, Vol. 2, Wiley, New York (1964), p. 192.
(166) B.A. Movchan and A.V. Demchishin, *Phys. Met. Metallogr. 28*, 83 (1969).
(167) J.A. Thornton, *J. Vac. Technol. 11*, 666 (1974).
(168) A.I. Shaldervan and N.G. Nakhodkin, *Sov. Phys. Solid State 11*, 2773 (1970).
(169) J.A. Thornton, *J. Vac. Sci. Technol. 12*, 830 (1975).
(169a) L.R. Gilbert, R. Messier and R. Roy, *Thin Solid Films 54*, 149 (1978).
(169b) R. Messier, S.V. Krishnaswamy, L.R. Gilbert, and P. Swab, *J. Appl. Phys. 51*, 1611 (1980).
(169c) S.V. Krishnaswamy, R. Messier, Y.S. Ng, T.T. Tsong, and S.B. McLane, "Atom Probe FIM Investigation of Voids in a-Ge," paper presented at 8th Intl. Conf. on Amorphous and Liquid Semiconductors, Cambridge, MA (August 1979).
(169d) P. Swab, S.V. Krishnaswamy, and R. Messier, *J. Vac. Sci. Technol. 17*, 362 (1980).
(170) D. Henderson, M.H. Brodsky, and P. Chaudhari, *Appl. Phys. Lett. 25*, 641 (1974).
(171) S. Kim, D.J. Henderson, and P. Chaudhari, *Thin Solid Films 47*, 155 (1977).
(172) R.E. Reed-Hill, *Physical Metallurgy Principles*, Van Nostrand, New York (2nd Edition, 1973).
(173) J.W. Patten, E.D. McClanahan, and J.W. Johnston, *J. Appl. Phys. 42*, 4371 (1971).
(174) M.H. Francombe, in *Epitaxial Growth*, Part A, J.W. Mathews, ed., Academic Press, New York (1973), p. 109.
(175) H.H. Wieder, *Intermetallic Semiconducting Films*, Pergamon Press, New York (1970).
(176) P.S. Vincett, W.A. Barlow, and G.G. Roberts, *J. Appl. Phys. 48*, 3800 (1977).
(177) D.H. Boone, T.E. Strangman, and L.W. Wilson, *J. Vac. Sci. Technol. 11*, 641 (1974).
(178) W. Grossklaus, N.E. Ulion, and H.A. Beale, *Thin Solid Films 40*, 271 (1977).
(179) J.W. Patten, *Thin Solid Films 54*, 325 (1978).
(180) J.G.W. van de Waterbeemd and G.W. Oosterhout, *Philips Res. Rep. 22*, 375 (1967).
(181) R.D. Bland, G.J. Kominiak, and D.M. Mattox, *J. Vac. Sci. Technol. 11*, 671 (1974).
(182) D.M. Mattox and G.J. Kominiak, *J. Vac. Sci. Technol. 9*, 528 (1972).
(183) D.M. Mattox and G.J. Kominiak, *J. Electrochem. Soc. 120*, 1535 (1973).
(184) J.A. Thornton, *Thin Solid Films 40*, 335 (1977).
(185) F.B. Koch, R.L. Meek and D.V. McCaughan, *J. Electrochem. Soc. 121*, 558 (1974).
(186) W.D. Sproul and M.H. Richman, *J. Vac. Sci. Technol. 12*, 842 (1975).
(187) J.J. Cuomo and R.J. Gambino, *J. Vac. Sci. Technol. 14*, 152 (1977).

(188) F. Vratny and N. Schwartz, *J. Vac. Sci. Technol. 1*, 33 (1964).
(189) L.I. Maissel and P.M. Schaible, *J. Appl. Phys. 36*, 237 (1965).
(190) A.G. Blackman, *Metall. Trans. 2*, 699 (1971).
(191) A.G. Blackman, *J. Vac. Sci. Technol. 10*, 299 (1973).
(192) R.W. Hoffman, in *Physics of Thin Films*, Vol. 3, G. Hass and R.E. Thun, ed., Academic Press, New York (1966), p. 211.
(193) C.J. Santoro, *J. Electrochem. Soc. 116*, 361 (1969).
(194) W.B. Pennebaker, *J. Appl. Phys. 40*, 394 (1969).
(195) S.K. Lahiri, *J. Appl. Phys. 41*, 3172 (1970).
(196) P.R. Stuart, *Vacuum 19*, 507 (1969).
(197) R.S. Wagner, A.K. Sinha, T.T. Sheng, H.J. Levinstein, and F.B. Alexander, *J. Vac. Sci. Technol. 11*, 582 (1974).
(198) D.W. Hoffman and J.A. Thornton, *Thin Solid Films 40*, 355 (1977).
(199) J.A. Thornton and D.W. Hoffman, *J. Vac. Sci. Technol. 14*, 164 (1977).
(200) D.W. Hoffman and J.A. Thornton, *Thin Solid Films 45*, 387 (1977).
(201) D.W. Hoffman and J.A. Thornton, *J. Vac. Sci. Technol. 16*, 134 (1979).
(201a) D.W. Hoffman and J.A. Thornton, *J. Vac. Sci. Technol. 17*, 380 (1980).
(202) D.W. Hoffman and P.I. Goodsmith, "Decorative Metallizing by Magnetron Sputtering: Effects of Geometry on Film Properties," paper #790217 presented at Society of Automotive Engineers winter meeting, Detroit, MI (1979).
(203) W.D. Kingery, *Introduction to Ceramics*, Wiley, New York (1960).
(204) A.S. Nowick and S. Mader, in *Basic Problems in Thin Film Physics*, R. Niedermayer and H. Mayer, ed., Vanderhoeck & Ruprecht, Göttingen (1966), pp. 212-217.
(205) S.D. Dahlgren, *Metall. Trans. 1*, 3095 (1970).
(206) R.F. Bunshah and R.S. Juntz, *Metall. Trans. 4*, 21 (1973).
(207) R.F. Bunshah, *Vacuum 27* 353 (1977).
(208) E.D. McClanahan, Battelle Pacific Northwest Laboratories (private communication).
(209) H.A. Beale, R.J. Hecht, P.R. Holiday, J.R. Mullaly, E. Thompson and C.T. Torrdey, *Thin Solid Films 54*, 326 (1978).
(210) J.A. Thornton and D.G. Cornog, *Thin Solid Films 45*, 397 (1977).
(211) J.A. Thornton and D.P. Ferris, *Thin Solid Films 40*, 365 (1977).
(212) W.T. Pawlewicz, *J. Appl. Phys. 49*, 5595 (1978).
(212a) T.D. Moustakas, D.A. Anderson, and W. Paul, *Solid State Comm. 23*, 155 (1977).
(212b) J.A. Thornton, D.G. Cornog, and W.W. Anderson, *J. Vac. Sci. Technol. 18*, 199 (1981).
(213) D.H. Grantham, E.L. Paradis, and D.J. Quinn, *J. Vac. Sci. Technol. 7*, 343 (1970).
(214) B.N. Chapman, *J. Vac. Sci. Technol. 11*, 106 (1974).
(215) C. Weaver, *J. Vac. Sci. Technol. 12*, 18 (1975).
(216) K.I. Mittal, *J. Vac. Sci. Technol. 13*, 19 (1976).
(217) H. Grat von Harrach and B.N. Chapman, *Thin Solid Films 13*, 157 (1972).
(218) D.S. Campbell, in *Handbook of Thin Film Technology*, L. Maissel and R. Glang, ed., McGraw Hill, New York (1970), p. 12-22.
(219) J.A. Thornton, U.S. Patent 4,126,530 (1978).
(220) J.L. Vossen, J.J. O'Neill, Jr., K.M. Finlayson, and L.J. Royer, *RCA Rev. 31*, 293 (1970).
(221) G.J. Kominiak and J.E. Uhl, *J. Vac. Sci. Technol. 13*, 170 (1976).
(222) J.E. Houston and R.D. Bland, *J. Appl. Phys. 44*, 2504 (1973).
(223) E.L. Hollar, F.N. Rebarchik, and D.M. Mattox, *J. Electrochem. Soc. 117*, 1461 (1970).
(224) J.L. Vossen, J.J. O'Neill, Jr., O.R. Mesker, and E.A. James, *J. Vac. Sci. Technol. 14*, 85 (1977).
(225) R.G. Frieser, *J. Electrochem. Soc. 113*, 357 (1966).
(226) W.W. Molzen, *J. Vac. Sci. Technol. 12*, 99 (1975).
(227) K.S. Keskar, Y. Yamashita, Y. Onodera, Y. Goto, and T. Aso, *J. Appl. Phys. 45*, 3102 (1974).
(228) T.K. Lakshmanan and J.M. Mitchell, Proceedings 10th National Vacuum Symposium (1963), p. 335.

(229) J.A. Thornton and A.J. Jonath, "Indium Phosphide Films Deposited by Cylindrical Magnetron Reactive Sputtering," Proceedings 12th IEEE Photovoltaic Specialists Conf., Baton Rouge, LA (1976), p. 549.
(230) S.F. Vogel and I.C. Barlow, *J. Vac. Sci. Technol. 10*, 381 (1973).
(231) D.J. Harra and W.H. Haywood, *Supplemento al Nuovo Cimento 5*, 56 (1967).
(232) J. Heller, *Thin Solid Films 17*, 163 (1973).
(233) S. Maniv and W.D. Westwood, *J. Appl. Phys. 51*, 718 (1980).
(234) E. Holland and D.S. Campbell, *J. Mater. Sci. 3*, 544 (1968).
(235) J. Harvey and J. Corkhill, *Thin Solid Films 6*, 277 (1970).
(236) A.J. Stirling and W.D. Westwood, *Thin Solid Films 7*, 1 (1971).
(237) T. Abe and T. Yamashina, *Thin Solid Films 30*, 19 (1975).
(238) F. Shinoki and A. Itoh, *J. Appl. Phys. 46*, 3381 (1975).
(239) L.F. Donaghey and K.G. Geraghty, *Thin Solid Films 38*, 271 (1976).
(240) S. Schiller, U. Heisig, K. Goedicke, K. Schade, G. Teshner, and J. Henneberger, *Thin Solid Films 64*, 455 (1979).
(241) S. Schiller, U. Heisig, K. Steinfelder, and J. Strümpfel, *Thin Solid Films 63*, 369 (1979).
(242) R.H. Glang, R.A. Holmwood, and J. Kurtz, in *Handbook of Thin Film Technology*, L. Maissel and R. Glang, ed., McGraw Hill, New York (1970), p. 2-1.
(243) E. Stern and H.L. Caswell, *J. Vac. Sci. Technol. 4*, 128 (1967).
(244) S. Maniv and W.D. Westwood, *J. Vac. Sci. Technol. 17*, 403 (1980).

BIBLIOGRAPHY FOR SUPPLEMENTARY READING

(1) K.L. Chopra, *Thin Film Phenomena*, McGraw Hill, New York (1969).
(2) L.I. Maissel and R. Glang, ed., *Handbook of Thin Film Technology*, McGraw Hill, New York (1970).
(3) W.D. Westwood, "Glow Discharge Sputtering," *Progress in Surface Science 7*, 71 (1976).
(4) J.A. Thornton, "High Rate Thick Film Growth," *Annual Review of Materials Sciences 7*, 239 (1977).
(5) J.L. Vossen and W. Kern, ed., *Thin Film Processes*, Academic Press, New York (1978).
(6) M.H. Francombe, "Growth of Epitaxial Films by Sputtering," in *Epitaxial Growth*, Part A, J.W. Mathews, ed., Academic Press, New York (1975).
(7) L.T. Lamont, "Some Special Pumping Problems Associated with the Sputtering Applications," *J. Vac. Sci. Technol. 10*, 251 (1973).
(8) L. Holland, *Vacuum Deposition of Thin Films*, Wiley, New York (1956).
(9) *Physics of Thin Films* (Series), Volumes 1-10, Academic Press, New York (1963-1978).
(10) B. Chapman, *Glow Discharge Processes; Sputtering and Plasma Etching*, Wiley, New York (1980).

6
Ion Plating Technology

Donald M. Mattox
Sandia National Laboratories
Albuquerque, New Mexico

INTRODUCTION

Ion plating is a generic term applied to atomistic film deposition processes in which the substrate surface and/or the depositing film is subjected to a flux of high energy particles sufficient to cause changes in the interfacial region or film properties compared to the nonbombarded deposition. Such changes may be in the adhesion of the film to the substrate, film morphology, film density, film stress, or the coverage of the surface by the depositing film material. This definition of ion plating refers only to processes which affect the film and/or substrate and does not specify the source of the depositing material nor the origin of the bombarding species. The ion plating technique was first reported in the technical literature in 1964.[1]

Variations of the ion plating technique have given rise to other terms such as Vacuum Ion Plating, Reactive Ion Plating, Chemical Ion Plating, Alternating Ion Plating and others which relate to specific environments or techniques, many of which are covered in the Bibliography Appendix.

Ion plating is typically done in an inert gas discharge system similar to that used in sputter deposition, except that the substrate is made of the sputtering cathode.[2-7] The substrate is subjected to inert gas ion bombardment for a time sufficient to modify the substrate surface prior to deposition of the film material. The film deposition is begun without interrupting the ion bombardment. For a film to form, it is necessary for the deposition rate to exceed the sputtering rate. Ion bombardment may or may not be continued after the interfacial region has been formed.

Ion plating has been used most often to provide good adhesion between a film and a surface. From the standpoint of adhesion, the principal benefits obtained from the ion plating process are its ability to:

(1) Modify the substrate surface in a manner conducive to good adhesion and maintain this condition until the film begins to form.

(2) Provide a high energy flux to the substrate surface, giving a high surface temperature, thus enhancing diffusion

Ion Plating Technology 245

and chemical reactions without necessitating bulk heating.

(3) Alter the surface and interfacial structure by introducing high defect concentrations, physically mixing the film and substrate material, and influencing the nucleation and growth of the depositing film.

The factors affecting the adhesion of coatings to substrates are reviewed in Reference 8.

In addition to modifying the substrate surface and interface formation, the ion bombardment of the growing film may cause a modification of the morphology of the deposited material, changes in the internal stress of the deposited film, or the modification of other physical and electrical properties (refer to Bibliography).

Another advantage of ion plating, in some situations, is the high "throwing power" or ability to cover a surface often found with ion plating as compared to that obtained by vacuum evaporation. This high throwing power results from gas scattering, entrainment and redeposition from the sputtered film surface. This allows coatings to be formed in recesses and areas remote from the source-substrate line of sight, thus giving more complete surface coverage.

The use of a partial pressure of reactive gas (oxygen, nitrogen, hydrocarbon) allows the deposition of compounds in *reactive ion plating*.

The application of ion plating has often been limited to situations where it is the only technique by which the desired results can be obtained. An exception to this are applications where the process is being used as one means to replace electroplating with "dry plating" processes in order to avoid water pollution problems.

When ion plating process was first introduced, the origin of the benefits derived from ion bombardment were the subject of much speculation and controversy. Since that time, a great deal of data and some understanding has been developed on the effects of low energy ion bombardment of surfaces and how ion bombardment affects adhesion and film properties.

METHODS OF ION BOMBARDMENT

Ion Gun (Vacuum Environment)

Figure 6.1 shows an ion gun system in which the ions are formed in a Kaufman ion source and then extracted into a vacuum chamber where they impinge on the substrate. The Kaufman ion source utilizes a magnetically enhanced low voltage discharge to create the ions which are then extracted at the desired energy (250 eV to 2,000 eV). After extraction, through a multiaperture screen grid, into the vacuum system, the ions impinge on the substrate. Deposition atoms are obtained by sputtering a target with a portion of the ion beam or by using a shielded evaporation source.

In this setup the bombarding ions do not suffer collisions and, therefore, impinge on the substrate surface with high energy. The charge compensation filament minimizes space change effects on the beam and change buildup on insulator surfaces. Commercially available ion gun systems are capable of providing argon ion beams up to 10 inches in diameter with energies to 2,000 eV and current densities to 1 mA/cm^2. These systems are most often used for ion beam etching.

The ion gun system has the distinct advantage that the ion bombardment parameters can be controlled independently of other deposition parameters. It has the disadvantage that some of the scattering mechanisms which operate at gas discharge pressures are not present since the substrate is in a vacuum.

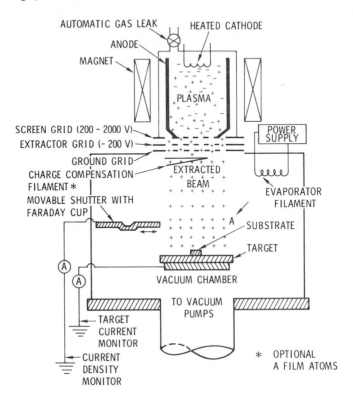

Figure 6.1: Ion gun ion-plating system with a resistive heater for the source of depositing material.

Gas Discharges (0.5–10 μ Gas Environment)

If one applies a DC voltage between two electrodes in a chamber containing a gas at a pressure greater than several microns, one finds that at low voltage little current is drawn and there is a uniform potential gradient across the chamber. As the voltage is increased, the current increases slightly until the breakdown potential is reached, at which point the current increases rapidly and the potential drops to a lower value. At this voltage and gas pressure, one enters the *normal glow discharge region* where, with a constant voltage, the current is a function of gas pressure and the cathode current density is constant over the cathode spot. The cathode spot has a velvety glow whose color is characteristic of the cathode material and is due to the de-excitation spectrum of sputtered material. This color often can be observed to change as the surface is sputter cleaned. At higher pressures it is found that the cathode spot covers the whole cathode and the current density becomes a function of the voltage at constant pressure. This is the abnormal glow discharge region and is the region in which ion plating and sputter deposition is done. To have a self-sustained DC diode gas discharge at several thousand volts DC, in argon requires a minimum of about

10 μ argon gas pressure. For neon, the pressure must be greater; but for a krypton discharge, the pressure can be less.

The *plasma region* is a low field region where there is an equal number of ions and electrons, thus the net space charge is zero. This field strength is low and the region is luminous. The plasma region in an ion plating gas discharge consists of a high percentage of neutral atoms and is thus a weakly ionized and low temperature plasma. Since the ions which are accelerated to the cathode originate primarily near the edge of the plasma region, it is important that the plasma density be kept constant over the cathode surface or the ion current density to the cathode will vary, causing variations in bombardment effects.

Of primary interest are the *cathode dark space* where most of the potential drop is found, and the *negative glow region* where most of the ionization occurs. The width d of the cathode dark space is inversely proportional to the gas pressure p such that

$$pd \simeq \text{constant}$$

This distance represents the number of electron mean ionization path lengths required for secondary electrons to produce enough ions to maintain the discharge and is determined by pressure, type, and temperature of the gas. If the discharge chamber windows are clear, you can see this relatively dark region vary in width with pressure during sputtering. For an argon discharge at 10 μ pressure, the cathode dark space is about 2 cm wide. It should also be pointed out that the color of a gas discharge depends on the gas present, as well as plasma density and temperature. An experienced operator can often tell when he has contaminant atoms in the discharge by its color. If the anode or a ground is moved into the cathode dark space, the discharge between the electrodes is extinguished.

Figure 6.2 schematically depicts the processes that occur in the cathode region of the discharge. (Numbers and letters in parentheses refer to portions of Figure 6.2.) The primary source for ionization of the discharge gas is electron-atom collision $e^- + G^0 \rightarrow G^+ + 2e^-$ (1). The gas ions that are produced may then become part of the plasma and ultimately be lost to the chamber walls; or they may be accelerated across the cathode dark space to the cathode (2). In traversing the cathode dark space, the accelerating ions may suffer charge exchange collisions with thermal neutrals, $G_1^+ + G_2^0 \rightarrow G_2^+ + G_1^0$, producing a spectrum of high energy neutrals and high energy ions. The high energy atoms/ions that impinge on the cathode may:

(1) produce secondary electrons (1)
(2) sputter cathode atoms (S) (4)
(3) sputter surface contaminants (sputter cleaning)
(4) become neutralized and reflect from the cathode surface as high energy neutrals or as metastable atoms (G^*) (3)
(5) or become incorporated into the cathode surface and change the surface properties.

In ion plating, the impinging particles may sputter film atoms (M) (5). High energy neutrals may cause sputtering from surrounding fixturing which is not at the cathode potential.

The secondary electrons that are produced are accelerated back across the cathode dark space causing ionization which sustains the discharge. These high energy electrons are also the source of much of the film/substrate heating during DC sputter deposition.

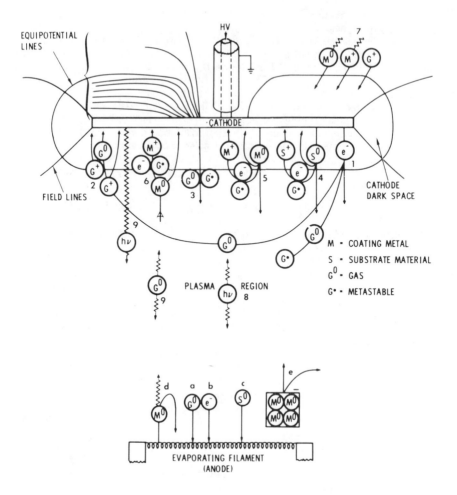

Figure 6.2: Processes which occur in a DC diode discharge using an inert gas.

The sputtered surface atoms (film, substrate, or contaminant) can be:

(1) Scattered back to the cathode (4).

(2) Ionized by electron-atom collision or metastable-atom collision $S^0 + G^* \rightarrow S^+ + G^0 + e^-$ (Penning Ionization) and accelerated back to the cathode or lost from the cathode region (4).

(3) Transported as high energy neutrals to some surface (sputter deposited) (a).

Penning ionization is probably the most important ionization process for sputtered cathode atoms. The energy distribution of sputtered atoms is higher (1-100 eV) than that of thermally vaporized atoms (0.2-1 eV). This may account for many of the interesting properties of sputter-deposited films though much of this energy is lost by thermalizing collisions except in a vacuum environment. The return of sputtered atoms to the cathode surface is a function of many discharge parameters, as well as the system throughput. This return is important in sputter cleaning, generation of surface features (such as cones and whiskers), surface coverage (ion plating), and the generation of the pseudodiffusion type of interface.[8] It has been estimated that at 0.1 torr gas pressure, 90% of the sputtered atoms are scattered back to the cathode. It should be noted that only neutral or negatively-charged particles can escape from the cathode surface.

In ion plating, film atoms are injected into the discharge from a source such as resistively-heated filament. Such a source is shown in Figure 6.2. Here the film atoms M may be:

(1) Scattered back to the filament (d).
(2) Deposited as neutrals on the substrate (cathode) (6).
(3) Ionized by electron-atom or metastable-atom collision and accelerated to the substrate (6), or
(4) Nucleated in the vapor phase to form small particles (gas evaporation) which become negatively charged, and may be collected on a positively-charged surface or lost to the system walls (e).

In any case, the film atoms act as foreign particles in the gas discharge and can cause changes in the discharge parameters.

The high energy gas atoms that are trapped in the cathode surface may be important. The incorporated gas atoms may be many atomic percent under some discharge conditions. Temperature is important in gas incorporation with higher temperatures giving less retention. The gas atoms that are incorporated into the cathode (film) may affect the film properties. The gases trapped in the surface during sputter cleaning may be released after film formation and cause poor film adhesion; therefore, care must be used in the application of sputter cleaning. Generally, heavy gas atoms are less likely to be incorporated into the surface than are light gas atoms.

If the gas in the system (or impurities in it) can react with the cathode surface, the ion bombardment can be used to form a compound or alloy surface. This makes sputter cleaning the surface of a reactive metal difficult if a reactive gas is present. Contaminant gases in the discharge are often more reactive than they are in the simple gaseous state since a portion of them are disassociated or ionized.

It should be noted that in an ion plating gas discharge only a small percentage of the gas atoms are ionized, so the cathode and other surfaces in the system are being continuously bombarded by thermal neutrals. In addition, the discharge acts as a rather intense source of ultraviolet radiation $h\upsilon_1$ (8) which can affect organic surfaces and possibly insulator surfaces. The cathode surface will become heated during sputtering since most of the energy of the bombarding ions is given up as heat and will radiate energy $h\upsilon_2$ (9). The current, as measured by the cathode, is not totally due to ion current, but is a combination of ion current and secondary electron current. It is, however, valuable to know the total cathode current density in order to establish reproducible sputtering conditions and in determining when a surface has been sputter-cleaned.

The *diode sputtering* configuration with a self-sustained discharge (cold cathode) is the most widely used ion plating technique. A typical system of this type is shown in Figure 6.3. Another configuration uses a hot electron-emitting filament as the electron source (hot cathode). This is called *triode ion plating*. In this technique, the plasma is generated between a hot electron-emitting filament and an anode. Ions are extracted from this plasma to bombard the negatively-biased cathode. This type of discharge can be maintained at a lower gas pressure than the diode configuration. The ions that bombard the cathode from such a low pressure plasma probably have an energy spectra in which most of the ions have energies close to the bias potential. A principal difficulty in triode ion plating is that of maintaining a uniform plasma density and composition.

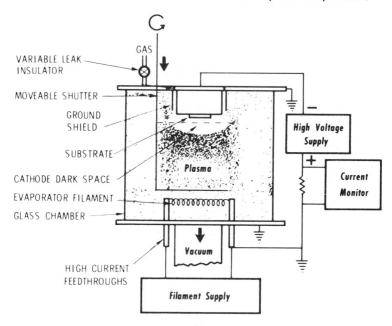

Figure 6.3: Typical ion-plating system using a DC diode discharge.

In any of the DC type discharges, an axial magnetic field of 50-200 Oe can be used over the substrate surface to increase the electron path length, thus increase the ionization probability as is done in magnetron sputtering. A problem with the use of the magnetic field is, again, maintaining plasma uniformity.

Since ion bombardment is necessary to sustain the discharge in DC ion plating, the above discussion applies only if the cathode is a conductor. If the cathode is an insulator, it builds up a surface charge which prevents bombardment by positively-charged particles. To permit ion bombardment of an insulator surface, such as glass, ceramic, or organic a setup such as shown in Figure 6.4 may be used.

In this case, a plasma of ions and electrons is generated either by the RF discharge or by an auxillary DC discharge. The insulator to be coated is placed close to the plasma with a metal electrode on the back. This arrangement is then a capacitor with capacitance C_g, with the plasma forming one electrode and the metal backing plate the other. When an RF field is applied to the capacitor, the insulator surface in contact with the plasma, is alternately polarized positive and negative. When negative it attracts ions. Figure 6.4 shows the potential generated at the surface. Since the electrons have a higher mobility than the ions, the surface does not attain a high positive bias but does attain a high negative

Ion Plating Technology   251

bias. If the RF supply has a high peak-to-peak voltage (typically >500 volts), the surface bias will be sufficient to accelerate ions from the plasma to a high enough energy to cause sputtering. This technique is called *RF ion plating* and may be used to coat insulators or metals which tend to form dielectric layers on their surfaces or when depositing a dielectric coating. A DC potential with a superimposed RF can also be used to increase the sputtering from materials which form insulating surfaces.

Whenever an electrically isolated surface is exposed to a DC plasma, it will assume a negative potential with respect to the plasma. This occurs because the electrons, with their higher mobility, are lost to the surface at a higher rate than are ions. The potential between the surface and the plasma is called the *wall potential* (or sheath potential) and will be about 10 volts negative with respect to the plasma for most ion plating arrangements. This wall potential accelerates ions from the plasma. These low energy ions are then capable of desorbing contaminants from the surface. This cleaning mode is often called *glow discharge cleaning* or "ion scrubbing" and is often used to clean insulator surfaces before vacuum deposition. This cleaning, along with heating from the plasma, is the source of much of the contaminant release during presputtering of a system. If the plasma is generated by an RF discharge, the surface may attain a positive, negative, or zero potential with respect to the plasma, depending upon its coupling to ground.

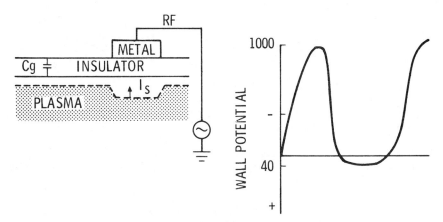

Figure 6.4: Ion bombardment of an insulator using an applied RF potential.

ION BOMBARDMENT EFFECTS: PRIOR TO DEPOSITION

Ion bombardment of a surface results in a number of effects including: (1) sputtering, (2) defect production, (3) crystallographic disruption, (4) surface morphology changes, (5) gas incorporation, (6) temperature rise, (7) changes in surface composition and (8) physical mixing of near-surface materials.

Sputtering

Physical sputtering, which is usually referred to as sputtering, is a momentum transfer process in which an incoming particle creates a collision cascade which intersects the surface causing the ejection of a surface atom. The sputtering yield is defined as the number of atoms ejected per incident particle (ion) and is a function of the relative masses of the impinging and surface atoms, incident particle energy, angle of incidence of the impinging particle, surface morphology and the bombardment history of the surface.

High mass and high energy impinging particles will give increasing sputtering yields up to the point where most of the energy is dissipated so deeply in the surface region that the major portion of the collision cascade produced does not intersect the surface. The sputtering yield will increase with off-normal angles of incidence up to 60°-70° and then decrease as the incident ions are reflected from the surface. Typically, angular dependence can account for a factor of two variations in the sputtering yield.

Sputtering yield is also bombardment dose dependent, increasing with bombardment dose to some nearly equilibrium valve. The increased sputtering yield with dose is attributed to the introduction of defects and foreign atoms into the near-surface region which reduces the local atom bonding. The variation of sputtering yield is also a function of the bombarding environment. For instance, if there is a high residual gas pressure, the sputtered atoms will be backscattered to the surface.

If materials having a low sputtering yield are mobile on the target surface, then the low yield material may agglomerate into an island structure giving cone formation on the surface. The surface mobility may allow appreciable amounts of bulk material to be removed without removing the initial surface material. This has led to problems with composition profile measurements by sputter-Auger and secondary ion mass spectrometry in some systems.

Chemical sputtering of a surface results when the energetic and activated bombarding species form a chemical bond with the surface species giving a volatile or more easily sputtered material. This effect enhances the apparent sputtering yield. It is difficult to separate the physical and chemical sputtering components of the sputtering of materials by reactive gases. Often chemical reaction with the cathode surface will reduce the apparent sputtering yield in DC sputtering by forming an insulator surface.

Sputtering of a surface results in the removal of surface material, hence may be called sputter cleaning: The use of a reactive gas gives reactive plasma etching or reactive plasma cleaning of the surface.

Defect Production

The amount of energy transferred to the lattice atom (E_t) by the bombarding particle depends on the relative masses of the particles and is given by:

$$E_t = \frac{4M_i M_t}{(M_i + M_t)^2} E$$

where E is the energy of the incident particle of mass, M_i and M_t is the mass of the target atom. If the transferred energy exceeds about 25 eV, the lattice atom may be displaced to an interstitial site and a point defect is formed (damage threshold). If less than 25 eV, the energy will all appear as heat.

The combination of thermal agitation and displacement will give migrating interstitials and vacancies which agglomerate and result in the formation of dislocation networks. Even with agglomeration of defects, there remains an extremely high residual concentration of point defects in the surface region of an ion bombarded surface. These defects have been studied by the trapping and desorption of low energy helium ions and show that the trapping sites are vacancies with varying binding energies.

Crystallographic Disruption

If the defects produced by ion bombardment are sufficiently immobile, the surface crystallography will be disrupted into an amorphous structure. Also, gas incorporation will tend to disrupt the surface crystallography. These effects are well known by investigations concerned with Low Energy Electron Diffraction (LEED) studies of surface crystallography. These studies normally require that ion bombarded surfaces be extensively annealed in order to restore the surface crystallography.

Surface Morphology Changes

Ion bombardment of a surface results in a wide variety of topographic changes in both crystalline and amorphous surfaces giving increased surface roughness and changes in sputtering yield. Generally, there is a strong dependence of surface morphology on ion bombardment dose, particle energy, species and angle of incidence as well as surface condition such as crystallography and impurity concentration.

Gas Incorporation

Incorporation of gases into surfaces and growing films by low energy ion bombardment has long been recognized and studied and is the basis for inert gas pumping by "getter ion pump" type vacuum pumps. Inert gases have been deliberately incorporated into surfaces and films by low energy ion bombardment for a variety of reasons including surface thermometry (^{85}Kr) and diffusion studies (H, He). An atom ratio of helium in a gold matrix as high as 40% atomic percent have been reported even though the solubility of helium in gold is zero. The ability to incorporate the insoluble gases into the lattice of the surface or depositing film depends on the mobility, trapping sites and temperature, as well as the energy of the depositing species. Generally, amorphous materials trap more gas than do crystalline materials. In bias sputter deposition done with argon, the gas concentration may be as high as several percent and may cause blistering on subsequent annealing. Gas incorporation increases with particle energy until heating by bombardment causes the release of the entrapped gas.

Temperature Rise

Most of the energy of bombarding particles appears as surface heating. The surface to mass ratio, thermal properties of the system and the energy input of the system determines the resulting bulk temperature of the bombarded material. The atoms in an ion bombarded surface are being displaced from their equilibrium positions with energies up to or greater than 25 eV (damage threshold) and the meaning of a surface temperature with such great agitation is uncertain.

Surface Composition

Without diffusion, atoms will be removed from a surface by sputtering in a stoichiometric ratio after an initial conditioning period. Because of differential sputtering yields of the components of a system, the surface composition may differ from that of the bulk. Ion bombardment itself may result in surface composition changes due to preferential "knock-on" or recoil effects sometimes called "recoil implantation."

Diffusion in the surface region can have a pronounced effect on composition. The presence of a high defect concentration and a high temperature will enhance diffusion such as is found with radiation enhanced diffusion. Since the surface acts as a sink for point defects, the defect flow will allow solute segregation with small ions segregating to the surface.

Physical Mixing

Physical mixing of the near-surface material results in what may be termed a "Pseudodiffusion" layer where the mixing occurs without the need for solubility or diffusion. Recoil implantation results in surface materials such as oxygen, or carbon being embedded in the surface region. If the sputtering atoms are returned to the surface by backscattering, there will be an intermixing. If the sputtered atoms are ionized and accelerated back to the surface, they will become embedded in the surface region.

ION BOMBARDMENT EFFECTS: INTERFACE FORMATION

Ion bombardment of a surface, while adding coating atoms, results in: (1) physical mixing, (2) enhanced diffusion, (3) modifying nucleation modes, (4) preferential removal of loosely bound atoms and (5) improved surface coverage.

Physical Mixing

Mixing of the near-surface region by nondiffusion type mechanisms will occur due to implantation of high energy particles, backscattering of sputtered atoms and recoil implantation of surface atoms. The high energy particle may arise from a variety of sources. The coating atoms may be ionized and accelerated while traversing the plasma in a glow discharge or may be ionized in an ion source then extracted and accelerated back to the surface. The net effect of this intermixing will be to form the aforementioned "pseudodiffusion" type interface.

Enhanced Diffusion

The high defect concentration and temperature of the near-surface region will enhance the diffusion rates. Diffusion into the bulk or surface segregation may be enhanced by solute segregation effects where small ions will tend to segregate to the surface which acts as a sink for point defects.

Nucleation Modes

The behavior of an atom condensing on the substrate surface is determined by its interaction with the surface. If there is no strong bonding between the condensing atom and the surface, the atom will diffuse over the surface until it nucleates at a high energy site or by collision with other diffusing atoms. This may be termed as a nonreaction nucleation mode and will lead to widely spaced nuclei. The growth of the nuclei will possibly lead to a "dewetting" type growth with interfacial porosity as the nuclei grow together. The nuclei will not grow together until they have reached an appreciable thickness. This type of nucleation and growth leads to the island-channel-continuous growth pattern often studied with electron microscopy.

If the condensing atom reacts strongly with the surface, the surface mobility will be limited and the nucleation density will increase and in the limit the depositing atoms will form a continuous monolayer on the substrate surface. If there is a chemical reaction or diffusion, the condensing atoms will react with the surface to form compound or alloy layers which extend both normally and laterally to the surface. This mode of growth will tend to decrease interfacial porosity and "grade" the properties of the interfacial region. Extensive interfacial reaction may degrade the interface by the formation of porosity, due to differential diffusion rates or by high intrinsic stresses generated by thick interfacial compound layers. Very thin contamination layers on the surface can convert the diffusion/reaction nucleation mode to the nonreaction mode with an attendant undesirable interfacial region.

An ion bombarded surface, with its surface morphology and disruption probably provides more nucleation sites than an undisturbed surface giving a higher nucleation density even for nonreactive systems. Ion bombardment of the surface will tend to remove and disrupt contamination and barrier layers which prevent the diffusion/reaction nucleation mode. Implanted atoms may also act as nucleation sites.

Preferential Removal

Since sputtering of surface atoms is dependent on the local bonding, it is not unexpected that ion bombardment of a surface will tend to sputter the more loosely bound atoms. This is particularly evident when the reaction/diffusion type of interface is being formed and has been used to deposit pure Pt-Si layers by sputter depositing platinum on silicon and backsputtering the excess platinum. It has also been noted that "high quality" SiO_2 films can be formed by bias sputtering and this is attributed to backsputtering of loosely bonded atoms, though other effects may be important.

Surface Coverage

The increased "throwing power" of the ion plating process over large areas and out of line-of-sight of the vapor source when used in a gas discharge, is most probably due to gas scattering. The effect has been seen at high gas pressures without the discharge present but ion bombardment is probably important in obtaining an adherent coating since evaporation into a high gas pressure will lead to vapor phase nucleation and the deposition of a fine powder as is done in "gas" evaporation. Vapor phase nucleated particles in a gas discharge will assume a negative potential and be repulsed by the cathodically charged ion plating substrate or may be used for "Ionized Cluster Beam" deposition.

On a microscopic scale, the enhanced surface coverage is probably due to sputtering of the deposited atoms into regions out of line-of-sight of the source. Backsputtering has been used to preferentially deposit materials on vertical sidewalls even to the extent that all deposited material on the surfaces normal to the impinging ion beam is completely removed.

ION BOMBARDMENT EFFECTS: FILM GROWTH

Deposited materials often have properties and behaviors which are appreciably different than those found in bulk material. These differing properties and behaviors may include: (1) small grain size, (2) high defect composition, (3) low recrystallization temperature (metals), (4) low strain point (glasses), (5) high intrinsic stresses, (6) metastable crystallography and phase composition and many others. Ion bombardment of a film during deposition may affect film morphology, crystallographic composition, physical properties and many other properties.

Morphology

The morphology of a depositing film depends on how the depositing adatoms are incorporated into the existing structure. On a smooth surface, preferential growth of one area over another may result from varying surface mobilities of the depositing atoms usually as a result of grain orientations. Preferential growth leads to a dominant grain orientation and a surface roughening with thickness. As the surface gets rough, geometrical shadowing will lead to preferential growth of the elevated regions giving a columnar morphology to the deposit. A pre-existing surface roughness or preferential nucleation will also lead to shadowing and a columnar morphology. An elevated substrate temperature will affect the morphology by increasing surface mobility, enhancing bulk diffusion and allowing recrystallization to occur. These effects have led to the structure zone model of deposited materials of Movchan and Demchishin. The model consists of the formation of three zones which depend on the ratio of the surface temperature (T) to the melting point of the deposited material (T_m) as is shown in Figure 6.5.

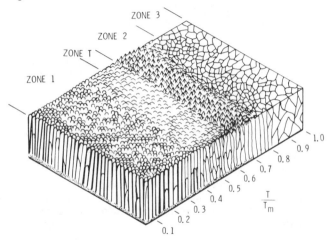

Figure 6.5: Movchan and Demchishin diagram of structural zones in vacuum condensates.

Zone 1 results when adatom diffusion is insufficient to overcome the effects of shadowing, giving a columnar structure with low density boundaries between the columns. The individual columns are polycrystalline and are usually highly defected with small grain size. The surface morphology is generally rounded.

Zone 2 is defined as the range of T/T_m where adatom diffusion dominates and the column structure consists of less defected and larger grains with higher density boundaries between the columns. The surface morphology of the Zone 2 material is generally more angular than that of the Zone 1 material.

Zone 3 is the T/T_m region where bulk diffusion and recrystallization dominate. The material is more equiaxed with high density grain boundaries and large grains.

Rough surfaces, high gas pressures, oblique adatom flux and reactive contaminant gases will tend to lower the T/T_m values for the zone boundaries. Ion bombardment during deposition may tend to raise the T/T_m values for the zone boundaries though this effect may be offset by gas incorporation, and smaller initial grain size.

Crystallography

Structural order is generally controlled by the surface mobility of the adatoms. Low mobility will lead to amorphous or small grained material. High mobility will lead to large grained material and more structural perfection. Bulk diffusion will determine the crystallography of materials deposited at elevated temperatures (Zone 3). Preferential crystallographic orientation will result from the differential surface mobilities at low temperatures or the recrystallization kinetics at high temperatures.

Ion bombardment of the depositing material may influence crystallography by enhancing diffusion and recrystallization or by altering the nucleation stage of the film growth.

Composition

As previously discussed, ion bombardment can alter the composition of the depositing material by preferentially sputtering loosely bound atoms or implanting atoms into the growing surface region to form metastable phases. In the extreme case, many atomic percent of insoluble gases can be incorporated into a depositing film by ion bombardment.

Physical Properties

Intrinsic stress is probably one of the most obvious properties which is affected by ion bombardment. Intrinsic stress is generated by atoms which are not in their lowest energy configurations. Generally, evaporative deposited films have a tensile stress while sputter deposited films have a compressive stress. The magnitude of the stresses often approaches the yield point of the bulk materials and in some cases may exceed the yield point of pure bulk materials because of incorporated impurity atoms. Compressive stresses develop when the atoms are in closer proximity to one another than they would be under more equilibrium circumstances. Tensile stresses develop when atoms are further apart than they would be under more equilibrium circumstances possibly due to large numbers of interstitial atoms. Some care must be taken with these interpretations since stress is often measured by x-ray analysis of lattice parameters which tell nothing about intergranular forces. Beam deformation measurements give an integrated stress measurement which may be more meaningful.

For sputter deposited films, an elevated gas pressure, an oblique adatom flex or an elevated temperature will reduce the compressive film stress. In some cases, stress may be reduced by a relief mechanism such as diffusion or recrystallization. Recrystallization temperature and rate is dependent on the intrinsic strain energy and defect concentration. In other cases, the structure may be unable to support the stress such as the low density structure of Zone 1 material where the weak columnar boundaries will not be able to transmit a stress. In such a case, the x-ray analysis may show high stress but a bending beam test will show little integrated stress.

Ion bombardment during deposition may increase the stress by forcing atoms into nonequilibrium properties and suppressing Zone 1 formation, or may decrease stress by enhancing stress relief mechanism such as diffusion and recrystallization. Generally, the more refractory the material, the more likely ion bombardment is to increase stress. A low intrinsic stress is usually desirable for good adhesion though, in some cases, a compressive stress may be beneficial since materials under compression are generally stronger than unstress materials because fractures will tend not to propagate.

Other physical properties such as hardness, yield strength, etc. are probably

258 Deposition Technologies for Films and Coatings

affected by ion bombardment during deposition, but there have been few studies in this area.

SOURCES FOR ION PLATING

Vapor Sources

Figure 6.6 shows a variety of methods of providing the vapor to be deposited in ion plating. Each technique has its advantages, disadvantages, and uses. Resistively heated filaments are the most commonly used vapor sources. The technique of using a metal-bearing gas or hydrocarbon (for carbon) as a vapor source may be compared to chemical vapor deposition and has been termed *chemical ion plating*. The electron beam heating technique (*e-beam ion plating*) provides the ability to deposit refractory materials or large quantities of materials easily. The use of a sputtering target as the vapor source in ion plating is essentially the same as bias sputtering except that a sputter cleaning step is introduced and often higher bias voltages are used in ion plating.

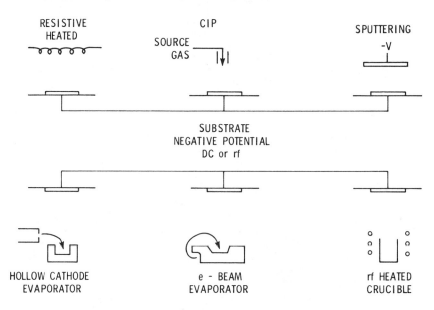

Figure 6.6: Vapor sources for ion plating.

Sometimes when using a sputtering target to provide the film material, it is necessary to continuously decrease the potential on the substrate so that at some point the deposition rate becomes greater than the removal rate. This is particularly true when depositing an easily sputtered material such as gold. *Vacuum ion plating* can be done using an ion gun and a vapor source in a relatively low pressure (10^{-5} torr).

Flash evaporation has also been used as a source for alloy deposition and plasma sources have been used to vaporize and partially ionize the film material. Arc discharges can also be used as a source of film material. The material ejected from vacuum arcs is often highly ionized. Multiple sources may be used to deposit multicomponent films.

Ion Plating Technology   259

An interesting variation on atom deposition is the use of ionized atom clusters obtained by adiabatic expansion of a vapor from a vaporization cell (ionized cluster beam deposition).[9] In this case, the atom clusters contain 10^3–10^5 atoms with only a few units of charge.

Enhanced Ionization Sources

The ion plating process generally depends on the ionization of a gaseous species to provide the ions needed for ion bombardment. The ionization in the gas phase can be enhanced by the application of a magnetic field to increase the path length of electrons as in magnetron sputtering. Enhanced ionization can be obtained by means of the configurations shown in Figure 6.7 where enhancement is obtained by the use of a hot cathode plasma normal to the deposition axis, by the use of an electron emitting filament between the source and substrate and in the case of an e-beam source, a positive electrode can be used to attract secondary electrons away from the region of the crucible to enlarge the plasma region as used in the ARE process (Chapter 1). In some cases, with easily ionized coating materials, such as chromium, there may be appreciable ionization of the material being deposited. In some instances, a glow discharge can be established using only metal vapor but the discharge is very unstable and this method is generally not practical.

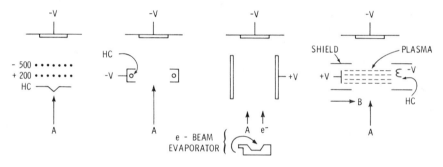

A ATOMS FROM SOURCE
B MAGNETIC FIELD
HC HOT CATHODE

Figure 6.7: Methods of enhancing the ionization in an ion-plating system.

By using the correct ion source, coating deposition can be done using only ions of the depositing species. Current densities are generally low and the ion energies must not be very high or self-sputtering will prevent the formation of a coating.

APPARATUS FOR ION PLATING

Figure 6.3 shows a typical batch ion plating system. In a batch-type operation, the system is cycled between atmosphere and deposition conditions. In some cases, this leads to an inconsistent process, for reasons which will be discussed later. If this problem is encountered, it would be well to use a system having a separate loading and unloading lock so that the bombardment chamber is subjected to as little change as possible. In the extreme case, a continuous (in-line) system can be used which allows the deposition system to be maintained in a clean condition. This is especially useful when using a sputtering source. The system shown in Figure 6.3 can be divided into subsystems and categorized as:

(1) Vacuum system
(2) Gas discharge chamber
(3) Gas handling system
(4) High voltage substrate
(5) Vapor source
(6) High voltage power supplies (RF, AC, or DC)

Vacuum System

The vacuum system associated with the ion plating process has two functions. The first is to reduce the level of undesirable gases to some tolerable amount and the second is to keep such undesirable gases at an acceptable level during deposition. A wide variety of pumping systems are available for this purpose and all have their disadvantages. A good vacuum system should be constructed and assembled using good vacuum practices and meet certain ill-defined criteria. These include:

(a) A good ultimate vacuum capability
(b) A low outgassing rate under heating and discharge conditions
(c) A low leak-up rate
(d) A high throughput of reactive gases at sputtered pressures
(e) Low backstreaming from the pumping system

All these requirements are relative to the process being used. For example, you could use air as the discharge gas and a mechanical pump for depositing gold on ceramics; however, if one were to attempt to deposit aluminum this way, the aluminum would be oxidized by residual oxygen. The message here is that for a given process and product, acceptable system capability limits should be established and then adhered to.

What then should be the reasonable capability limits of a good batch type vacuum system? The system should be capable of being evacuated to the high 10^{-7} torr range in a few hours without baking. With bakeout to 300°C, the vacuum should be in the low 10^{-7} torr range when cold. Bakeout reduces the potential for contamination by desorbing gases from the internal surfaces of the system. This kind of ultimate vacuum can be obtained with diffusion pumps, ion pumps, turbomolecular pumps, cryopumps, etc. In our laboratory, turbomolecular pumps have been found to be very satisfactory for batch-type operations as long as the foreline pressure is kept low to prevent backstreaming. We often use inert gas and oxygen discharges to clean up vacuum systems which cannot easily be baked. Flushing with the discharge gas several times after short discharge cycles is very effective in reducing the contamination level.

Good ultimate vacuum capability is not the whole story. For instance, increasing the pumping speed to compensate for a leaky system does not take the place of a tight vacuum system. Outgassing and leak-up rates are important. Outgassing usually comes from:

(1) Poorly designed or constructed system
(2) Dirty system
(3) Poor process procedures

Under Item (1) comes things like heating organic seals, no relief on threaded holes, tight mechanical fits, etc. Item (2) occurs when there are too many gadgets in the system, components improperly cleaned before assembly, or there is too much deposit buildup on fixturing. Item (3) is often difficult to pinpoint, but can come from such things as humid air as the bleedback gas, water cleaning of fixturing, condensing moisture in the system, etc.

Leak-up rates, in which you watch the pressure rise after closing off the pumping system, allow you to measure the real and virtual leaks in the system. Again, an acceptable value must be established for each system and product.

A real problem with a sputtering or ion plating system is that the throughput of system is usually greatly reduced in order to allow the pressure of the sputtering gas to be raised to the necessary level to establish a gas discharge. This can be disastrous! If the throughput is too low, the system will be unable to flush contaminants out of the processing chamber. If the throughput is too high, the pumps may operate inefficiently and backstreaming can occur. Pay attention to the foreline pressure and keep it below the maximum pressure quoted by the manufacturer.

The presence of contaminant gases can be a very real problem in most cases. These contaminants can come from the chamber walls, real or virtual leaks, substrates, targets, etc. Usually a *presputtering or preconditioning* time is used before deposition to rid the system of as much of these contaminant gases as possible. The preconditioning causes desorption of gases from the walls, as well as cleans the cathode. The throughput of the system is very important in determining the length of this presputtering time. An equilibrium value is taken to be when the discharge current reaches an equilibrium at a given gas pressure and voltage. This presputtering is used in ion plating to sputter clean the substrate surface and to desorb gases from the system walls.

The problem of a low throughput can be alleviated when using an inert gas discharge by incorporating a getter pump (such as a titanium sputtering, sublimation, or evaporation pump) in the deposition chamber. This pump will have a high pumping speed for reactive gases, but a low pumping speed for the inert gases. Of course, this will not work when using reactive sputtering or sputtering some types of compounds such as oxides.

Backstreaming can be kept low through good vacuum system design and by keeping the foreline pressures so low that the pump can work efficiently and baffling/cold trapping the diffusion pumped systems. Generally, system performance is an empirically determined parameter for each process. This is not always the case, however; and some process control techniques will be discussed in a separate section (Process Controls).

Gas Discharge Chamber

The primary function of the gas discharge chamber is to provide as uniform an environment as possible. The most important features of this environment are:

(a) Uniform electric field distribution
(b) Uniform plasma density

(c) Uniform plasma composition
(d) Uniform deposition
(e) Uniform temperature

In order to have a uniform field distribution and flow pattern, it is desirable to construct the system as axially symmetric as possible. If the system is not symmetric, it means that some portion of the cathode will be closer to ground than another, the electric field will be distorted, and there will be a plasma density variation due to wall losses. This effect can be felt for many times the cathode dark space distance. For this reason, all surfaces not at cathode potential should be 5-10 times the cathode dark space distance from the surface to be coated. Shields around the cathode should be kept well away from the active area of the substrate.

The gas inlet should be axially symmetric to the substrate holder and opposite the pumping port. Particularly in the case of reactive deposition, there should be no stagnation zones in the gas flow. Gases should be premixed before being injected into the chamber to ensure compositional uniformity. It should be kept in mind that vacuum pumps may pump different gases at different rates which may affect gas composition.

To prevent dust and debris from settling on the substrates and causing pinholes, it is usually best to deposit up or horizontally. This may present some problems in holding the substrates on the substrate holder, but this can be solved with a little ingenuity.

All systems should be equipped with a shutter so that the surfaces can be sputter cleaned and the system preconditioned without depositing contaminants on critical surfaces. When opened, this shutter should be axially symmetric and well away from the plasma region of the discharge (if possible). It is helpful if the shutter can be moved close to critical surfaces while sputter cleaning in order to prevent scattering of the contaminants behind the shutters.

Gas Handling System

In the usual ion plating system, inert gases or gas mixtures are used. Gases are normally provided from commercial sources whose purity is known. In general, the highest purity gas that is economically feasible should be used. Inert gases may be purified by flowing them over a hot active metal (e.g., Ti, Zr, etc.) chip bed or by the use of a *getter deposition* system where the incoming gases are exposed to a freshly deposited film before they enter the discharge region. The gases should be introduced into the system in a noncontaminating way. In reactive sputter deposition, it is usually only necessary to have a few percent reactive gas and the balance inert gas. Relative composition of gas mixtures can be very important to the product characteristics.

Because the gas pressure is higher in the inlet tube than in the chamber, an intense discharge may be present which causes undesirable heating. This discharge can be extinguished using a grounded grid in the inlet opening. A discharge may also be found in the throat of the vacuum system because of the hollow geometry.

The gas flow rate can be monitored using a suitable flow meter and the flow rate can be controlled using a variable leak. This leak can be set manually or can be controlled by a servo system from a gauge in the discharge system or foreline.

In addition to the use of variable leaks to control gas flow, it is usually necessary to throttle the pumping speed of the vacuum system to maintain efficient pumping and to prevent backstreaming. This is normally done by throttling the high vacuum valve. Since an appreciable throughput is necessary to flush contaminant gases out of the discharge chamber, excessive throttling should be avoided. One technique is to "bleed back" the system to a certain pressure (say 1 μ). After the deposition is completed, the system should be cooled and then let up to atmospheric pressure using dry nitrogen or dry air. Moisture condensation on cooled surfaces should be avoided. Flowing hot water through the cooling lines while the system is open will prevent moisture condensation.

High Voltage Substrate

In the ion plating process, the substrate and holder is made in the high voltage cathode. If the substrate to be coated is planar, it will have a uniform ion bombardment over the surface, but geometrical effects can cause some of the problems if the surface is not planar. Figure 6.8 illustrates some of the problem areas encountered with a complex geometry and also when using a mask. The presence of a ground shield in the vicinity of the substrate (Region A) may cause field distortion and shadow a portion of the substrate from ion bombardment, giving poor sputter cleaning. Also high energy neutrals may sputter material from the shield on to the substrate. Through holes, blind holes, and reentrant corners (Regions B, C, D) can give a hollow cathode effect in which secondary electrons are trapped and focused, giving localized regions of high plasma density causing local heating and high sputter erosion. Corners and edges (Region E) are high field regions where high sputtering rates are to be found. This may prevent these areas from forming a thick coating because of the high sputtering or may give excessive erosion. Points (Region F) or other high surface-to-volume geometries may become overheated because of poor thermal transfer properties. Poorly fitting or thick masks (1, 2, 3) act much like holes, giving electron trapping, high plasma densities, and heating. The mask should be tapered, as shown in Figure 6.8 (4), to minimize this effect. Also, material from the mask may contaminate the coating. If this is a problem, the mask can be made of (or coated with) the material to be deposited.

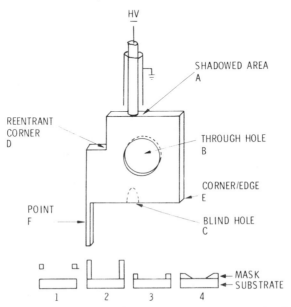

Figure 6.8: Geometries where obtaining a uniform ion bombardment is difficult.

Often the substrate to be ion plated must be held in some type of holder. This holder must often be larger than the substrate to avoid edge effects, and care must be taken to ensure that material sputtered from the holder does not contaminate the substrate or film. The holder may often be cooled by chilled water or liquid nitrogen. A liquid nitrogen cooled substrate/substrate holder may be in the form of a cold finger; and by ensuring good thermal contact between the holder and the substrate, the bulk temperature can often be kept very low. In order to reduce contamination, the holder and nearby surfaces may be coated with the material to be deposited.

Sometimes when depositing on an insulating substrate, a high transmission negative grid is placed in front of the substrate. This allows more effective bombardment of the insulator surface. Care must be taken to ensure that the substrate surface is not contaminated by material sputtered from the grid. Insulator surfaces can also be ion plated by applying an RF bias to the substrate holder. Another technique is to deposit a very thin metal film to make the surface conductive, then ion bombard the surface before continuing the deposition.

High Voltage Power Supplies (RF, AC, or DC)

In DC applications for both sputtering and ion plating, the power supply should be capable of supplying 3–10 kV at 1 mA/cm^2 to the cathode. Filtered or regulated power supplies are not required. The output of the supply should have a high reactance to prevent overloading if an arc occurs in the system.

Commercial RF generators have a capacitance output with several thousand volts peak-to-peak amplitude and operate at the FCC allowed frequency of 13.56 MC. Usually this frequency is crystal-controlled. Impedance matching between the power supply and the target in order to get a maximum power into the target is often a problem. Two very important considerations in impedance matching are the transmission lines and target design. As for power requirements, 5–10 watts/cm^2 seems to give good sputtering rates.

PROCESS CONTROL

Most thin film processing is developed on an empirical base. This is normally done by characterizing the system for a given product requirement. Once the processing is defined, then the process is used repeatedly. In this approach, there are several questions: Is each substrate position equivalent? Are all portions of the surface treated equally? Is the process definition sufficient to provide uniformity from run to run? Obviously, these questions are related to the sensitivity of the product to the processing. If the product is process sensitive, then care must be used to establish that the system design and process and product specifications are adequate to meet the requirements.

To determine whether substrate positions are equivalent means that you must understand the product requirements. These requirements may include such things as adhesion, electrical properties, stability, aging properties, bondability, composition, etc. For instance, electrical stability may be a function of film morphology, which may depend upon the substrate position (center or edge) and may vary with bombardment parameters. As in the case of reactively-sputtered film, the film composition may be a function of substrate position (center or edge) and may depend on gas flow.

To determine that substrate positions are equivalent means that important properties must be determined at enough positions for the results to be statistically meaningful. Meaningful limits have to be established. Once position equivalency has been established, the processing must remain constant. If the processing is changed, then it is necessary that equivalency be reaffirmed.

Some deposition variables are:

>System geometry
>Initial system vacuum
>Substrate/source distance
>Target and system preconditioning
>Substrate/source geometry
>Sputtering gas purity (in chamber)
>Sputtering gas pressure
>Sputtering gas flow rate
>Substrate voltage and current
>Substrate temperature
>System gas throughput
>Sputtering time
>Surface conditioning of substrate
>Cleanliness of vacuum system
>Source deposition rate

If these are carefully duplicated, you should get reproducible results and the process will be reproducible from run to run, as well as all substrate positions being equivalent.

In order to be certain that properties are constant, it is desirable to do batch sampling to ensure batch-to-batch uniformity. This may require the use of monitor slides with some property (such as surface finish) which makes testing easier. During system characterization where position equivalency is determined, certain positions will probably be found to consistently represent the extremes in film properties (say, the center versus an edge). These positions should be monitored from lot to lot for film properties. Some of the film properties which can be investigated are:

>Resistivity (sheet and specific)
>TCR (temperature coefficient of resistivity)
>Resistivity stability
>Seebeck coefficient
>Hall coefficient and Hall mobility
>Electrical noise
>Composition (compositional uniformity)
>Intrinsic stress
>Thickness
>Density
>Crystallography (structure, grain size, orientation, fiber axis)

Contamination (condensable elements, compounds, dissolved gases, insoluble gases)

Adhesion

Current-voltage behavior on anodization

Pinholes

Bumps

Morphology

Grain boundary composition

Boundability

Obviously, not all of these properties are of interest in all products.

So far this discussion seems to imply a black art, and in many cases it is. A good process engineer understands his machine and knows how to make the fine adjustments necessary to maintain a high quality product. In some cases, this type of control is not sufficient. If the product is very process-sensitive or demands particularly high reliability, or if the production personnel are not careful, it may be necessary to generate better process controls. It is in this area that much development work needs to be done. To change ion plating processing from a black art to a science will require an understanding of the processing which, in general, we do not have now.

Several techniques exist for process control and are either being used or could be used to the advantage where sophisticated process controls are required.

Rate and Deposition Monitors

Crystal rate monitors exist for use in gas discharge systems. If an RF system is used, then the detector and circuit must be well shielded electrically. A big problem with rate monitoring is position of the detector. Ideally, the detector should be in a substrate position and the readings are only relative. In ion plating there may be quite a variation betweeen deposition on the substrates and monitor.

Pressure Monitoring

Capacitance manometers have been found to give reliable pressure measurements, though they may be sensitive to vibration. If they are used in the presence of an RF discharge, they must be well shielded electrically to give reliable results. Pirani gauges are also reliable pressure gauges. All gauging should be stable and the calibration should be checked periodically.

Gas Composition Monitoring

Often the composition of the gas in the chamber is an important variable. Commercially, there are available differentially-pumped mass spectrometers which allow continuous monitoring of the gases in the discharge. Unfortunately, this monitoring technique does not indicate the composition of the gases striking the cathode surface. This variable may be very important, particularly in the reactive deposition techniques. Research instruments have been assembled which monitor the species striking the cathode by means of a small orifice in the cathode. Perhaps some day such a monitoring system will be commercially available.

PREPARATION OF SUBSTRATE SURFACES

Even though the cleaning of the substrate surface is an inherent part of ion plating, it is important that the surface be initially as clean as possible to minimize system contamination and reduce cleaning time. Reference 10 gives an overview of the various substrate cleaning techniques which may be used prior to ion plating.

APPLICATIONS OF ION PLATING

Adhesion

The improved adhesion obtained by the ion plating process has been the basis of most of the applications of ion plating. Most often these applications have been developed when no standard deposition technique could be found which could be used. A major exception to this is the use of ion plating to replace electrodeposition by the Japanese because of their concern with water pollution from "wet" deposition processes. There are four principal categories under adhesion application: (1) wear and erosion, (2) corrosion inhibition, (3) joining, and (4) incompatible coating-substrate systems.

Wear and Erosion

Coatings may be used to form hard surfaces which exhibit good wear characteristics or may be used as lubrication layers. Titanium carbide sometimes with hafnium nitride is a good wear coating. The hafnium nitride provides a low coefficient of friction for sliding surfaces and the titanium carbide provides a diffusion barrier and a surface resistant to deformation and bonding. Several types of lubricants are deposited by ion plating. The low shear metals are of particular interest in vacuum environments, and good adhesion is a requirement in order to prevent fretting and galling.

Corrosion Inhibition

In corrosion protection, one is concerned with forming a coating which will not be corroded or corrode instead of the surface to be protected. Interfacial corrosion can cause coating failure if the adhesion is poor. An interesting example of corrosion protection is the use of ion-plated carbon coatings on surgical implants. Also, the use of ion plated aluminum for the corrosion protection of uranium, steel and other such materials is of great interest, particularly to replace cadmium plating.

Joining

Ion-plated coatings, because of their excellent adhesion, are being used more and more as joining surfaces for materials where there are joining problems. Ion-plated silver is being used to join exotic materials such as uranium and beryllium.

Incompatible Coating-Substrate Systems

Systems which have little solid-solid solubility or have surface barrier layers or other restrictions to the type or extent of interfacial region formation, often show poor coating adhesion. Ion plating can assist in forming a desirable type of interfacial region for adhesion. Sometimes ion-plated coatings are used as a "strike" for subsequent coating techniques such as electrodeposition.

Table 6.1 summarizes some of the applications of ion plating which are covered in the Bibliography Appendix.

Table 6.1: Some Applications of Ion Plating

Optical	Electrical	Mechanical	Corrosion
Cr-Plastics	Contact	Lubrication	Gas/Liquid
In(Sn)O-Glass	Pt,Al,Au,Ag-Si	Au,Ag,MoS$_2$-Metals	Al-U,Ti,Steel
-Plastics	Al-GaAs	Wear/Erosion	Cr-Steel
Tin-Metals	In(Ga)-CdS	Ti(C,N), Zr(C,N), Cr(C,N)	Al$_2$O$_3$-Steel
ZrC-Metals	Cu-Al$_2$O$_3$ (VIAS)	-Metals, Carbides	Si$_3$N$_4$-Mo
	Passivation	Bonding	Cd-Steel
	Si$_3$N$_4$-Si	Ag-Be	Ti-Steel
		Cu-Ta,W,Nb,Oxides	Biological
		Au,Cr-Mo	C-Metals
		Creep	Ti-Metals
		Pt-Ti	Chemical
			Ta-Steel

SUMMARY

Ion plating has been accepted as another deposition technique as is evidenced by the increased usage of the term without a definition or reference being used. The technique joins other coating techniques as a possible process of choice for certain applications. It is not a universal panacea for coating, but has its advantages and disadvantages. The economics of ion plating seems to be comparable to those of other deposition techniques such as electrodeposition, vacuum deposition, and sputter deposition. In certain applications it seems to be the only technique which can achieve results; in other applications, it provides a better system.

REFERENCES

(1) D.M. Mattox, "Film Deposition Using Accelerated Ions," Sandia Corporation Development Report No. SC-DR-281-63 (1963); *Electrochem. Tech.* 2, 295 (1964).
(2) D.M. Mattox, "Design Consideration for Ion Plating," Sandia Corporation Report No. SC-R-65-997 (1966).
(3) D.M. Mattox, "Sputter Deposition and Ion Plating Technology," AVS-Thin Film Division Monograph (1974); also Sandia Laboratories Report SLA-73-0619 (July, 1973).
(4) D.M. Mattox, *Jpn. J. Appl. Phys.*, Suppl. 2. Pt. 1, p. 443 (1974).
(5) D.M. Mattox, *J. of Vac. Sci. Technol.*, 10, 47 (1973).
(6) D.M. Mattox, "Mechanisms of Ion Plating," Proc. IPAT-79 Conf. Available from CEP Consultants, 26 Albany St., Edinburgh EH1 3QH, United Kingdom.
(7) D.M. Mattox, "Fundamental Processes in Ion Plating," Presented to the 109th AIME Annual Meeting, Feb. 24-28, 1980, Las Vegas, NV.
(8) D.M. Mattox, "Thin Film Adhesion and Adhesive Failure—A Perspective," *Adhesion Measurement of Thin Films, Thick Films and Bulk Coatings*, ASTM STP 640, K.L. Mittal ed. American Society for Testing and Materials 1978, pp. 54–62.
(9) Toshinori Takagi, Isao Yamada and Akio Sasaki, "An Evaluation of Metal and Semiconductor Films Formed by Ionized-Cluster Beam Deposition," *Thin Solid Films* 39, 207 (1976).
(10) D.M. Mattox, *Thin Solid Films* 53, 81 (1978).

BIBLIOGRAPHY

1963

D.M. Mattox, "Interface Formation during Thin Film Deposition," *J. Appl. Phys. 34*, 2493 (1963).

1964

D.M. Mattox, "Film Deposition using Accelerated Ions," Sandia Corp. Development Report No. SC-DR-281-63 (1963); *Electrochem. Tech. 2*, 295 (1964).

1965

D.M. Mattox, "Metallizing Ceramics using a Gas Discharge," Sandia Corp. Report No. SC-R-64-1330 (1964); *J. Amer. Cer. Soc. 48*, 385 (1965).

L.I. Maissel and P.M. Schaible, *J. Appl. Phys. 36*, 237 (1965) (First Reference to "Bias Sputtering").

"Ion Plating," *Electronics 38*, No. 1, 35 (1965).

D.M. Mattox, "Interface Formation and the Adhesion of Deposited Thin Films," Sandia Corp. Report No. SC-R-65-852 (1965).

D.M. Mattox and R.D. Bland, "Ion Plating of SPRF Reactor Parts," Sandia Corp. Development Report No. SC-R-65-530 (1965).

R.D. Bland, J.E. McDonald, and D.M. Mattox, "Ion Plated Coatings for the Corrosion Protection of Uranium," Sandia Corp., Report No. SC-R-65-519 (1965).

R.L. Jacobson and G.K. Wehner, *J. Appl. Phys. 36*, 2674 (1965).

1966

D.M. Mattox, "Design Consideration for Ion Plating," Sandia Corp. Report No. SC-R-65-997 (1966).

E.F. Krummel and A. Gordon, "Deposition of Thin Films Under the Influence of Fast Ion Irradiation," *Zeit. Angew. Physik 22*, 1 (1966).

T. Spalvins, J.S. Przbyszewski, and D.H. Buckley, "Deposition of Thin Films by Ion Plating on Surfaces Having Various Configurations," NASA Lewis, TN-D3707 (1966).

D.M. Mattox, "Chromizing Molybdenum for Glass Sealing," *Rev. Sci. Instr. 37*, 1609 (1966).

R. Culbertson and D.M. Mattox, "High Strength Ceramic-Metal Seals Metallized at Room Temperature," 8th Conference on Tube Technology, IEEE Conf. Record, pp. 101–107 (1966).

R.J. MacDonald and D. Haneman, "Depths of Low-Energy Ion Bombardment Damage in Germanium," *J. Appl. Phys. 37*, 1609 (1966).

R.J. MacDonald and D. Haneman, "Low-Energy Ion Bombardment Damage in Germanium," *J. Appl. Phys. 37*, 3048 (1966).

1967

D.M. Mattox and R.D. Bland, "Aluminum Coating of Uranium Reactor Parts for Corrosion Protection," *J. Nucl. Mat. 21*, 349 (1967).

D.M. Mattox, Patent No. 3,329,601 (1967).

C.F. Schroeder and J.E. McDonald, "Adherence and Porosity in Ion Plated Gold," *J. Electrochem. Soc. 114*, 889 (1967) (Au on Cu/Be).

E.L. Hollar and R.D. Bland, "A Process Control for Ion Plating in Glow Discharge Cleaning," Paper presented at the American Vacuum Society Midwest Regional Meeting, Boulder, Colorado (May, 1967).

J.F. Carpenter, "Ion Vapor Deposition of Aluminum," ASME 67-WA/AV-2 (1967).

T. Spalvins and D.H. Buckley, "Importance in Lubrication of Various Interface Types Formed during Vacuum Deposition of Thin Metallic Films," Trans. Int. Vac. Met. Conf. (1967), pp. 247–259 (Au-steel).

R. E. Jones, C.L. Standley and L.I. Maissel, "Re-emission Coefficients of Si and SiO_2 Films Deposited Through RF and DC Sputtering," *J. Appl. Phys. 38*, 4656 (1967).

1968

D.M. Mattox and F.N. Rebarchik, "Sputter Cleaning and Plating Small Parts," *Electrochem. Tech. 6*, 374 (1968).

R.D. Bland, "A Parametric Study of Ion Plated Aluminum Coatings on Uranium," *Electrochem Tech. 6*, 272 (1968).

L.E. McCrary, J.F. Carpenter, and A.A. Klein, "Specialized Applications of Vapor-Deposited Films," Proceedings of the 1968 Vacuum Metallurgy Conference, American Vacuum Society (1968) (Al-Steel).

T.W. Wood, "Thick Adherent Metal Coatings by Means of Two Cathodes in a Glow Discharge," Mound Laboratory Report No. MIM-1453 (1968).

J.F. Carpenter, A.A. Klein, and L.E. McCrary, "Aluminum Coatings for Aerospace Fasteners," Paper presented at the Materials Engineering Exposition and Congress of the American Society for Metals (October, 1968), Detroit, Mich. (Al-Steel).

R.T. Johnson, Jr. and D.M. Darsey, "Resistive Properties of Indium and Indium-Gallium Contacts to CDS," *Solid State Elec. 11*, 1015 (1968).

J. Starkovich, "A Mass-Production Method for Metal Plating a Ferrite Core by Ion Plating Techniques," SC-DR-68-188 (May, 1968).

D.M. Mattox, "Ion Plating," Sandia Corp. Report No. SC-R-68-1865 (1968) (Review).

E. Stern and T.B. Light, *Appl. Phys. Lett. 13*, 381 (1968).

J.L. Vossen and J.J. O'Neill, Jr., "DC Sputtering with RF-Induced Substrate Bias," *RCA Review 29*, 566 (1968).

K. Erents, G. Farrell and G. Carter, *Proc. 4th Int. Vac. Congress* (Inst. of Physics and Physical Society, London, p. 145, 1968).

1969

F.G. Peters, "Bond Strength of Metallic Films Deposited on Polycrystalline Garnets," *Ceramic Bull. 48*, 228 (1969) (Al, Cu on Garnet).

T. Spalvins, "Energetics in Vacuum Deposition Methods for Depositing Solid Film Lubricants," *J. Amer. Soc. Lub. Eng.*, pp. 436–441 (November, 1969) (MoS_2 on Steel).

D.M. Mattox, "Interface Formation, Adhesion, and Ion Plating," Trans. of SAE (1969), pp. 2175–2181 (Review).

G.N. Jackson, "Substrate Bombardment in RF Sputtering Systems," *Vacuum 19*, 493 (1969).

1970

T. Spalvins, "Deposition of Alloy Films on Complex Surfaces by Ion Plating with Flash Evaporation," NASA-Lewis Report No. N70-32006 (June, 1970).

R.T. Bell, "Aluminum Ion Plating of Uranium-Molybdenum Alloy Fast-Burst Reactor Elements," *J. Vac. Sci. Technol. 7*, S125 (1970).

J.L. Vossen and J.J. O'Neill, Jr., "Evaporation of Aluminum with RF-Induced Substrate Bias," *RCA Review*, pp. 276–292 (June, 1970) (Al-Si).

L.I. Maissel, R.E. Jones and C.L. Standley, "Re-emission of Sputtered SiO_2 during Growth and Its Relation to Film Quality," *IBM J. Res. Dev., 14*, 176 (1970).

J.L. Vossen, J.J. O'Neill, Jr., K.M. Finlayson and L.J. Royer, "Backscattering of Materials Emitted from RF-Sputtering Targets," *RCA Review* 293 (1970).

V.P. Belevsku and V.M. Koptenko, "Device for Investigation of Material Vaporized by an Electron-Beam," *Pribory i Tehlinka Eksperimenta 4*, 188 (July–August, 1970).

E.V. Kornelsen, "Entrapment of Helium Ions at (100) and (110) Tungsten Surfaces," *Can. J. Phys. 48*, 2812 (1970).

G. Carter, "Ion Reflection, Penetration and Entrapment in Solids," *J. Vac. Sci. Technol. 7*, 31 (1970).

1971

T. Spalvins, "Bonding of Metal Lubricant Films by Ion Plating," *J. Amer. Soc. of Lub. Eng.*, pp. 40–46 (February, 1971).

D.L. Chambers and D.C. Carmichael, "Development of Processing Parameters and Electron-Beam Techniques for Ion Plating," Proceedings of the 1971 Conference of the Society of Vacuum Coaters, Miami, Florida (1971) (Technique).

D.L. Chambers and D.C. Carmichael, "Electron-Beam Techniques for Ion Plating," *R/D Magazine 22*, No. 5, 32 (1971) (Techniques).

C.T. Wan, D.L. Chambers, and D.C. Carmichael, "Investigation of Electron-Beam Techniques for Ion Plating," Proceedings of the 1971 Vacuum Metallurgy Conference, American Vacuum Society, New York City (June, 1971) (Techniques).

Kurt D. Kennedy, Glen R. Scheuermann and Hugh R. Smith Jr., "Gas Scattering and Ion Plating Deposition Methods," *R/D Magazine*, p. 40 (November, 1971) (Technique).

J.L. Vossen, "Control of Film Properties by RF Sputtering Techniques," *J. Vac. Sci. Technol. 8*, S12 (1971) (Induced Bias).

F.L. Schuermeyer, W.R. Chase and E.L. King, "Self-Induced Sputtering during Electron-Beam Evaporation of Ta," *J. Appl. Phys. 42*, 5856 (1971).

R. Carpenter, "Ion Plating: Increasing Adhesion of Deposited Films," *EPP Internat'l 7*, 5 (1971).

A.G. Blackman, *Metall. Trans. 2*, 699 (1971) (Bias Sput. = Peening).

1972

Gerald E. White, "A High Rate Source for Ion Plating," 1972 Vacuum Metallurgy Conference, Pittsburgh, PA (June 19-22, 1972).

A. Smith, "High Rate Ion Plating," 1972 Vacuum Metallurgy Conference, Pittsburgh, PA (June 19-22, 1972).

Howard R. Harker and Russel J. Hill, "The Deposition of Multi-Component Phases by Ion Plating," *J. Vac. Sci. Technol. 9*, 1395 (1972).

J.C. Withers, "High Temperature Coatings," *Am. Ceramics Soc. Bul. 51*, 350 (1972).

Hugh R. Smith, Jr., "Current Developments in Ion Plating," Meeting of Society of Vacuum Centers, Detroit, Mich., March 23, 1972.

John R. Morley and Hugh R. Smith, "High Rate Ion Production for Vacuum Deposition," *J. Vac. Sci. Technol. 9*, 1377 (1972).

Larry Fritz, "Ion Plating: The Best of Sputtering and Evaporation," *Circuits Manuf. 12*, 10 (1972) (Review).

J.S. Philbin and F. Gonzalez, "Procedures for Ion Plating–SPR II Fuel," Sandia Lab. Report SC-M-72-0458 (September, 1972) (Al-Uranium Alloys).

H. Venher, and C.C.R. Euratom, "The Vapor Deposition of High Work Function Materials in a Gas Discharge," 3rd International Conference on Thermionic Electrical Power Generation, Julich, Federal Republic of Germany, June 5–9, 1972.

T. Terada, "Aluminum Ion Impression into Gallium Arsenide," *J. Appl. Phys. D 5*, 756 (1972).

J.L. Vossen, "Method of Metallizing Semiconductor Devices," U.S. Patent 3,640,811 (February 8, 1972).

J.W. Patten and E.D. McClanahan, *J. Appl. Phys. 43*, 4811 (1972).

R.F. Bunshah and R.S. Juntz, "The Influence of Ion Bombardment on the Microstructure of Thick Deposits Produced by High Plate Physical Vapor Deposition Process, *J. Vac. Sci. Technol. 9*, 1404 (1972).

F.L. Schuermeyer, W.R. Chase and E.L. King, "Ion Effects during E-Beam Deposition of Metals," *J. Vac. Sci. Technol 9*, 330 (1972).

J.W. Coburn and Eric Kay, "Positive-Ion Bombardment of Substrates in RF Diode Glow Discharge Sputtering," *J. Appl. Phys. 43*, 4965 (1972).

D.M. Mattox and G.J. Kominiak, "Structure Modification by Ion Bombardment during Deposition," *J. Vac. Sci. Technol. 9*, 528 (1972.

R.D. Bland, J.K. Maurin and S.F. Driliere, "The Effect of DC Substrate Bias on Thick RF-Sputtered Chromium," Sandia Labs Report SC-DR-72-0922 (1972).

P.T. Stroud, Helen M. Lindsay and J.G. Perkins, "Some Preliminary Studies of the Structure of Ion Bombarded Thin Films," *Vacuum 23*, 125 (1972).

E.V. Kornelsen, "The Interaction of Injected Helium with Lattice Defects in a Tungsten Crystal," *Rad. Eff. 13*, 227 (1972).

J.J. Bellina, Jr., and H.E. Farnsworth, "Ion Bombardment Induced Surface Damage in Tungsten and Molybdenum Single Crystals," *J. Vac. Sci. Technol. 9*, 616 (1972).

1973

B. Swaroop and J. Adler, "Ion Plated Copper-Steel Graded Interface," *J. Vac. Sci. Technol. 10*, 503 (1973).

Gerald W. White, "New Applications of Ion Plating," *R/D Magazine*, p. 43 (July, 1973).

Gerald W. White, "Applications of Ion Plating," SAE Meeting, Detroit, Mich., (May 14–18, 1973) Paper No. 730546.

J.E. Varga and W.A. Bailey, "Evaporation, Sputtering and Ion Plating: Pros and Cons," *Solid State Technol.* p. 79 (December, 1973) (Review and Comparison).

S. Aisenberg and R.W. Chabot, "Physics of Ion Plating and Ion Beam Deposition," *J. Vac. Sci. Technol. 10*, 104 (1973) (Theory-Technique).

Vance Hoffmann, "Thin Film Vacuum Equipment," *Solid State Technol.* p. 93 (December, 1973) (Technique).

F.C. Douglas, E.L. Paradis and R.D. Veltri, "Application of Diffusion Barriers to the Refractory Fibers of Tungsten, Niobium, Carbon and Aluminum Oxide," NASA-CR-134466 (September, 1973) (Ion Plating).

T. Spalvins, "Ion Plating," Paper 730545, Meeting of Society of Automotive Engineers, Detroit, Mich. (May 16, 1973).

B.C. Strupps, "Sputtering and Ion Plating as Industrial Processes," Paper 730547, Meeting of Society of Automotive Engineers, Detroit, Mich. (May 16, 1973).

R.T. Bell and J.C. Thompson, "Applications of Ion Plating in Metals Fabrication," Oak Ridge, Y-12 Plant, Y-DA-5011 (1973).

D.M. Mattox, "Sputter Deposition and Ion Plating Technology," AVS-Thin Film Division Monograph (1974); also Sandia Laboratories Report SLA-73-0619 (July, 1973) (Theory and Technique).

V.G. Babear and M.B. Guseva, "Adsorption of Metal Vapors in the Presence of Ion Irradiation," *Isv. Nauk SSSR Ser. Fiz. 37*, 2596 (1973).

A.G. Blackman, "DC Bias-Sputtered Aluminum Films," *J. Vac. Sci. Technol. 10*, 299 (1973).

W. Heiland and E. Taglauer, "Bombardment Induced Surface Damage in a Nickel Single Crystal Observed by Ion Scattering and LEED," *Ion Surface Interaction, Sputtering and Related Phenomena*, ed., R. Behrisch, W. Heiland, W. Poschenrieder, P. Stoib and H. Verbeik, Gordon and Beach 1973, p. 117.

R.S. Nelson and D.J. Mazey, "Surface Damage and Topography Changes Produced during Sputtering," *Rad. Effects 18*, 127 (1973).

J.E. Houston, R.D. Bland and D.M. Mattox, "The Role of Surface Cleanliness on the Properties of Films Deposited by Ion Plating Techniques," Proc. 1973 Internat'l. Conf. on Sputtering and Its Applications (Montpellier, France, October, 1973) (Sputter Cleaning).

Preter Davidse, Joseph S. Logan and Fred Maddocks, "Method for Sputtering a Film on a Rough Surface," U.S. Patent 3,755,123 (August 28, 1973) (Negative Bias).

D.M. Mattox, "Fundamentals of Ion Plating," *J. Vac. Sci. Technol. 10*, 47 (1973).

J.W. Patten and E.D. McClanahan, "Effects of Deposition Temperature and Substrate Bias on Orientation and Hardness of Thick Sputter Deposited Beryllium Foils," *J. Less Common Metals 30*, 351 (1973).

1974

W.R. Stowell and D. Chambers, "Investigation of the Hollow Cathode Effects in the Structure of Bulk Films," *J. Vac. Sci. Technol. 11*, 653 (1974).

J.J. Cuomo, R.J. Gambino and R. Rosenberg, "The Influence of Bias on the Deposition of Metallic Films in RF and DC Sputtering," *J. Vac. Sci. Technol. 11*, 34 (1974).

J.L. Vossen, "A Sputtering Technique for Coating the Inside Walls of Through Holes in a Substrate," *J. Vac. Sci. Technol. 11*, 875 (1974).

R.S. Nowicki and W.D. Buckley, "Effects of Deposition Parameters on Properties of RF Sputtered Molybdenum Films," *J. Vac. Sci. Technol. 11*, 675 (1974).

E.A. Fagen, R.S. Nowicki and R.W. Segvin, "Effects of DC Substrate Bias on the Properties of RF-Sputtered Amorphous Germanium Ditelluride Films," *J. Appl. Phys. 45*, 50 (1974).

B.E. Nevis and T.C. Tisone, "Low Voltage Triode Sputtering and Backsputtering with a Combined Plasma, Part IV Heat Transfer Characteristics," *J. Vac. Sci. Technol. 11*, 1177 (1974) (Low Voltage Triode Backsputtering—LVBS 50-250 eV).

D.W. Hoffman and R. Nummagadda, "Depth Profiling of an Ion Plated Interface by Ion Scattering Spectrometry," *J. Vac. Sci. Technol. 11*, 657 (1974).

R.D. Bland, G.J. Kominiak and D.M. Mattox, "Effect of Ion Bombardment during Deposition and Thick Metal and Ceramic Deposits," *J. Vac. Sci. Technol. 11*, 671 (1974) (Properties).

D.C. Carmichael, "Effects of the Bias Voltage on the Properties and Structure of Sputtered and Ion Plated Coatings," *J. Vac. Sci. Technol. 11*, 639 (1974) (Properties).

R.D. Bland, "Substrate Heater for Bias Sputtering and Ion Plating," *J. Vac. Sci. Technol. 11*, 906 (1974) (Technique).

T. Takagi, I. Yamade, and A. Sasaki, "ZnSiMn DC Electroluminescent Cells by Ion Implantation Techniques," (Vaporized Metal Cluster Ion Source) Proceedings of International Conf. on Ion Implantation in Semiconductors and Other Materials (O$_s$aka, Japan, August 26–30, 1974) (Technique).

W.R. Stowell and D. Chambers, "Investigation in Hollow Cathode Effects on the Structure of Bulk Film," *J. Vac. Sci. Technol. 11*, 653 (1974).

R.F. Janninck, C.R. Heiden and A.E. Guthensohn, "A Rapid Loading and Unloading Cohesion Type Fixture for Ion Plating Reedblade Contacts," *J. Vac. Sci. Technol. 11*, 535 (1974) (Au-Metal).

J. Gordon Davy and J.J. Havak, "RF Bias Evaporation (Ion Plating) of Nonmetal Thin Films," *J. Vac. Sci. Technol. 11*, 43 (1974).

P.S. McLeod and G. Mah, "The Effect of Substrate Bias Voltage on the Bonding of Evaporated Silver Films," *J. Vac. Sci. Technol. 11*, 119 (1974) (Ag-Metal).

K. E. Sturbe and L.E. McCrary, "Thick Ion-Vapor-Deposited Aluminum Coatings for Irregularly Shaped Aircraft and Spacecraft Parts," *J. Vac. Sci. Technol. 11*, 362 (1974) (Al-Metal).

Duane G. Williams, "Vacuum Coating with a Hollow Cathode Source," *J. Vac. Sci. Technol. 11*, 374 (1974).

N. Ohmae, T. Nakai and T. Tsukizoe, "Prevention of Fretting by Ion Plated Film," *Wear 30*, 299 (1974) (Tribology Adhesion).

Nobuo Ohmae, Tetsuo Nakai and Tadasu Tsukizoe, "Ion Plated Film and Its Properties Against Friction and Wear," Proceedings of the JSLE-ASLE International Lubrication Conference (Tokyo, June, 1975), to be published. (Vacuum Ion Plating, Tribology, Interfact Formation.)

D.G. Teer, "Ion Plating," (Abstract Only) *Vacuum 24*, 482 (1974).

Hajime Watanbe, Masatoshi Takao and Akira Tasaki, "Ion Plating Applied to Preparation of Magnetic Thin Films," *Jpn J. Appl. Phys. 13*, 727 (1974).

William R. Stowell, "Ion Plated Titanium Carbide Coatings," *Thin Solid Films 22*, 111 (1974) (TiC in Steel).

Toshimori Takayi, Isao Yamada, Kiruchi Yanagawa, Masatoshi Kunori and Shigemitsu Kobiyama, "Vaporized-Metal Cluster Ion Source for Ion Plating," Proc. Sixth International Vacuum Congress (Kyoto, Japan, 1974) (Technique).

L.B. Leder, "Fundamental Parameters of Ion Plating," *Met. Finish 72*, 41 (1974).

Bernard C. Strupp, "Industrial Applications of Ion Plating and Sputtering," *Plating 61*, 1090 (1974) (Cu-Ti, Au-BeO, Undercoat for Electrodeposition).

C.T. Wan, D.L. Chambers and D.C. Carmichael, "Effect of Processing Conditions on Characteristics of Coatings Vacuum Deposited by Ion Plating," Proc. 4th Int. Conf. of Vac. Met., T. Saito, Ed., The Iron and Steel Institute of Japan, to be published.

D.H. Boone, T.E. Strangman and L.W. Wilson, "Some Effects Structure of Composition in the Properties of Electron Beam Vapor Deposited Coatings for Gas Turbine Superalloys," *J. Vac. Sci. Technol. 11*, 641 (1974) (Ion Plating).

Chikara Hayashi, "Vacuum Industry in Japan," Proc. 6th Internat'l Vac. Congress. 1974 (*Jpn J. Appl. Phys.* Suppl. 2, Pt. 1, 1974) p. 241.

Toshinori Takagi, Isao Yamada, Kovichi Yanagawa, Masatoshi Kunori, and Shigemitsu Kobiyama, "Vaporized-Metal Cluster Ion Source for Ion Plating," Proc. 6th Internat'l Congress 1974 (*Jpn J. Appl. Phys.* Suppl. 2, Pt. 1, 1974) p. 427.

D.M. Mattox, "Recent Advances in Ion Plating," Proc. 6th Internat'l Vac. Congress 1974 (*Jpn J. Appl. Phys.* Suppl. 2, Pt. 1, 1974) p. 443.

Nobuo Ohmae, Tetsuo Nakai and Talasu Tzukizoe, "On the Application of Ion Plating Technique to Tribology," Proc. 6th Internat'l Vac. Congress 1974 *Jpn J. Appl. Phys.* Suppl. 2, Pt. 1, 1974) p. 451.

Kiyoshi Matsubara, Yuji Enomoto, Go Yaguchi, Makuto Watanabe, and Ryutaro Yanazaki, "Study of Deposition Process and Film Property in Ion Plating," Proc. 6th Internat'l Vac. Congress 1974 (*Jpn J. Appl. Phys.* Suppl. 2, Pt. 1, 1974) p. 455.

Yoichi Murayama, "Structures of Gold Thin Films Formed by Ion Plating," Proc. 6th Internat'l Vac. Congress. (*Jpn J. Appl. Phys.* Suppl. 2, Pt. 1, 1974) p. 459.

T. Narusawa and S. Komiya, "Composition Profile of Ion Plated Au Film on Cu Analyzed by AES and Sims during Xe Bombardment," *J. Vac. Sci. Technol. 11*, 312 (1974).

D.M. Mattox, "Ion Plating—Review and Update," Proc. of 6th Internat'l Conf. on Electron and Ion Beam Science and Technology. Edited by Robert Bakish, The Electrochemical Society (1974) p. 254.

T. Nakai, N. Ohmae and T. Tzukizoe, "Some Comments on the New Technique of Ion Plating," *Bull. Jpn Sol. Precision Eng. 8*, 17 (1974).

T. Spalvins, "Horizons in Ion Plated Coatings," *Metal Finishing 72*, 38 (1974).

B.E. Nevis and T.C. Tisone, "Low Voltage Triode Sputtering and Backsputtering with a Confined Plasma. Part IV," *J. Vac. Sci. Technol. 11*, 1177 (1974).

R. Carpenter, "The Basic Principles of Ion Plating," in Science and Technology of Surface Coatings. Brian N. Chapman and J.C. Anderson, editors, Academic Press (1974). Proc. of NATO Advanced Study Institute, p. 393.

L. Holland, "Substrate Treatment and Film Deposition in Ionized Gases," (Abstract Only) *Vacuum 24*, 481 (1974).

J.A. McHugh, "Ion Beam Sputtering—The Effect of Incident Ion Energy on Atomic Mixing in Subsurfaced Layers," *Rad. Eff. 21*, 209 (1974).

K.H. Ecker, "Transmission Sputtering of Gold by Heavy Ion in the Lower KeV Energy Region," *Rad. Eff. 23*, 171 (1974).

1975

T. Takagi, I. Yamada and A. Sasaki, "Ionized Cluster-Beam Deposition," *J. Vac. Sci. Technol. 12*, 1128 (1975).

Geoffrey Beardmore and Hugh N. Evans (Smiths Ind., England), "Methods of Sputter Deposition of Materials," U.S. Patent 3,869,368 (3/4/75) Sputter Cleaning-Graded Interface.

Y. Miyakawa, M. Nishimura and M. Nosaka, "Friction and Wear Performance of Gold and Silver Films," Proceedings of the JSLE-ASLE Internat'l Lubrication Conf. (Tokyo, Japan, 6/9-11/75).

G.J. Hale, G.W. White and D.E. Meyer, "Ion Plating Using a Pure Ion Source: An Answer Looking for Problems," *Electronic Packaging & Production 15*, 39 (May, 1975).

Y. Murayama, "Thin Film Formation of In_2O_3, TiN and TaN by RF Reactive Ion Plating," *J. Vac. Sci. Technol. 12*, 818 (1975).

J. Chin and N.B. Elsner, "Preparation of Si_3N_4 Coatings by Ion Plating," *J. Vac. Sci. Technol. 12*, 821 (1975).

Y. Enomoto and K. Matsubara, "Structure and Mechanical Properties of Ion Plated Thick Films," *J. Vac. Sci. Technol. 12*, 827 (1975).

S. Schiller, U. Heisig and K. Goedicke, "Alternating Ion Plating—A Method of High-Rate Ion Vapor Deposition," *J. Vac. Sci. Technol. 12*, 858 (1975).

T. Spalvins, "Growth Defects in Thick Ion Plated Coatings," Report No. NASA-TN-D-7895 E-8130 (January, 1975) also *Sci. Tech. Aerospace Reports 13*, 638 (3/23/75).

L. Holland (Review Paper), "Substrate Treatment and Film Deposition in Ionized and Activated Gas," *Thin Solid Films 27*, 185 (1975).

J. Chin and N.B. Elsner, "Preparation of Silicon-Aluminum-Nitrogen Compounds by Reactive Ion Plating," 5th Internat'l Conf. on Chemical Vapor Deposition, Sept. 21-26, 1975, Fulmer, England.

K.D. Kennedy and G.R. Schuermann, "Method for Coating a Substrate to Provide a Titanium- or Zirconium-Nitride or Carbide Deposit Having a Hardness Gradient Which Increases Outwardly from the Substrate," U.S. Patent 3,900,592, *Offic. Gaz. U.S. Pat. Off. 937*, 1002 (8/19/75).

K. Matsubaro, "Method for Control of Ionization Electrostatic Plating," U.S. Patent 3,900,585 *Offic. Gaz. U.S. Pat. Off. 937*, 1000 (8/19/75).

S. Schiller and U. Heisig, "Bedampfungstechnik-Verfahren, Einrichtungen und Anwendungen," p. 62 VEB Verlag Technik, Berlin (1975) Broschure-Forschungs-institut Manfred von Ardenne Dresden.

M. Sokolowski, A. Sokolowska, E. Rolinski and A. Michalski, "Ion Treatment of Silicon in a Glow Discharge," *Thin Solid Films 30*, 29 (1975).

D. Meyer, et al, "Solving Microelectronics Adhesion Problems with Ion Plating," *J. Electrochem. Soc. 122*, 361C (1975).

N.N. Engel, "Ion Plating Method and Products Therefrom, U.S. Patent 3,915,757, *Offic. Gaz. U.S. Pat. Off. 939*, 2064 (10/28/75).

Joseph E. Berg, Randolph E. Brown, Jr., "High Vacuum Deposition Apparatus," (Ionizing Element in System), U.S. Patent 3,913,520 (10/75).

R.R. Hart, H.L. Dunlap and O.J. Marsh, "Ion Induced Migration of Cu into Si," *J. Appl. Phys. 46*, 1947 (1975).

W.W. Carson, "Sputter Gas Pressure and DC Substrate Bias Effects on Thick RF Dioxide Sputtered Films of Ti Oxycarbide," *J. Vac. Sci. Technol. 12*, 845 (1975).

Y. Enomoto and K. Matsubara, "Structure and Mechanical Properties of Ion Plated Thick Films," *J. Vac. Sci. Technol. 12*, 827 (1975).

D. Edwards, Jr., and E.V. Kornelsen, "Vacancy Creation during Low Energy Ion Bombardment," *Rad. Eff. 26*, 155 (1975).

P. Blank and K. Wittmaack, "The Influence of Implanted Xenon on the Sputtering Yield of Silicon," *Rad. Eff. 27*, 29 (1975).

G.J. Kominiak, "Target and Substrate Interactions in Bias Sputter Deposition," *J. Vac. Sci. Technol. 12*, 689 (1975).

K.E. Steube (McDonnell Douglas) "Glow Discharge-Tumbling Vapor-Deposition Apparatus," U.S. Patent 3,926,147 *U.S. Pat. Off. Gaz. 941*, 1128 (12/16/75).

R. Ueda, "Synthesis and Epitaxial-Growth of CdTe Films by Neutral and Ionized Beams," *J. Cryst. Growth 31*, 331 (1975).

E. Tadokoro and T. Kitamoto, "Method for Forming Protective Film by Ionic Plating (on Magnetic Tape)," (Fuji Photo), U.S.P. 3,928,159 *U.S. Pat. Off. Gaz. 941*, 1771 (12/23/75).

T. Tsukizoe and N. Ohmae, "Friction Properties of Advanced Composite Materials," Proceedings ASME-JSME Joint Conf. on Applied Mechanics, Hawaii, March, 1975. Jpn Soc. Mech. Eng. Tokyo, p. 407.

N. Ohmae, T. Nakai and T. Tsukizoe, "Ion Plated Film and Its Properties Against Friction and Wear," Proceedings ASLE-JSLE Internat'l Lubr. Conf. Tokyo, June, 1975, Jpn Soc. Lubr. Eng. (In Print).

T. Spalvins and W.A. Brainard, "Induction-Heating Simplified Metal Evaporation for Ion Plating," NASA Tech. Brief B75-10,288, 12/75.

D.G. Teer, "Ion Plating," *Tribol. Internat. 8*, 247 (1975).

John A. Thornton and Virgle L. Hedgcoth, "Tubular Hollow Cathode Sputtering onto Substrates of Complex Shapes," *J. Vac. Sci. Technol. 12*, 93 (1975) (Bias-Morphology).

1976

Y. Namba and T. Mori, "Epitaxial Growth of Ag Deposited by Ion Deposition on NaCl," *J. Vac. Sci. Technol. 13*, 693 (1976).

G.J. Kominiak and J.E. Uhl, "Substrate Surface Contamination during Ion Plating," Sandia Report, SAND75-0134 (1975).

J. Chin and N.B. Elsner, "Preparation of Silicon Nitride Coatings by Ion Plating," *Sci. Tech. Aerospace Rpts. 14*, (7) 854 (4/8/76).

N. Ohmae, T. Nakai and T. Tsukizoe, "MoS$_2$ Ion Plated Carbon Fibre-Reinforced Epoxy Resin as an Advanced Self-Lubricating Composite," *Wear 38*, 181 (1976).

S. Schiller, U. Heisig and K. Goedicke, "Ion Plating—A New Process in Vacuum Coating," (In German) *Vak. Technik 25*, (3) 65 (1976).

Yoichi Murayama, Masayuki Matsumoto and Kunohiro Kashinagi, "Ion Plating Apparatus Having an H. F. Electrode for Providing an H. F. Glow Discharge Region," U.S. Patent 3,962,988 (6/15/76).

K. Matsubara, U.S. Patent 3,935,619 (4/27/76).

T. Spalvins and W.A. Brainard, "Ion Plating with an Induction Heating Source," NASA Rpt. TM-X-3330, Jan. 1976 also *Sci. Tech. Aerospace Rpts. 14*, (8) 979 (4/23/76).

D.E. Carlson and C.E. Tracy, "Metallization of Glass using Ion Injection," *Ceram. Bul. 55*, (5) 530 (1976).

J.R. Clarke, A.K. Weiss, J.L. Donovan, J.E. Greene and R.E. Klinger, "Ion Plated Lead Oxide—An X-Ray Sensitive Photoconductor," Post Paper 23rd Nat'l Symp. of American Vacuum Soc. 9/21-24/76.

A.W. Morris, "The Ion Vapor Deposited Aluminum Coatings," *Plating and Surface Finishing 63*, (10) 42 (1976).

D.G. Teer, "The Energies of Ions and Neutrals in Ion Plating," *J. Phys. D, 9*, L187 (1976).

D.G. Teer, "Ion Plating," *Trans of the Inst. of the Metal Finishing 54*, (4) 159 (1976).

R.I. Sims, "The Ion Plating Process and Its Applications," *The Metallurgist and Materials Technologist*, January, 1976.

Kaizo Kuwahara and Yukito Matsui, "Thickness Uniformity of Ion Plated Films Deposited Through a Mask Hole," *Bull. Jpn Soc. Press. Eng. 10*, 115 (1976).

Nobuo Ohmae, Tetsuo Nakai and Tadsau Tsukizoe, "Ionenstrahlbedampfung in Hochvakuum," Technology Report of Osaka Univ., V. 21 No. 1302 (1976).

Talivadas Spalvins, "Industrialization of the Ion Plating Process," *R/D Magazine 27*, (10) 45 (Oct., 1976).

G.J. Kominiak and J.E. Uhl, "In-Situ Investigation of Substrate Surface Recontamination during Glow-Discharge Sputter Cleaning," *J. Vac. Sci. Technol. 13*, 170 (1976).

E.R. Naimon, D. Vigil, J.P. Villegas and L. Williams, "Adhesion Study of Silver Films Deposited from a Hot Hollow-Cathode Source," *J. Vac. Sci. Technol. 13*, 1131 (1976).

A. Cavalera, C. Dinea and B. Delcea, "A Study of Ion Plated Gold Film on Silicon Substrates by Means of Auger-Electron Spectroscopy," *Stud. Cercetari. Fiz. 28*, 113-119 (1976, Romanian).

S. Schiller, V. Heisig and K. Steinfelder, "A New Sputter Cleaning System for Metallic Substrates," *Thin Solid Films 33*, 331 (1976).

P. Glaser, A.R. Herman and G. Vajo, "Ion Evaporation," *Thin Solid Films 32*, 69 (1976).

J. Kominiak and J.E. Uhl, "Substrate Surface Contamination from Dark-Space Shielding during Sputter Cleaning," *J. Vac. Sci. Technol. 13*, 1193 (1976).

Russell Messier, Takeshi Takamori and Rustrum Roy, "Structure Composition Variation in RF-Sputtered Films of Ge Caused by Process Parameter Changes," *J. Vac. Sci. Technol. 13*, 1060 (1976) (Possibly due to self-bias).

J. Kominiak, "Metallization Adhesion Layer Formed by Backscattering during DC Sputter Etching," *J. Vac. Sci. Technol. 13*, 1100 (1976).

J. Amano, P. Bryce and R.P.W. Lawson, "Thin Film Deposition using Low-Energy Ion Beams, I System Specification and Design," *J. Vac. Sci. Technol. 13*, 591 (1976).

N. Ohmae, "Recent Work on the Tribology of Ion Plated Thin Films," Published *J. Vac. Sci. Technol. 13*, 82 (1976).

B. Swaroop, D.E. Meyer and G.W. White, "Ion Plated Aluminum Oxide Coatings for Protection Against Corrosion," *J. Vac. Sci. Technol. 13*, 680 (1976).

Editor, "Versatile Aluminum Coating Controls Corrosion," *Iron Age*, p. 48, 2/9/76.

R. Vela, "Film Growth of Semiconductors and Metals by Ion Plating Method and Surface Imperfections," Internat'l Conf. on Metallurgical Coatings, San Francisco, CA, 4/5-9/76.

Y. Namba and T. Mori, "Epitaxial Growth of ZnTe on NaCl Deposited by Ion Deposition," International Conf. on Metallurgical Coatings, San Francisco, CA, 4/5-9/76.

J.M. Walls and H.N Southworth, "The Optimum Ion Species for Sputter-Cleaning on Ion Profiling Tungsten Surfaces," *Surf. Technol. 4*, 255 (1976).

Boris Navinsek, "Sputtering—Surface Changes Induced by Ion Bombardment," *Prog. in Surf. Sci. 7*, 49 (1976).

D.G. Teer, "The Role of Neutrals in Ion Plating," *J. Adhesion 8*, 171 (1976).

Tushinori Takagi, Isao Yamada and Akio Sasaki, "An Evaluation of Metal and Semiconductor Films Formed by Ionized-Cluster Beam Deposition," *Thin Solid Films 39*, 207 (1976).

S. Komiya and K. Tsuruoka, "Physical Vapor Deposition of Thick Cr and Its Carbide and Nitride by Hollow-Cathode Discharge," *J. Vac. Sci. Technol. 13*, 520 (1976).

1977

Tadasu Tsukizoe, Tetsuo Nakai, and Nobuo Ohmae, "Ion Beam Plating using Mass Analyzed Ions," *J. Appl. Phys. 48*, 4770 (1977).

Masoo Fukutomi, Masahiro Kitajima, Masatoshi Okada, and Ryoji Watanabe, "Silicon Nitride Coatings on Molybdenum by RF Reactive Ion Plating," *J. Electrochem. Soc. 124*, 1420 (1977).

McDonnell Aircraft Co., "IVADIZE-Aluminum Plating by Ion Vapor Deposition," IVD-080-07, January, 1977.

N. Ohmae, T. Nakai, T. Tsukizoe, "Ion Plated Thin Films for Anti-Wear Applications," To be published in Proceedings of the Internat'l Conf. of Wear of Materials, St. Louis, MO, April 25, 1977.

E. Tsunasawa, K. Inagaki and K. Yamanaka, "Process Parameters and Some Properties of Ion Plated Titanium Coatings on Steel Substrates," To be published *J. Vac. Sci. Technol.*, January, 1977.

M.G.D. El-Sherbiney and J. Hallin, "An Experimental Study of the Hertzian Contact of Surfaces Covered with Ion Plated Films," *Wear 41*, 365-371 (1977).

C.W.B. Martinson, P.J. Nordlander and S.E. Karlson, "AES Investigations of the Interface between Substrate and Chromium Films Prepared by Evaporation and Ion Plating," *Vacuum 27*, (3) 119 (1977).

S. Fujishiro, D. Eylon, "Effect of Environment and Coating on the Creep Behavior of Commercial Ti Alloys," *Scripta Met. 11*, 1011 (1977).

Gerald W. White, "Thin Film Treated Drilling Bit Cones," U.S. Patent 4,054,426 (October 18, 1977).

Gerald W. White, "High Rate Ion Plating Source," U.S. Pat. 4,016,389 (4/5/77).

E.P. Eernisse, "Stress in Ion-Implanted CVD Si_3N_4 Films," *J. Appl. Phys. 48*, 3337 (1977) (Compressive Stress).

J.S. Colligon, G. Fischer and M.H. Patel, "Recoil Implantation of Materials," *J. Mat. Sci. 12*, 829 (1977).

D.W. Hoffman and John A. Thornton, "Internal Stresses in Sputtered Chromium," *Thin Solid Films 40*, 355 (1977) (Peening = Compression).

IPAT-77 (Conf. on Ion Plating & Allied Techniques) Edinburgh, June, 1977. Proceedings published by CEP Consultants Ltd., 14a Henderson Row Edinburgh EH3 5DH, United Kingdom.

"The Terminology of Ion Plating," P.A. Walley.
"The Ion Plating Technique," D.G. Teer.
"Ion Plating—Atomic Processes Occurring Within the Surface Layers," R.S. Nelson.
"An Investigation into the Factors Affecting the Cathodic Etching for Cleaning Metal Surfaces Prior to Ion Plating," D.W. Tomkins, D.S. Coleman.
"The Role of Ions and Neutrals in Ion Plating," D.G. Teer, B.L. Delcea.
"Metal Ion Sources for Plating, R. Clampitt, L. Gowland, K.L. Aitken.
"Ion Plating with Electron Beam Evaporation," B. Heinz, G. Kienel.
"An Assessment of Present and Future Applications for Ion Plating," R.I. Sims.
"Corrosion Resistant Coatings of Cadmium by an Ion Plating Method," J.R. Smith, B.J. Williams.
"Ion Plating of Carbon," D.G. Teer, M. Salama.
"Ion Plating Metals on Plastics," K. Jones, A.J. Griffiths, E.W. Williams.
"Optical Applications of Ion Plating," E. Henderson.
"The Formation of Low Friction, Wear Resistant Surfaces on Titanium by Ion Plating," D.G. Teer, F. Salem.
"A Preliminary Evaluation of Ion Plating for the Deposition of High Temperature Corrosion Resistant.
"The Electron Beam Coating of Turbine Components and Ion Plating," D.H. Boone, D. Lee, J.M. Shafer.
"Sputter Ion Plating," R.A. Dugdale.

Tadasu Tsukizoe, Tetsuo Nakai and Nobuo Ohmae, "Ion Beam Plating using Mass Analyzed Ions," *Appl. Phys. Lett. 48,* 4770 (1977).

Proceedings of the 7th International Vacuum Congress and the 3rd International Conference on Solid Surfaces, Vienna, 1977, Ed. R. Dobrozemsky, F. Rudenauer, F.P. Viehbock and A. Breth.

"Trends in Thin Film Deposition Methods," Chr. Weissmantel, p. 1533 (Sources).
"Ion Deposition Techniques for Industrial Applications," S. Schiller, U. Heiseg and K. Goedicke, p. 1545.
"Production of Thin Films by PVD Techniques and Their Applications," R.F. Bunshah, p. 1553.
"Ion Plating in Industrial Thin Film Production," H.K. Pulker, R. Buhl and E. Moll, p. 1595.
"Ion Plating and Vapor Deposition of Steel Substrates with Copper in a Technical Vacuum Equipment," R. Wilberg, G. Blasek, D. Schulze and W. Teubner, p. A-2740.
"Mass Spectrometer-Controlled Ion Beam Plating," N. Ohmae, T. Nakai, T. Tsukizoe, p. 1607.
"Influence of Ion Implantation into the Source of Evaporation or Growing Structure on the Properties of Vacuum Deposited Thin Films," L. Pranevicius, K. Bilius, J. Dudonis, A. Meskauskas.
"Structure of Evaporated, Ion Plated and RF Sputtered Chromium Films," K.A.B. Andersson, H.T.G. Nilsson, P.J. Nordlander, S.E. Karlsson, p. 1753.

"On the Properties of Ion Plated Ag Layers on Si," A. Cavaleru, G. Dinca, B. Dunare.

I. Roikh, L.N. Koltunova and O.V. Lebedinskii, "Protective Coatings Produced by Ion Plating in a Vacuum (A Review)," *Zashch. Met. 13*, 649 (1977).

H. Yoshihara, "Properties of Ion Plated Silver Films," *J. Jpn Int. Metals 41*, 999 (1977).

E. Williams, "Ion Plating: Coat of Many Colors," *New Sci. 74*, 588 (1977).

No Author, "Aluminum Plating by Ion Vapor Deposition," *Finish Highlights 9*, 26 (1977).

T. Spalvins, "New Applications of Sputtering and Ion Plating," ASME Paper No. 77-DE-21 Am. Sci. Mech. Engin. New York, NY, 1977.

W.A. Brainaid and D.R. Wheeler, "X-Ray Photoelectron Spectroscopy Study of RF Sputtered Titanium Carbide, Molybdenum Carbide and Titanium Boride Coatings and Their Friction Properties," NASA Technical Report NASA-TP-1033, October, 1977.

B.C. Swartz, R.D. Silkensen and G. Steving, "RF Bias Sputtering Method for Producing Insulating Films Free of Surface Irregularities," U.S. Patent 4,036,723.

M. Okada, et al, "Silicon-Nitride Coatings on Molybdenum by RF Ion Plating," *J. Electrochem. Soc. 124*, 297C (1977).

J.A. Thornton, "The Influence of Bias Sputter Parameters on Thick Copper Coatings Deposited using a Hollow Cathode," *Thin Solid Films 40*, 335 (1977).

J.P. Coad and R.A. Dugdale, "Some Applications of Sputter-Ion Plating, A Coating Method for Forming Dense Well-Adhered Coatings," *Wire Ind. 44*, 771-773 (December, 1977).

E. Tsunasawa, K. Inaguak and K. Yamanaka, "Properties of Ion Plated Titanium Coatings on Steel," *J. Met. Finish Soc. Jpn 28*, 454-458 (September, 1977).

A.G. Dirks and H.J. Leamy, "Columnar Microstructures in Vapor-Deposited Thin Films," *Thin Solid Films 47*, 219 (1977).

E. Tsunasawa, K. Inagaki and K. Yamanaka, "Properties of Ion Plated Chromium Coatings on Steel," *J. Met. Finish. Soc. Jpn 23*, 159 (1977).

H. Yoshihara, "Silver Ion Plating on Copper Substrates," *J. Met. Finish. Soc. Jpn 23*, 156 (1977).

E. Tsunasawa, K. Inagaki and K. Yamanaka, "Influence of Deposition Parameters on Formation of Thin Films by the Ion Plating Process," *J. Met. Finish. Soc. Jpn 28*, 140 (1977).

H.E. Hintermann and C. Menound, "Ionic Deposition," *Oberflache Surf. 18*, 227 (1977).

N. Hosokown, T. Tsukadu and T. Missioni, "Self Sputtering Phenomena in High Rate Coaxial Cylindrical Magnetron Sputtering." *J. Vac. Sci. Technol. 14*, 138 (1977).

D.G. Teer, "Adhesion of Ion Plated Films and Energies of Deposition," *J. Adhesion 8*, 289 (1977).

P.K. Haff and Z.E. Switkowski, "Ion Beam-Induced Atomic Mixing," *J. Appl. Phys. 48*, 3383 (1977).

H.W. Lehman, L. Krausbauer and R. Widner, "Redeposition a Serious Problem in RF Sputter Etching of Structure with Micrometer Dimensions," *J. Vac. Sci. Technol. 14*, 281 (1977).

B.F.T. Bolker, "Effects of Temperature and Bias on the Composition of RF-Sputtered Te-As-Ge Ternary Films," *J. Vac. Sci. Technol. 14*, 254 (1977).

J.L. Vossen, Jr., "Method of Low Stress Hafnium Thin Film," U.S. Pat. 4,056,457 (Bias Effect on Stress).

N. Ohmae, T. Nakai and T. Tsukizoe, "Ion Plated Thin Films for Anti-Wear Applications," pp. 350-357, *Wear of Materials*, 1977.

S. Komiya et al, "Characteristics of Thick Chromium-Carbon and Chromium-Nitrogen Films Deposited by Hollow-Cathode Discharge," *Thin Solid Films 45*, 433-445 (1977).

D.G. Teer and F.B. Salem, "The Formation of Low-Friction, Wear Resistant Surfaces by Ion Plating," *Thin Solid Films 45*, 583 (1977).

J. L. Vossen, J.J. O'Neill, O.R. Mesker and E.A. James, "Extremely High Stress in Graded Interfacial Layers: Thin Films of Hf on Sapphire," *J. Vac. Sci. Technol. 14*, 85 (1977).

J. Chevallier, "Platinum Metal Coatings by Vacuum Deposition Processes," *Thin Solid Films 40*, 223 (1977).

S. Schiller, U. Heisig and K. Goedicke, "On the Use of Ring Gap Discharges for High Rate Vacuum Coating," *J. Vac. Sci. Technol. 14*, 815 (1977).

1978

W. Patz and A. Flamig, "The Dependence of the Porosity of Ion Plated Films on the Process Parameters," *Thin Solid Films 51*, 265 (1978).

P.D. Parry, "Localized Substrate Heating during Ion Implantation," *J. Vac. Sci. Technol. 15*, 111 (1978).

Susumu Takeuchi, "Method of Manufacturing a Chromium Oxide Film," U.S. Patent 4,096,026 (June 20, 1978) (RF Ion Plating).

E. Eser, R.E. Ogilivie and K.A. Taylor," Structural and Compositional Characterization of Sputter-Deposited WC + Co Films," *J. Vac. Sci. Technol. 15*, 396–400 (1978).

N.S. Plakakis and L. Missel, "Wet and Vacuum Processes—A Comparison Technology and Products, Part III," *Metal Finishing 76*, 50 (1978).

E.R. Fannin and D.E. Muehlberger, "Ivadizer-Applied Aluminum Coating Improves Corrosion Protection of Aircraft," Proc. 14th Ann. Airlines Plating Forum (1978).

A.E.T. Kuiper, G.E. Thomas and W.J. Schouten, "Thin Film Deposition from Beams of Ionized Atoms and Clusters," *J. Cryst. Growth 45*, 332 (1978).

E. Eser, R.E. Ogilivie and K.A. Taylor, "Friction and Wear Results from WC + Co Coatings by DC-Biased RF Sputtering in a Helium Atmosphere," *J. Vac. Sci. Technol. 15*, 401–406 (1978).

J. Amano and R.P.W. Lawson, "Thin Film Deposition using Low-Energy Ion Beams (4) Ion-Source Modification," *J. Vac. Sci. Technol. 15*, 118 (1978).

H. Yano, H. Hashimoto and Y. Toyama, "Damage Caused by Ar-Ion-Beam Etching," *J. Electrochem. Soc. 125*, 155C (1978).

G. Fischer, A.E. Hill and J.S. Colligon, "A Two-Gun Ion Beam System for Dynamic Recoil Implantation," *Vacuum 28*, 277 (1978).

B. Zega, "Process for Coating Insulating Substrates by Reactive Ion Plating," U.S. Patent 4,112,137.

T. Spalvins, "Morphology of Gold and Copper Ion Plated Coatings," NASA TP-1262 (June, 1978).

D.E. Muehlberger, "Plating with Aluminum by Ion Vapor Deposition," *Plating and Surface Finishing 65*, 20 (1978) also Society of Automotive Engineers Reprint No. 780252.

E.H. Hirsch and I.K. Varga, "The Effect of Ion Irradiation on the Adherence of Germanium Films," *Thin Solid Films 52*, 445 (1978) (Recoil Implantation: Film Stress).

M.S. Raven, "ESCA Interface Study of Ion Plated and Thermally Evaporated Selenium Films," *J. Physics D 11*, 631–642 (1978).

George R. Thompson, Jr., "Ion Beam Coating: A New Deposition Method," *Solid State Technology 21*, 73 (1978).

D.G. Teer and B.L. Delcea, "Grain Structure of Ion Plated Coatings," *Thin Solid Films 54*, 295 (1978).

J.M. Walls, D.D. Hall, D.G. Teer and B.L. Delcea, "A Comparison of Vacuum-Evaporated and Ion Plated Thin Films using Auger Electron Spectroscopy," *Thin Solid Films 54*, 303 (1978).

S. Fujishiro and D. Eylon, "Improved High Temperature Mechanical Properties of Titanium Alloys by Platinum Ion Plating," *Thin Solid Films 54*, 309 (1978).

M. Lardon, R. Buhl, H. Signer, H.K. Pulker and E. Moll, "Morphology of Ion Plated Titanium and Aluminum Films Deposited at Various Temperatures," *Thin Solid Films 54*, 317 (1978).

M.K.H. Shorshorov, O.V. Gusev, T.L. Roslyakova and M.V. Grusin, "Peculiarities of Formation and Structure in Films Produced by Condensation from a Partially Ionized Vapor Flow," (ABS) *Thin Solid Films 54*, 323 (1978).

M.A. Bayne, R.W. Moss and E.D. McClanahan, "Krypton Entrapment in Pulse-Biased Sputter-Deposited Metals," *Thin Solid Films 54*, 327 (1978).

Joseph E. Berg, Randolph E. Brown, "Method for Depositing Film on a Substrate," U.S. Patent 4,107,350 (August 15, 1978).

Nobuhiko Iwao, Tadashi Suzuki, Masaofukutomi Okada and Ryoji Watanabe, "Effect of Ion Plated Molybdenum on Corrosion of V/Mo Alloys in Liquid Sodium," *J. Nuc. Mat. 73*, 105 (1978).

Eric Kay and Gunther Heim, "Model of Bias Sputtering Applied to the Control of Nb Film Properties," *J. Appl. Phys. 49*, 4862 (1978).

E.W. Williams, "Gold Ion Plating," *Gold Bulletin 11*, 30 (1978).

Toshinori Takagi, "Ion Plating Method," U.S. Patent 4,082,636 (4/4/78) (Electron Emitter Above Vapor Source).

K.G. Stephens and I.H. Wilson, "Properties and Applications of Ion-Implanted Films," *Thin Solid Films 50*, 325 (1978).

M.A. Nicolet, "Diffusion Barriers in Thin Films," *Thin Solid Films 52*, 415 (1978).

Mitsunori Kobayashi and Yoshihako Doi, "TiN and TiC Coating on Cemented Carbides by Ion Plating," *Thin Solid Films 54*, 67 (1978).

T. Sato, M. Tada and Y.E. Huang, "Physical Vapor Deposition of Chromium and Titanium Nitrides by the Hollow Cathode Discharge Process," *Thin Solid Films 54*, 61 (1978).

G. Dearnaley, "Ion Implantation for Improved Resistance to Wear and Corrosion," *Mater. Eng. Appl. 1*, 28 (1978).

1979

J.W. Mayer and P.T. Clogston, "Metal Semiconductor Interfaces," To be published in *J. Vac. Sci. Technol. 16*, (1) 1979.

R. Wang, F. Baron and R. Reihl, "Induced Interface Interaction in Ti/Si System by Ion Implantation," *J. Vac. Sci. Technol. 16*, 130 (1979).

D.W. Hoffman and J.A. Thornton, "Effects of Substrate Orientation and Rotation in Internal Stresses in Sputtered Metal Film," *J. Vac. Sci. Technol. 16*, 134 (1979).

S.A. Schwartz and C.R. Helms, "Model of Ion Knock-on Mixing with Application to Si-SiO$_2$ Interface Studies," *J. Vac. Sci. Technol. 16*, 781 (1979).

Z.L. Liau, B.Y. Tsaur and J.W. Mayer, "Influence of Atomic Mixing and Preferential Sputtering in Depth Profiles and Interfaces," *J. Vac. Sci. Technol. 16*, 121 (1979).

D.M. Mattox, "Bibliography on Ion Plating and Related Ion Bombardment Effects," Sandia Laboratories Report SAND-0524.

D.M. Mattox, "Ion Plating Technology," Course Notes UCLA Extension Course on *Deposition Technologies and Applications*, R.F. Bunshah, Organizer.

E.W. Williams, "Gold Ion Plating: A Recently Developed Coating Process," *Solid State Technol. 22*, 80 (1979).

International Conference on Metallurgical Coatings, San Diego, CA, April 23-27, 1979, Proceedings to be published in *Thin Solid Films*.

> B.Y. Tsaur, Z.L. Liau, S.S. Lau and J.W. Mayer, "Ion Induced Intermixing of Surface Layers," *63*, 37 (1979).
>
> J.R. Hirvonen, C.A. Carosella, R.A. Kant, I. Singes, R. Vardiman and B.B. Ruth, "Improvement of Metal Properties by Ion Implantation," *63*, 5 (1979).
>
> L. Panevicius, "Structures and Properties of Deposits Grown by Ion Beam Activated Vacuum Deposition Technology," *63*, 77 (1979).
>
> M.A. Bayne, R.W. Moss and E.D. McClanahan, "Krypton Entrapment in Continuous Biased Sputter-Deposited Glassy Metals," *63*, 137 (1979).
>
> R.P. Howson, J.N. Avaritsiotis, M.I. Ridge, C.A. Bishop, Formation of "Transparent Heat Mirrors by Ion Plating onto Ambient Temperature Substrates," *63*, 163 (1979).
>
> L.D. Hartsough, "Resistivity of Bias-Sputtered Ti-W Films."
>
> T. Spalvins, "Characterization of Defect Growth Structures in Ion Plated Films by Scanning Electron Microscopy."
>
> J.W. Coburn, "The Influence of Ion Sputtering on the Elemental Analysis of Solid Surfaces."
>
> R. Clampitt, "Ion Plating by Field Emission Deposition."

"Ion Plating," *Vacuum 29*, (1979).

> A.R. Goode and M. St. J. Burden, "Inert Gas Handling in Ion Plating Systems," p. 9.
>
> M. St. J. Burden and K.B. Cross, "Radio Frequency Plasma Excitation," p. 13.
>
> K.B. Cross and J. O'Donnell, "An Evaporation Source Pellet or Slug Feeding System," p. 15.

Michael J. Mirtich and James S. Sovey, "Adhesive Bonding of Ion-Beam-Textured Metals and Fluoropolymers," *J. Vac. Sci. Technol. 16*, 809 (1979).

D.G. Welkie and M.G. Lagally, "Correlation of Short-Range Order and Sputter Dose in GaAs (110) using a Vidicon-Based LEED System," *J. Vac. Sci. Technol. 16*, 784 (1979).

Ion Plating and Allied Techniques, London, July 3-5, 1979, Proceedings available from CEP Consultants, Ltd., 26 Albany Street, Edinburgh EH1 3QH, UK.

> D.M. Mattox, "Mechanisms of Ion Plating."
>
> D.G. Teer, A. Matthews, A.J. Kirkham, "Wear Resistant Coatings By Reactive Ion Plating."
>
> M.I. Ridge, R.P. Howson, J.N. Avaritsiotis, "Reactive Ion Plating with a Plasma Magnetron Sputtering Source."
>
> Y. Enomoto, Y. Tsuya, G. Yaguchi, T. Ohgami, R. Takagi, "Frictional Properties of Gold and Gold/Molybdenum Thin Films onto Stainless Steel in Vacuum at Elevated Temperatures."
>
> D.G. Teer, A.J. Kirkham, "Factors Influencing Current Density in Ion Plating."
>
> S. Husa, "Ion Plating with an Inverted Biasing Arrangement."
>
> L. Pranevicius, K. Klimasauskas, "The Significance of Induced Stress in Thin Films Grown by Ion Beam Activated Vacuum Deposition Technology."
>
> J. Franks, "Ion Source for High Vacuum Ion Plating."
>
> R.P. Howson, J.N. Avarisiotis, M.I. Ridge, "Solar Heat Mirrors Produced by Ion Plating."
>
> B. Raicm, G. Dinca, "Ion Plated Optical Layers."
>
> M.J. Carter, H.A. MacLeod, I.M. Reid, E. Henderson, "Ion Plating of Optical Thin Films for the Infra-red."

O.S. Heavens, "Surface Analytical Methods for the Study of Ion Plated Surfaces."

T.A. Eckler, B.A. Manty, S. Fujishiro, "Evaluation of Ion Plated Coatings on Titanium Alloys."

R.W. Young, R.E. Hurley, "The Deposition of Aluminum Layers on Glass using Ion Plating."

D.W.L. Tolbree, D.G. Teer, M. Salama, "The Preparation of Self Supporting Ion Plated Carbon Foils for Lifetime Measurements in an Accelerated Heavy Ion Beam."

W.B. Nowak, B.A. Okorie, "Electrochemical (Corrosion) Behavior of Ion Plated Fe-Cr Films."

B.L. Delcea, B. Raicm, G. Dinca, "Ion Plated Films of Al, Au and Cu for Microwave Applications."

T. Takayi, I. Yamada, K. Matsubara, "Ion Beam Epitaxial Techniques and Applications."

J.P. Coad and R.A. Dugdale, "Recent Applications of Sputter Ion Plating."

B. Zega, "Reactive Deposition—A Challenge for Ion Plating and Allied Techniques."

C. Weissmantel, "Recent Developments in Ion-Activated Film Preparation."

Proceedings of the Third Symposium on Ion Sources and Applications Technology (February 19-22, 1979) Inst. of Electrical Engineers of Japan.

Toshinori Takagi, Kakuei Matsubara, Hiroshi Takaoka and Isao Yamada, "Some Investigations of the Role Played by Ions in Ion-Based Film Formation," p. 119.

Masayuki Itakura and Fumikazu Chira, "Electrical Properties of Ion Plated Al Contacts to Si," p. 127.

Kakvei Matsubara, Naoto Kondo, Kenichi Fujii, Hiroshi Takaoka and Toshinori Takagi, "Domain Growing Mechanism and Curie Point Writing of MnBi Films Prepared by Multitype Ionized-Cluster Beam Technology," p. 131.

Sunji Misawa, Sadafumi Yoshida, Shun-ichi Gonda, "Doping of Impurities into AlN using Partially Ionized Beam in Reactive Molecular Beam Technology," p. 135.

Kazuyuki Toki and Kazutoshi Kosakabe, "Electron Bombardment Ion Source for Ion Beam Deposition," p. 137.

G. Shimaoka, "Epitaxial Growth of CdSi Thin Films on Rock Salt by Ion Beam Deposition," p. 141.

R. Busch, "Effects of Sputtered Surface Structure on Adherence of Thermal Barrier Coatings," Battelle PNW Labs Conf. Report 781018, pp. VI 37-VI 39, *Energy Abs. 4*, 3198 (1979).

L. Holland and S.M. Ojha, "The Growth of Carbon Films with Random Atomic Structure from Ion-Impact Damage in a Hydrocarbon Plasma," *Thin Solid Films 58*, 107 (1979).

P.A. Higham and D.G. Teer, "Influence of the Deposition Parameters on the Structure of Ion Plated Chromium," *Thin Solid Films 58*, 121 (1979).

C. Weissmantel, "Film Preparation using Plasma or Ion Activation," *Thin Solid Films 58*, 101 (1979).

T. Shioyama, S. Ogawa and K. Yamanaka, "Electrical Properties and Structure of Tantalum Thin Films Deposited by Bias-Ion-Beam Sputtering," *Thin Solid Films 58*, 127 (1979).

R.P. Howson, J.N. Avaritsiotis, M.I. Ridge and C.A. Bishop, "Reactive Ion Plating of Metal Oxides onto Insulating Substrates," *Thin Solid Films 58*, 379 (1979).

Masahiro Kitajima, Masao Fukutomi, Masatoshi Okada and Ryoji Watanabe, "Electrographic Detection of Pinholes in Ion Plated Silicon Nitride Films on Molybdenum," *Denri Kagaku 47*, 214 (1979).

J. Dudonis et al, "The Influence of Ion Bombardment during Condensation on the Properties of Vacuum Deposited Thin Films (ABS)," *Thin Solid Films 58*, 106 (1979).

T. Spalvins, "Industrial Potential, User, and Performance of Sputtered and Ion Plated Films," NASA-TM-79, 107.

J. Franks, P.R. Stuart and R.B. Withers, "Ion-Enhanced Film Bonding," *Thin Solid Films 60*, 231 (1979).

T. Takagi, I. Yamada and K. Matsubara, "Ionized-Cluster-Beam Deposition and Epitaxy," *Thin Solid Films 58*, 9 (1979).

R. Clampitt, "Field Emission Deposition Sources," *Thin Solid Films 58*, 129 (1979).

D.J. Dimara, L.M. Ephrath and R.R. Young, "Radiation-Damage in Silicon-Dioxide Films Exposed to Reactive-Ion Etching," *J. Appl. Phys. 50*, 4015 (1979).

F.A. Otter et al, "Ion-Recoil Plating of Ultrathin Films," *Bull. Am. Phys. Soc. 24*, 316 (1979).

S. Fujishiro and D. Eylon, "Titanium and Titanium Alloys Ion Plated with Noble Metals and Their Alloys," U.S. Pat. 4,137,370, *Offic. Gaz. U.S. Pat. Off. 978*, 1822 (1979).

E.N. Williams, K. Jones, D.G. Teer, "Structural and Resistivity Studies of Ion Plated Metal Films on Plastics." *Mat. Res. Bull. 14*, 59 (1979).

3rd International Congress on Cathodic Sputtering and Related Applications, Sept. 11-14, 1979, Nice, France.

> "Thin Film Growth under Ion Bombardment," C. Weissmantel.
> "New Ion Plating and Sputtering Units," C. Hayashi.
> "Substrate Sputter Etching during Plasma-Deposition of Diamond-Line Carbon Films," H. Norstrom, R. Olaison, L.P. Andersson, S. Berg.
> "The Effects of Substrate Ion Bombardment in Relation to Coating Adhesion and Other Properties," P. Coad, M. Gettings, L. Martindale, R. Dugdale.
> "Reactive Ion Plating with a Plasma Magnetron Sputtering Source," M. Ridge, R. Howson, J. Avaritsiotis, C. Bishop.

R. Kirchheim and S. Hofmann, "Coverage of Foreign Atoms on Surfaces as a Function of Absorption, Sputtering and Diffusion Rates," *Surf. Sci. 83*, 296 (1979).

J. Franks, P.R. Stuart and R.B. Withers, "Ion Enhanced Film Boriding," *Thin Solid Films 60*, 231 (1979).

Hideo Yoshihara and Hidefumi Mori, "Enhanced ARE Apparatus and TiN Synthesis," *J. Vac. Sci. Technol. 16*, 1007 (1979).

T. Spalvins, "Characterization of Defect Growth Structures in Ion Plated Films by Scanning Electron Microscopy," NASA TM-79110 (1979).

Chr. Weissmantel, K. Bewiloqua, C. Schurer, K. Breuer and H. Zscheile, "Characterization of Hard Carbon Films by Electron Energy Loss Spectroscopy," *Thin Solid Films 61*, L-1 (1979).

Chr. Weissmantel, C. Schurer, F. Frohlich, P. Grau and H. Lehman, "Mechanical Properties of Hard Carbon Films," *Thin Solid Films 61*, L-5 (1979).

G. Reisse, U. Ebersbach, F. Henny and Chr. Weissmantel, "Ion Beam Nitriding of Iron," *Thin Solid Films 61*, L-9 (1979).

M. Marinov, "Influence of Ion Bombardment on the Surface Structure of Silicon Single Crystals," *Thin Solid Films 61*, 363 (1979).

H.R. Kaufman, "Focused Ion-Beam Designs for Sputter Deposition," *J. Vac. Sci. Technol. 16*, 899 (1979).

K. Morimoto, Y. Utamora and T. Takagi, "Ionized Cluster-Beam Deposition Process for Fabricating P-N Junction Semiconductor Layers," U.S. Patent 4,161,418, *Offic. Gaz. U.S. Pat. Off. 984*, 656 (1979).

R.P. Howson et al, "Properties of Conducting, Transparent Oxide Films Produced by Ion-Plating onto Room Temperature Substrates," *Appl. Phys. Lett. 35*, 161 (1979).

W.B. Nowak and B.A. Okorie, "Transient Electrochemical (Erosion) Behavior of Ion-Plated Fe-Cr-Films," *J. Electrochem. Soc. 126*, 333C (1979).

E.R. Zannin, "Ion Vapor Deposited Aluminum Coatings for Improved Corrosion Protection," *Sci. Tech. Aerospace Reports 17*, 1824 (1979).

G. Mezey, S. Matteson and M.A. Nicolet, "Comparison of Ion-Beam-Induced Intermixing of Two Elements in Bilayer or in Thin, Buried-Layer Configuration," *J. Electrochem. Sci. 126*, 345C (1979).

Donald T. Hawkins, "Ion Milling (Ion Beam Etching) 1975–1978: A Bibliography," *J. Vac. Sci. Tech. 16*, 1051 (1979).

Masato Nakajima, Kenji Kusao, Takashi Hirao, Kaoru Inoue and Shigetoshi Takayanagi, "Enhanced Interdiffusion in the Al-Si System During Ar Ion Bombardment," *Jpn J. Appl. Phys. 18*, 1869 (1979).

William A. Brainard and Donald R. Wheeler, "Use of a Nitrogen-Argon Plasma to Improve Adherence of Sputtered Titanium Carbide Coatings on Steel," *J. Vac. Sci. Technol. 16*, 31 (1979).

Masao Fukutomi, Masahiro Kitajima, Masatoshi Okada and Ryoji Watanabe, "Silicon Carbide Coating on Molybdenum by Chemical Vapor Deposition and Its Stability Under Thermal Cycle Conditions," *J. Nucl. Mat. 87*, 107 (1979) (IP-Boron, Carbon).

Paul R. Forant, "Vacuum Metallizing," *Metal Finishing 77*, (11) 17 (1979).

H. Mori and H. Yoshihara, "Ion-Effect on γ-Al_2O_3 Films by Enhanced ARE Process," *Jpn J. Appl. Phys. 18*, 837 (1979).

B.Y. Tsaur, S.S. Lau and J.W. Mayer, "Ion-Beam-Induced Formation of the PdSi Silicide," *Appl. Phys. Lett. 35*, 225 (1979).

K.E. Steube, "Fabrication and Optimization of an Aluminum-Ion Vapor Deposition System," McDonnell Douglas Corp. Report AFML-TR-78-132, Contract F33615-76-5209, June, 1978, MCIC-404120.

M.A. Bayne, R.W. Moss and E.D. McClanahan, "Krypton Entrapment in Continuously Biased Sputter Deposited Glassy Metals," *Thin Solid Films 63*, 137 (1979).

S Spooner, H. Solnick-Legg and E.A. Starke, Jr., "The Effect of Ion Plating and Ion-Implantation on the Cyclic Responses and Fatigue-Crack Initiation of Metals and Alloys," Georgia Institute of Technol. Report AD-A066547, Contract N00014-78-C-0270 (March, 1979).

Bernard Bourdon, "Method for Depositing Thin Layers of Materials by Decomposing a Gas to Yield a Plasma," U.S. Patent 4,173,661 (Nov. 6, 1979).

C. Weissmantel, H.J. Erler and G. Reisse, "Ion Beam Techniques for Thin and Thick Film Deposition," *Surf. Sci. 86*, 207 (1979).

K. Bewiloqua, D. Dietrich, L. Pagel, C. Schurer and C. Weiss Mantel, "Structure and Properties of Transparent and Hard Carbon Films," *Surf. Sci. 86*, 308 (1979).

Peter Sigmund, "Recoil Implantation and Ion-Beam-Induced Compositional Changes in Alloys and Compounds," *J. Appl. Phys. 50*, 7261 (1979).

C. Weissmantel, G. Reisse, N.J. Erler, F. Henn, K. Bewiloqua, U. Ebersbach and C. Schurer, "Preparation of Hard Coatings by Ion Beam Methods," *Thin Solid Films 63*, 315 (1979).

W. Fleischer, D. Schulze, R. Wilberg and A. Lunk, "Reactive Ion Plating (RIP) with Auxiliary Discharge and the Influence of the Deposition Conditions on the Formation and Properties of TiN Films," *Thin Solid Films 63*, 347 (1979).

A.K. Weiss and J.R. Clarke, "Plasma Plating," (PbO) U.S. Patent 4,170,662, *Offic. Gaz. U.S. Pat. Off. 987*, 457 (Oct. 9, 1979).

L. Pranevicius, "Structure and Properties of Deposits Grown by Ion Beam Activated Vacuum Deposition Techniques," *Thin Solid Films 63*, 77 (1979).

"Industrial Applications of Ion Plating," *Galvano-Organo (Supl) 487*, 45 (1978).

1980

L.R. Gilbert, S.V. Krishnaswamy and R. Messier, "Resputtering Effects in Ba(Pb,Bi)O$_3$ Perovskites," *J. Vac. Sci. Technol. 17*, 389 (1980).

T. Spalvins, "Survey of Ion Plating Sources," *J. Vac. Sci. Technol. 17*, 315 (1980).

H.J. Erler, G. Reisse and C. Weismantel, "Nitride Film Deposition by Reactive Ion Beam Sputtering," *Thin Solid Films 65*, 233 (1980).

J.N. Avaritsiotis and R.P. Howson, "The Ion Plating of Desirable Optical Coatings onto Plastics," *Thin Solid Films 65*, 101 (1980).

I.N. Evdokimov and G. Fischer, "Surface Hardness Improvement by Dynamic Recoil Implantation," *J. Mater. Sci. 15*, 854 (1980).

E. Eser, R.E. Ogilvie and K.A. Taylor, "The Effect of Bias on DC and RF Sputtered WC-Co Coatings," *Thin Solid Films 67*, 265 (1980).

Tallivaldis Spalvins, "Survey of Ion Plating Sources," *J. Vac. Sci. Technol. 17*, 315 (1980).

S.M. Myers, "Properties and Application of Ion Implanted Alloys," *J. Vac. Sci. Technol. 17*, 310 (1980).

D.W. Hoffman and M.R. Gaerttner, "Modification of Evaporated Chromium by Concurrent Ion Bombardment," *J. Vac. Sci. Technol. 17*, 425 (1980).

S. Shimizu and S. Komiya, "Epitaxial Growth of Si on (T012) Al$_2$O$_3$ by Partially Ionized Vapor Deposition," *J. Vac. Sci. Technol. 17*, 492 (1980).

E.H. Hirsch and I.K. Varga, "Thin Film Annealing by Ion Bombardment," *Thin Solid Films 69*, 99 (1980).

I.N. Eudokimov and G. Fischer, "Surface Hardness Improvement by Dynamic Recoil Implantation, *J. Mat. Sci. 15*, 854 (1980).

7

Microstructures of PVD-Deposited Films Characterized by Transmission Electron Microscopy

Birgit E. Jacobson

Linköping University
Department of Physics and Measurement Technology
Linköping, Sweden

1. INTRODUCTION

This article will discuss the microstructures of thick coatings deposited on metal substrates. The microstructural features play a central role in the relationship film processing/film structure/film properties and is therefore worth more attention since there is lack of knowledge in this field, especially in the area of thick coatings, i.e., coatings that are thick enough to develop and exhibit their own bulk properties. More fundamental studies have been performed previously concerning the nucleation and growth of thin films, and in this case the substrate material has usually been nonmetallic or semiconducting. The goal here is to show how important and useful information about these microstructures can be obtained by careful characterization using transmission electron microscopy (TEM). It will be shown how the growth morphology and structure can be varied over wide ranges by the variation of the processing parameters and composition during film deposition. It will also be shown how these structure variations will affect the film properties. This will be done by the discussion of selected examples representing a variety of film compositions and deposition methods.

Why Study the Microstructure?

Many opportunities exist for developing alloys and compounds with preferred combinations of unique properties not previously obtained with conventional materials processing. A variety of deposition methods can be used for this purpose, each exhibiting advantages and limitations depending on the selection of coating material, specified film properties and final processing costs. In order to optimize the result of the coating process it is important to learn more about the potential of these opportunities, i.e., over what ranges the mechanical, physical and chemical properties of the films can be varied by proper selection of deposition process and deposition conditions. To avoid the trial and error method it is important to study in more detail, first, the fundamental relationships between the processing parameters and the resulting microstructural features and, next, the relationships between these features and the film properties.

It should also be pointed out that the well known structure/property relationships valid for conventionally produced coarse grained structures not a priori can be extrapolated to include also these often extremely fine grained deposited structures.

What to Learn About the Microstructure?

The film structure is characterized by the film composition, by the grain morphology, by the internal orientation relationships between the individual grains (texture), and by the defect structure, i.e., the distribution of dislocations, stacking faults, porosities and second phase particles. The characteristics of the film composition include, besides the overall composition, the chemical segregation of alloying elements in various scales, e.g., in a long range scale as a function of the distance from the film/substrate interface or in a short range scale as a distance within the grains from the grain boundaries. All these features have to be considered in a complete film structure characterization.

The most important prerequisite for meaningful studies of the microstructure and related film properties is that the films are prepared under extremely well controlled deposition conditions, and that subsequently they are well characterized in all aspects that might have any influence on the film properties. Then these studies allow more general conclusions to be drawn for the purpose of getting a better understanding of the process/structure/property relationships.

In general, deposited film structures are much more finely dispersed than those of conventionally produced materials; at moderate deposition temperatures, the grain sizes vary typically from the micron range down to the scale of some hundreds or tens of an Å, alternatively, the structure can be synthesized in the amorphous state. The presence of these high densities of grain boundaries or, respectively, the absence of grain boundaries will each affect the film properties, specifically in situations where diffusion is an important mechanism. The same is valid for the distribution of voids and pores of various size ranges that are frequently present in deposited films. Furthermore, unexpectedly or purposely nonequilibrium phases can be produced by proper selection of the deposition condition, and these have to be characterized as well to get a proper understanding of the film structure. These defect distributions are thus all important features of deposited films, and they promise possibilities of unique combinations of properties to be obtained, i.e., deformation, corrosion, magnetic and electrical properties. It is thus important to learn in detail about the production of these defect distributions, how they can be stimulated or inhibited during film processing.

Why TEM?

In the structural investigation of deposited films the transmission electron microscopy (TEM) has many advantages over the more traditional x-ray diffraction and optical microscopy techniques. The most powerful features of the TEM are the high spatial resolution capability (< 10 Å) and the variety of operational modes that can be used for the structure analysis. Both are particularly crucial in the case of fine crystalline structures with high imperfection densities. With the various modes of imaging, contributions to the contrast from each individual object can be sorted out and studied in detail. Besides the capability of real-space image formation and electron diffraction pattern formation, contemporary TEM instruments possess the additional capability of chemical x-ray analysis from small selected volumes (30-50 Å in diameter) where other traditional methods give average compositional information only. This is achieved by an additional lens, located above the object and used solely to form an electron probe which can be scanned across the object or placed to illuminate a selected stationary spot, and in addition an energy dispersive detector for the analysis of excited X-rays from this illuminated spot.

Content

Section 2 is a condensed review of the possibilities and limitations that are rendered in a TEM instrument. Various ways to prepare thin transparent foils for the TEM study are discussed.

Section 3 is concerned with the variations in microstructure that can be achieved by electron beam evaporation under various conditions. Creep deformation and superconducting properties are considered.

Section 4 deals with the microstructure of an important class of materials, refractory metal compounds prepared by activated reactive evaporation.

Section 5 shows an example of magnetron sputter deposited coatings where the specific opportunity of low temperature deposition is utilized to synthesize a common alloy (stainless steel) in the amorphous state. The crystalline to amorphous transition zone is studied.

2. THE POSSIBILITIES IN A TEM/STEM INSTRUMENT

The basic reasons for the utilization of the electron microscope are (a) the superior space resolution and (b) the versatility of the instrument allowing characterization of the structure in a variety of modes.

Transmission
Electron
Microscope

Images	Diffraction	Chemical Analysis
Morphology	Phase Identification	Particle Composition
Defect Structure	Orientation Relationship	Compositional Gradients

The combination of bright field and dark field and other imaging modes with diffraction pattern analysis and chemical x-ray analysis is essential in the characterization procedure. It is therefore convenient to be able to carry out all these analyses in one and the same instrument.

Imaging

The information obtained in the image is derived from the electron scattering processes, elastic and inelastic, that take place when the electron beam passes through the specimen. The contrast of the image depends on the intensity distribution leaving the bottom surface of the specimen, and the image of the microstructure can be formed, alternatively, by phase contrast or amplitude contrast. After suitable increase in magnification, utilizing the projector lenses, the image can be obtained and focused either on a fluorescent screen or photographically recorded. If the transmitted and scattered beams are allowed to recombine, then the lattice image of the planes which are diffracting can be obtained by the phase contrast with a resolution of about 2 Å. The more conventional technique of imaging is by amplitude contrast, either in the bright field (BF) or in the dark field (DF) mode. These images are obtained by inserting an aperture in the back focal plane of the objective lens which, alternatively, removes all diffracted beams (but not the primary beam) from the image or allows only one strong diffracted beam to form the image. In the BF case, all regions in the specimen that are out of diffraction orientation appear bright and those observed under diffraction conditions dark. In the DF case, all regions appear dark except those under exact Bragg diffraction conditions, which contribute to the intensity of the selected diffracted beam and thus appear bright.

Image contrast will thus arise from changes in local diffraction conditions: variations in lattice orientation of the grains; presence of lattice defects such as point defects, line defects, planar defects and volume defects; variations in composition and variations in structure. Also, by tilting and rotation of the specimen under the incident electron beam the image will continuously vary in contrast, revealing different features in the microstructure. All these contrast effects have to be considered during the interpretation of the images and the intended characterization of the microstructure. The situation can thus be rather complicated especially when the defect density is high as in most materials produced by deposition methods. Besides the BF and DF imaging modes, however, there is a whole range of more recently developed, so-called nonconventional imaging methods, that can be utilized as complements in various situations. These are, for instance, lattice imaging, 3D-stereo imaging, 2½D imaging, weak beam imaging, out-of-focus imaging, and imaging of magnetic domain structures. It would be too voluminous to cover them all in this place; the reader is instead referred to standard books in electron microscopy, e.g., Hirsch's book, *Electron Microscopy of Thin Crystals* (1965)[1] which is the standard textbook; J.W. Edington's book, *Practical Electron Microscopy in Materials Science* (1976),[2] which is the users handbook; or to the lecture notes from the 3rd course of the International School of Electron Microscopy (1975), *Electron Microscopy in Materials Science,* Parts I, II, III and IV,[3] which covers more recent theories and instrumentational questions.

However, one of these imaging techniques has been found particularly useful in the study of deposited structures. This is the out-of-focus imaging technique which allows the revealment and identification of microcavities (10-50 Å in diameter) that are common in this kind of materials. The cavities represent a weak phase object which is invisible in focus, i.e., when the objective lens is focused on the bottom surface of the foil. However, when imaged under appropriate out-of-focus conditions, it produces a contrast that in under-focused condition consists of a bright dot surrounded by a dark Fresnel fringe and in over-focused condition of a dark dot surrounded by a bright Fresnel fringe. Furthermore, the outer diameter of the central spot, or the inner diameter of the first Fresnel fringe, corresponds fairly well to the diameter of the cavity. This contrast effect from small strain-free cavities was treated by Rühle and Wilkens (1972).[4]

The resolution of the TEM images is generally below 10 Å, down to 2-3 Å. The maximum foil thickness, transparent to the electron beam in a 100 kV electron microscope, is about 1000-2000 Å, and the maximum area observable in one single TEM-foil is about 0.03 mm^2. For fine grained structures, with a grain diameter of a couple of hundred Å, this corresponds to a volume containing many thousands of grains.

Electron Diffraction

Both the image plane and the diffraction pattern plane can alternatively be focused onto the fluorescent screen. If an aperture is placed in the image plane only electrons passing through the corresponding area in the specimen will contribute to the diffraction pattern, and they will thus produce a selected area diffraction pattern (SADP). Due to spherical aberration the minimum selected area in a 100 kV microscope is about 1 μm. This is apparently a problem for the study of fine grained microstructures in deposited materials; the orientation and lattice d-spacing of individual grains cannot be studied in this way. A 1 μm diameter polycrystalline specimen area will instead produce a ring pattern of more or less densely distributed spots. The breadth and the spottiness of the rings depend on the size and number of the crystals contributing to the pattern; the finer the grain size, the more continuous and broader is the pattern. Very diffuse broad rings may finally be referred to as originating from material in the amorphous state. Each ring diameter (2r) is related to a specific lattice spacing d-value, and this can be evaluated from the relation $\lambda L = r \cdot d$, where λL is the camera constant of the microscope for an electron wavelength of λ. The intensity and continuity of the rings give information about texture, i.e., preferred growth orientations of the grains. However, a complete characterization of the texture can best be performed utilizing a texture goniometer stage attached to a conventional x-ray diffraction apparatus.

The limitation of the conventional SAD method is experienced when phase identification or orientation relationships between individual grains are searched for in a fine grained material. There are two ways to decrease the size of the diffracting area. In contemporary TEM instruments the incident beam can be narrowed to cover a 300-400 Å specimen only and thus produce a convergent beam diffraction pattern. Still higher resolution can be obtained with a special lens (the STEM lens, located above the specimen) that can be used to focus the beam and form an electron probe of desirable size down to 30 Å diameter. This can be used for illumination of the individual grains of that size and thus produces single crystal diffraction patterns.

The final thinning of the 3 mm disc can be achieved by either electropolishing (jet polishing) or ion beam thinning. Electropolishing is preferred when possible because as soon as the proper polishing conditions are found (the composition and temperature of the electrolyte, the jet flow rate and the voltage-current settings), then a thin, damage-free foil can be produced in a couple of minutes. Lists of recipes and polishing conditions for various materials are found in the literature.[1,6] However, differences in alloying and deformation will affect the conditions so these can be used as guidelines only for thin foil preparation of the actual material. Electropolishing is, however, not suitable for nonconducting materials or materials containing second phase structures exhibiting different polishing conditions. Instead, ion beam milling can be utilized, and on all types of materials: metals, ceramics, polymers and composites. In this method a beam of ionized argon gas bombards the center area of the 3 mm sample disc at an incident angle of between 15 and 30°. The beam current is typically 20-80 μA and the accelerating voltage 2-6 kV. Another advantage of ion beam milling is that the size of the center hole can be allowed to grow large (1 mm in diameter) still maintaining thin surrounding edges and thus allowing large transparent areas to be studied, whereas electropolishing requires rapid interruption of the process at perforation to give smallest possible center hole. This is achieved by a photo-sensing unit. The thinning rate of ion milling is low, however, about 1 μm per hour, so each sample thus requires typically several days of milling. Another disadvantage is the radiation damage introduced on the foil surface. The ion beam current should be kept low to avoid heating of the sample which will enhance this effect. On the other hand, a decrease in current effects a decrease in milling rate, and a compromise between the two thus has to be made when selecting proper milling data. This is also the case for the angle setting; 30° gives 3-4 times higher milling rate than 15°, but the lower angle gives a smoother surface and a thinner damaged surface layer. It seems that soft metals, like Ni,

Cu and Au, are hard to prepare without severe damage to the structure, whereas hard materials, like TiC and Nb_3Sn compounds, are suitable for this preparation method. In the following sections, examples will be given from materials prepared with alternative methods. The Nb_3Sn-Cu compounds discussed contained a two phase structure of both hard and soft phase, the two thus representing different electropolishing properties, sputtering-yields and radiation damage sensitivity. Ion beam milling was used for preparing TEM foils of this structure, and the milling data were selected for the harder phase (Nb_3Sn). The Cu phase was thus subject to both radiation damage and preferential etching. This situation is not always a disadvantage, however; in this study it could be utilized to assist the identification of the two different phases, which would otherwise be difficult when the Cu concentration was low and the grain size of both phases fine (200 Å).

TEM foil preparation of thick films (0.5 to 100 μm in thickness), deposited on a substrate, can be prepared in a similar way by disc cutting and subsequent thinning of the disc from the substrate side either by electropolishing or ion beam milling. When the interface is reached, the thinning can be continued, alternatively, from the substrate side only, from the film side only, or simultaneously from both sides. This allows the study of, respectively, the interface structure, the film surface structure, or the center bulk structure of the film. The ion beam milling is in this case easier to utilize, because the slow milling rate allows a better control of the process.

Deposited films on flat substrates can be studied in cross-section as well. For this purpose several pieces of the sample are sandwiched together held by an epoxy. This piece of laminate is subsequently sliced and disc-shaped and finally ion beam milled in the usual way. A similar technique can be used for preparing TEM samples of thin coated wires. The wires are densely packed in a 3 mm diameter tube where they are held together by an epoxy. The whole package is then sliced to form 3 mm discs which are subsequently thinned by the ion beam miller. Many wires can be studied if the disc is repolished several times, allowing wires further and further out from the disc-center to be thinned.

There are alternative thinning procedures to those discussed above. For information about these and for more details about the ones discussed, the reader is referred to more comprehensive literature such as references 1, 2, 5 and 6. As a conclusion it should be pointed out that the foil preparation is an important step in the structure characterization by TEM analysis, because a good foil is the prerequisite for a careful and detailed TEM analysis. It involves, however, a tedious work that requires patience and some experience.

3. MICROSTRUCTURE AND PROPERTIES OF ELECTRON BEAM EVAPORATED FILMS

The growth morphology and microstructure of thick films of single phase composition prepared by electron beam evaporation have been studied extensively for a variety of materials, as is previously discussed and reviewed by Bunshah.[7] In most cases, however, these studies are limited to SEM and optical microscopy observations and thus in structure resolution. With the following examples it will be shown how TEM studies can contribute with additional information that is essential for a proper understanding of the film properties. Two examples are given of single phase structures, one pure metal, Ni, and one compound, Nb_3Sn. The influence of deposition rate and substrate temperature on the microstructural features is discussed. Then the effect of small amounts of co-evaporated second phase, Al_2O_3 and ZrO_2 on the growth morphology is considered. Finally, it will be shown how the growth morphology can be varied over wide ranges by co-evaporation of large amounts of second phase, Nb_3Sn-Cu, over the whole composition range. The related electrical (Nb_3Sn) and mechanical (Ni) properties of these structures are compared to those of conventionally produced materials.

Chemical Analysis

Chemical analysis can be achieved in the microscope if a suitable detector is appropriately placed either above the specimen to analyze the excited x-rays from the illuminated specimen area or below the specimen for energy-loss spectroscopy analysis of the electron beam after it has penetrated through the specimen. For the x-ray analysis, an energy dispersive detector is normally used, and this allows quantitative analysis of all elements above $Z = 11$. The presence and distribution of the light elements ($Z < 11$) is, however, also very important to the material properties, and the energy loss detector, which allows detection of these elements, is thus an important new attachment to the TEM instrument. The size and localization of the analyzed area can be selected by focusing the electron beam probe diameter ($\geqslant 30$ Å) using the STEM lens, and the analyzed depth is determined by the actual foil thickness of that area. Composition gradient profiles can be obtained with extremely high spatial resolution by a linear or X-Y scan of the probe over the specimen surface.

Heating, Cooling and Deformation

Besides image formation, electron diffraction and chemical analysis, the TEM instrument allows dynamical experiments to be performed in situ in the microscope. The structural changes resulting from heating, cooling or tensile deformation of the specimen can be simultaneously observed and recorded. The temperature covers the whole range from liquid-N_2 temperature up to 1000°C. The same 3 mm disc specimen can be used as is usual for the room temperature studies. It should be pointed out that these annealing experiments of conventional coarse grained materials have a limited value, because the presence of the thin foil surfaces prevents grain growth and enhances dislocation annihilation. This surface influence is less dominating, however, when the grain size is only a small fraction of the foil thickness, as is often the case in deposited materials. The tensile deformation experiments require a special, rather tedious, sample preparation procedure.

Limitations and Complementary Techniques

In spite of the versatility of the electron microscope, there are limitations where complementary techniques should be utilized. The spatial resolution is not a limitation for the purpose of characterization of fine grained deposited film structures; the limitation is instead in the low magnification end and is determined by the limited area that can be prepared transparent to the electron beam. It is therefore important to observe morphological variations on a coarser scale (magnifications <1000 X) by utilizing conventional optical microscopy and scanning electron microscopy prior to the more detailed and tedious TEM analysis.

X-ray diffraction measurements of lattice d-spacing for phase characterization is more accurate than electron diffraction and is performed over considerably larger areas and volumes of the specimen than the electron diffraction analysis. The two techniques are therefore complementary and should be combined for a more complete structure characterization. X-ray diffraction should also be used for a complete quantitative texture analysis. The sample is then placed in a texture goniometer and the result is projected on a stereogram, which shows the preferred orientations of a selected set of planes (normally the close-packed planes) in relation to the film surface and a specific direction along the surface, e.g., a direction that is related to the geometry of the deposition process set-up.

Chemical analysis can be performed using a variety of methods. An averaging of the composition over large volumes is obtained by wet chemical analysis, optical spectroscopy or x-ray fluorescence analysis. Analysis of smaller volumes in the micron size range is obtainable by the electron microprobe utilizing wavelength dispersive x-ray analysis or by the scanning electron microscope (SEM) utilizing energy dispersive x-ray analysis. Chemical analysis over large specimen surface areas (mm^2) and extremely thin depth, in the range of some Å up to tens of an Å, is obtained in cases where excited electrons are utilized for the elemental analysis in electron spectroscopy for chemical analysis (ESCA) and ultraviolet photoelectron spectroscopy (UPS). This is also the case for Scanning Auger spectroscopy (SAM) which exhibits the same depth resolution and besides a better resolution (in the μm range) in surface area analyzed. Chemical analysis in the way it can be performed in the TEM/STEM instrument cannot be done with any other technique. The specific feature is the small analyzed area (30 Å or larger), which allows composition gradient profiling with extremely high resolution. If only the variations of the average composition in one direction is searched for, e.g., the profile over a substrate/film interface, similar results could be obtained by sputter-etch depth profiling and simultaneous Auger analysis. However, the sputter-etching procedure can effect a change in composition by preferentially etching multiphase structures.

Specimen Preparation for TEM

It is nowadays possible to prepare TEM-foil specimens of almost any type of material. The most conventional shape of this specimen is a disc, 3 mm in diameter and a few tenths of a mm in thickness, thinned in the center region until perforation. The smooth edges of this center-hole are electron transparent, and the thicker rim is useful for the handling of the specimen, thus preventing deformation of the thin area. This foil preparation procedure thus involves two main operations, the 3 mm disc shaping and the final thinning.

The disc shaping from bulk and sheet materials can be achieved by a variety of methods. A slicing machine with a rotating diamond wheel is employed for slicing hard and brittle materials, a spark erosion machine for slicing metals, and a chemical string saw for slicing soft materials when minimum disturbance of the structure is of great importance. The subsequent 3 mm disc cutting from the sliced material is done by mechanical punching, spark erosion punching or ultrasonic cutting. The disc is then preliminarily thinned by grinding (on 600 grade SiC paper) to a thickness of 0.1-0.2 mm. The depth of damage caused by these procedures varies with method and material. It is therefore important to select the proper method to assure that the defect structure in the final thin foil truly represents that of the bulk material.

Growth Morphology and Cavity Formation in Single Phase Structures

Structure and Properties of Superconducting Nb_3Sn Films:[8] High critical temperature superconducting compounds such as Nb_3Ge (T_c = 23 K) can only be obtained by vapor phase quenching. It has been difficult, however, to characterize such Nb_3Ge films, particularly their composition and long-range order. Therefore, a study of the more stable structure of evaporated Nb_3Sn has been undertaken to be used as a model system in order to understand in detail the relationships between structure, composition and superconducting properties of such A15 compounds in the course of getting a better understanding of the physics behind these properties. For instance, a uniformly small grain size and other defects of proper size and distribution are important as magnetic flux pinning centers and thus lead to high critical current densities at high magnetic fields in the superconducting stage; atomic ordering and stoichiometric composition are the important factors for obtaining the highest possible critical temperature.

A wide variety of Nb_3Sn films were produced for this purpose by variation of the deposition parameters, including rate, substrate temperature and chemical composition. The electron beam evaporator used was equipped with three colinear element sources, positioned to produce a linear spread in the composition along the substrate, from under stoichiometric (24 at%Sn) to excess tin (31 at%Sn) composition. A careful feedback control of the evaporation rates was an important feature of this system; the atomic fluxes arriving at the substrate were kept constant to within 1% with both short and long term stability. Six rate monitors were used for this purpose. The substrate temperature was kept within ±5°C during deposition, and the vacuum enclosure system had a pressure capability, in moderately baked (100°C) and fully equipped condition, of the low 10^{-8} Torr range.

The microstructural features of the 4-7 μm thick deposits were studied in transverse section using SEM and in cross section, perpendicular to the growth direction using TEM; thin foils were prepared at locations of selected composition using ion beam milling. This preparation technique allowed typically 10^6 grains (400 Å in diameter) to be observed in one single foil thus ensuring that the observations were statistically sound. A large number of observations were recorded in the magnification range 20,000-400,000x, and the micrographs shown in the following are selected as being typical of each structure observed.

Low Rate-Low Temperature Deposition: An extremely fine columnar structure of grains 250 Å in diameter is obtained when Nb and Sn are codeposited at a low rate (32 Ås^{-1}) and a low substrate temperature (570°C), (Figure 7.1a). The individual grains are defined by surrounding high angle grain boundaries. The characteristic contrast distribution indicates, however, that neighboring columns are clustered together within 0.2 μm areas having near equal growth orientations. This grain configuration is a rather typical feature of low temperature deposits in general. This kind of structure is strongly textured, in this case with a preferred growth orientation in the <200> direction according to X-ray diffraction and in agreement with the appearance of the SADP having an enforced (200) reflection ring and a missing (211) ring. The rotary orientation is random, however, as is reflected in the diffraction rings having even contrast (Figure 7.1b).

Another characteristic feature of this kind of structure in general is the distribution of grain boundary cavities. In this structure they are 50-100 Å in diameter when at grain boundary triple points and 50 Å in width and 200 Å in length when extending along the boundaries. They occupy approximately 10-15 vol. % of the structure.

This microstructure can be related to the Zone I structure in the classification of zone models for growth of pure metals proposed by Movchan and Demchishin[9] and Thornton.[10] However, the actual homologous temperature for growth of this compound structure of Nb_3Sn is high (0.41 T_m) compared with the T_1 transition temperature that was suggested for pure metals (0.30 T_m) or oxides (0.36 T_m).

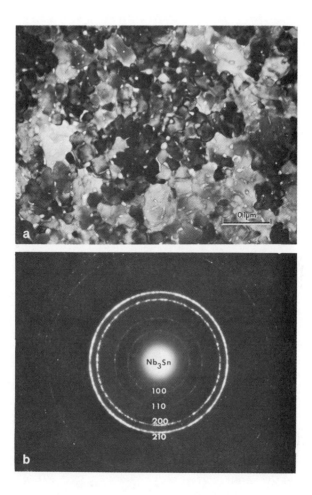

Figure 7.1: (a) Fine grained (d = 250 Å) columnar Nb_3Sn structure prepared at low rate (32 Ås^{-1}) and low temperature (570°C). Cavities are frequently dispersed along the grain boundaries. (b) An SAD ring pattern from the area above shows a strongly preferred <200> growth orientation—the <211> reflection is missing.

Low Rate-High Temperature Deposition: An increase of the deposition temperature to 750°C (T_s = 0.49 T_m) effects a dramatic change of the microstructure (Figure 7.2a). First of all, the grain size increases about 10 times, d = 2000 Å, and, secondly, there are no grain boundary cavities produced. The structure is still columnar, however, (Figure 7.3). The grains are separated by wavy high angle grain boundaries, and each grain exhibits its own specific contrast; i.e., there is no clustering effect of grains having near equal growth orientations. Consequently, the degree of texture is less pronounced; a weak preference of <200> and <211> growth orientations is seen by X-ray diffraction and is also reflected in the SADP (Figure 7.2b), where the (210) reflection is dominating in consistance with preferred (211) growth.

The grain size distribution is not homogeneous in this structure, however; large grains, 1-2 μm in diameter, are dispersed occasionally throughout the otherwise fine grained structure of grains 0.2 μm in diameter (Figure 7.4), thus reminding of abnormal grain growth. As will be shown later, this growth behavior can be prevented by the presence of second phase particles as Al_2O_3.

Figure 7.2: (a) Coarse grained (d = 2000 Å) columnar Nb_3Sn structure prepared at low rate (32 Ås^{-1}) and high temperature (750°C). No grain boundary cavities are produced. (b) An SAD pattern from the area above shows a weak <200> and a weak <211> growth orientation.

Microstructures of PVD-Deposited Films   299

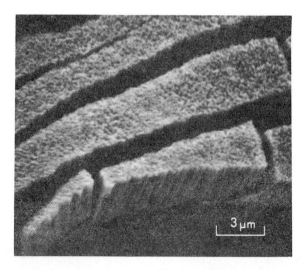

Figure 7.3: An SEM micrograph of the Nb_3Sn structure, same as in Figures 7.2a and 7.4, shows the columnar structure in side view.

Figure 7.4: Abnormal grain growth has occurred during deposition at low rate (32 $Å s^{-1}$) and high temperature (750°C) of single phase Nb_3Sn thus affecting a bimodal grain structure of large grains (d = 2 μm) dispersed in a finer grain structure of 0.2 μm size grains.

There are few defects in this structure, except for the grain boundaries, that will have any influence on the superconducting properties. Occasionally, clusters of bubbles are observed in the larger grains, probably originating from grown-in and coalesced voids, but these are too rare to have any influence on the superconducting properties.

This structure can be classified as a Zone 2 structure in Movchan and Demchishin's model.

High Rate-High Temperature Deposition: Deposition at a five times higher rate (160 Ås^{-1}), at the same substrate temperature (750°C), produces a columnar structure of two times smaller grain size, d = 950 Å (Figure 7.5a). Besides, the structure is now homogeneous with no abnormally large grains present. The high angle grain boundaries are still wavy in the characteristic way that seems to be connected to the columnar growth. The growth texture is similar to that produced at low rate and low temperature; the SADP (Figure 7.5b) exhibiting strong (200) and (210) reflections and a missing (211) ring, consistent with preferred <200> growth.

Columnar cavities along grain boundaries are again frequent. They are approximately 200 Å wide and elongated sometimes up to 1000 Å. They occupy about 6-8 vol % of the structure. No other defects are present.

This structure should also be related to a Zone 2 grain model. However, it is not fully dense in this case as an effect of the high deposition rate.

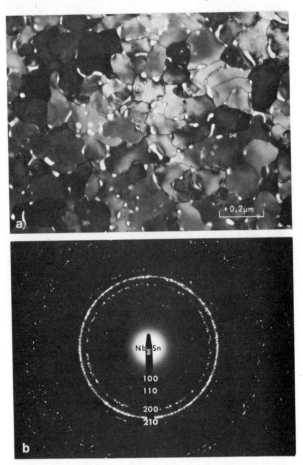

Figure 7.5: (a) A medium fine grained (d = 950 Å) columnar Nb$_3$Sn structure prepared at high rate (160 Ås^{-1}) and high temperature (750°C). Grain boundary cavities are produced because of the high deposition rate. (b) An SAD pattern from the area above.

The superconducting critical temperature shows the highest value (T_c = 18.0 K) for this last structure (750°C - 160 Ås^{-1}) and was 17.7 K for the (750°C - 32 Ås^{-1}) structure and only 17.0 K for the (570°C - 32 Ås^{-1}) structure. The resistivity above the T_c, $\rho(T_c)$, and the slope of the upper critical field versus temperature at zero field, H'_{c2}, were those of an ideal Nb$_3$Sn structure.

Structure and Mechanical Properties of Fine Grained Nickel Films: It is well known that grain size exerts a major influence on the mechanical properties of materials at both room temperature and elevated temperatures. First, the higher density of grain boundaries can act as barriers to dislocation motion and can thus impede plastic flow, leading to grain size strengthening at low temperatures, $T < 0.3 T_m$. Second, at elevated temperatures, $T > 0.3 T_m$, where the diffusion of atoms to or from grain boundaries is important, deformation at a given stress is expected to be faster for fine grained structures. This reduction of strength at high temperatures frequently involves a change in the deformation mechanism, from dislocation controlled plastic flow for coarse grains to grain boundary sliding, and sometimes superplasticity, for fine grains. A small grain size is a characteristic and thus important feature of evaporated films in general compared to grain sizes typically obtained in conventionally produced materials. It is therefore of importance to investigate more in detail the deformation behavior of such film structures. An example will here be given on the creep properties of electron beam evaporated freestanding sheets of pure Ni.

Thick films of Ni were prepared[11] on either side of the T_1 transition temperature,[9] at $0.26 T_m$ and at $0.35 T_m$. The deposition rate was 5 μm min^{-1}, and the final thickness varied from 200 to 250 μm over a 10 cm by 10 cm area. The deposits were released from the substrate, and samples prepared for mechanical testing and for TEM studies by electrolytical thinning. Creep testing was in vacuum at temperatures between 270° and 650°C and under constant stress testing conditions. The creep data were analyzed using an empirical expression for the steady state strain rate ($\dot{\epsilon}$):

$$\dot{\epsilon} = K L^a (\sigma/E)^n \exp(-Q/RT)$$

where K is a structure constant, L the grain size, σ the applied stress, E the elastic modulus, R the gas constant and T the absolute temperature. The quantities α, n and Q are the grain size sensitivity, the stress exponent and the activation energy for creep, respectively. Their values indicate the operative mechanisms of deformation.

The low temperature deposit had a uniform distribution of columnar grains, about 0.5 μm in diameter, and was remarkably free of internal defects. The key feature of this structure was a large number of columnar cavities distributed throughout the film along the grain boundaries and occupying about 1-2% of the volume. This structure exhibited a hardness of 185 DPH and a 0.2% yield strength of 265 MPa, in good agreement with room temperature properties of conventionally processed Ni hardened by the grain size effect. The creep properties, however, were determined by the presence of cavities. Grain growth during high temperature deformation was suppressed by cavity pinning, and the columnar grains transformed upon loading to a microstructure consisting of fine equiaxed grains, with an ultimate grain size depending on the test temperature. The activation energy Q for creep of this structure was 116 ± 8 kJ mol^{-1}, a value which coincides with the activation energy for grain boundary self diffusion, whereas that of conventionally produced coarse grained Ni in the same temperature range is Q = 170 kJ mol^{-1} and at higher temperatures Q = 276 kJ mol^{-1}. These latter energies correspond to dislocation core diffusion and lattice self diffusion respectively. The corresponding stress exponent was n = 3 for the PVD-Ni whereas n = 7 for dislocation controlled creep, and the creep ductility was clearly superior to that of conventional Ni; steady state elongations of 120% were achieved under certain conditions. All these features are characteristic of deformation controlled by grain boundary sliding, accomodated by grain boundary diffusion.

By contrast, the film prepared at $0.35T_m$ had a larger grain size, averaging 1.8 μm in diameter and occasionally showing 3 μm grains and no grain boundary cavities. The hardness was 125 DPH and the yield strength 200 MPa as expected, but the creep properties were substantially different to those of the $0.26T_m$ deposits. The behavior was highly erratic because of concurrent abnormal grain growth. The absolute creep rates and a stress exponent of n = 7 suggested a deformation mechanism involving dislocation climb in agreement with the results of coarse grained conventional Ni.

There is one result of this work which may have an immediate application. In coarse grained materials, voids produced in radiation environment have a strong embrittling effect on creep ductility. These studies show, however, that under some circumstances voids or cavities in fine grained materials affect an increased ductility, although the creep strength is decreased.

Influence of Fine Dispersed Particles

The relationship between structure and mechanical properties of thick freestanding coatings of dispersion strengthened materials prepared by electron beam evaporation has been studied extensively for many years by Movchan and Demchishin, and their work is reviewed in Ref. 13. They show that the relative effect of increasing the volume fraction of particles, on the mechanical properties, is largest at low volume fractions and then levels off. The type of particles used is also of importance, and it is shown both by them and others[8,12] that Al_2O_3 particles are very effective changing the growth morphology and properties of evaporated films. This will be shown in the following two examples. For comparison, the influence of ZrO_2 particles will also be discussed.

Structure and Properties of Superconducting Nb_3Sn-Al_2O_3 Films:[8] Codeposition of Al_2O_3 during evaporation affects the Nb_3Sn structure in three ways: the grain boundary morphology produced is equiaxed instead of columnar, the grain size is refined, and a new kind of fine intragranular voids are homogeneously dispersed throughout the A15 lattice. The SEM image (Figures 7.3 and 7.6) does not show any significant difference; however, the TEM micrographs reveal these distinct changes in the structure. Compare Figures 7.2a and 7.4 with Figure 7.7 showing structures without and with Al_2O_3, respectively. The structure is refined and abnormal grain growth during deposition is suppressed.

A structure prepared at high rate (160 $Ås^{-1}$) and high temperature (700°C) with 4.1 vol % Al_2O_3 is typified by Figure 7.8a. The grain size has decreased by the presence of alumina from 950 Å diameter columnar to 570 Å equiaxed. The grain orientations are random as observed by X-ray diffraction. The texture free SAD pattern of the A15 structure is shown in Figure 7.8b. Two different types of voids are produced under these deposition conditions (Figures 7.9a and 7.9b). One type is caused by the high deposition rate and is thus of the same kind as produced with no alumina present, i.e., cavities located preferentially at grain boundary triple points. They are, however, smaller and more spherical in shape (100-200 Å diameter), and they occupy approximately 6 vol % of the structure.

These cavities are not produced at lower rates (Figure 7.7). The second type of voids, however, is caused by the presence of Al_2O_3. These are intragranular voids, 25 Å in diameter and dispersed homogeneously, typically 75-100 Å apart, throughout the structure. Proof that those fine dots observed by the TEM truly represent the contrast from voids within the lattice is given by the Figures 7.9a and 7.9b; the former is exposed in underfocused condition ($\Delta f = -3.9\ \mu m$) showing white spots surrounded by dark rims, and the latter is taken in overfocused condition ($\Delta f = +3.9\ \mu m$), showing dark spots with white border lines—the grain boundary cavities show corresponding Fresnel fringes.

Figure 7.6: An SEM micrograph of an Nb_3Sn-Al_2O_3 deposit shows the equiaxed structure in side view.

Figure 7.7: A fine grained (d = 540 Å) equiaxed Nb_3Sn-2.5 vol % Al_2O_3 structure prepared at 32 Ås^{-1} and 700°C. No grain boundary cavities were produced.

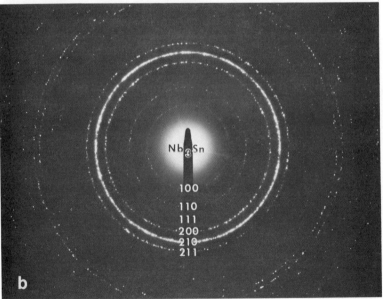

Figure 7.8: (a) A fine grained (d = 570 Å) equiaxed Nb_3Sn-4.1 vol % Al_2O_3 structure prepared at 160 Ås^{-1} and 700°C have grain boundary cavities because of the high deposition rate. (b) The SAD pattern from the structure above represents that of a completely randomly oriented Nb_3Sn grain structure.

Microstructures of PVD-Deposited Films   305

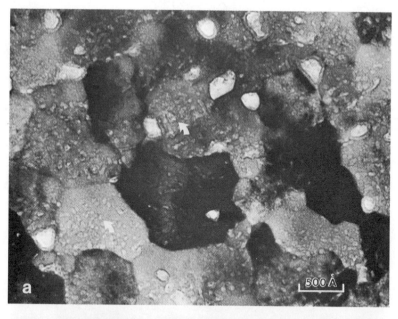

Figure 7.9: Intragranular voids, 25 to 55 Å in size, are produced by the codeposition of Al_2O_3. These are revealed by the trough-focus-imaging technique (a) is exposed in underfocused condition and (b) in overfocused condition.

The presence of Al_2O_3 was not easily revealed by TEM, however. No secondary particles were actually observed, in spite of the fact that the growth morphology was strongly affected. Chemical analysis was therefore performed using a TEM/STEM instrument equipped with an energy dispersive X-ray detector. This revealed a segregation of Al to grain boundary cavities when these were formed at high deposition rates. The resolution of this technique was not good enough to reveal increased Al content in the intragranular voids, however, though it is believed that these voids were formed enclosing fine clusters of alumina not wet by the Ni matrix lattice; the high T_c values obtained in these structures had proved that Al was not present as free Al metal atoms. See Figures 7.10a and 7.10b. This type of void formation can be expected to occur during growth of films also by other methods and compositions whenever fine clusters of non-wetting impurities or second phase particles are present, and these will affect the film properties. In this case their presence was of great importance because they affected an increase of the critical current value while not lowering the high T_c value.[8,14]

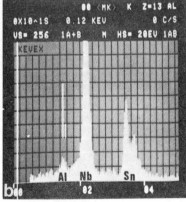

Figure 7.10: (a) STEM micrograph of an Nb_3Sn-4.1 vol % of Al_2O_3 structure with grain boundary cavities (bright spots). (b) An energy dispersive X-ray analysis at these grain boundary cavities shows a significant Al signal.

Structure and Mechanical Properties of Ni-Al$_2$O$_3$ Films: In an effort to further refine and stabilize the Ni structures discussed above, low amounts of Al$_2$O$_3$ (f = 0.20 - 0.76 vol %) were codeposited with Ni in the temperature range T_s = 0.32 - 0.45T_m.[12] The presence as well as small variations in the alumina content had, as in the case of Nb$_3$Sn, a drastic influence on the growth morphology and thus the mechanical properties. The as-deposited structure was equiaxed (not columnar), homogeneous (no abnormal grain growth) and refined to a grain diameter of 0.3 μm. In spite of the rather high deposition rate, r = 4.5 μm min^{-1}, the films were in all cases fully dense (Figure 7.11).

Figure 7.11: An Ni-Al$_2$O$_3$ structure stabilized against grain growth; annealed at 625°C for 2 hr.

The expected strengthening effects from alumina were obtained both in room temperature hardness and creep deformation at increased temperatures. The hardness varied from 220 VHN for the Ni-0.20 vol % Al$_2$O$_3$ structure to 335 VHN for the Ni-0.76 vol % Al$_2$O$_3$ structure compared to the 130-150 VHN values that are obtained in pure Ni.

The corresponding increase in the creep strength with increasing amounts of alumina was accompanied by a decrease in ductility from above 40% to less than 1% in the actual composition range. However, the deformation temperature is an important factor in this consideration, because the stabilized fine grained structure affects a change in deformation mode at about 550°C. Below this temperature the deformation was controlled by dislocation core diffusion as is the case for conventional Ni; the activation energy for deformation was Q = 178 kJ mol^{-1} and the stress exponent n = 5 (Figure 7.12). Above this temperature a change occurred in the stress dependence of the creep rate at $\dot{\epsilon}$ = 4 x 10^{-4} s^{-1}. Below this limit the stress exponent was high, n = 5, and the creep deformation dislocation controlled, and above this limit the stress exponent was low, n = 2.5, the ductility high, $\epsilon > 41\%$ (the limit set by the creep testing machine) and the deformation homogeneous, all suggesting a grain boundary sliding mechanism involved. In order to quantify the contributions from the various deformation mechanisms when superimposed on one sample, the TEM was used for measuring the mean grain elongation, i.e., the mean width and length of about 150 grains perpendicular to and along the strain axis (Figure 7.13a). The difference between total elongation and mean grain elongation was attributed to strain developed due to grain boundary sliding,[12] and these values were in all cases in agreement with the results obtained from the creep testing curves, strain rate versus stress. A characteristic feature of this creep deformed structure was a distribution of very fine twins preferentially oriented in the direction of the strain axis, and these were attributed to the accommodation process during grain boundary sliding (Figure 7.13b).

Figure 7.12: Elongated grains and a high dislocation density is typical of an Ni-Al$_2$O$_3$ structure creep deformed at low temperature (T = 345°C). ϵ = 20%.

These properties were drastically changed, however, when the Al$_2$O$_3$ content was increased to above approximately 0.4 vol % where fine alumina particles actually could be observed in the structure (Figure 7.14a), increasing in size with the alumina content from 20 to 100 Å in the actual range. These particles then inhibit the motion of dislocations and propagation of twins (Figure 7.14b) thus affecting a hard and brittle structure.

No voids were produced around alumina particles in this case, thus indicating a better wetting between the Ni matrix and alumina than obtained by the Nb$_3$Sn matrix. These interface properties still play an important roll, however, as will be discussed in the following section.

Microstructures of PVD-Deposited Films 309

Figure 7.13: A high density of fine twins, expanding from one grain to the next in the stress direction, and a low dislocation density is typical of an Ni-Al$_2$O$_3$ structure creep deformed at intermediate temperatures (T = 535°C), (a) and (b). ϵ = 41%.

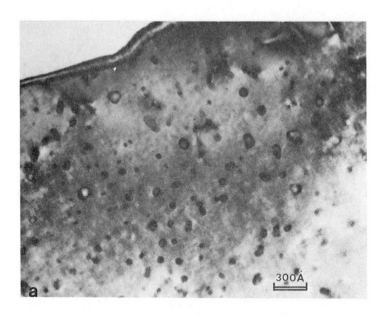

Figure 7.14: (a) Al_2O_3 particles, 20 – 70 Å in diameter, are randomly dispersed in the Ni matrix at high alumina concentrations. (b) These particles inhibit dislocation motion and the development of twins during creep deformation thus affecting a brittle material. $\epsilon = 0.06\%$.

Structure and Mechanical Properties of Ni-ZrO₂ Films: Thick condensates of electron beam evaporated Ni-ZrO$_2$ composition with increasing amounts of second phase, from 0.3 to 2.6 vol % ZrO$_2$, were studied by TEM and tensile testing after cold rolling.[15] The deposition temperature was 850°C and the deposition rates 1-10 μm min^{-1}. The ZrO$_2$ phase was stabilized by 6 wt % CaO. The sheets were annealed at 1200°C (Figure 7.15) and subsequently cold rolled to reductions of various degrees from 20 to 78%. The deformation properties of these structures were tested by tensile testing.

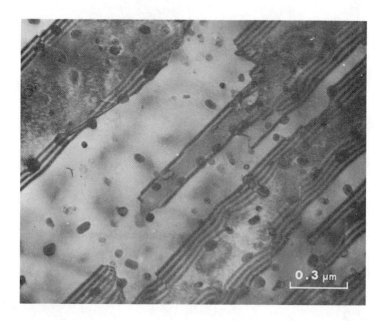

Figure 7.15: As-deposited Ni-1.75 vol % ZrO$_2$ structure with a distribution of ZrO$_2$ particles and twins.

The TEM studies revealed a specific kind of cavity formation at the particle/matrix interfaces after cold rolling as is shown by Figure 7.16a, with the dislocation cell boundaries in contrast, and Figure 7.16b, with the dislocations out of contrast. (This is not a foil preparation effect.) There is a common specific orientation relationship between the ZrO$_2$ lattice and the Ni lattice where the cavities first appear,[15] and they are created by the interaction between dislocations, sliding on (111) Ni planes, and the particle/matrix interface.

An interesting effect of these cavities is that they limit the degree of work-hardening that can be attained in this material, because they act as relaxation centers during cell formation. This effect is more pronounced the higher the particle content and gives rise to a large minimum cell size upon increased deformation.

Variations in Growth Morphology with Composition in Two-Phase Structures

Structure and Properties of Superconducting Nb$_3$Sn-Cu Films: The following is an example of the wide variations in growth morphology that can be

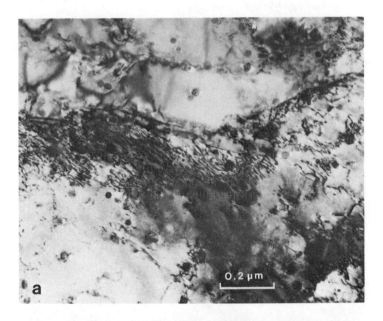

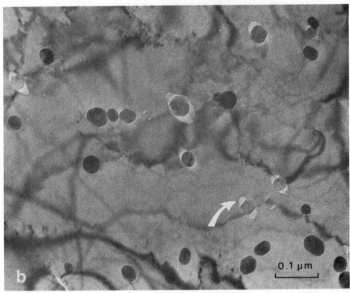

Figure 7.16: Cold rolled Ni-1.75 vol% ZrO_2 structure, reduced 69%, exhibits (a) a dislocation cell structure of 0.7 μm cell diameter and (b) cavities at the particle/matrix interfaces and extending along the rolling direction.

obtained by variations of the relative amount of two phases in a deposit. This relationship was explored for codeposited Nb-Sn-Cu structures in which the important superconducting phase is Nb₃Sn and the Cu phase is chosen because of its good mechanical properties. The α-Cu(Sn) solid solution is the most stable Cu-Sn phase, and Cu does not react with Nb nor with Nb₃Sn; thus the Nb₃Sn-Cu form a unique system for the study of a non-reacting metal dispersion.

A variety of deposits were obtained by deposition under various conditions of chemical composition (4 to 85 vol % Cu), substrate temperature (400°, 500° and 690°C) and deposition rate (16 and 32 Ås^{-1}).[16,17] Figure 7.17 is a schematic presentation of the results, and the corresponding TEM micrographs are shown in the following.

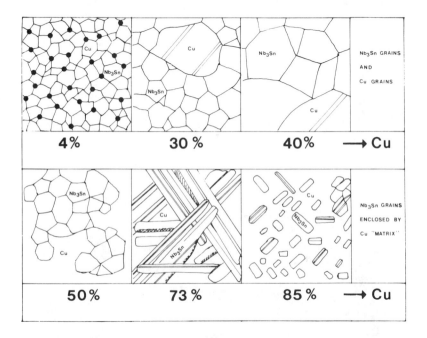

Figure 7.17: A schematic representation of the variations in microstructure of Nb₃Sn-Cu composites with increasing amounts of Cu.

The codeposition of Cu even in small amounts (4 vol%) affected a drastic change in the growth morphology, from columnar to equiaxed (Figure 7.18). An extremely fine grained (180 Å diameter) and homogeneous structure was thus produced at 400°C deposition temperature, both phases having the same grain size. A frequent formation of grain boundary cavities (50 Å in diameter) was a typical effect of the low temperature, and these were not obtained at 500°C and above.

The grain size of Nb₃Sn increased somewhat, to typically 300 - 500 Å, with deposition temperature. However, the most important grain growth effect was obtained by increasing the relative Cu content; the grain size of the Cu phase increased with Cu concentration affecting an uneven grain size distribution (Figure 7.19a). Below 50 vol% Cu, both phases were randomly distributed and oriented as is shown by the SAD pattern (Figure 7.19b). Above this limit, however, the Cu phase grew epitaxially with the Hastelloy substrate and thus adopted a 25 μm large grain size (Figure 7.20), thereby enclosing the still fine grained Nb_3Sn phase. Under these circumstances the two lattices exhibited a preferred growth orientation relationship, schematically represented by Figure 7.21 and also shown by the SADP in Figures 7.22b, 7.23b and 7.24b, exposed under the same selected area aperture as Figure 7.19b. These patterns all show a single crystalline Cu-pattern and a polycrystalline discontinuous Nb_3Sn ring pattern. Above 50 vol% Cu, the growth morphology of the Nb_3Sn phase changed with Cu concentration: from equiaxed grains, grown in clusters surrounded by the Cu phase (Figure 7.22a); to a continuous network of rods, grown parallel to the <224> Cu directions on the close packed Cu planes (Figure 7.23a); to, finally, isolated bean-shaped particles, dispersed in the Cu matrix (Figure 7.24a). The growth relationship between the two lattices was $\{111\}$ Cu // $\{001\}$ Nb_3Sn and $<0\bar{1}1>$ Cu // $<010>$ Nb_3Sn.

The superconducting T_c values were high and comparable with those of pure Nb_3Sn even up to 85 vol% Cu, which is an interesting result in this case where the aim is to produce a structure with optimized good superconducting and mechanical properties simultaneously. However, the critical current density values were not in the range obtained for Nb_3Sn-Al_2O_3 composits, and the reason for that has to be found by extended and more detailed TEM studies of the interface boundaries.

Figure 7.18: A fine grained (d = 180 Å) equiaxed structure of Nb_3Sn-4 vol % Cu deposited at 32 Ås^{-1} and 400°C. Grain boundary cavities are typical of the low deposition temperature.

Microstructures of PVD-Deposited Films

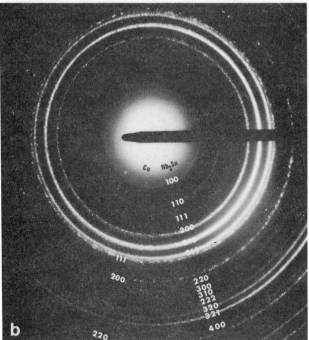

Figure 7.19: The grain diameter of the Cu phase increases with the Cu concentration. (a) Nb_3Sn-30 vol % Cu deposited at 32 $Ås^{-1}$ and 500°C. (b) The SADP from the structure above is that from a fine grained two-phase structure, Nb_3Sn and Cu. Both phases are poly-crystalline and randomly oriented.

316 Deposition Technologies for Films and Coatings

Figure 7.20: An optical micrograph of the coarse grained (d = 25 µm) Cu phase, epitaxially grown on the Hastelloy substrate at high relative Cu deposition rates.

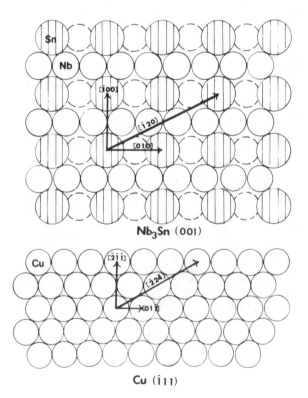

Figure 7.21: A schematic diagram shows the growth orientation relationship between the two lattices, Nb_3Sn and Cu.

Microstructures of PVD-Deposited Films 317

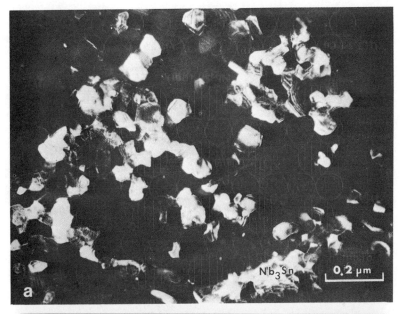

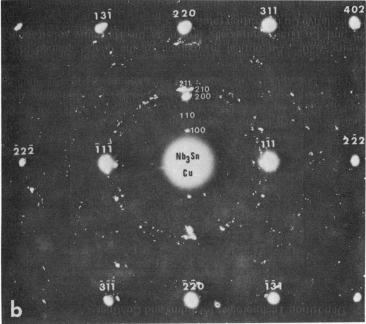

Figure 7.22: (a) A dark field micrograph of Nb_3Sn grains clustered in the Cu matrix of an Nb_3Sn-60 vol % Cu structure deposited at 32 $Ås^{-1}$ and 690°C. (b) The SAD pattern from this structure shows a single-crystal ($\bar{1}12$)Cu pattern and a poly-crystal Nb_3Sn ring pattern of fine spots at preferred locations.

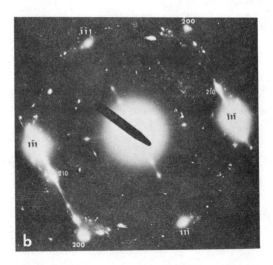

Figure 7.23: (a) At low deposition rate (16 Ås^{-1}) and high substrate temperature (690°C) a rod-like structure of the Nb$_3$Sn phase grows preferentially in $<\bar{1}20>$Nb$_3$Sn // $<\bar{2}2\bar{4}>$Cu directions of a Nb$_3$Sn-82 vol % Cu structure. (b) In this case the Nb$_3$Sn rods are typically faulted along their growth directions and this causes a streaking effect on the Nb$_3$Sn spots in the SAD pattern. The strong spots represent a single-crystal (011)Cu pattern.

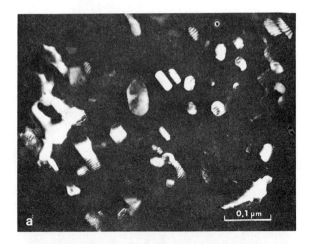

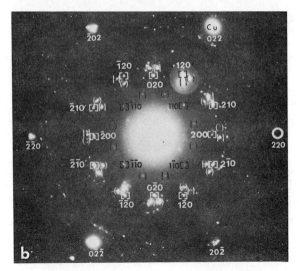

Figure 7.24: At high Cu concentrations, Nb_3Sn-92 vol % Cu, the Nb_3Sn phase can no longer grow as a continuous structure throughout the Cu lattice; instead bean shaped isolated particles are formed (a), and these grow still in the preferred lattice orientations as is revealed by the SAD pattern in (b). The strong spots represent a $(\bar{1}11)$Cu pattern and the fine ones represent three equivalent growth orientations of the Nb_3Sn grains.

4. MICROSTRUCTURE AND PROPERTIES OF FILMS PREPARED BY ARE

Activated reactive evaporation (ARE) is a promising technique for processing refractory compound coatings.[18] The microstructure and thus the mechanical properties of these coatings can be varied over wide ranges by variation of the processing parameters and the composition of the film. The carbides and nitrides, for instance, exhibit extremely good high temperature strength properties, and these can be improved by proper alloying. In the following examples the microstructures of TiC, (V,Ti)C and TiN coatings are characterized and discussed in relation to the composition, deposition temperature, metal evaporation rate and gas pressure. The related film properties are characterized by hardness testing.

Microstructure and Hardness of Single Phase Carbide Films, TiC

The metal was evaporated by an electron beam from a rod-fed source and the carbide formed when this metal vapor was allowed to react with an C_2H_2 atmosphere between the source and the substrate, Ti sheet. The substrate temperature was varied from 400° to 1000°C, the deposition rate was 1.2-3.0 μm per minute and the reactive gas pressure 6.5-11 x 10^{-4} Torr.[19] Samples for TEM studies were prepared by ion beam milling.

The ARE process produced single phase TiC structures at all deposition conditions under consideration. The Ti:C ratio was within the range x = 0.78 to 0.93.[19] The SADP (Figure 7.25b) was the same for all deposits, showing randomly oriented grains in the film plane (even ring patterns) and a preferred <111>TiC and sometimes <220>TiC growth direction (X-ray analysis).

Two characteristic and completely different grain morphologies were produced, one extremely fine grained at low deposition temperatures (<700°C) and one coarse grained at higher deposition temperatures. Figure 7.25a shows a structure that is typical of that produced at very low deposition temperatures in general, i.e., also for other film compositions and PVD deposition methods. It is exposed in overfocused condition to reveal the faint network of microporosities irregularly dispersed in the structure (dark lines). Extremely fine fibers of near equal growth orientations are clustered together in areas exhibiting the same dark contrast in the image (other areas will appear dark when the sample is tilted in relation to the incident beam in the microscope). The spottiness of the SADP and dark field imaging (see later Figure 7.29b) reveals the true size of the individual fibers. In this specific case the deposition temperature varied between 235° and 400°C, and the resulting fiber diameter was about 100 Å.

At 600°C, a similar structure was obtained (Figure 7.26; overfocused image) with 100 Å diameter fibers and a somewhat more regularly distributed network of cavities. At 700°C and above (Figure 7.27), a normal grain structure was produced with micron-sized grains separated by high angle grain boundaries. No microcavities were observed, but, eventually, large cracks along the grain boundaries were produced. The image contrast also reveals a rather high dislocation density and internal stresses within the grains.

The hardness of the TiC deposits does not vary significantly with grain size. The fine fibrous structures are in fact less hard than those with normal grain morphologies. It is therefore likely that the hardness values are influenced by the fine networks of cavities; in fact, structures with similar network configurations have similar hardness values. However, hardness varies also with stoichiometry, and as this is likely to be higher at higher deposition temperatures it could account for the increased hardness obtained at higher temperatures (KHN = 4880 at 1000°C and KHN = 2700 at 700°C) where the grain size is considerably larger.

Effect of Carbide Alloying, VC-TiC[21]

At equilibrium, VC and TiC have complete solubility in each other, and the hardness of this alloy passes through a maximum at the (V-70at%Ti)C composition. Improved ductility at high temperatures has also been observed in (V- vol % Ti)C structures.[20]

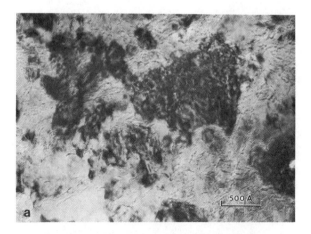

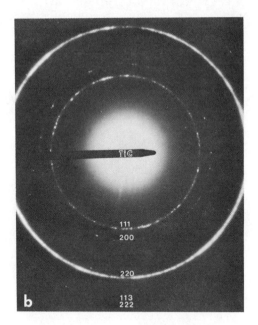

Figure 7.25: Clusters of fine grains, 100-500 Å in diameter, and a fine network of cavities are typical of a low temperature deposit. (a) This shows a TiC structure prepared between 235° and 400°C. The micrograph is exposed in overfocused condition. (b) This SAD ring pattern is typical of those from TiC structures evaporated under various conditions. KHN = 2375.

322 Deposition Technologies for Films and Coatings

Figure 7.26: A fine fibrous TiC structure deposited at 2.0 μm min^{-1} and 600°C. The micrograph is exposed in overfocused condition, and a fine feather-like network of cavities is revealed (dark lines). KHN = 2500.

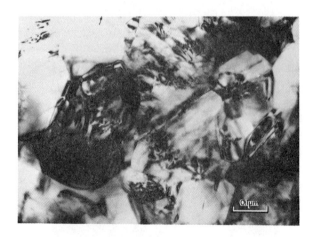

Figure 7.27: A drastic change occurs in growth morphology when the deposition temperature is increased from 600°C (see above) to 700°C. A normal grain structure of TiC is produced with a grain size of 0.1-0.2 μm. The grain interior exhibits a contrast from internal stresses and dislocations, but no cavity network. KHN = 2700.

Microstructures of PVD-Deposited Films   323

Evaporation of V-Ti in a reactive C_2H_2 atmosphere produces, however, a non-equilibrium structure of VC-TiC-free graphite phases when the deposition temperature is low, 550°C (Figures 7.28a and 7.28b). The defect density is then extremely high, i.e., regions of heavily disturbed structure are separated by wide dislocation cell boundaries. The SADP shows that the TiC and free graphite are present as large grains, 300-500 Å in diameter, whereas the VC phase is extremely fine grained or amorphous. It is interesting to note that no microcavities are produced in this structure.

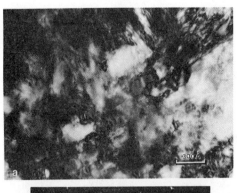

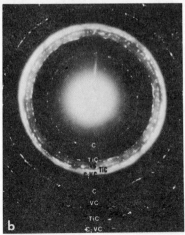

Figure 7.28: (a) A non-equilibrium structure of mixed VC, TiC and free graphite phases is produced during deposition at 2.8 μm min^{-1} and 550°C. No cavities are produced, but the structure is heavily distorted, reminding of a dislocation cell structure. (b) The SAD pattern from this structure shows that TiC and graphite have a larger grain size (distinct spots) than the VC (diffuse rings). KHN = 1935.

At 700°C and in the presence of excess carbon, the equilibrium single phase (V,Ti)C-structure is formed (Figure 7.29a). The structure exhibits fine, 50 Å size, well-defined grains, which are clustered in larger regions where the individual grains have near equal growth orientations, as is revealed by dark field imaging (Figure 7.29b). A network of fine cavities is a dominating feature. Less carbon—a lower C_2H_2 gas pressure—effects a coarser (100 Å size) and less well defined structure (Figure 7.30). At high deposition temperatures, 1000°C, a normal grain structure is produced with 0.3-1 µm size grains separated by high angle grain boundaries (Figure 7.31). No internal defects or stress fields are produced.

The hardness of these structures is determined by three factors: the presence of free graphite and microcavity networks effect a low hardness (KHN = 1935 at 550°C and KHN = 1385 at 700°C with excess carbon), and high hardness values are achieved by increasing the stoichiometry (temperature) (KHN = 2900 at 1000°C).

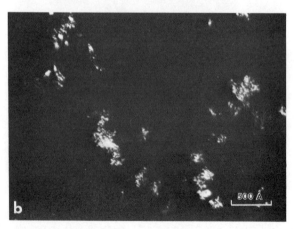

Figure 7.29: (a) An extremely fine grained (d = 50 Å) (V,Ti)C structure prepared at a higher gas pressure (10^{-3} Torr) than the structure in Figure 7.30. (5×10^{-4} Torr), both at 700°C. The micrograph is exposed in underfocused condition to reveal the faint network of cavities present. (b) The dark field image from the same structure shows that grains of near equal growth orientations are clustered together. KHN = 1385.

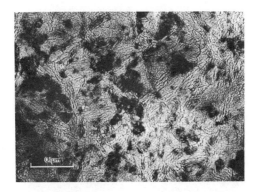

Figure 7.30: A 'low temperature' deposit of (V,Ti)C structure prepared at 1.5 μm min^{-1} and 700°C. Compare Figure 7.27 showing TiC prepared under the same conditions. The micrograph is exposed in overfocused condition to reveal the cavity network present (dark lines). KHN = 2435.

Figure 7.31: A 'high temperature' deposit of (V,Ti)C structure prepared at 1.2 μm min^{-1} and 1000°C. KHN = 2900.

Effect of in Situ TEM Annealing of Fine Grained Carbide Films

An interesting general question is how these structures with extremely fine grains and microcavity networks will behave during high temperature annealing. Will the cavities anneal out and thus effect a fully dense structure? Will the fine grain size be preserved after recrystallization? To answer those questions the TEM was utilized for in situ annealing of selected carbide films up to 1000°C under continuous observation and imaging in the various TEM modes to reveal any structural changes.[21] Annealing was performed in a stepwise manner with 100°C intervals and about 30 minutes at each temperature and ending at 1000°C prolonged for 4 hours.

No changes occurred in structures of equilibrium phase compositions until 1000°C was reached. Then the cavity network started to coarsen and new recrystallized grains to appear after about 10 minutes. However, the network mesh size seemed to stabilize after growth from typically 100 to 150 Å and thus inhibit grain growth after recrystallization (Figures 7.32a and 7.32b). The structure thus transformed from a fine grained fibrous to an equiaxed of only slightly larger grain diameter.

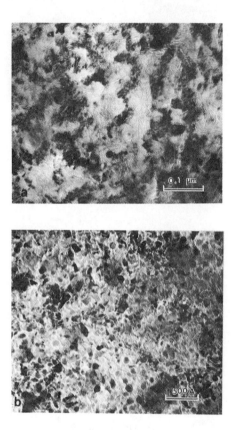

Figure 7.32: In situ TEM annealing of a TiC structure prepared at 600°C: (a) an underfocused micrograph of as-deposited structure at room temperature; (b) an underfocused micrograph exposed at 1000°C after 4 hours annealing. The recrystallization is completed and well defined equiaxed grains are seen. The cavity network has coalesced but has not annealed out.

The non-equilibrium phase structure was more sensitive to transformation, and the dominating process was the growth of a ribbon-like network of faulted grains of HCP structure. The SADP revealed a new superimposed ring pattern from Ti_3O that originated from these grains (Figures 7.33a, 7.33b, 7.33c and 7.33d). The oxygen was probably present in the film in the as-deposited condition, and this might also have been the case for low amounts of free Ti with reflections too faint to be resolvable in the complex SAD pattern.

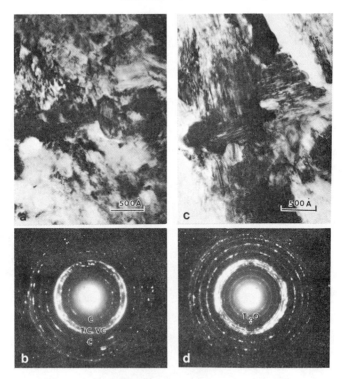

Figure 7.33: In situ TEM annealing of a non-equilibrium poly-phase structure of mixed VC,TiC and free graphite: (a and b) bright field image and corresponding SADP of as-deposited structure exposed at room temperature; (c and d) image and SADP exposed at 1000°C after annealing for 30 min. Rod-like grains of a faulted HCP Ti_3O structure are growing in distinct directions. The Ti_3O ring pattern of large d-spacings appear in the center of the SADP.

Microstructure and Hardness of Poly-Phase Nitride Films, Ti-Ti_2N-TiN

Titanium nitride is chemically extremely stable and is therefore used commercially as a deposit on cutting tools. The CVD process is at present the deposition method used for this purpose. However, the ARE process, as well as various other techniques, have also been used for synthesizing such nitride coatings. For comparison with the above discussed carbide films a similar study was done on the microstructure and hardness of ARE processed titanium nitride films. The processing variables were: the deposition temperature, the evaporation rate of metal, and the reactive gas pressure. As the nitride structures turned out to be considerably more coarse grained than the carbide structures and inhomogeneous on the scale that limits examination by TEM, it was important to combine the SEM and TEM studies to get a proper characterization of the deposits.[22]

These results showed that as the ratio of the evaporation rate of Ti to the partial pressure of N_2 decreases, then the composition changes from Ti to Ti_2N to TiN including two phase mixtures. A titanium nitride coating deposited at 550°C under a high Ti/N_2-gas pressure rate thus exhibits a bimodal structure of 1-2 μm diameter Ti grains interspersed with areas of a subgrain structure of 0.1 μm Ti_2N phase. The Ti grains are free from internal defects whereas the Ti_2N phase exhibits a TEM contrast of internal stresses and high dislocation densities

(Figure 7.34a). The corresponding SADP (Figure 7.34b) shows a ring pattern from randomly oriented Ti_2N grains and superimposed a few strong spots from an adjacent Ti grain. No cavities are observed in this structure. The surface topography is rather smooth (Figure 7.34c).

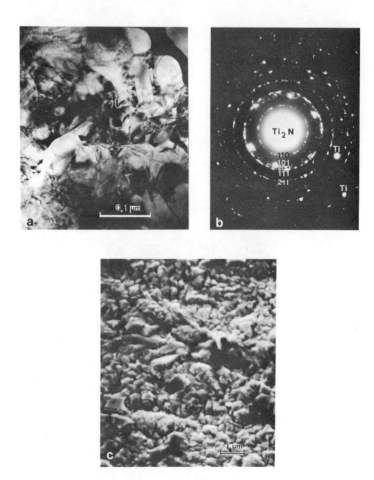

Figure 7.34: A $Ti+Ti_2N$ deposit prepared at 550°C and a high relative Ti/N_2-gas-pressure deposition rate: (a) the TEM micrograph of the Ti_2N phase reveals high dislocation density and internal stresses; (b) the corresponding SADP shows a ring pattern from fine grained Ti_2N and a few isolated spots from a large adjacent Ti grain; (c) A SEM micrograph of the surface topography.

A decreasing Ti/N_2-gas-pressure ratio produces a structure of faceted Ti_2N and TiN grains of about 0.5 μm diameter. Figures 7.35a, 7.35b and 7.35c show a TiN structure deposited at the same temperature as the previous one, 550°C. The TEM (Figure 7.35a) reveals that these facets extend through the depth of the film, which thus includes a distribution of cavities between the grains.

Microstructures of PVD-Deposited Films   329

Figure 7.35: A TiN deposit prepared at 550°C under a low relative Ti/N$_2$-gas-pressure deposition rate: (a) the TEM micrograph of this structure reveals faceted grains with a featureless internal structure and open cavities between the grains; (b) the corresponding SADP shows a ring pattern from randomly oriented TiN grains. (c) The SEM micrograph shows the faceted topography of this film.

An increase of the deposition temperature effects an increase in grain size (20-30 μm at 1000°C) and a transition to an increasingly smoother grain morphology. Figure 7.36 shows a Ti$_2$N+TiN structure deposited at 735°C.

It was found that the hardness was always the highest when the Ti$_2$N phase was present and in the largest volume fraction, as was the case for the structure shown in Figure 7.36, thus having a hardness value of KHN = 2800. The faceted TiN structure shown in Figure 7.35 had a very low hardness, KHN = 625, and this should be explained by the presence of cavities.

Figure 7.36: A Ti_2N+TiN deposit prepared at 735°C exhibits a smoother and more continuous grain structure.

5. MICROSTRUCTURE AND PROPERTIES OF SPUTTER DEPOSITED FILMS

The sputtering technique allows low temperature deposition of almost any kind of material and composition. The low substrate temperature affects low surface mobility of arriving species and thus rapid vapor quenching rates, typically of the order of $10^{15} - 10^{18}$ Ks^{-1}; i.e., five to nine orders of magnitude larger than those obtainable by mechanical splat cooling. This technique is thus suitable for the synthesis of extremely fine grained or amorphous structures, non-equilibrium phases, and non-reacting multilayered structures. The synthesis of amorphous metals have become of increasing importance; one application being their superior corrosion resistance properties. A TEM study of the crystalline-to-amorphous transition of a 304 Stainless Steel will thus be discussed in the following as an example of sputter deposited films.[23]

Crystalline-to-Amorphous Transition in a Magnetron Sputter Deposited Film

Thin film samples of 304SS composition and varying amounts of excess carbon were deposited using magnetron sputtering techniques. Individual sources were used for the carbon and 304SS, the composition thus varying in a monotonic fashion along the substrate (float glass) due to the deposition profile overlap of the individual sources. Deposition was at room temperature and the deposition rates directly under the sources were 0.6 $Ås^{-1}$ for carbon and 8.1 $Ås^{-1}$ for 304SS. Film thickness varied from 2 μm to 10 μm along the substrate due to the large difference in the C and 304SS rates. The structures were characterized by X-ray diffraction, differential thermal analysis (DTA) and TEM. Samples for TEM were prepared by ion beam milling.

The chemical composition of the deposits was exactly the same as that of the target material (18.1%Cr-7.6%Ni-1.6%Mn-0.9%Si-0.02%C-Fe) except for the carbon concentration. The structure was, however, ferritic BCC whereas that of the target material was austenitic FCC. Figure 7.37a typifies the pure 304SS, showing uniform equiaxed grains of approximately 400 Å diameter and a moderate internal dislocation density. The corresponding SADP (Figure 7.37b) shows a sharp ring pattern of even distribution of individual spots.

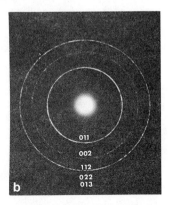

Figure 7.37: (a) Grain structure of a ferritic 304SS film with no excess carbon. (b) The corresponding SADP shows a sharp ring pattern from the BCC phase.

With excess carbon the structure is changed dramatically (Figures 7.38a and 7.38b). The grain boundaries are not easily identified because of the high internal dislocation densities. In fact, the 300 to 400 Å large regions of uniform diffraction contrast are composed of small grains varying in size between 100 and 150 Å as was revealed by dark field imaging—this structure thus reminds of those of electron beam evaporated at low temperatures. The grain refinement is also reflected in the SADP; the spottiness is not resolved anymore, and the rings are less sharp—all SADP in this study were exposed under the same selected area aperture. A mean grain diameter of 120 Å was determined by X-ray analysis using the Scherrer equation.

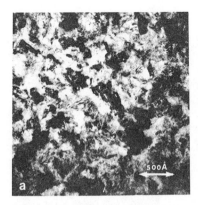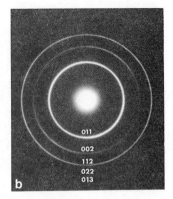

Figure 7.38: (a) Grain structure of a 304SS-4.3 at % Carbon film with high dislocation concentrations. (b) The grain refinement and lattice distorsion are reflected in the less sharp SADP with no resolvable spots.

The crystalline-to-amorphous transition occurs with increasing carbon concentration at approximately 5 at% C, i.e., at a considerably lower metalloid concentration than can be achieved by liquid quenching (15-30 at% C, B, P or Si) of amorphous ferrous alloys and reflects the effect of the large quenching rate inherent to vapor deposition. Figures 7.39a, 7.39b and 7.40a, 7.40b show the bimodal transition zone structure, where the grey areas represent a completely featureless amorphous structure and the black-and-white areas a structure of fine crystallites 10 to 30 Å in size. The characteristic shape of these areas derives from the topography of the substrate surface, causing a shadowing effect relative to the carbon source and thus a local decrease in the carbon content. The size of these areas decreases and the distance between them increases with increasing carbon content; i.e. nearer the carbon source. The fully amorphous structure is shown in Figures 7.41a and 7.41b; the rings in the SADP are broadened, forming halos typical for an SADP from an amorphous phase.

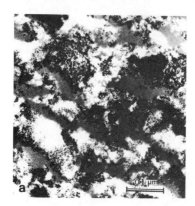 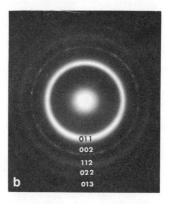

Figure 7.39: (a) The structure of 304SS-5at%Carbon at the crystalline-to-amorphous transition zone. The black and white areas consist of fine crystallites of 10-30 Å diameter, and these are embedded in a amorphous phase (grey contrast). (b) The SADP shows a superimposed sharp and diffuse ring pattern.

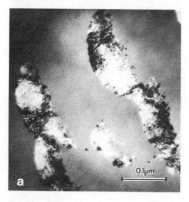 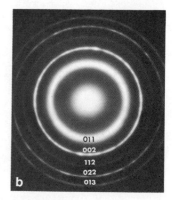

Figure 7.40: (a) The crystalline areas are fewer and smaller at the carbon-rich end of the transition zone. (b) The SADP transforms towards a more diffuse ring pattern.

 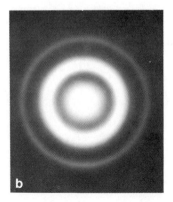

Figure 7.41: (a) The fully amorphous 304SS-22.4at%Carbon structure is completely featureless. (b) The corresponding SADP shows a diffuse halo pattern.

Through-focus imaging was utilized for the purpose of revealing networks of cavities or fine voids, which are common in low-temperature deposits. No evidence was found for void formation, however, in either the crystalline or amorphous regions.

The crystallization temperatures of these amorphous structures increases with the carbon content, from 377°C close to the transition zone at 6 at% C to between 450° and 500°C at 15 at%C. A transition from ferritic to austenitic structure occurs at about 575°C.

REFERENCES

(1) *Electron Microscopy of Thin Crystals,* Eds. P. Hirsch, A. Howie, R.B. Nicholson, D.W. Pashley and M.J. Whelan, R.E. Krieger Publishing Company, Hungington, N.Y., (1977).
(2) J.W. Edington: *Practical Electron Microscopy in Materials Science,* Van Nostrand Reinhold Company, N.Y., (1976).
(3) *Electron Microscopy in Materials Science,* Eds. U. Valdrè and E. Ruedl, Third Course of the Int. School of Electron Microscopy, Bologna, January, 1975. Publ. by The Commission of the European Communities, Luxembourg, (1976).
(4) M. Rühle and M. Wilkens: Proc. 5th Eur. Conf. Electron Microscopy, Manchester, 1972, London: The Inst. of Physics, p. 416.
(5) P.J. Goodhew in *Practical Methods in Electron Microscopy,* Vol. 1, Part 1, Ed. M. Glauert, North-Holland Publ. Co., London, (1972).
(6) K.C. Thompson-Russel and J.W. Edington: *Practical Electron Microscopy in Materials Science,* Monograph Five: Electron Microscope Specimen Preparation Techniques in Materials Science, Philips Technical Library, Eindhoven (1977).
(7) R.F. Bunshah, ibid.
(8) B.E. Jacobson, R.H. Hammond, T.H. Geballe and J.R. Salem: *J. Less Common Metals, 62* (1978) pp 59-87.
(9) B.A. Movchan and A.V. Demchishin: *Phys. Met. Metallogr. USSR, 28* (1969) p. 83.
(10) J.A. Thornton: *Annu. Rev. Mater. Sci., 7* (1977) p. 239.
(11) J.R. Spingarn, B.E. Jacobson and W.D. Nix: *Thin Solid Films, 45* (1977) pp 507-515.
(12) B.E. Jacobson, J.R.Spingarn and W.D. Nix: *Thin Solid Films, 45* (1977) pp 515-526.
(13) B.A. Movchan, G.F. Badilenko and A.V. Demchishin: Presented at The Sixth Int. Vac. Met. Conf., San Diego, California, April 23-27 (1979).

(14) R.H. Hammond, B.E. Jacobson, T.H. Geballe, J. Talvacchio, J.R. Salem, J.C. Pohl and A. Braginski: IEEE Trans. Magn., *MAG-15*, January (1969) pp 169-1972.
(15) B.E. Jacobson, A.V. Demchishin and B.A. Movchan: Presented at The Sixth Int. Vac. Met. Conf., San Diego, California, April 23-27 (1979).
(16) B.E. Jacobson, R.H. Hammond, T.H. Geballe and J.R. Salem: *Thin Solid Films, 54* (1978) pp 243-258.
(17) B.E. Jacobson and R. Sinclair: *New Developments and Applications in Composites,* Ed. D. Kuhlman-Wilsdorf, Symposium at the TMS-AIME Fall Meeting, St. Louis, Missouri, Oct. 15-18, 1978.
(18) R.F. Bunshah and A.C. Raghuram: *J. Vac. Sci. Technol. 9* (1972) p. 1385.
(19) B.E. Jacobson, R.F. Bunshah and R. Nimmagadda: *Thin Solid Films, 54* (1978) pp 107-118.
(20) G.E. Hollox: *Mater. Sci. Eng., 3* (1968) p. 121.
(21) B.E. Jacobson, R.F. Bunshah and R. Nimmagadda: *Thin Solid Films, 63* (1979) 357-362.
(22) B.E. Jacobson, R. Nimmagadda and R.F. Bunshah: *Thin Solid Films, 63* (1979) 333-339.
(23) T.W. Barbee, B.E. Jacobson and D.L Keith: *Thin Solid Films, 63* (1979) 143-150.

8

Chemical Vapor Deposition

John M. Blocher, Jr.*
Consultant
Oxford, Ohio

INTRODUCTION

A large number of chemical reactions are available for the deposition of solid films or coatings up to (e.g.) 1 inch thick from gaseous precursors on a heated substrate.[1] These include the thermal decomposition or reduction of fluorides, chlorides, bromides, iodides, hydrides, organometallics (including carbonyls), hydrocarbons, phosphorus trifluoride complexes, and ammonia complexes. The process, which we call chemical vapor deposition and abbreviate CVD, is useful for the deposition of metals, alloys, intermetallics, boron, silicon, carbon, borides, silicides, carbides, nitrides, oxides, sulfides, etc.

Although in most cases, CVD reactions are activated thermally, in some cases, notably in exothermic chemical transport reactions (to be discussed later), the substrate temperature is held below that of the feed material to obtain deposition. Activation by other than thermal means has been receiving considerable attention recently as reviewed by Yee[2]. By electric discharge plasma activation, for example, it is possible to obtain deposition at lower substrate temperatures. In some cases, unique materials are produced in plasma-activated CVD, such as "amorphous silicon"[3] from silane where 10 to 35 m/o hydrogen remains bonded in the solid deposit.[4] This material is of interest for thin-film solar cells. Passivating films of SiO_2 or SiO_2/Si_3N_4 deposited by plasma-activated CVD are of interest in the semiconductor industry.

In his review, Yee[2] includes reactive evaporation, reactive sputtering, reactive ion plating, and activated reactive evaporation within the definition of CVD as involving deposition of solids by chemical reaction of gas-phase precursors; however, except for this reference to plasma-activated CVD and a brief discussion of chemical transport reactions, this presentation will include only the more conventional CVD technology and will emphasize the so-called "open tube" process where the reactants flow through the CVD reactor. Before a more detailed discussion of this type of CVD process is undertaken, the closed tube or chemical transport system will be reviewed briefly.

In chemical transport processes, the transfer of material occurs usually by

---
*Retired from Battelle-Columbus Laboratories, Columbus, Ohio.

gas-phase interdiffusion of the reactants and reaction products of a reversible reaction, for example,

$$CrI_2(g) = Cr(s) + 2I(g).$$

This reaction is endothermic as written and shifts to the right with increased temperature. This is the basic of the deBoer-VanArkel "iodide process" for metal purification in which iodine reacts with the crude metal at low temperature, forming a volatile iodide which diffuses to a heated substrate where deposition of the metal takes place and the iodine is released to diffuse back to the crude and repeat the cycle. Elements in the crude that do not form iodides or are not released as gases are not transported to the deposited metal. High-purity chromium is prepared commercially in a reactor of special design[5] by this principle for use in metallurgical research and for the preparation of thin-film chromium masks for microelectronic applications.

Professor Harald Schäfer[6] has studied chemical transport reactions extensively not only for metals but for oxides and many other compounds. In chemical transport, the driving force for deposition is generally a temperature differential. Pack cementation processes involve chemical transport reactions in an isothermal system where the driving force is not a temperature differential, but the difference in activity between a metal in its free state and in solution with another metal as an alloy coating. For example, iron objects packed in a mixture of chromium powder, ammonium iodide as an "activator", and aluminum oxide as an inert material added to form a porous mass, will become coated with a chromium alloy or "chromized" by the chemical transport reaction noted above. Figure 8.1 illustrates a pack chromizing system, which can be several feet in diameter and height. Turbine blades and vanes for jet aircraft are chromized, aluminized, and siliconized for oxidation resistance in this way.

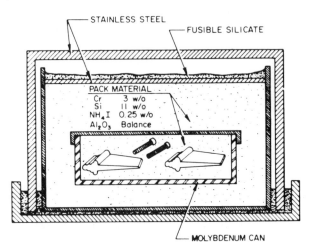

Figure 8.1: Cutaway view of retort used in diffusion coating.[1]

Now turning to the more general aspects of CVD with flowing reactants, the greatest dollar volume of CVD products lies in the electronics industry where semiconducting, insulating, resistive, and conductive coatings are applied extensively by CVD; however, other applications are of industrial importance, and still others are being developed rapidly. In general, the current applications of CVD include coated particles, extended-surface coatings, vapor-formed free-standing shapes, high-purity materials, powders, vapor-impregnated parts, and filaments, fibers and whiskers for reinforcement in structural composites.

The advantages of CVD include the fact that a large variety of materials can be deposited at near-theoretical density (or controlled lower density) with good adherence, and with throwing power in excess of that achieved by electrodeposition or by vacuum evaporation, sputtering, or ion plating. Grain structure and orientation may be controlled to give unique products, and CVD is uniquely applicable to the coating of particles or granules by carrying out the process in a fluidized bed.

The process is not without limitations, however, in that not every material is obtainable in the desired structure at temperatures compatible with substrate limitations. That is, suitable chemical reactions must be available, the substrate must be chemically stable toward the coating reactants, and it must not deteriorate as the result of the inherent heat treatment received in processing. Further, the substrate and coating must be matched sufficiently with regard to coefficient of thermal expansion to avoid prohibitive stresses in cooling from deposition temperature to room temperature.

The problems of successfully engineering a CVD process become evident on brief consideration of the fundamentals. In most cases, one wants a uniform deposit of material over an extended area of substrate, and since the rate of CVD reaction as well as the morphology of the deposited material depend primarily on temperature and reactant composition, one seeks to provide the same conditions over the entire area of the substrate. This is not as simple as it may sound. Since the mass transport from an initially "cold" gas must occur through a boundary layer at the substrate surface, which in turn varies in thickness and diffusion characteristics with gas temperature (a function of position in the reactor), this objective can only be partially achieved. Compounding the problem is the variation of reactant/product composition due to reactant depletion as the gas travels downstream.

As if this were not enough to worry about, gas-phase precipitation must be considered. Given enough dwell time at high enough gas temperatures (for example, in eddy currents or, indeed, in the main stream), a smoke of the reaction product or of an intermediate chemical species can form. Although sometimes desirable, gas-phase precipitation is generally undesirable as, for example, in the deposition of epitaxial silicon from silane. On the other hand, for some deposition reactions, the formation in the gas phase of an intermediate is necessary, in which case the time-temperature-composition history of the gas phase is particularly important. For example, microcrystalline boron is formed on tungsten wire for fiber reinforcement by the reaction

$$\tfrac{3}{2}H_2(g) + BCl_3(g) \rightarrow B(s) + 3HCl(g) \text{ at } \sim1025°C.$$

However, an intermediate, $HBCl_2(g)$, must form in the gas phase before it is finally reduced at the substrate surface.[7] If insufficient dwell time at temperature is provided in the gas phase for appreciable concentrations of $HBCl_2$ to form, the deposition rate of boron is prohibitively slow. This deficiency has been observed in attempting to coat particles with boron in a fluidized bed fed directly with hydrogen and $BCl_3(g)$. The problem is solved by preheating the mixed reactants short of the boron deposition temperature before they enter the fluidized bed.[8]

Clearly, the characteristics of the reactants, the chemical reaction, and the gas-flow dynamics for a given substrate geometry are of prime importance in the successful use of CVD. Because of the great differences in these factors across the broad spectrum of CVD applications, there can be no universally applicable CVD reactor system. Each application, or, at best, class of applications, requires its own engineering.

COATING PROCEDURES

Some examples of reactions of interest in the deposition of metals are:

$$WF_6(g) + 3H_2(g) \xrightarrow{350-1000°C} W(s) + 6HF(g)$$

$$WCl_6(g) + 3H_2(g) \xrightarrow{500-1100°C} W(s) + 6HCl(g)$$

$$TaCl_5(g) + \tfrac{5}{2}H_2(g) \xrightarrow{700-1100°C} Ta(s) + 5HCl(g)$$

$$(C_9H_{12})_2Cr(g) \xrightarrow{400-600°C} Cr(s) + 2C_9H_{12}(g)$$

(With, e.g., HI(g) as a catalyst to suppress codeposition of carbon.)

$$Ni(CO)_4(g) \xrightarrow{150-200°C} Ni(s) + 4CO(g)$$

(Deposition possible at 100°C with a trace of H_2S to catalyze the substrate surface.)

Reactions such as the following can be used to deposit compounds:

$$Al_2Cl_6(g) + 3CO_2(g) + 3H_2(g) \xrightarrow{800-1400°C} Al_2O_3(s) + 3CO(g) + 6HCl(g)$$

$$BCl_3(g) + NH_3(g) \xrightarrow{500-1500°C} BN(s) + 3HCl(g)$$

$$TiCl_4(g) + CH_4(g) \xrightarrow{800-1100°C} TiC(s) + 4HCl(g)$$

$$Ga(g) + AsCl_3(g) + \tfrac{3}{2}H_2(g) \xrightarrow{750-900°C} GaAs(s) + 3HCl(g)$$

Since we are dealing usually with reactants that are corrosive, toxic, or hygroscopic, an enclosed system is usually required.

Apparatus for CVD consists generally of three sections:

  (1) Reactant supply system
  (2) Deposition system
  (3) By-product recycle or disposal system.

The techniques used in each of these areas will be discussed in turn.

Reactant Supply System

Feed materials that are gases at room temperature such as nickel carbonyl (diluted), WF_6, methane, acetylene, hydrogen, ammonia, nitrogen, etc., can be purchased in tanks and their flows can be monitored by simple flowmeters and pressure gages.

Hydrogen can be purified to better than 1 ppm residual impurities by diffusion through commercially available palladium-silver barriers. Argon and helium are best gettered by passing through zirconium chips at $\geqslant 900°C$.

Where the feed materials are volatile liquids, they can be boiled if they do not decompose at or below their boiling points. Silicon tetrachloride and titanium tetrachloride are common examples.

It is frequently convenient to bubble a carrier gas through a volatile liquid reactant to saturate the carrier gas with a desired partial pressure of the reactant. If the carrier gas bubbles are small or dwell time in contact with the liquid is long enough, the gas can become practically saturated and the composition of the resultant gas mixture can be calculated from the equilibrium vapor pressure of the liquid at the temperature of the vaporizer. That is, the mole fraction fraction of the volatile reactant in the saturated carrier gas is the ratio of the vapor pressure of the liquid to the total pressure in the gas phase above the liquid. If the vaporizer is not replenished and the liquid level drops, one runs the danger of the gas becoming less saturated, with accompanying loss of composition control.

In cases where mixed reactants are being fed from a vaporizer, it is necessary to avoid preferential pick-up of the more volatile component by the carrier gas. This would result in a change of composition as the liquid is vaporized. Known compositions of mixed reactants (or single reactants for that matter) can be obtained by using the "flash vaporizer" technique, wherein the liquid is metered to a chamber in which it is rapidly volatilized or actually boiled. The hysteresis of gas composition in such a system in response to change of liquid flow rate depends upon the inventory of liquid in the flash vaporizer relative to the gas flow rate. This inventory can be made small, but cannot be zero. To avoid undesirable compositional changes or uncertainties on either start-up or shut-down of a deposition reaction, it is common practice to provide for by-passing the deposition chamber until the desired feed gas composition has been established.

If the carrier gas technique is used to introduce the vapor from a volatile liquid, the carrier gas may be an inert gas such as argon, helium, nitrogen, or carbon dioxide if non-reactive. Frequently one of the reactants can be used as the carrier gas if the mixture is stable at temperatures approaching that of the substrate upon which deposition is to take place.

Hydrogen is a frequently used carrier gas since it is a convenient reductant for many reactions. Its low density can be a problem in mixing it with higher molecular weight reactants [to get good mixing of gases, flows in the area of mixing should be in the turbulent range (Reynolds number >2000)].

Hydrogen is but one of many materials used in CVD that require special handling. In the case of hydrogen, the explosion hazard must be considered and accommodated. Some reactants are toxic. For example, nickel carbonyl is much more toxic than its equivalent carbon monoxide content. Fortunately, its odor can be detected at low concentrations, and a gas flame burning in the room will give a characteristic gray luminescence in the presence of 5 ppm of $Ni(CO)_4$. However, since th exposure limit is 0.01 ppm[9], special precautions must be taken in its use and disposal (to be discussed later).

Deposition System

CVD reactions can be carried out on wafers, for example, in apparatus of the type shown schematically in Figure 8.2. Tubular shapes can be vapor formed by deposition on a rotated, induction-heated removable mandrel in apparatus such as that for deposition of tungsten shown in Figure 8.3. Apparatus for the deposition of nickel from the carbonyl on extended surfaces is shown in Figure 8.4.

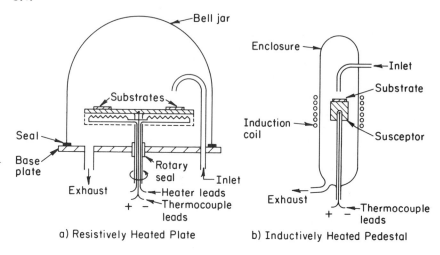

Figure 8.2: Experimental CVD reactors.

Figure 8.3: Apparatus for vapor forming 120-cm x 2.5-cm diameter tungsten tubing by hydrogen reduction of tungsten hexafluoride.[10]

Chemical Vapor Deposition 341

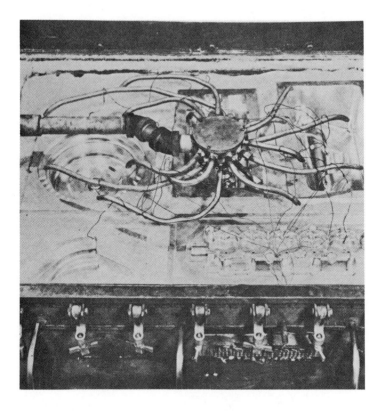

Figure 8.4: Carbonyl nickel deposition chamber with gas distributor, thermocouples, and molds in place.[11]

In general, the design of equipment for substrates of various configurations seeks to optimize heating of the substrate and exposure to the coating atmosphere. As pointed out earlier, in the case of small particles (5 μm to 1000 μm), uniform exposure of the substrate to the coating reactants is obtained by carrying out the reaction in a fluidized bed such as that pictured in Figure 8.5[12] used experimentally for coating nuclear fuel particles with pyrolytic carbon at temperatures of from 900° to 2000°C. This laboratory reactor is a noteworthy example of a thoroughly reliable piece of high-temperature CVD equipment that can be assembled in a minimum time with readily available materials.

Gas Dynamics

For uniformity of reactant composition across a substrate promoted by good mixing, one would like to operate a CVD reactor in the turbulent flow range (Reynolds number >2000). However, this advantage of operating in the turbulent flow range is outweighed in most cases by the disadvantages:

342 Deposition Technologies for Films and Coatings

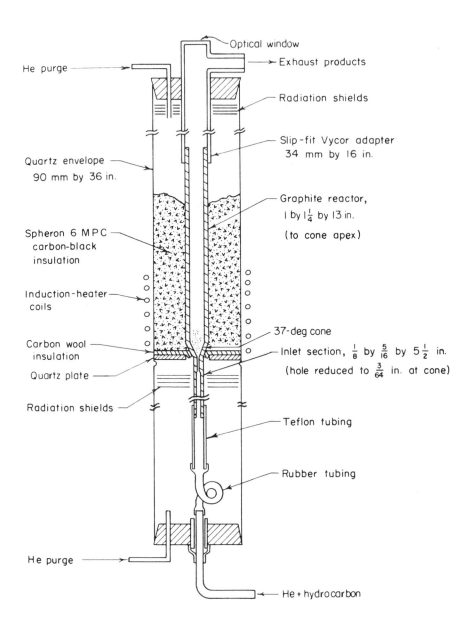

Figure 8.5: Apparatus for coating powders with pyrolytic carbon.[12]

(1) The shortened "dwell time" of the reactant gases in contact with substrate at the higher flow rates generally decreases the efficiency of reactant utilization per pass.

(2) Where a gas-phase reaction to form an intermediate (e.g., the $HBCl_2$ mentioned earlier), little or no deposition may result.

(3) Very large volumes of gas would have to be handled to operate in the turbulent range in most cases, particularly with reactants high in hydrogen, e.g., 20 to 50 scfm through a 4-cm diameter tube, depending upon the average gas temperature.

(4) At the high rate of gas flow, convective cooling of the substrate is increased and adjusting the heat input to compensate for uneven cooling is more of a problem.

Accordingly, most of CVD is done in the laminar flow region, with considerable attention being given to the gas-flow patterns around the object being coated. In this connection, it is useful for any new geometry to monitor the flow patterns with a visible tracer such as the smoke formed on mixing a gas containing $TiCl_4$ or $SiCl_4$ with one containing $H_2O(g)$.

To level out the rate of deposition, one can compensate for the downstream depletion of reactant composition by gradual constriction of the reactor toward the downstream end to increase the gas velocity past the substrate and decrease the thickness of the boundary layer through which the reactant and reaction-product gases must diffuse. Alternatively, one may simply reverse the direction of gas flow, accept a tapered deposition profile in the first stage, and compensate for it in the second. This device is useful only in cases where the morphology of the coating is either uninfluenced by the reactant composition/reaction rate or is relatively unimportant, and thus it is of limited utility. Operation at reduced pressure, e.g., a few torr, is used to improve uniformity, as will be explained later.

Rotating the substrate is frequently used to compensate for non-uniform gas dynamics. Translation of the substrate can also be used. As a last resort, appropriate movement has been applied to either the reactant inlet system, the reaction by-product exit system, or both.

The use of a fan or blower in the deposition chamber has also been used for gas circulation as, for example, in the gas-phase chromizing of coils of sheet steel by the reaction

$$CrCl_3(g) + \tfrac{3}{2}H_2(g) = Cr\ (in\ Fe) + 3HCl(g).[13]$$

Substrate Heating

Four major means of substrate heating are employed in CVD:

(1) A "hot plate" in direct contact with the substrate, as illustrated in Figure 8.2a. It may be either resistively or inductively heated. For hollow cylinders being coated on the outside, the "hot plate" can take the form of a bayonet heater.

(2) Radiant heat (plus some gaseous convection/conduction) from a tube furnace surrounding the substrate. This arrangement has the disadvantage of deposition on the reactor walls, usually at a higher rate than on the substrate.

344 Deposition Technologies for Films and Coatings

(3) Radiant heating of the substrate through the transparent wall of the reactor (prevention of deposition on reactor wall is achieved by air cooling, water cooling, etc.).

(4) Direct inductive coupling to the substrate. This is a very useful technique as it puts the heat where it is needed. It is most useful for axially symmetrical forms which can be rotated in the induction field for uniformity of exposure.

(5) Resistive heating of a conductive substrate as, for example, in the deposition of high-modulus, high-strength boron for fiber reinforcement onto a tungsten core wire by hydrogen reduction of boron trichloride.

Figure 8.6 is an example of a commercially available reactor system in which the substrates, in this case wafers of silicon, are heated by radiation through an air-cooled bell jar. Automatic or manual controls are provided for all of the required functions such as substrate heating, reactant choice (for HCl etching or for deposition from $SiCl_4$ with choice of dopant), and reactant flow rate. It will be noted that the substrate assembly is rotatable to improve uniformity and that the downstream cross section is constricted to increase the gas flow rate to compensate for reactant depletion. In this system, wafer-to-wafer uniformity of ±4 percent in epi-layer thickness and ±5 percent in resistivity can be assured.

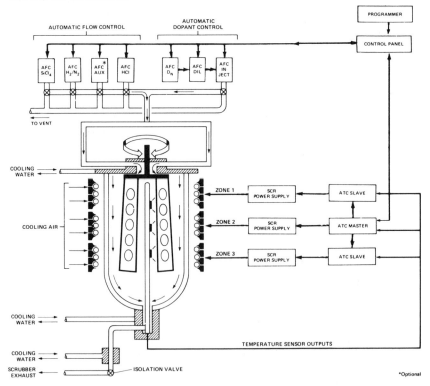

Figure 8.6: Block diagram of system for epitaxial deposition of silicon (courtesy of Applied Materials, Inc.)

By-Product Recycle or Disposal

Since CVD is carried out mostly with corrosive, toxic, or inflammable gases, disposal of the by-products and partially converted reactants must be considered. Obviously, the cost of the feed material, the scale of the operation, and the efficiency of conversion per pass determine whether the feed material should be recovered or discarded. It is probably uneconomical to try to recover the 80 percent of unreacted $SiCl_4$ from an epitaxial silicon deposition system. However, BCl_3 is relatively more expensive and the 80 percent of unreacted BCl_3 in a boron filament deposition system is routinely condensed from the by-product HCl and hydrogen reductant and recycled with fresh hydrogen.

Before discharge to the atmosphere, the effluent gas from a CVD reactor must be treated to remove undesirable products. This is frequently accomplished by scrubbing with aqueous solutions containing suitable neutralizing agents. Commercial CVD effluent scrubbers are available for discharge of gas mixtures containing HCl, $SiCl_4$, BCl_3, PCl_3, etc.

Residual nickel carbonyl and iron carbonyl are generally passed through "decomposers", i.e., heated tubes with (e.g.) copper turnings to provide surface where most of the remaining nickel or iron content is deposited. As a final step, the resulting CO is passed through a hydrocarbon-supported "burner" to remove the last traces of the carbonyl.

Obviously, each CVD reaction has its own recycle/disposal problems whose solution must be engineered into the system.

Surface Preparation

The interface between a CVD deposit and the substrate is of prime importance in determining the adhesion of the deposit, and sometimes its morphology. It is generally desirable to have the substrate surface clean of oxides or other common contaminating surface films, and attention must be paid to the chemistry of the substrate surface with reference to the coating reaction, lest undesirable intermediate products form, or lest the substrate be corroded away before a coating can be deposited. Absolute prevention of substrate corrosion is not always mandatory if it can just be delayed enough for the coating to get started, after which the substrate is protected from further corrosion.

Substrates of active metals such as titanium or aluminum are particularly difficult to coat by hydrogen reduction of halides because the active metal itself is a better reducing agent, and the surface can be eroded away or a nonvolatile intermediate halide can be formed which invariably results in a nonadherent coating. Bryant and Meier of the University of Pennsylvania have classified such reactions and substrate materials in terms of the free energy potential for reaction.[14]

The use of interlayers is frequently advisable to improve coating/substrate compatibility or to protect the substrate from the coating atmosphere. In the author's laboratory, for example, a nickel coating via $Ni(CO)_4$ is used on ferrous alloys prior to coating with tungsten. Titanium carbide is a good interlayer for tantalum coatings on ferrous alloys.[15]

It is not only important that the substrate be compatible with the CVD coating reaction, but with the coating itself at the temperature of deposition. The flexibility of CVD as a coating technique decreases greatly at temperatures below 800°C; that is, the choice of reactions useful below 800°C is limited. Obviously, the situation is ripe for substrate/coating interdiffusion to form brittle interlayers or for possible interaction (usually with a volume decrease and resultant weakness at the interface). For example, tungsten and silicon readily form WSi_2 at 1200°C, so if a silicon carbide coating is desired on tungsten for service below, say, 1000°C, and a choice is available, it is better to deposit tungsten on silicon carbide at 500°C by the reaction

$$WF_6(g) + 3H_2(g) = W(s) + 6HF(g)$$

than silicon carbide on tungsten at 1200°C by the reaction

$$CH_3SiCl_3(g) \xrightarrow{H_2} SiC(s) + 3HCl(g).$$

Substrate surface preparation methods vary with the substrate surface and the coating reaction being used. After cleaning by pickling, grit blasting, degreasing, etc., surfaces of metals whose oxides are reducible by hydrogen (e.g., iron) are conveniently exposed to hydrogen at or above the coating temperature for a time before the CVD reactants are introduced. Substrates such as titanium, whose oxides, nitrides, and carbides are soluble in the metal, can be cleared of these surface contaminants by simple soaking in a gettered inert gas at 1000°C or above.

Oxide-contaminated silicon surfaces to be coated epitaxially with doped silicon are generally etched in HCl(g) at ~1000°C:

$$SiO_2(s) + 4HCl(g) = SiCl_4(g) + 2H_2O(g).$$

DIAGNOSTIC TOOLS

Characterization of CVD products is carried out by conventional metallography, scanning electron microscopy, x-ray diffraction, election diffraction, scratch tests, bend tests, etc. However, some less conventional techniques have been developed to monitor the CVD reaction locally as reviewed by Sedgwick.[16] The work in this area on silicon deposition is largely due to Sedgwick by Raman and fluorescence spectroscopy and to V. Ban by mass spectroscopy.[17]

Although crystal-oscillator deposition-rate monitors are used extensively in deposition by vacuum evaporation, etc., their use is impractical at the temperatures of most CVD reactions. Where the wall of the deposition chamber is transparent, laser interference monitoring (λ = 6328 Å for SiO_2 on Si and 1.15 μm for epitaxial Si) has been used by Sugarawa, et al.[18] Film thickness control of ±5 percent in the range of 5000 Å for both SiO_2 and epitaxial silicon films is obtained.

To obtain a given thickness of CVD coating, where the substrate surface is not optically accessible, one relies on experience or on reaction-product monitoring [e.g., by absorbing and titrating HCl(g) from a chloride + hydrogen reaction]. Control to ±20 percent by this route is characteristic.

COST OF CVD

Before the advent of recent inflation, a brief analysis of the cost of CVD materials was made by the author.[19] An update of this information is introduced here not to provide reliable cost projections, but to indicate how widely the cost of CVD materials and processing can vary with the nature of the material and the processing technique.

Table 8.1 shows the cost of typical starting materials, which may be high relative to the processing cost in cases where the chemical is in low demand. A greater demand would bring the costs of boron trichloride, nickel carbonyl, and tungsten hexafluoride more in line with the larger-volume chemicals listed above them.

Table 8.1: Cost of CVD Reactants

Product	Reactant(s)	Cost (March, 1980) $/lb Reactant*	$/cm^3 Product**
C	CH_4	0.08	0.0005
Si	$SiCl_4$	0.40	0.018
TiC	$TiCl_4$	0.35	0.025
	CH_4	0.08	
Al	$Al(C_4H_9)_3$	1.40	0.09
B	BCl_3	3.80	0.32
Ni	$Ni(CO)_4$	6.00	0.50
W	WF_6	30.00	2.95

*In "large" lots.
**At 70% utilization.

As shown in Table 8.2, the processing cost varies widely with the coating reaction used and with the nature and configuration of the substrate. By far the most economical coating is obtained on particles in a fluidized bed. At the other end of the scale, the individual handling involved in coating silicon wafers and tool inserts has limited the economy with which these coating operations have been performed.

Table 8.2: Cost of CVD Processing

Process	Material Deposited	March, 1980 Cost per Unit* $/cm^3, ($/cm^2)
Fluidized-Bed (60-cm diameter) Coating of Particles 1200°C	Carbon from CH_4	0.005
Deposition on Mandrel, 1 atm, 160°C	Nickel from $Ni(CO)_4$	0.60
Deposition on Mandrel, 0.01 atm 2300°C	Pyrolytic Graphite from CH_4	1.50 to 2.00
Deposition on Mandrel, 1 atm, 550°C	Tungsten from WF_6	2.00 to 5.00
Pack Cementation 50 μm Diffusion Coating 1 atm, 950–1150°C	Aluminum or Chromium (Halide Carrier)	1.00 to 10.00** (0.005 to 0.05)**
Silicon Wafer Coating 0.5 μm Polycrystalline Si on 7-to 10-cm diameter Wafers, 0.0005 atm, 650°C	Poly Si from SiH_4	25.00 to 50.00 (0.001 to 0.0035)
Retort coating (8 μm) of 1.25 x 1.25-cm tool inserts 0.1 atm, 1000°C	Titanium Carbide from $TiCl_4 + CH_4 + H_2$	50.00 (0.05)

*Exclusive of reactant cost; varies widely with scale, conditions, application; values obtained in conversation with commercial processors.
**Depends strongly on packing density of parts.

348   Deposition Technologies for Films and Coatings

Although CVD costs are expected to rise, an across-the-board correction for inflation is not warranted, as technology will improve in some cases and not in others. Further, market changes can quickly change the relative costs of starting materials. Thus, Tables 8.1 and 8.2 should be viewed as only representative of the spectrum of CVD costs.

FUNDAMENTALS OF CVD

As in other chemical systems, the progress of CVD reactions is governed by *thermodynamics* which determines the driving force and *kinetics* which defines the available rate control mechanisms. Although properly included in kinetics, the subject of *mass transport* is of such importance in CVD that it can override the significance of the other aspects of chemical kinetics. So many CVD reactions are carried out at high temperatures where surface reaction kinetics is not limiting, that the assumptions of chemical equilibrium at the surface and diffusional mass transport of reactants and reaction products through a boundary layer at the surface will often define the deposition rate.

THERMODYNAMICS

With the increasingly large body of thermodynamic data and computer technology that has become available within the last 20 years, it is now practical to check the feasibility of many proposed CVD reactions and determine the preferred range of operating conditions. Data for the free energy of formation of compounds of interest in CVD are available as a function of temperature, principally in the JANAF Tables.[21] What one does is to identify all of the possible reaction products of a given starting mixture, relate them by stoichiometry and minimize the total free energy of the system. What was a tedious operation is now routine with computer programs to do the job such as that of Gordon and McBride.[22]

An example of the power of this tool is given by the work of Hunt and Sirtl on the Si-H-Cl system.[23] The various compounds that must be considered in this system are $SiCl_4$, $SiCl_3$, $SiCl_2$, $SiCl$, $SiHCl_3$, SiH_2Cl_2, $SiClH_3$, and HCl. After making estimates where data were not available and using the computer program EQUICA developed at Battelle by Hall and Broehl[24], Hunt and Sirtl determined the equilibrium composition over a wide range of temperature, composition, and pressure. The results are most directly useful in the form of diagrams such as that of Figure 8.7 in which the equilibrium Cl/H ratio is plotted against the Si/Cl ratio in the gas phase. Thus, one can start with any Si-H-Cl combination, e.g., pure $SiHCl_3$ where the Cl/H ratio is 3, and inquire as to the equilibrium Si/Cl ratio, which is found to be, e.g., 0.27 at 1450 K. The difference between this ratio and that of 0.333 in the feed material divided by the latter is the fraction of the silicon fed that will leave the gaseous "solution" at equilibrium, i.e., the "deposition efficiency", 19 percent in this example.

Frequently one would like to know the conditions under which a particular phase will deposit to the exclusion of others, for example, NbC in the Nb-Cl-C system. It can be shown[25] that where reactions such as

$$NbCl_5(g) = Nb(s) + \frac{5}{2}Cl_2(g)$$

and

$$NbCl_5(g) + C = NbC(s) + \frac{5}{2}Cl_2(g)$$

are competing with

$$\left(\frac{5}{x}-1\right)Nb(s) + NbCl_5(g) = \frac{5}{x}NbCl_x(g)$$

and

$$\left(\frac{5}{x}-1\right)NbC(s) + NbCl_5(g) = \frac{5}{x}NbCl_x(g) + C(s),$$

the equation

$$\ln P_{MC} - \ln P_M = \frac{-\Delta F_{MC}}{(x-1)RT}$$

(where ΔF_{MC} is the free energy of formation of the carbide) will define a pressure-temperature region between two straight lines where elemental niobium will not deposit from $NbCl_5(g)$ but NbC will form in the presence of carbon. This principle was used in the Rover nuclear rocket program[25] to deposit NbC on the inside bore of long graphite tubes to make them resistant to attack by high-temperature hydrogen. By limiting the deposition to the pressure-temperature zone defined by the above equation, NbC can form only as rapidly as carbon can diffuse through the coating to meet the coating atmosphere. Thus, with a sufficient supply of $NbCl_5$ and uniform temperature, the coating growth rate is self-controlling and thus uniform.

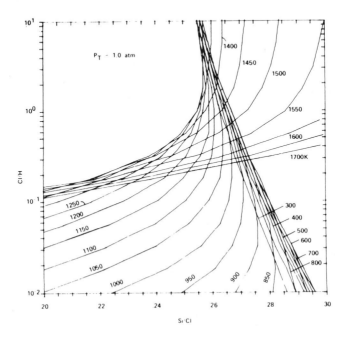

Figure 8.7: Compositions in equilibrium with elemental silicon at 1 atm.[23]

Unfortunately, the thermodynamic data were not available at the time to define the system, and it was necessary to obtain the data of Figure 8.8 experimentally.

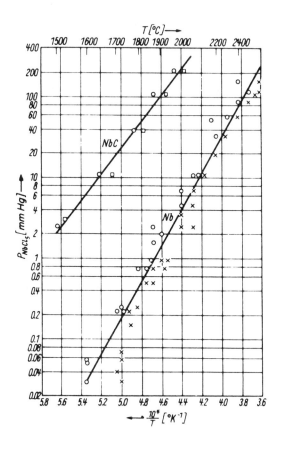

x = metal deposited
open squares = carbide formed
open circles = no deposition, or attack of metal or carbide

Figure 8.8: Critical decomposition temperature of $NbCl_5$ as a function of the initial pressure of $NbCl_5$.[25]

Schäfer and Kahlenberg[26,6] were later able to confirm the lower line thermodynamically [see Figure 8.9 ("Berechnung" = calculation)] and data are now available[27,28] to complete the confirmation. However, the opportunity has not yet arisen to justify the work.

Probably the most detailed phase stability calculations for CVD systems are those of Wan and Spear on the Nb-Ge-H-Cl system.[29]

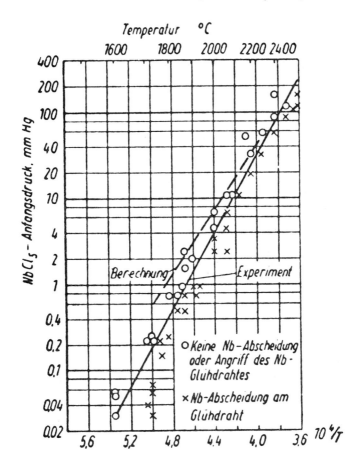

x = metal deposited
open circles = no deposition, or attack of metal

Figure 8.9: Comparison of the experimental data of Figure 8.8 with the critical decomposition temperature line calculated by Schäfer and Kahlenberg.[6,26]

Mass Transport

Figure 8.10 shows a typical log-reaction-rate-versus-reciprocal-temperature graph for a CVD system.[30] In this case, the decomposition of iron carbonyl at 20 torr, the plot is linear below about 200°C, indicating that in that region, the reaction rate is limited by kinetics at or near the substrate. Above 200°C, this reaction is limited by interdiffusion of reactants and reaction products through a boundary layer at the substrate surface as shown in Figure 8.11 from Oxley's treatment of Transport Processes in Reference 1. The fact that the rate varies with flow conditions (N_{Re} = 10 to 200 in Figure 8.10) below 200°C indicates a significant diffusional transport contribution in the kinetically controlled region.

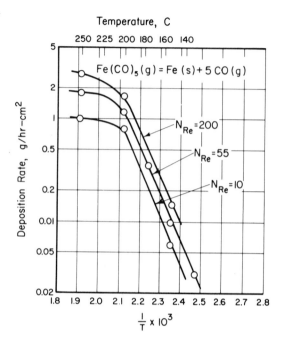

Figure 8.10: Effect of temperature and Reynolds number on deposition rate of iron by thermal decomposition of iron carbonyl at a pressure of 20 torr.[30]

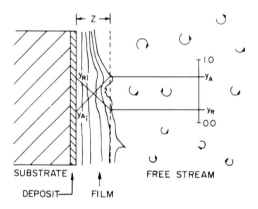

Figure 8.11: Simplified diffusion model for deposition.[1]

A useful quantitative model of the diffusion-limited epitaxial growth of silicon from silane in hydrogen is given by Eversteyn, et al.[31] Based on diffusional transport of silane between the substrate surface (where its partial pressure is essentially zero at temperatures >1000°C) through a boundary layer to a convectively mixed bulk gas whose silane concentration decreases downstream due to deposition, they were able to predict the deposition rate as a function of gas velocity and position in a reactor with a 2.9-degree substrate tilt as shown in Figure 8.12.

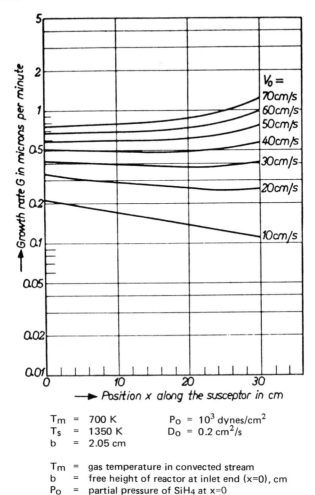

T_m = 700 K	P_0 = 10^3 dynes/cm^2	
T_s = 1350 K	D_0 = 0.2 cm^2/s	
b = 2.05 cm		

T_m = gas temperature in convected stream
b = free height of reactor at inlet end (x=0), cm
P_0 = partial pressure of SiH$_4$ at x=0
D_0 = diffusion coefficient for SiH$_4$ at 300 K (corrected for thermal diffusion)

Figure 8.12: Calculated values of the growth rate versus position along the susceptor at different gas velocities for an angle of tilting of 2.9°.[31]

From these calculations, it was predicted that relatively uniform deposition rates should occur from one end of the reactor to the other at a gas velocity between 30 and 40 cm sec^{-1}. Experimental verification of this prediction is shown at 34 cm sec^{-1} in Figure 8.13. Note that not only is the trend of deposition rate with flow confirmed, but the predicted deposition rate of 0.27 μm min^{-1}* at P_0 = 639 dyne cm^{-2} (0.063 percent SiH$_4$ in H$_2$ at 1 atm) is also confirmed.

---

*0.43 μm min^{-1} (from Figure 8.12 at P_0 = 1000) x $\frac{639}{1000}$ = 0.27 μm min^{-1}.

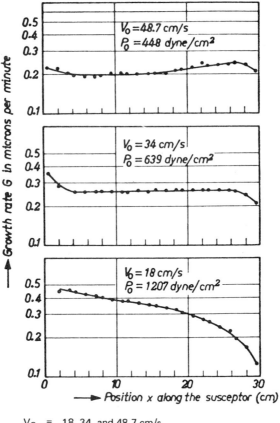

V_0 = 18, 34, and 48.7 cm/s
b = 2.05 cm
T_s = 1350 K

T_m = gas temperature in convected stream
b = free height of reactor at inlet end (x=0), cm
P_0 = partial pressure of SiH_4 at x=o
D_0 = diffusion coefficient for SiH_4 at 300 K (corrected for thermal diffusion)

Figure 8.13: Experimental growth rate versus position along the susceptor for an angle of tilting of 2.9°.[31]

A useful concept relating kinetically controlled and diffusion-controlled CVD has been introduced by van den Brekel[32,33] in which a dimensionless "CVD number" is defined:

$$CVD = \frac{P_b - P_s}{P_s - P_{eq}} = \frac{k_D \delta T_o^2 \ln(T_s/T_b)}{D_o(T_s - T_b)T_s}$$

where P_b and P_s = partial pressures of the reactant gas in the bulk and at the surface respectively

P_{eq} = equilibrium partial pressure of reactant at surface
k_D = mass transfer coefficient at the surface, cm sec^{-1} (rate of coating thickness increase)
δ = thickness of boundary layer, cm
T_s, T_b = temperatures of surface and bulk gas stream, respectively
D_O = diffusion coefficient, cm^2 sec^{-1}, at reference temperature, T_O; $D_T = D_O (T/T_O)^2$.

Where the surface reaction is not at equilibrium, $P_s \neq P_{eq}$ and P_s/P_{eq} is the supersaturation at the surface which determines the ultimate rate of deposition as well as (with temperature) the morphology.

In the simplest case, where $T_s \cong T_b$,

$$(CVD)_{T_s=T_b} = \left(\frac{P_b - P_s}{P_s - P_{eq}}\right)_{T_s \cong T_b} = \frac{k_D \delta}{D_{T_s}}$$

which is the equivalent of the Nusselt number hd/k in heat transfer where

h = film coefficient for heat transfer, cal cm^{-2} sec^{-1} °C^{-1}
d = length parameter, cm
k = thermal conductivity of gas, cal cm^{-1} sec^{-1} °C^{-1}.

In fact, the equivalence of mass transfer and heat transfer has been clearly established (see, for example, Treyball[34]) and the correlations of heat transfer with fluid flow parameters have been useful in establishing mass-transfer coefficients.

In the equation for the CVD number, where the diffusional resistance is relatively large, $(P_b - P_s) > (P_s - P_{eq})$ and CVD > 1. Where the surface reaction is low $(P_s - P_{eq}) > (P_b - P_s)$ and CVD < 1, the process is surface-reaction limited. By properly evaluating the CVD number, the equivalence or lack of equivalence of different CVD systems can be established using this tool. Van den Brekel has been particularly successful in explaining quantitatively the phenomena of surface leveling and the development of protrusions in CVD coatings.[35-37]

By operating at the lowest practical temperature and reducing the total pressure to increase diffusivity and subject the system more to kinetic control, uniformity of deposition is enhanced. The data of Christin et al[38] on the penetration of CVD silicon carbide into porous carbon bodies illustrate the principle well. A cylinder 2.0 cm in diameter (with a connected pore size of a few micrometers) was exposed (transport by diffusion only; no ΔP across specimen) to flowing 10% CH_3SiCl_3 –90% H_2 at temperatures of 900-1100°C. As shown in Figure 8.14, at 1 atm and at a substrate temperature of 1100°C, the reaction occurred on the outer surface (x-ray microprobe analysis). At lower temperatures, incomplete reaction at the outer surface allowed for some diffusion of CH_3SiCl_3 and deposition of silicon carbide on the interior. However, uniformity was still poor. At 900°C the deposition rate was too low to be of interest. As shown in Figure 8.15, a reduction in pressure from 1 atm to 0.5 atm improved the uniformity considerably. Below 0.5 atm, improvement in uniformity was reported to have been less marked.

Before the subject of vapor-phase mass transport is concluded, the work of Takahashi, Sugarawa, et al[39] on the effects of convection currents in CVD reactions should be mentioned, along with that of Curtis and Dismukes[40] who were able to explain the growth striae in expitaxial silicon by thermal convection instability in the CVD "barrel coater" reactor.

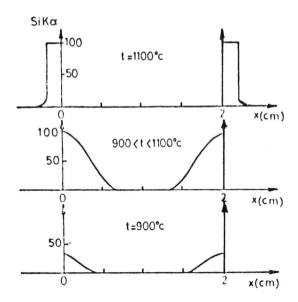

Figure 8.14: Effect of temperature on the deposition of silicon carbide from CH_3SiCl_3-H_2 on the interior of a porous carbon body[38]: (a) t = 1100°C; (b) 900°<t<1100°C; (c) t = 900°C.

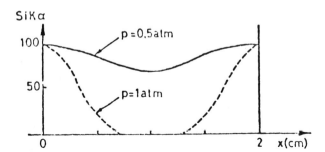

Figure 8.15: Effect of pressure on the deposition of silicon carbide from CH_3SiCl_3-H_2 on the interior of a porous body[38]: ———, p = 0.5 atm; - - - - -, p = 1 atm; temperature, 900°<t<1100°C.

Chemical Kinetics

Although it is possible to write reaction rate constants that fit the data for CVD reactions reasonably well in the kinetically controlled range, it is often difficult to define the reaction mechanism. In the simple reaction

$$SiH_4(g) = Si(s) + 2H_2(g),$$

Joyce and Bradley[41] concluded that the rate-controlling step is reaction between molecules absorbed on neighboring sites followed by deposition of the product, rather than chemisorption of the reactant.

Chemical Vapor Deposition

By contrast, despite the large amount of work that has been done, the rate-limiting mechanism of the reaction

$$3H_2(g) + WF_6(g) = W(s) + 6HF(g)$$

is still in doubt, as reviewed by Brecher.[42]

Other CVD kinetics studies have been made with varying degrees of success, depending upon the complexity of the system, for example, the kinetics of the hydrogen reduction of $BCl_3(g)$.[7,43]

STRUCTURE OF CVD COATINGS

As in condensation from the vapor phase without chemical reaction, the structure of the deposited material depends on the temperature and supersaturation roughly as pictured in Figure 8.16; however, in the case of CVD, the effective supersaturation (i.e., the local effective concentration in the gas phase of the material to be deposited, relative to its equilibrium concentration) depends not only on concentration, but on temperature, as the reaction is thermally activated. Since the effective supersaturation for thermally activated reactions increases with temperature, the opposing tendencies can lead in some cases to a reversal of the sequence of crystalline forms listed in Figure 8.16, as temperature is increased.

Figure 8.16: Effect of supersaturation and temperature on the structure of vapor deposited materials.

Growth of columnar grains such as those pictured in Figure 8.17 is characteristic of many materials in certain ranges of conditions. This structure results from uninterrupted grain growth toward the source of supply. Where growth in one crystallographic direction is preferred over others, grains having that orientation engulf those of other orientations. A fibered texture results, the grains having facets at the surface characteristic of the preferred growth, e.g., in a cubic system, tetragonal pyramids with (111) facets for the [100] growth direction, and trigonal pyramids with (100) facets for the [111] growth direction.

The structure/property/process relationships in CVD have been reviewed by the author.[44] Examples of potential uses of deposits with preferred growth are (1) for uniformity of emission in thermionic emitters[45] and (2) for black-body absorption in a forest of tungsten dendrites.[46] Although preferred growth can be useful, the accompanying faceted surface is undesirable where a smooth surface is required. Frequently the as-grown surface is ground off, as in the CVD vapor fabrication of ZnS IR windows.[47]

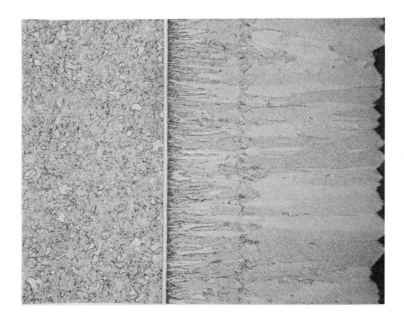

Figure 8.17: Characteristic columnar grains of tungsten deposited by hydrogen reduction of tungsten hexafluoride at 500°C.[8] (Tungsten thickness = 0.65 cm.)

In addition to promoting surface irregularity, columnar structures tend to be weak perpendicular to the growth direction. Thus, for maximum strength and smooth surface finish, deposition conditions should be chosen to give a fine-grained coating. The choice of a deposition reaction that can develop a high effective supersaturation at low temperatures is desirable. As the choices of substrate and deposition reaction are frequently limited, it is well to be aware of two additional means of promoting a fine-grained deposit, (1) codeposition of a grain-growth inhibitor, and (2) mechanical deformation of the growing surface. In some cases, the codeposition is inherent in the available CVD reactions, e.g., the codeposition of silicon and silicon carbide from methyltrichlorosilane. This effect can also be obtained by carrying out two deposition reactions simultaneously. For example, the codeposition of carbon from a hydrocarbon and beryllium oxide from hydrolysis of beryllium chloride was used in the author's laboratory to obtain a fine-grained beryllium oxide and to limit its grain growth in heat treatment.[48] The potential of inhibiting grain growth by codeposition of second-phase materials in CVD has been explored only superficially, and constitutes a potentially highly productive area for research.

Grain refinement during deposition is illustrated by Figure 8.18 which shows a niobium coating on a thermocouple well which was partially immersed in a fluidized bed of uranium dioxide nuclear fuel particles during their coating with niobium for matrix compatibility.[8] A fine-grained deposit developed in the area of contact with the fluidized particles. Not only is the gaseous boundary layer broken at the point of contact of particles, thus increasing the effective supersaturation locally, but the mechanical deformation of the growth surface on contact provides high-energy sites for renucleation of crystallites.

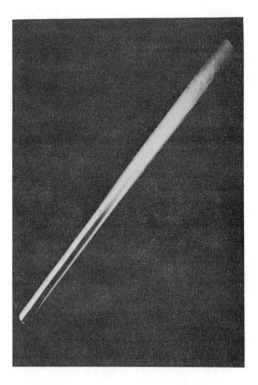

Figure 8.18: Niobium coating on thermocouple well partially immersed in a fluidized bed.[8]

W. Holman and his associates at Lawrence Livermore Laboratories obtain a grain refining effect in tungsten-rhenium alloys by "rubbing" the surface with a tungsten brush during deposition. This leads to tensile strengths equal to those of the wrought material.[49]

In general, if conditions and techniques can be adjusted to give a dense fine-grained material, its mechanical properties approximate those of its wrought counterpart.

CURRENT CVD PRODUCTS

In the following section, we will consider a number of commercially significant CVD products. In noting the reason for the unique applicability of CVD processing in each case, we will be better able to appreciate the opportunities for new applications of CVD in the future.

The use of CVD in the semiconductor industry was mentioned above. There, a use of great importance has been the growth of doped epitaxial layers of silicon on single crystal substrates, although polycrystalline films are of increasing utility. Films of compound semiconductors such as $GaAs_xP_{1-x}$ deposited by CVD are used for light emitting diodes, and other CVD films are being developed for radiation detectors, and piezoelectric transducers. All of the high-purity silicon used in the semiconductor industry is currently made by CVD, either the hydrogen reduction of trichlorosilane, $SiHCl_3$, or the thermal decomposition of silane, SiH_4. As one of the several CVD processes being investigated for the Department of Energy to provide low-cost silicon for solar power, the zinc reduction of silicon tetrachloride in a fluidized bed of seed particles to yield a granular product is being studied in the author's laboratory. Eventual construction of a 50 metric ton per year experimental facility is contemplated.[50]

Another application in which the ability to control crystal growth during deposition is of importance is the vapor forming of tungsten emitter elements for thermionic converters. There, a (110) preferred grain orientation, obtained by hydrogen reduction of WCl_6, gives maximum emitter efficiency. Unfortunately, development work on a nuclear thermionic converter has been dropped in the United States, although progress has continued in France and Russia.

Control of grain growth habit is also exercised in the deposition of pyrolytic graphite, as reviewed by Smith and Leeds.[51] When methane is decomposed at pressures near 10 Torr on a substrate at 2300° to 2400°C, the basal graphitic planes of the deposited carbon line up nearly parallel to the deposition surface. Preferential orientation in the c direction results, although the orientation in the ab directions is nearly random. This so-called "turbostratic" structure is highly anisotropic, with thermal conductivity in the ab direction exceeding that in the c direction by a factor of 100, a property that makes pyrolytic graphite useful in rocket nozzles, atmosphere re-entry vehicles, and as a bowl liner for smoking pipes.

Other forms of pyrolytic carbon formed at lower temperatures, higher pressures, and with different hydrocarbon precursors are useful in coated-particle nuclear fuels for gas-cooled reactors. A strong dense isotropic variety is used to encapsulate these 200-μm-diameter particles to prevent escape of fission gases, and another weak and porous variety is used as an inner cushion to accommodate swelling of the nuclear fuel and avoid cracking the outer shell. An intermediate layer of silicon carbide prevents the migration of solid fission products.

Structural shapes of some materials are made by CVD simply because it is most economical. Thus, vapor-formed crucibles and tubing of tungsten are commercially available for use in fused-salt chemistry, etc.; vapor-formed furnace tubes of silicon are also commercially available, along with various items of hardware for supporting silicon wafers during treatment.

CVD is not the cheapest method of obtaining nickel shapes; however, the superior throwing power of CVD recommends its use over the less expensive electroforming in some cases. Figure 8.19 illustrates the more uniform cross section of CVD deposits at corners which can be sources of weakness in electroformed structures. This improved structure is particularly important in some pressure molding operations where the longer life of the molds justifies their 50 to 100 percent greater cost.[52]

In the case of some coatings for wear and corrosion resistance, or for prevention of reaction with the substrate, CVD is used as the coating method simply because no other method will give a dense impermeable coating of the desired coating material. Silicon carbide-coated graphite is used as a pedestal heater for silicon wafer processing in the semiconductor industry, where the activity of carbon in the furnace must be kept low.

Chemical Vapor Deposition

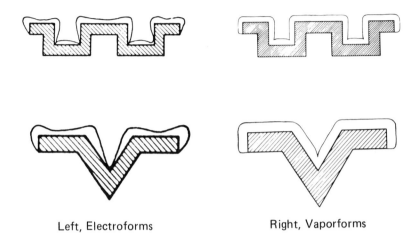

Left, Electroforms   Right, Vaporforms

Figure 8.19: Comparison of electroformed and CVD-formed structures (courtesy of Pyrolytic Company).

Reduction of the loss of carbon moderator at 2500°C in the hydrogen propellant of the nuclear rocket reactor was accomplished by niobium carbide and zirconium carbide coating of the bore holes in the graphite reactor core. Improved coatings were being studied at the time the Rover/NERVA program was terminated.

Boron nitride and titanium boride coatings on graphite are being used to form containers for the vacuum evaporation of liquid aluminum. Titanium nitride, and alumina coatings on cemented carbide tool bits provide increased wear resistance.[53,54]

The CVD coating of steel with tantalum recently promised to make a major impact on the chemical processing industry. By using an intermediate titanium carbide diffusion layer, Glaski[55] was able to ensure good bonding to steel of an impermeable tantalum coating several mils in thickness, on pipe, thermocouple wells, and valve parts. Pipes up to 10 feet in length and 12 inches in diameter (without flanges) could be coated in production equipment. However, it was found that complex economic factors made it impossible to compete with glass-lined steel pipe, which the CVD product promised earlier to displace. In spite of that set-back, it is believed that the CVD tantalum coating of steel will have a significant future role in the chemical processing industry, in the coating of other equipment, if not ultimately in the coating of pipe.

FUTURE APPLICATIONS OF CVD

It is evident from the above information that CVD will never displace electroplating and others of the established coating techniques. However, it is equally evident that a large number of CVD products are here to stay. As new applications are devised for coatings whose structure is critical, so will the use of CVD increase because of its good throwing power and ability to provide control of growth habit. Another major area for increased utility of CVD is in the deposition of the lesser known refractory coatings, such as silicon nitride and silicon oxynitride.

CONCLUSION

Some of the complexity of CVD can be appreciated from the many factors considered in this chapter. It is obvious that CVD is not a universally applicable technique; however, the right "fit" of substrate properties and coatng requirements has resulted in a large number of CVD products, and the opportunities are there for greatly increased utilization in the future.

ACKNOWLEDGEMENT

Permission to reproduce copyrighted material is appreciated. Figures 8.7, 8.12, and 8.13 from the *Journal of The Electrochemical Society* are reprinted by permission of the publisher, The Electrochemical Society, Inc. Their permission has also been granted for use of Figures 8.1, 8.4, and 8.11 from the book *Vapor Deposition,* published by John Wiley and Sons, Inc., as one of the ECS Monograph Series.

Figure 8.9 is reproduced from the *Zeitschrift für anorganische und allgemeine Chemie* with the permission of the publisher, Johann Ambrosius Barth, Leipzig, DDR.

Figure 8.6 is a schematic diagram of the Series 7600 reactor manufactured by Applied Materials, Inc., Santa Clara, California; system design covered by U.S. Patents 3,523,712, 3,796,182, and 3,836,751.

The Pyrolytic Company (Figure 8.19) is now known as Vapor Form Products, New Kensington, Pennsylvania.

REFERENCES

(1) Powell, C.F., Oxley, J.H., and Blocher, J.M., Jr., Eds., *Vapor Deposition,* J. Wiley and Sons, New York (1966).
(2) Yee, K.K., *International Metals Review,* (1), 19-41 (1978).
(3) Carlson, D.E., and Wronski, C.R., *Appl. Phys. Lett., 28* 671 (1976).
(4) Fritsche, H, Tsai, C.C., and Persaus, P., *Solid State Tech.,* 55-60 (January, 1978).
(5) Blocher, J.M., Jr., and Loonam, A.S., Can. Pat. 620,235 (May 16, 1961), U.S. Pat. 3,116,144 (December 31, 1963).
(6) Schäfer, H., *Chemical Transport Reactions,* Academic Press, New York, New York (1964).
(7) Carlton, H.E., Oxley, J.H., Hall, E.H., and Blocher, J.M., Jr., *Proceedings of the Second International Conference on Chemical Vapor Depostion,* The Electrochemical Society, Inc., Pennington, New Jersey, 209-225 (1970).
(8) Browning, M.F., et al, unpublished work, Battelle's Columbus Laboratories.
(9) OSHA standard.
(10) Unpublished work, Battelle's Columbus Laboratories.
(11) Whittington, G.A., *Research Devel., 11* (5), 86 (1960).
(12) Blocher, J.M., Jr., Browning, M.F., and Oxley, J.H., "Fluidized-Bed Processing of Particulate Nuclear Fuels", *Chem. Eng. Progress Symposium Series, 62,* 64-78 (1966).
(13) *Iron Age, 188* (16), 133 (October 19, 1961).
(14) Bryant, W.A., and Meier, G.H., *J. Vac. Sci. Tech., 11* (4), 719-724 (1974).
(15) Glaski, F., *Proceedings of Fourth International Conference on Chemical Vapor Deposition,* G.F. Wakefield and J.M. Blocher, Jr., Eds., The Electrochemical Society, Inc., Pennington, New Jersey, 521-535 (1973).
(16) Sedgwick, T., "In Situ Detection of Gas-Phase Species in Chemical Vapor Deposition", *Proceedings of the Sixth International Conference on Chemical Vapor Deposition,* The Electrochemical Society, Inc., Pennington, New Jersey, 59-78 (1977).
(17) Ban, V., *J. Crystal Growth, 31* 248 (1975).
(18) Sugarawa, K., et al, "Monitoring of CVD Film Thickness by Laser System", *Proceedings of the Fourth International Conference on Chemical Vapor Deposition,* The Electrochemical Society, Inc., Pennington, New Jersey, 205-217 (1973).

(19) Blocher, J.M., Jr., *Proceedings of the 16th Annual Conference of the Society of Vacuum Coaters,* Chicago, Illinois, 82 (March 28-29, 1973).
(20) Oxley, J.H., Browning, M.F., Blocher, J.M., Jr., and Dryden, C.E., *Nuclear Science and Engineering, 16,* 280 (1963).
(21) Stull, D.R., et al, *JANAF Thermochemical Tables,* The Dow Chemical Company, Midland, Michigan.
(22) Gordon, S., and McBride, B.J., NASA SP273 (1971).
(23) Hunt, L., and Sirtl, E., *J. Electrochem. Soc., 119,* 1771-1745 (1972).
(24) Hall, E.H., and Broehl, J.H., "A Computer Program for the Rapid Solution of Complex Chemical Equilibrium Problems", Eighth Quarterly Report to Chemical Vapor Deposition Research Group II, Battelle's Columbus Laboratories (April 15, 1967).
(25) Blocher, J.M., Jr., and Campbell, I.E., "Carbide Coatings for Graphite", *Proceedings, 2nd International Conference on Peaceful Uses of Atomic Energy,* Geneva, 7, 374 (1958).
(26) Schäfer, H., and Kahlenberg, F., "Thermochemistry of the Niobium Chlorides", *Z. Anorg. Allgem. Chem., 305,* 291-326 (1960).
(27) Storms, E., Calkin, B., and Yencha, A., "The Vaporization of the Defect Carbides, Part I: The Nb-C System", *High Temperature Science, 1* (4), Academic Press, New York and London (1969).
(28) Keneshea, F.J., and Cubiciotti, D., "The Thermodynamic Properties of Gaseous Niobium Chlorides", *J. Phys. Chem., 73* (9), 3054-3062 (1969).
(29) Wan, C.F., and Spear, K.E., "Thermodyanmic Equilibrium in the Nb-Ge-H-Cl System for the Chemical Vapor Deposition of Nb_3Ge", *Proceedings of the 6th International Conference on Chemical Vapor Deposition,* L.F. Donaghey, et al, Eds., The Electrochemical Society, Inc., Pennington, New Jersey, 47-58 (1977).
(30) Carlton, H.E., and Oxley, J.H., *AIChE Journal, 11* (1), 79 (1965).
(31) Eversteyn, F.C., et al, *J. Electrochem. Soc., 117* (7), 925-931 (1970).
(32) Van den Brekel, C.H.J., "Characterization of Chemical Vapor Deposition Processes, Part I", *Philips Res. Repts., 32,* 118-133 (1977).
(33) Van den Brekel, C.H.J., and Bloem, J., "Characterization of Chemical Vapor Deposition Processes, Part II", *Philips Res. Repts., 32,* 134-146 (1977).
(34) Treyball, R.E., *Mass Transfer Operations,* McGraw-Hill Publishing Company, Inc., New York, New York (1955).
(35) C.H.J. Van den Brekel, and A.K. Jansen, *J. Crystal Growth, 43,* 364-370 (1978).
(36) A.K. Jansen, and C.H.J. Van den Brekel, *J. Crystal Growth, 43,* 371-377 (1978).
(37) C.H.J. Van den Brekel, *Acta Electronica,* 21 (3), 209-220 (1978).
(38) F. Christin, L. Heraud, J.J. Choury, R. Naslain and P. Hagenmuller, in H.E. Hintermann, Ed., *Proc. 3rd Eur. Conf. on Chemical Vapor Deposition,* 1980, Laboratoire Suisse de Recherches Horlogeres, Neuchatel, Switzerland, 1980.
(39) Takahaski, R., Sugarawa, K., et al, *J. Electrochem. Soc., 119* (10), 1406 (1972).
(40) Dismukes, J.P., and Curtis, B.J., "A Survey of Convective Instabilities in Silicon CVD Systems", *Semiconductor Silicon—1973,* H.R. Huff and R.R. Burgess, Eds., The Electrochemical Society, Inc., Pennington, New Jersey 258-270 (1973).
(41) Joyce, B.A., and Bradley, R.R., *J. Electrochem. Soc., 110* (12), 1235-1240 (1963).
(42) Brecher, L.E., "A Comparison of Models Advanced to Explain the Kinetics of Tungsten Hexafluoride Reduction", *Proceedings of the Fourth International Conference on Chemical Vapor Deposition,* G.F. Wakefield and J.M. Blocher, Jr., Eds., The Electrochemical Society, Inc., Pennington, New Jersey (1973).
(43) Gruber, P.E., "On the Kinetics of Chemical Vapor Deposition", *Proceedings of the Second International Conference on Chemical Vapor Deposition,* J.M. Blocher, Jr., and J.C. Withers, Eds., The Electrochemical Society, Inc., Pennington, New Jersey, 25-36 (1970).
(44) Blocher, J.M., Jr., "Structure/Property/Process Relationships in Chemical Vapor Deposition", *J. Vac. Sci. Tech., 11* (4), 680-686 (1974).
(45) Huberman, M.N., and Holzl, R.A., *J. Appl. Phys., 35,* 1357 (1964).
(46) Cuomo, J.J., Zieger, J.F., and Woodall, J.M., *Appl. Phys. Lett.* (May 15, 1975).
(47) Donadio, R., Connolly, J., Kohane, T., and Pappis, J., "Zinc Sulfide IR Domes", AFML-TR-77-197 (November, 1977).
(48) Wilson, W.J., Browning, M.F., Secrest, V.M., and Blocher, J.M., Jr., USAEC Report, BMI-1718 (March 16, 1965).

(49) Holman, W.R., and Huegel, F.J., *Proceedings of the Conference on Chemical Vapor Deposition of Refractory Metals, Alloys, and Compounds,* Gatlinburg, Tennessee, A.C. Schaffhauser, Ed., The American Nuclear Society, Hinsdale, Illinois, 127 (September 12-14, 1967).
(50) Blocher, J.M., Jr., and Browning, M.F., "The Role of Chemical Vapor Deposition in the Production of High-Purity Silicon", *Proceedings of the Sixth International Conference on Chemical Vapor Deposition,* L.F. Donaghey, et al, Eds., The Electrochemical Society, Inc., Pennington, New Jersey, 129-143 (1977).
(51) Smith, W.H., and Leeds, D.H., "Pyrolytic Graphite", *Modern Materials,* B.W. Gonser, Ed., Academic Press, New York, *7,* 139-218 (1970).
(52) Anon., *Plastics Machinery and Equipment* (April, 1974).
(53) Münster, A., and Ruppert, W., *Z. Elektrochem., 57,* 558 (1953).
(54) Various authors, *Proceedings of the Third International Conference on Chemical Vapor Deposition,* Salt Lake City, Utah, F.A. Glaski, Ed., The American Nuclear Society, Hinsdale, Illinois, 340-415 (April 24-27, 1972).
(55) Glaski, F.A., *Proceedings of the Fourth International Conference on Chemical Vapor Deposition,* G.F. Wakefield and J.M. Blocher, Jr., Eds., The Electrochemical Society, Inc., Pennington, New Jersey, 521 (1973).

GENERAL REFERENCES

(1) Powell, C.F., Oxley, J.H., and Blocher, J.M., Jr., Eds., *Vapor Deposition,* John Wiley and Sons, New York, New York (1966).
(2) Schäfer, H., *Chemical Transport Reactions,* Academic Press, New York, New York (1964).
(3) *Proceedings of the International Conferences on Chemical Vapor Deposition:*
  CVD I — A.C. Schaffhauser, Ed., American Nuclear Society (ANS), Hinsdale, Illinois (1967).
  CVD II — J.M. Blocher, Jr., and J.C. Withers, Eds., The Electrochemical Society, Inc. (ECS), Pennington, New Jersey (1970).
  CVD III — F. Glaski, Ed., ANS (1972).
  CVD IV — G.F. Wakefield and J.M. Blocher, Jr., Eds., ECS (1973).
  CVD V — J.M. Blocher, Jr., H.E. Hintermann, and L.H. Hall, Eds., ECS (1975).
  CVD VI — L.F. Donaghey, P. Rai-Choudhury, and R.N. Tauber, Eds., ECS (1977).
  CVD VII — T.O. Sedgwick and H. Lydtin, Eds., ECS (1979).
  CVD VIII — J.M. Blocher, Jr., G.E. Vuillard, and G. Wahl, Eds., ECS (1981).
(4) Vapor Deposition Literature Survey (monthly abstract service), Battelle's Columbus Laboratories, Columbus, Ohio (1970-).
(5) Yee, K.K., "Protective Coatings for Metals by Chemical Vapor Deposition", Review No. 226, *International Metals Reviews,* (1), 19-42 (1978).
(6) Blocher, J.M., Jr., Soc. Auto. Eng., 1780 (May, 1973); [Reprint No. 730543; *Mach. Des., 43* (16), 58 (1971)].
(7) Hintermann, H.E., and Gass, H., *Oberfläche Surface, 12* (10), 177 (1971); *Schweizer Archiv., 33* (6), 157 (1967).
(8) Haskell, R.W., and Byrne, J.G., *Treatise on Materials Science and Technology,* H. Herman, Ed., Academic Press, New York and London, *1,* 293 (1972).
(9) Béguin, C., *Metall, 1,* 21 (1974).
(10) Archer, N.J., *Chem. Eng.* (London), 780 (December, 1974); Anon., Des. Eng., 76 (November, 1971).
(11) Bryant, W.A., "The Fundamentals of Chemical Vapor Deposition", *J. Mater. Sci., 12,* 1285-1306 (1977).
(12) Hintermann, H.E., Ed., Proc. 3rd Eur. Conf. on Chemical Vapor Deposition, Laboratoire Suisse de Recherche Horlogeres, Neuchatel, Switzerland, 1980.

9
Plasma Assisted Chemical Vapor Deposition

Thomas D. Bonifield
Texas Instruments Incorporated
Dallas, Texas

INTRODUCTION

Scope

Plasma assisted chemical vapor deposition (PACVD) consists of the techniques of forming solid deposits by initiating chemical reactions in a gas with an electric discharge. These techniques are commonly divided into two categories depending on the thermodynamics of the plasma: *thermal plasmas,* such as arcs at atmospheric pressure, where the electrons, ions and neutral gas molecules are in local thermodynamic equilibrium, and *non-equilibrium* or *"cold" plasmas,* such as low pressure glow discharges, where the electrons and, to a lesser extent, the ions are much more energetic than the background gas molecules. The theoretical ideas and the experimental practices used to describe thermal and "cold" plasmas are very different and at present they are quite distinct techniques for PACVD (see for instance Ref. 17). In recent years, the latter, non-equilibrium glow discharge PACVD, has found important applications in the microelectronics, optics and solar energy industries and is the subject of this chapter. In what follows, PACVD will refer only to non-equilibrium glow discharge CVD.

Many of the phenomena of glow discharge PACVD are similar to those of conventional high temperature CVD. Likewise, much of the physical description of glow discharge plasmas for sputtering carries over to plasma CVD. Even the reactor designs resemble planar diode sputtering equipment or low pressure CVD reactors. However, plasma assisted chemical vapor deposition is not nearly as well understood as either of these related technologies. General ideas of the phenomena of PACVD will be presented here and are necessary to guide our thinking, but the emphasis will be on empirical descriptions of processes and materials deposited to date. For more information, the references supplied here should be consulted, especially the excellent review articles by Reinberg,[1] Hollahan and Rosler,[2] Rand,[3] Yasuda[4] and Hollahan and Bell.[5]

Advantages and Limitations

The chief advantage of plasma CVD over thermal CVD is the ability to deposit films at relatively low substrate temperatures (typically less than 300°C). Instead of requiring thermal energy, the energetic electrons in the plasma can activate almost any reactions among the gases in the discharge. At the same time, the bulk of the gas and the substrates do not reach high temperatures because of the non-equilibrium nature of the glow discharge plasma.

As a result, CVD films, which are mismatched in thermal expansion coefficient with the substrate, can be deposited with plasma assistance without severely stressing the film upon cooling to room temperature. Also, high temperature CVD processes are often unacceptable for deposition over other films or structures which would either vaporize, melt (e.g., CVD silicon nitride over aluminum), flow, diffuse or undergo a chemical reaction. Some materials, such as some polymers, are unstable at the temperatures required to activate the deposition reaction. Another way of stating the temperature advantage is that at reasonable or acceptable temperatures, PACVD often has much higher deposition rates than conventional thermal CVD.

In terms of complexity of operations, PACVD is more complex than thermal CVD. Although direct comparisons are few, it is generally more complex, but often more forgiving, than other vacuum deposition techniques such as evaporation or sputter depositions.

The major limitation of PACVD is that deposition of pure materials is virtually impossible. Because of the reduced substrate temperatures, desorption of product gases is ineffective resulting in the incorporation of these elements in the film. For example, amorphous silicon deposited from silane is heavily hydrogenated. However, as in the case of hydrogenated amorphous silicon for solar cells, this can be an advantage.

Another major disadvantage of plasma CVD is the strong interaction of the plasma with the growing film. Although, with proper understanding, the plasma interaction can be used to advantage, it generally means that deposition rates and some film properties will depend on the spatial profile of the plasma. The result is often a non-uniform deposition across the reactor so that the process as well as the equipment has to be adjusted for uniformity.

PHYSICAL AND CHEMICAL CONCEPTS OF PACVD

Plasma assisted CVD is a highly complex chemical process. The discharge produces a wide variety of chemical species, free radicals, electrons and ions, and it can also induce chemical changes at the surface of the growing film. Modeling is impaired not only by the fact that the most important physical and chemical processes have not been singled out, but also by the ignorance of boundary conditions.

Glow Discharge Plasmas

The glow discharges used in plasma assisted CVD are basically of the same type used in glow discharge sputtering, fluorescent lamps and some kinds of gas discharge lasers. Radio frequency (RF) discharges are most common, since the deposition of insulating films destroys the conduction path necessary for DC or very low frequency discharges. Microwave discharges are also less common, because of the complication of using resonant cavity structures. The following discussion, adapted from published analyses of sputtering glow discharges, describes the aspects of RF discharges important to PACVD. Two excellent articles for further reference are the papers of Koenig and Maissel[6] and of Coburn and Kay.[7]

At low frequencies (low kHz range), some degree of secondary electron emission from the electrodes is necessary in order to maintain the discharge. At higher frequencies (MHz range), enough electrons gain the energy required to ionize molecules in the gas to maintain the discharge without secondary emission. Electron energy is lost mainly through inelastic collisions such as ionization, electronic and vibrational excitation and molecular dissociation, which are very sensitive to the particular chemical species in the gas. The major mechanisms removing electrons from the discharge are recombination with ions, mainly at surfaces, and attachment to form slow negative ions.

Many of the gases used in PACVD are polyatomic molecules and many have low ionization potentials. Gases used for sputtering tend to be just the opposite, for example, argon. Also, the gas pressures are usually higher in PACVD (0.1 - 1 Torr) than in sputtering (approximately 0.01 Torr), resulting in higher collision frequencies and shorter mean free paths for the electrons. As a result, electron energies are lower in PACVD and sensitive to the chemical nature of the gases, and product gases, in the discharge. Larger molecules generally result in lower electron energies. Many of the reactive gases in PACVD, particularly halides, are also good electron attachers. Quite often, these gases adversely affect discharge stability at high pressures or low powers.

Positive ions are much less mobile than electrons and gain less energy from the RF field. They also transfer more of their energy to the background gas during elastic collisions. Since there is seldom a single dominant positive ion, resonant charge transfer is not as important a process in PACVD as it is in glow discharge sputtering. However, as in sputtering, positive ions do bombard the substrates during film growth.

Figures 9.1a and 9.1b illustrate the space charge potential of the plasma with and without a developed DC bias on the powered electrode. In each, the two curves represent the extremes of the positive and negative cycle of the RF voltage. Figure 9.1b is typical of planar diode sputter discharges while Figure 9.1a is more typical of PACVD discharges. For diode sputtering, the powered electrode (target) is capacitively coupled and is purposely kept small so that a larger voltage develops across the sheath between the target and the plasma than at the sheath between the substrate and the plasma. On the other hand, most plasma assisted CVD reactors have nearly equal powered and grounded surface areas. Therefore, large voltages develop across the sheath at the grounded surface. Floating, unpowered surfaces are closer to the plasma potential.

Evidently, positive ions will bombard both powered, grounded or floating surfaces. Plasma potentials of several hundred volts with respect to ground have been measured in some plasma CVD processes.[8]

Several factors will determine the energy of the ions impinging on the film during growth. Besides factors such as discharge power, gas type and pressure, which affect the plasma density, RF voltage, target bias, RF frequency and pressure determine what fraction of the plasma potential the ions can pick up as they cross the plasma sheath. At low pressure (less than 0.05 Torr) and low frequency ($\ll 1$ MHz), the ions can follow the RF field and are not seriously scattered in the sheath. They will then strike the surface with the full energy available to them. At high frequencies the ions cannot respond to the RF fields and will drift through the sheath with less energy, seeing only an average plasma potential. At increased pressure scattering in the sheath will reduce that energy further.

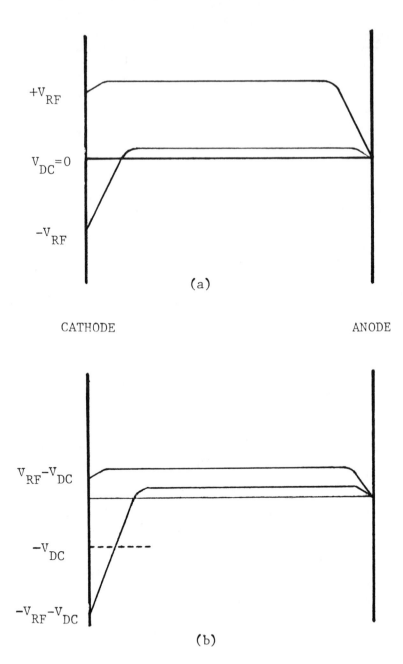

Figure 9.1: A schematic illustration of the space charge potential in an RF glow discharge plasma at the positive and negative extremes of the RF voltage of the cathode: (a) no DC cathode bias and (b) a large negative DC cathode bias.

Collisional Dissociation by Electrons

The most characteristic phenomenon of plasma assisted chemical vapor deposition is the decomposition in the discharge of reactants which are normally stable and unreactive at the deposition temperature. The major decomposition mechanism is collisional dissociation by energetic electrons in the plasma:

$$RX + e^- \rightarrow R + X + e^-.$$

The process is entirely non-thermodynamic; equilibrium constants cannot be used. The densities of dissociated products are determined by the kinetics of this and other reactions leading to deposition.

As discussed in the previous section, the small mass of the electron makes it more responsive to the RF field. Electrons are therefore the most energetic species in the plasma. Since many electrons have enough energy to produce ionization, even more will have enough energy for dissociation processes. The high mobility of the electrons further enhances their reaction rates over chemical processes involving atoms and molecules. In general, the efficiency of dissociation is very high, resulting in decomposition of 10 to 100 percent of the original gas introduced into the discharge. The relatively slow recombination of even very reactive radicals results in radical densities which are a significant percentage of the net molecule density.

Although there are no equilibrium constants to determine the relative levels of dissociation of different gases in the discharge, a qualitative measure of dissociation can be obtained by comparing bond strengths. For example, two common oxidizers in SiO_2 deposition (with SiH_4) are N_2O and CO_2. The energies for dissociating one oxygen atom are 40 kcal/mole for N_2O and 127 kcal/mole for CO_2. As a result, silicon oxides made with CO_2 tend to be silicon-rich unless much higher fractions of CO_2 in the gas are used than are necessary for N_2O. Another aspect of this problem is the relative bond strengths of N–NO, 115 kcal/mole, versus $O-N_2$, 40 kcal/mole. Since oxygen is much more easily dissociated from N_2O than nitrogen, SiH_4/N_2O discharges produce silicon oxide with very little nitrogen in the film.

Transport Kinetics

The pressure and flow conditions used in plasma assisted CVD are best described as slow, viscous, laminar flow. Accept at the very lowest pressures used (less than 0.1 Torr), the mean free path of molecules in the gas, a small fraction of a millimeter, is much smaller than the dimensions of the flow channel of a reactor. Reynolds numbers are very small so that diffusion dominates over turbulent or convective mass transfer. Typical diffusion times are a few milliseconds, whereas the residence time of a reactor (pressure x channel volume/ total flow) is closer to a second, because of the low flows used (a few standard liters per minute).

The transport kinetics of plasma CVD bear some resemblance to thermal, low pressure CVD. The net flow velocity of the gas decreases to zero at the surface of the substrates and mass transport of reactants to the surface relies on diffusion. However, the idea of a boundary layer just above the substrate surface defined by diffusion of reactants out of the gas stream is less appropriate in PACVD where the real reactants are the dissociation fragments produced by the discharge. Instead, the spatial distribution of free radical production in the discharge should be a dominant factor affecting the film growth kinetics. Unfortunately, this kind of kinetic information for plasma chemistry has been very difficult to obtain.

Surface Chemical Effects

Plasma assisted CVD has been described thus far as chemical vapor deposition by free radicals created in the discharge. The formation of a film proceeds by adsorption of these radical species to the substrate or film surface, chemical bonding to other atoms on the surface and ablation (desorption) of by-products of the surface reaction. While dissociation is a gas phase reaction involving free electrons, chemical bonding in the film and ablation of product gases are surface phenomena.

The temperature of the substrate can play a role in the surface reactions and ablation. An example is the incorporation of hydrogen in plasma CVD silicon nitride from SiH_4 and NH_3. Varying the substrate temperature from room temperature to 350°C greatly reduces the hydrogen content in the film, and increases its density and chemical stability as seen in Figure 9.2. This indicates that the species which arrive at the substrate are subhydrides of silane and ammonia (SiH, SiH_2, NH, etc.). Removal of the hydrogen is necessary before a highly crosslinked silicon nitride glass can be formed.

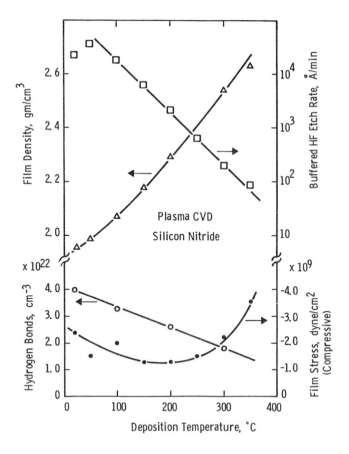

Figure 9.2: Properties of silicon nitride deposited from a gas mixture of $SiH_4/NH_3/N_2$ as a function of the substrate temperature during deposition. (From the author's lab.)

Plasma Assisted Chemical Vapor Deposition 371

Other mechanisms which can activate the surface chemistry involve radiation from the plasma: electrons, ions and photons. Light from the discharge can be very intense, but unless it is strongly absorbed by the film, it will only produce a slight rise in substrate temperature.

Electrons will produce some local heating of the surface of the film and may be an important activation of the surface chemistry. However, electron energies at surfaces are essentially the same as in the glow. They are mobil enough to follow the field and are always accelerated away from the surface in the plasma sheath (see Figure 9.1).

Positive ions, unlike electrons, are accelerated in the plasma sheath toward the film surface as described in the glow discharge section. At low enough pressures, the mean free path of the ions will become a large fraction of the sheath thickness, and ions will impact the film with significant energy. Sputtering of the film surface will remove atoms, aiding in ablation of product gases. Film atoms are removed, but they are usually more likely to be redeposited than the by-product atoms.

PRACTICAL GLOW DISCHARGE REACTORS

A variety of reactor designs have been used for PACVD in laboratory studies. However, only parallel plate reactors have been used for production applications. A typical deposition reactor system is shown in Figure 9.3. The three most important subsystems are described below based on the experiences of designing equipment for the semiconductor industry. Other peripheral equipment, such as flow, temperature and pressure control systems, are common to most chemical processing equipment and are not discussed here.

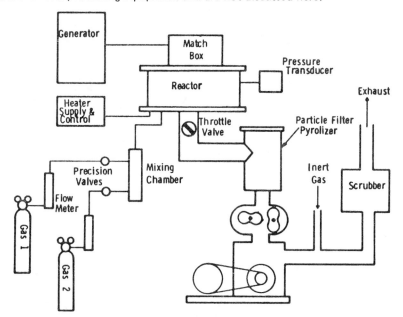

Figure 9.3: Overall system block diagram for glow discharge plasma assisted CVD (from Ref. 1, reprinted with the permission of Annual Reviews Inc.).

372   Deposition Technologies for Films and Coatings

Reactor Designs

Parallel plate reactors are most often categorized according to gas flow patterns. The earliest production scale reactor was the radial flow reactor invented by Reinberg[9,10] and used extensively by several semiconductor manufacturers. Shown in Figure 9.4, the gas flows from the periphery of a circular, grounded substrate plate, across the substrates and out the center of the reactor. As the initial gas is decomposed in the discharge, the direction of flow is such that the deposition occurs on decreasing areas of the substrate plate. Also, the flow velocity increases toward the center of the reactor since the effective channel area of the flow is decreased. As in a tilted susceptor CVD reactor, the higher flow velocity compensates for the reduced reactant concentration. These features, together with the cylindrical symmetry of the reactor, aid in maintaining deposition uniformity over large areas.

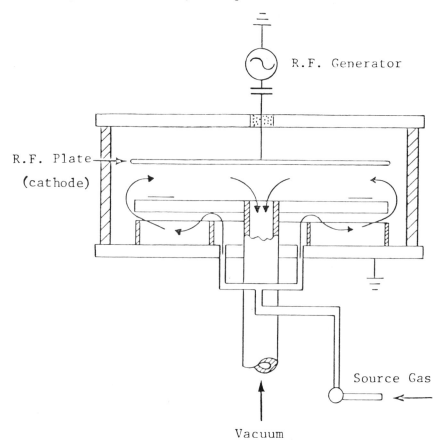

Figure 9.4: Radial flow parallel plate reactor.

Flow effects are important to uniformity control, but discharge plasma conditions are the dominant effects. Egitto recently published a detailed description of uniformity control in a radial flow reactor.[11] The plasma density is also a function of the radial position in the reactor, increasing toward the center, due to electron diffusion to the reactor walls. As a result, the reaction rate increases toward the center compensating for reduced reactant concentration. However, the flow pattern can be turned around from inside to outside and uniformity can still be obtained,[12] but uniformity control differs from the outside-to-inside radial flow reactors.

Another major reactor design, pictured in Figure 9.5, uses longitudinal flow between long parallel plates inside a quartz tube.[13] The plates are alternately driven positive and negative by the RF voltage. Uniform depositions are achieved by maintaining a uniform discharge at low power. Apparently, operation at low power results in a deposition rate that is limited by the power density instead of the gas concentration. These and other reactor designs used in the semiconductor industry have been compared in the trade literature.[14]

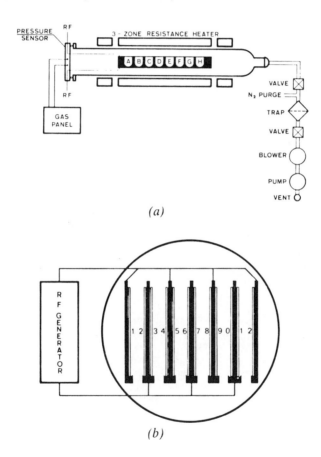

Figure 9.5: Longitudinal flow parallel plate reactor: (a) side view cross section of system, (b) front view cross section of reactor internals (from Ref. 13, reprinted with the permission of Cowan Publishing Corp.).

Discharge Power Supplies

Plasma CVD reactors vary in discharge power frequency from 50 KHz to 13.56 MHz (the industrial frequency band). Matching networks are needed to couple the RF power into the high impedance discharge. At high frequencies, matching networks consist of L–C networks with adjustable components, while low frequencies are matched through transformers. High frequency networks often require retuning with changes in power and pressure; low frequency impedance matching is less sensitive to changes in the discharge, but only a limited range of operating conditions is possible for a given matching network.

If DC bias is desired on the powered electrode, the final coupling to the reactor must be a series capacitor. RF coupling can be either to the substrate plate itself or to a counter electrode with the substrates floating or grounded. These factors determine the extent of ion bombardment seen by the film as described previously.

Vacuum Requirements

Operating pressures are in the range of moderate vacuums, 0.1 to 5 Torr. Although this is not a very high vacuum, high pumping speeds (2000-5000 liter/min) are usually required in order to achieve these pressures at reasonable flow rates. Roots type blowers backed by vacuum pumps are most common.

Many of the gases used or made in PACVD are flammable (SiH_4, H_2, CH_4, etc.) or toxic. Scrubbers or "burn-offs" are required for safe handling of these gases in the exhaust of the vacuum pump.

DEPOSITION OF VARIOUS MATERIALS

Amorphous Silicon

The most common reactant for plasma CVD of silicon has been SiH_4, either pure or diluted in an inert gas. The resultant material is a hydrogenated amorphous silicon (often written a-Si:H). Recent interest in a-Si:H films is for solar cells due to its characteristically low density of mid band gap defect states. The hydrogen in the film apparently passivates dangling bonds in the glass like structure.

Plasma CVD and reactive sputtering in Ar/H_2 plasmas are the major techniques for making a-Si:H. These have been reviewed recently by Brodsky.[15] Nonreactive sputtering and evaporation are other low temperature deposition techniques which result in nonhydrogenated amorphous silicon films. High temperature CVD depositions are used to produce polycrystalline or epitaxial silicon films, and recently, low temperature (230°C) PACVD films have been prepared with a significant polycrystalline character.[16] Large areas of polycrystalline silicon films have also been produced on hot substrates (1100°C) by thermal plasmas.[17]

Other PACVD techniques include the reduction of $SiCl_4$[18] in H_2 which results in amorphous silicon with variable degrees of halogenation and hydrogenation. Webb and Veprek[19] demonstrated an unusual technique whereby silicon is transported from a solid silicon charge by plasma chemical reaction in an H_2 discharge. However, the most common technique remains the decomposition of SiH_4, pure or diluted in argon.

A good illustration of the kinetics of PACVD from SiH_4/Ar mixtures is shown in Figure 9.6, from the work of Street, et al.[20] Deposition rates vary with power and SiH_4 concentration. At high concentrations, the deposition rate increases with power indicating that the kinetics become limited by reactant supply at increased powers.

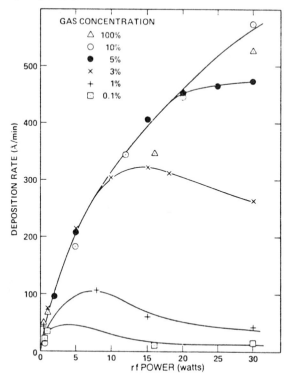

Figure 9.6: Dependence of the deposition rate of a-Si films on RF power for various SiH_4 concentrations in argon (from Ref. 20, reprinted with the permission of the American Physical Society).

Brodsky characterized the deposition kinetics with the distinction between gas phase and surface chemistry.[15] Infrared absorption spectra showed that the films prepared at low pressure contained hydrogen bonded as SiH, while films prepared at high pressure had infrared absorptions characteristic of hydrogen in the form of SiH_2 and SiH_3. His interpretation is that high pressures result in gas phase decomposition and perhaps some polymerization, while at low pressures, decomposition is able to take place at the surface of the film as well. Lucovsky, et al.,[21] have distinguished SiH_2 and $(SiH_2)_n$ chains in the IR absorption spectra. They find that the SiH_2 in the film gives way to $(SiH_2)_n$ chains with increasing power for samples prepared on the grounded anode, but films formed on the capacitively coupled powered cathode have very little $(SiH_2)_n$ at any power. They also found that increasing the substrate temperature enhances the formation of SiH and decreases the SiH_2 and $(SiH_2)_n$ densities.

The mechanism of these plasma interactions has not been identified. Brodsky suggests surface decomposition by electrons. Probably ion bombardment is also important since more bond breaking occurs on the negatively biased cathode than on the grounded anode.

Amorphous Carbon

Plasma decomposition of hydrocarbons is a rich technology. An abundance of starting materials are available, and the resulting films can either be hydrogen rich polymers, "graphite-like" or "diamond-like" amorphous carbon. Interest in the "diamond-like" carbon films has focused recently on their use as hard optical coatings. Although actually amorphous glasses, they are "diamond-like" in the character of their chemical bonding[22] and in their physical qualities of exceptional hardness, high electrical resistivity and optical transmission. Besides plasma decomposition of alkanes[22-24] (usually butane) two ion beam techniques have produced "diamond-like" amorphous carbon films: argon ion beam sputtered carbon with similtaneous argon ion beam bonbardment of the film,[25] and low energy carbon ion deposition.[26,27]

Plasma CVD "diamond-like" carbon films are produced under plasma conditions that promote significant ion bombardment of the film. Under these conditions, low pressures and cathode coupled substrates, the choice of starting gas affects film growth rates but does not seem to be very important to the film properties. Higher carbon to hydrogen ratio in the gas results in higher deposition rates.[23] The physical properties of the film depend instead on the extent of ion bombardment of the film during deposition.[24] Conditions of low ion impact, low power/high pressure, produce polymers (high hydrogen content). High power/low pressure discharges (especially on capacitively coupled cathodes) produce hard, high resistivity (10^{12} ohm-cm) "diamond-like" carbon films. Further increase of ion bombardment, very high power/low pressure, results in graphite-like depositions with low resistivity (0.1 ohm-cm). The transition to graphite, which is thermodynamically favored, may be due to excessive heating of the film by ion impact.

Silicon Nitride

Silicon nitride was the first material to be deposited on a large production scale by plasma assisted chemical vapor deposition. It is a hard, chemically stable glass that is used extensively as a final protective coating and passivation for integrated circuits because it is an excellent diffusion barrier against moisture and alkali ions. The advantage of plasma deposited silicon nitride is the low deposition temperature, typically 300°C, which is acceptable for use over metal leads on the circuit. Silicon nitrides, prepared by a variety of plasma depositions, were reviewed at length by Reinberg.[1]

The most common reactants for PACVD of silicon nitride are SiH_4 and NH_3. These are often diluted in argon[10,11,28-39] or nitrogen.[12,31-34] Since ammonia decomposes in the plasma at a smaller rate than silane, ammonia flows range from 1.4 to 6 times greater than silane flows. Plasma silicon nitride has also been deposited without ammonia, using only SiH_4 and N_2.[12,35,36] However, nitrogen is much more difficult to dissociate than ammonia. The dissociation energies (in kcal/mole) are: 227 for N–N, 110 for H–NH_2, 90 for H–NH, and 79 for H–N. While the total energy required per nitrogen atom is higher from NH_3 than from N_2, the kinetics are more favorable with NH_3 because of the lower energy per step. As a result, SiH_4/N_2 mixtures for plasma deposition contain only 0.1 to 1% SiH_4 in order to get acceptable film compositions (Si/N approximately 1) and resistivity.

Deposition rates are usually lower with N_2 than NH_3 because of the lower SiH_4 flows required. One way to overcome the kinetic limitations with N_2 is to pre-dissociate the N_2 feed gas in a separate discharge just prior to entry into the reaction chamber.[65] Deposition of nitrides from SiH_4/N_2 plasmas are motivated by an interest in lower hydrogen concentrations.

The stoichiometric form of silicon nitride is Si_3N_4, but PACVD nitrides have variable Si/N ratios, usually silicon-rich, and contain 15 to 30 atomic percent hydrogen.[37,38] Several simple measurements have been devised to monitor the composition of plasma nitrides. Changes in the Si/N ratio are reflected in the ultraviolet absorption edge[39] and refractive index[40] of the film as shown in Figure 9.7. Both of these measurements however, are sensitive to the density of the film as well. The Si/N ratio is primarily a function of the SiH_4/NH_3 or SiH_4/N_2 gas ratio. The hydrogen content can be determined from the infrared absorption spectrum of the film.[37] Deposition temperature is the dominant factor for hydrogen content. The effect of hydrogen on most other film properties has not been determined except for a correlation to density and etch rate in hydrofluoric acid.[37] (See Figure 9.2.)

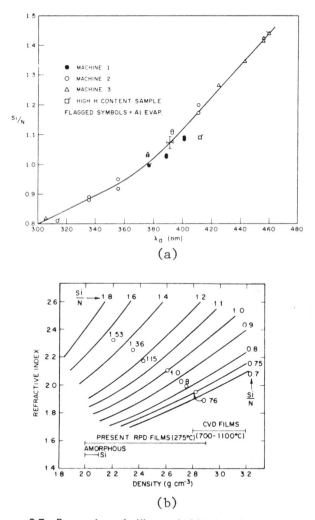

Figure 9.7: Properties of silicon nitride deposited from SiH_4 and NH_3: (a) U.V. absorption edge dependence on Si/N ratio in the film (from Ref. 39, reprinted with the permission of The Electrochemical Society, Inc.), (b) refractive index (5461 Å) as a function of Si/N ratio and film density (from Ref. 40, reprinted with the permission of the American Institute of Physics).

Film stress, the most important mechanical property of the film, is very sensitive to the discharge conditions during deposition. Nitrides deposited at low RF frequency tend to be compressively stressed,[38] while at high frequency (13.56 MHz), the film stress can be either tensile or compressive depending on the power and pressure in the discharge.[28] High powers and low pressures make compressively stressed films. No mechanism has been determined for these effects, but we can note that compressive films are made under conditions of increased ion bombardment of the film during growth. Extremes of either tensile or compressive stress promote cracking and delamination of the film; a mild compressive stress is best.

Other film properties—density, electrical resistivity, and stoichiometry, for instance, depend strongly upon discharge conditions. Figures 9.8 and 9.9, from the work of Sinha, et al,[28] illustrate these effects. When the control of these film properties is necessary, careful control of power, pressure and cathode-to-anode spacing is usually necessary.

Silicon Oxide and Oxynitride

The deposition of silicon oxides by PACVD has not gained the wide spread use that silicon nitride has gained. Other low temperature oxide depositions are available,[41] most notably, the atmospheric pressure CVD reaction of SiH_4 and N_2O at 450°C. Future requirements for lower temperatures (less than 300°C) and conformal silicon oxide coatings will be able to take advantage of PACVD for silicon oxide.

As with silicon nitride, the most common source of silicon is SiH_4, but several oxiders have been used, N_2O,[10,42] O_2[43] and CO_2.[10] Their dissociation energies (in kcal/mole) are: 40 for $O-N_2$, 119 for $O-O$ and 127 for $O-CO$. In one study,[10] films with a stoichiometry near SiO_2 required a N_2O/SiH_4 gas ratio of 7, but a CO_2/SiH_4 gas ratio of 200. The removal of one oxygen atom from either N_2O or CO_2 leaves a very strongly bound diatomic molecule, N_2 or CO, which is only incorporated into the film to a small degree.[42]

Plasma silicon oxides from silane and nitrous oxide have been best characterized. The silicon to oxygen ratio is variable by varying the SiH_4/N_2O gas ratio, as shown in Figure 9.10. Nitrogen contents are typically 5 atomic percent for films with a silicon to oxygen ratio of 1:2.[42] Hydrogen content has not been determined. Plasma silicon oxides can be phosphorus doped during deposition, for lowering the flow temperature of the glass, by adding PH_3 to the reactor.[42] (Note: PH_3 and AsH_3, another common additive to SiO_2, are highly toxic, and even arsenic oxide powders can be toxic.) The film density and dielectric constant of plasma silicon oxide, when saturated with oxygen, are similar to thermally grown oxides. The film stress is mildly compressive, in part due to the thermal compression of cooling from the deposition temperature (200-350°C).

Silicon oxynitride, $SiN_xO_y(H_z)$, can be formed by combinations of any of the techniques described for silicon nitride and silicon oxide deposition. The optical refractive index of these films can be varied continuously (n = 1.45 to 2.0) together with their composition by varying the relative amounts of nitriding and oxidizing reactants.[44] Silicon oxides are generally found to be permeable to external contaminants, such as moisture, while nitrides are impermeable. Silicon oxynitride with nitrogen to oxygen ratios in the film greater than 3 are still good moisture barriers.[45]

Silicon Carbide

Silicon carbide films can be deposited by PACVD from mixtures of gases used to deposit amorphous silicon and carbon films. The composition of the

Plasma Assisted Chemical Vapor Deposition 379

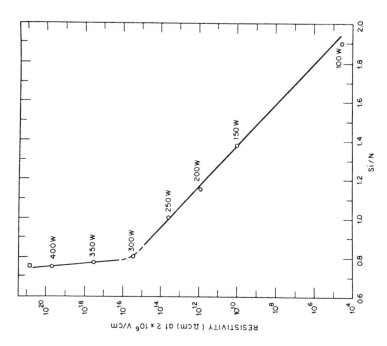

Figure 9.9: Variation of silicon nitride film resistivity at 2×10^6 V/cm with composition. O - plasma CVD silicon nitrides deposited at different RF powers, □ - thermal CVD Si_3N_4 film (from Ref. 29, reprinted with the permission of the American Institute of Physics).

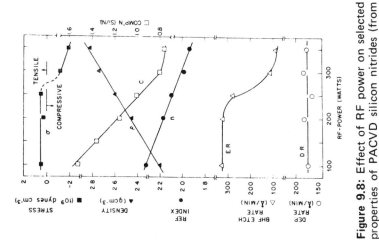

Figure 9.8: Effect of RF power on selected properties of PACVD silicon nitrides (from Ref. 28, reprinted with the permission of The Electrochemical Society, Inc.).

380 Deposition Technologies for Films and Coatings

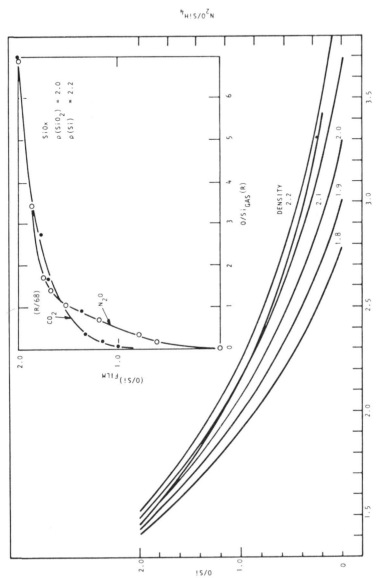

Figure 9.10: Oxygen to silicon ratio for plasma CVD silicon oxide films of varying density and refractive index (6328). Inset is the dependence of O/Si ratio in film to O/Si ratio in gas for SiH$_4$ and N$_2$O or CO. Note: The curve for CO$_2$ has been compressed by dividing the O/Si ratio in the gas by 68. (From Ref. 10, reprinted with the permission of the American Institute of Mining, Metallurgical and Petroleum Engineers, Inc.)

film can be varied continuously between "pure" carbon and "pure" silicon, with a large concentration of hydrogen bonded to both silicon and carbon.[46-48] The optical band gap and refractive index of the film varies smoothly with the Si/C ratio in the film. In one system, using SiH_4 and CH_4,[46] the band gap peaked at 2.8 eV at a Si/C ratio of 2/3; the refractive index (6328 Å) increased from 1.9 to 2.6 with increasing silicon contents. All of the SiC films formed with SiH_4 and CH_4 had a higher Si/C ratio than the corresponding SiH_4/CH_4 ratio.

Yoshihara, et al.,[48] measured film properties for SiC formed on a negatively biased substrate holder. Decreasing the substrate holder bias from 0 to -1500 volts (which increases ion bombardment of the film) decreased the hydrogen content in the film, increased the hardness of the films and decreased the magnitude of the stress, which for all samples was compressive. Plasma CVD silicon carbides produced at high RF powers and high negative substrate voltages had similar resistance to etching in KOH solutions as high temperature SiC. An entire silicon substrate can be etched away without attacking the SiC film.

Plasma Polymers

Literally, the term polymer implies that a monomer exists and can be identified as a building block component in the film, but the term is also used to refer to a characteristic bonding structure of amorphous solids. A polymer-like film differs from a glass by having less cross bonding of the major chemical components of the film and instead, a great deal of hydrogen or halogens saturating the dangling bonds. Polymerization is therefore often used to refer to the formation of polymer-like amorphous films even if no monomer exists. An example from the earlier discussion in this article is the deposition of polysilane—very hydrogen rich, low density amorphous silicon—at high discharge pressure in silane, where gas phase reactions are thought to form chains such as $(SiH_2)_n$ which survive the discharge and incorporate in the film.[15]

Some plasma polymers have been formed which appear to consist of linked monomers of the starting gas.[60] The more common observation, however, is that glow discharge plasmas decompose the starting gas into fragments. These fragments either nucleate into a polymer film at the surface or polymerize, forming chains and clusters, in the gas phase. The relative magnitude of gas phase or surface polymerization are sensitive to the power, pressure and choice of starting material.[4] Fluorocarbon plasma polymerization is especially sensitive to the interaction of the plasma with the polymer. For instance, plasma conditions which promote ion bombardment of the film, enhance etching of silicon substrates by the fluorine, instead of polymerization by fluorocarbon radicals in the discharge.[61]

Plasma polymerization has been reviewed well by Yasuda,[4] Shen,[62] Millard,[63] and Kay and Dilkes.[64] Because of the enormous number of different starting materials and because of the difference in the character of polymer films compared to the glass-like films discussed here, no attempt will be made to review plasma polymerization in this article.

Miscellaneous Materials

Table 9.1 is a list of other inorganic materials, mostly dielectrics, which have been deposited by PACVD in laboratories around the world. Often the reactors are quite different from the parallel plate type reactors described here. Besides making comparisons of process conditions more difficult, the limitations of the reactor reduce the range of film properties which can be achieved. Materials which have not been successfully deposited may become possible with improved reactor designs.

Table 9.1

Material	Reactants	References
Germanium	GeH_4	49
Germanium Oxide	$Ge(OC_2H_5)_4/O_2$	50
Germanium Carbide	GeH_4/CH_4	51
Arsenic	AsH_3	52
Boron	BCl_3/H_2	53
Boron Oxide	$B(OC_2H_5)_3/O_2$	50
Boron Nitride	B_2H_6/NH_3	54
Iron and Iron Oxide	$Fe(CO)_5$	55
Aluminum Oxide	$AlCl_3/O_2$	56
Aluminum Nitride	$AlCl_3/N_2$	57
Titanium Oxide	$TiCl_4/O_2$	58
Phosphorus Nitride	$P + N_2$	59

SUMMARY

Plasma assisted chemical vapor deposition has been used to produce a large number of thin films. Some of these materials, such as silicon nitride and amorphous silicon, are being well characterized because of the present applications of these films. Other materials, such as silicon oxides and "diamond-like" amorphous carbon films, are receiving growing attention for future applications. The chief motivation has been the possibility of much lower deposition temperatures than thermal CVD processes, but some applications, for example, hydrogenated silicon, are beginning to exploit unique properties of plasma CVD films.

Several important process considerations have been identified. Some can only be stated as empirical rules, while others can be understood, in at least a qualitative way, by the use of ideas borrowed from the theories of glow discharges, sputtering and thermal CVD. Application of these ideas depends upon the material and reactor design being considered, but they are stated below, by way of conclusion, in abstract form.

(1) The plasma deposition technique described here is a chemical deposition from reactive species in the vapor above the film. The plasma assistance can only overcome kinetic limitations of otherwise thermodynamically favorable deposition reactions. Film composition and deposition rate are sensitive to the reactant gas and free radical concentrations at the region of deposition, but diffusion dominates local reactant transport and coatings conform well to substrate topography.

(2) The distinguishing feature of plasma assisted CVD is electron impact dissociation of reactants. Electron energies are in the range of one electron volt (a corresponding thermal activation would require temperatures above 10,000°K), with enough electrons in the high energy tail of the distribution to dissociate any gas molecule. Dissociation rates, however, increase with decreasing bond strength.

(3) In most cases, deposition rates are highest where the plasma density and reactant concentrations are highest. These are dominant factors controlling the uniformity of the deposition. Often, no deposition will occur outside of the discharge plasma.

(4) Reactor surfaces in contact with the plasma, even grounded or floating surfaces, are bombarded by ions which pick up energy as they cross the plasma sheath above the surface. Ion bombardment can have a large effect on physical properties of the films—stress and density—and on the ablation of product gases (e.g., hydrogen) from the film during growth. Increasing power and decreasing pressure increases the flux and energy of the ions.

(5) Plasma assisted CVD has many process variables—power, frequency, powered or grounded substrates, pressure, cathode/anode spacing, substrate temperature, gas flows and flow ratios. While the number of variables makes a process complicated, it allows a process to be adjusted for control of film properties and for uniform depositions.

REFERENCES

(1) A.R. Reinberg, *Ann. Rev. Mater. Sci. 9,* 341 (1979).
(2) J.R. Hollahan and R.S. Rosler in *Thin Film Processes* (J.L. Vossen and W. Kern, eds.), Academic Press, New York, 1978.
(3) M.J. Rand, *J. Vac. Sci. Technol. 16,* 420 (1979).
(4) H. Yasuda in *Thin Film Processes* (Same as Ref. 2).
(5) J.R. Hollahan and A.T. Bell, eds., *Techniques and Application of Plasma Chemistry,* Wiley, New York, 1974.
(6) H.R. Koenig and L.I. Maissel, *IBM J. Res. Develop.,* 168, March 1970.
(7) J.W. Coburn and E. Kay, *J. Appl. Phys., 43,* 4965 (1972).
(8) J.L. Vossen, *J. Electrochem. Soc., 126,* 319 (1979).
(9) A.R. Reinberg, U.S. Patent No. 3,757,733, Sept. 1, 1973.
(10) A. R. Reinberg, *J. Elect. Materials, 8,* 345 (1979).
(11) F.D. Egitto, *J. Electrochem. Soc. 127,* 1354 (1980).
(12) R.S. Rosler, W.C. Benzing, J. Baldo, *Solid State Technology,* p. 45, June 1976.
(13) R.S. Rosler and G. Engle, *Solid State Technology,* p. 88, December 1979.
(14) P.S. Burggraaf, *Semiconductor International,* p. 23, March 1980.
(15) M.H. Brodsky, *Thin Solid Films, 50,* 57 (1978).
(16) Z. Igbal, A.P. Webb and S. Veprek, *Appl. Phys. Lett. 36,* 164 (1980).
(17) K.R. Sarma, *Solid State Technology,* 143, April 1980.
(18) G. Bruno, P. Capezzuto, F. Cramarossa and R. D'Agostino, *Thin Solid Films 67,* 103 (1980).
(19) A.P. Webb and S. Veprek, *Chem. Phys. Lett. 62,* 173 (1979).
(20) R.A. Street, J.C. Knights and D.K. Bieglesen, *Phys. Rev. B., 18,* 18, 1880 (1978).
(21) G. Lucovsky, R.J. Nemanich and J.C. Knights, *Phys. Rev. B, 19,* 2064 (1979); See also J.C. Knights, *J. Non-Crystalline Solids, 35,* 159 (1980).
(22) S. Berg and L.P. Andersson, *Thin Solid Films, 58,* 117 (1979).
(23) L.P. Andersson, S. Berg, H. Norstrom, R. Olaison and S. Towta, *Thin Solid Films, 63,* 155 (1979).
(24) L. Holland and S.M. Ojha, *Thin Solid Films, 58,* 107 (1979).
(25) C. Weissmantel, K. Bewilogua, C. Schmer, K. Bremer and H. Zscheile, *Thin Solid Films, 61,* L1 (1979).
(26) S. Aisenberg and R. Chabot, *J. Appl. Phys. 53,* 2953 (1971).

(27) E.G. Spencer, P.H. Schmidt, D.J. Joy and F.J. Sanssalone, *Appl. Phys. Lett. 29,* 118 (1976).
(28) A.K. Sinha, H.J. Levinstein, T.E. Smith, G. Quintana and S.E. Haszko, *J. Electrochem. Soc. 125,* 601 (1978).
(29) A.K. Sinha and T.E. Smith, *J. Appl. Phys. 49,* 2756 (1978).
(30) A.K. Sinha, H.J. Levinstein and T.E. Smith, *J. Appl. Phys. 49,* 2423 (1978).
(31) E.A. Taft, *J. Electrochem. Soc. 118,* 1341 (1971).
(32) W. Kern and R.S. Rosler, *J. Vac. Sci. Technol. 14,* 5 (1977).
(33) K.M. Mar and G.M. Samuelson, *Solid State Technology, 137,* April 1980.
(34) H.J. Stein, V.A. Wells and R.E. Hampy, *J. Electrochem. Soc. 126,* 1750 (1979).
(35) R. Gereth and W. Scherber, *J. Electrochem. Soc. 119,* 1248 (1972).
(36) Y. Kuwano, *Jpn. J. Appl. Phys. 8,* 876 (1969).
(37) W.A. Lanford and M.J. Rand, *J. Appl. Phys. 49,* 2473 (1978).
(38) R. Chow, W.A. Lanford and W. Ke-Ming, *Electrochemical Society Extended Abstracts, 80-2,* 770 (1980).
(39) M.J. Rand and D.R. Wonsidler, *J. Electrochem. Soc. 125,* 99 (1978).
(40) A.K. Sinha and E. Lugujjo, *Appl. Phys. Lett. 32,* 245 (1978).
(41) See for instance, B. Mattson, *Solid State Technology,* January 1980, 60.
(42) J.R. Hollahan, *J. Electrochem. Soc. 126,* 930 (1979).
(43) D. Kuppers, *J. Electrochem. Soc. 123,* 1079 (1976).
(44) R. Kirk, Chapter 9 in Reference 5.
(45) K. Takasaki, K. Koyama and M. Takagi, *Electrochemical Society Extended Abstracts, 80-2,* 767 (1980).
(46) Y. Catherine and G. Turban, *Thin Solid Films, 60,* 193 (1979).
(47) Y. Catherine, G. Turban and B. Grolleau, *Thin Solid Films, 76,* 23 (1981).
(48) H. Yoshihara, H. Mori and M. Kiuchi, *Thin Solid Films, 76,* 1 (1981).
(49) R.C. Chittick, *J. Noncrystalline Solids, 3,* 255 (1970).
(50) D.R. Secrist and J.D. Mackensie, *Bull. Am. Ceram. Soc., 45,* 784 (1966).
(51) D.A. Anderson and W.E. Spear, *Phil. Mag. 35,* 1 (1977).
(52) J.C. Knightz and J.E. Mahan, *Solid State Commun. 21,* 983 (1977).
(53) A.E. Hultquist and M.E. Sibert, in "Chemical Reactions in Electrical Exchanges," *Adv. Chem. Series,* 80, p. 182, Washington, D.C.: Am. Chem. Soc.
(54) S.B. Hyder and T.O. Yep, *J. Electrochem. Soc. 123,* 1721 (1976).
(55) D.M. Wroge, "Plasma Enhanced Deposition of Iron and Iron Oxide Thin Films," M.S. Thesis, USGLBL-9879, Dept. of Energy, Washington, D.C. (1979).
(56) N. Kaho and Y. Koga, *J. Electrochem. Soc. 118,* 1619 (1971).
(57) J. Bauer, *Phys. Status Solidi* (a) *39,* 173 (1977).
(58) J.H. Alexander, et al., in *Thin Film Dielectrics,* F. Vranty, ed. (Electrochem. Soc., Princeton, N.J. 1970).
(59) S. Veprek and J. Roos, *J. Phys. Chem. Solids, 37,* 554 (1976).
(60) See for instance: M. Stuart, *Nature 199,* 59 (1963); N. Paul and M.G.K. Pillai, *Thin Solid Films 48,* 319 (1978).
(61) J.W. Coburn and H.F. Winters, *J. Vac. Sci. Technology 16,* 391 (1979); J.W. Coburn and E. Kay, IBM *J. Res. Develop. 23,* 33 (1979).
(62) M. Shen, ed., "Plasma Chemistry of Polymers," Dekker, New York 1976.
(63) M. Millard in Reference 5.
(64) E. Kay and A. Dilks, *J. Vac. Sci. Technology 18,* 1 (1978).
(65) M. Shiloh, B. Gayer, F.E. Brinckman, *J. Electrochemical Soc. 124,* 295 (1977); M. Shibagaki, Y. Horiike, T. Yamazaki, M. Kashawagi, *Extended Abstracts, Electrochemical Society 77-2,* 416 (1977).

10
Deposition from Aqueous Solutions: An Overview

Morton Schwartz
Consultant
Los Angeles, California

INTRODUCTION

Electrodeposition, also called electroplating or simply plating has both decorative and engineering applications. The emphasis is on engineering applications and the structures and properties of deposits. It should be noted, however, that these are also of interest in decorative plating which constitutes the greatest commercial use of plated coatings, finding wide application in automotive hardware and bumpers, household articles, furniture, jewelry and others. Since the purpose of decorative plating is to provide a durable, pleasing finish to the surfaces of manufactured articles, the corrosion characteristics of the deposits and their ability to protect the substrate are important factors. These and other deposit characteristics involved in the selection and performance of decorative coatings including hardness, wear resistance, ductility, stress are also important to engineering applications of plated coatings.

Engineering applications of plated coatings involve imparting special or improved properties to significant surfaces of a part or assembly and/or protecting or enhancing the function of a part in its operating environment. Other applications include salvage of mismachined or worn parts and other types of reworking as well as material savings, and use of less expensive materials. Special technologies such as electroless deposition, electroforming, anodizing, thin films (for electronics), magnetic coatings, and printed circuit boards have been selected for discussion as representing specific engineering applications.

GENERAL PRINCIPLES

Figure 10.1 represents a simplified plating cell. A DC source, usually a rectifier or motor generator, supplies current flowing in one direction through the external portion of the circuit when a potential difference is imposed across the system. The current flow is that of electrons in the external conductors. The mechanism of electrical transfer in the solution is by means of electrically charged "particles" called ions. Positive ions (cations) travel toward the negative electrode (cathode) and negative ions (anions) travel toward the positive electrode (anode) when the potential is applied, thus completing the electrical circuit. The electrolyte usually contains other components which influence the process (see Figure 10.7 later in this chapter.).

386   Deposition Technologies for Films and Coatings

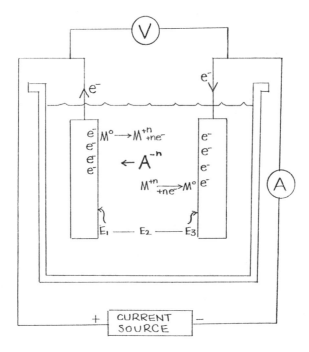

Figure 10.1: Plating Cell

The cathode or deposition reactions are characterized as reduction reactions since electrons are consumed, and the valence states of the ions involved are reduced. The anodic reactions are oxidation reactions wherein electrons are liberated, and the valence states are increased. Each set of reactions represents half-cell reactions and proceeds independently of the other, limited by a condition of material balance, i.e., electrons liberated in the anode reactions must equal the number of electrons consumed in the cathode reactions.

The above describes systems such as nickel or copper deposition from acidic solutions* of their simple ions. Since these are divalent ions (Ni^{2+}, Cu^{2+}), the equations shown in Figure 10.1 would involve two electrons.

Deposition from solutions in which the metallic ions are combined with other ions or ligands as complex ions involves more complicated mechanisms. The cyanide containing electrolytes represent the largest group of such systems. Some of these complex ions are so tightly constituted, i.e., the ionization constant releasing the simple ion is so small, that reduction or deposition of the metal atoms at the cathode occurs directly from the complex ions. This appears to be the mechanism involved with copper, silver and gold cyanide complex ions:

$$[Cu(CN)_3]^= + e \rightarrow Cu + 3\ CN^- \qquad (1)$$

$$[Ag(CN)_2]^- + e \rightarrow Ag + 2\ CN^- \qquad (2)$$

$$[Au(CN)_2]^- + e \rightarrow Au + 2\ CN^- \qquad (3)$$

---

*Most acidic plating solutions fall into this category, involving the simple ions.

The stability of the gold cyanide complex ion is such that it exists in acidic gold plating solutions.

It should be noted that the complex ions described above and other types are anionic and would migrate to the anode during electrolysis. Yet, deposition still takes place at the cathode, indicating that mechanisms other than simple electron reactions are involved in the cathode film. These complex anions approach the cathode by convection and/or diffusion where specific adsorption effects can occur in the double layer (q.v.) as discussed by Wagner, citing Frumkin and Florianovich.[1] The influence of simple cations present in the film are also involved in the reduction process.

Faraday's Laws of electrolysis (1833) are basic to electrodeposition. They relate the current flow, time, and the equivalent weight of the metal with the weight (or thickness) of the deposit and may be stated as follows:

1. The amount of chemical change at an electrode is directly proportional to the quantity of electricity passing through the solution.

2. The amounts of different substances liberated at an electrode by a given quantity of electricity are proportional to their chemical equivalent weights.

Faraday's Laws may be expressed quantitatively by the following equation:

$$W = \frac{ItA}{Fz} \quad (4)$$

where: W = weight of deposit in grams
I = current flow in amps
t = time in seconds
A = atomic weight of element deposited
z = number of charges (electrons) involved (valence change)
F = faraday, a constant = 96,500 coulombs

$I \cdot t$ is the quantity of electricity consumed (coulombs = amp-seconds) and A/z is the equivalent weight of the element. If the current is not constant, then $I \cdot t$ must be integrated:

$$\int_{t_1}^{t_2} I \, dt$$

The Faraday,* F, can be experimentally determined by rearranging the equation,

$$F = \frac{ItA}{Wz} \quad (5)$$

Rearranging equation (5) to

$$It = \frac{WFz}{A} \quad (6)$$

---

*The Faraday can be derived from the fact that 1 gram-atomic weight of an element contains 6.023×10^{23} atoms (Avogadro's Number, N). If the charge on the ion is Z, then $Z \times N$ electrons are required to deposit 1 gram-atomic weight and $Z \times N/Z = 6.023$ electrons are required to deposit (or dissolve) 1 equivalent weight of an element. Since the charge on an electron is 1.602×10^{-19} coulombs, then 6.023×10^{23} electrons $\times 1.602 \times 10^{-19}$ coul. = 96,500 coulombs.

permits the determination of the charge passing through a circuit by the known deposition or dissolution of an element, usually silver. Devices which utilize this application of Faraday's Laws are known as coulometers. Coulometers are used to determine the efficiency of a deposition process. They are also employed as either timers or integrators possessing "electrochemical memory" and producing sharp potential "end-points," i.e., significant changes in electrode potentials which activate electronic circuits. Figure 10.2 schematically illustrates such a device—an electrochemical cell called an E-cell—as part of an electronic circuit.

Figure 10.2: Microcoulometer-E-cell (Courtesy Plessey Electro-Products)

Faraday's Laws are absolute laws, and no deviations or exceptions from them have been found. Apparent exceptions can be shown to be incorrect or explained by the failure to take into account all the chemical or electrochemical reactions involved at the electrode. Thus, the efficiency of an electrochemical reaction can be determined:

$$\% \text{ Electrode Efficiency} = 100 \times \frac{\text{actual weight of deposit}}{\text{theoretical wt. of deposit}} \qquad (7)$$

With knowledge of the actual efficiency, predicted (average) thickness of deposit can be obtained, limited by the control of the current distribution.

The current flowing through a conductor is driven by a potential difference or voltage, the magnitude of which is determined by the relationship expressed as Ohm's Law (1826-7):

$$E = IR \qquad (8)$$

where E = volts, I = amps, and R = ohms. This law regulates both the current flow and its paths in an electrodeposition cell. A commercial electroplating installation and operation involves a multiplicity of series and parallel electrical circuits with only the total current and applied voltage controlled. The current distribution on each individual part or portion of a part (and resulting deposit thickness and properties) depends on the electrode potentials and resistances involved in the "mini-circuits" as well as the geometry and spacing of parts. Since the resistances of the solid, metallic conductors in the circuit are several orders of magnitude lower than the electrolytic (solution) resistances, they can usually be neglected. The potentials within the electrolyte and, more importantly, the electrode-electrolyte interfaces are fundamental, controlling factors and are not as straightforward as suggested by Ohm's Law.

When a metal is immersed in a solution containing its ions, an equilibrium condition is set up between the tendency for the metal to go into solution and the tendency for the metallic ions in solution to deposit on the metal:

$$M° \rightleftarrows M^{n+} + ne^-$$

However, before this equilibrium is established (i.e., the exchange currents or current densities are equal $|i_+| = |i_-| = i_o$), one of the reactions may be faster than the other, resulting in a "charge separation." If the reaction proceeding to the right is faster than to the left, the metal surface would be negatively charged. If the deposition reaction (to the left) is faster, then the surface would be positively charged. This resulting potential between the metal and the solution (at unit activity*) is called the single or standard electrode potential. Since this is a half-cell reaction, a reference electrode, the saturated hydrogen electrode (SHE) is used to complete the circuit and is given the arbitrary value of zero potential. In many instances other reference electrodes such as the calomel electrode are substituted with appropriate corrections. Potential measurements made in this manner (or values derived thermodynamically) result in a series known as the Electromotive Force (EMF) Series.

This origin of the electrode potential was first formulated by W. Nernst (1889). The magnitude of the potential difference between the metal and its ionic solution is given by the Nernst equation:

$$E = E° + \frac{RT}{nF} \ln \frac{a^x \text{ (products)}}{a^y \text{ (reactants)}^{**}} \qquad (9)$$

where: E = observed emf, potential difference (volts)
E° = standard emf
R = gas constant, 8.314 ($J \cdot °K^{-1}\ mol^{-1}$)
T = absolute temperature, °K
n = valence change (electron transfer)
F = Faraday, 96,500 coulombs ($A \cdot sec.\ mol^{-1}$)
a = activity (apparent degree of dissociation)

If the natural logarithm is converted to logarithm to base 10, and T is 298K (25°C), then equation (9) becomes:

---

*Activity is the "corrected" or effective ion concentration, i.e., the deviation from ideal, complete ionization of a 1 molar ion concentration. Activity = Molar Concentration x Activity Coefficient.

**Since the metal (solid) is the reactant in a plating cell, its activity is considered = 1 for all practical purposes and can be neglected. Also, as a practical approximation, the concentration in moles/ℓ can be substituted for activities.

$$E = E^o + \left(\frac{0.059}{n}\right)^* \log a \text{ (or log c approx.)} \quad (10)$$

Thus, a tenfold change in ion concentration changes the electrode potential by 59 mV/n (a negative change makes the electrode potential less positive). This is significant when complexing agents are present since the ionic concentration can be reduced drastically with the accompanying change in electrode potential.** For example, $E_o = -0.76v$ for zinc. But, when zinc is complexed with cyanide:

$$Zn^{2+} + 4CN^- \rightarrow [Zn(CN)_4]^= \quad (13)$$

the electrode potential shifts to approximately $-1.1v$. The standard electrode potential for the Cu^{1+}/CuM half cell is $+0.52v$ which shifts to approximately $-1.1v$ when complexed with cyanide:

$$Cu^{1+} + 3(CN)^- \rightarrow [Cu(CN)_3]^= \quad (14)$$

The practical significance is that a copper cyanide strike provides the best undercoat on a zinc surface since the potentials are essentially the same. Attempts to use an acid copper(II) ($E^o = +0.34v$) solution would provide a potential difference of 1.1v, resulting in an immersion or displacement deposit with poor adhesion. The similarity of electrode potentials for the $[Zn(CN)_4]^=$ and $[Cu(CN)_3]^=$ complexes also permits these metals to be deposited simultaneously as a brass alloy deposit from cyanide solutions.

For electrodeposition reactions to occur, an external source of current and potential is required to overcome the equilibrium conditions discussed above, i.e., to provide a non-equilibrium, irreversible condition. Referring to Figure 10.1, the total plating voltage is the sum of three components. E_2 represents the potential required to overcome the resistance of the electrolyte and obeys Ohm's Law; it would be the only potential required if only the single electrode potentials were involved in the electrodeposition process. E_1 and E_3 are the potentials at the electrodes required to sustain the electrolysis process when the current is flowing and exceed the single electrode potentials. The additional voltage is called polarization which usually increases as the current increases. The electrical energy is converted to heat according to Joules Law:

$$E_{heat} = IEt = I^2Rt \quad (15)$$

resulting in increased temperatures of the electrolytic solutions.

Polarization, also called overvoltage or overpotential, is an important, con-

---

$$*0.059 = \frac{2.303 \times 8.316 \times 298.1}{96,496}$$

**When complex ion reactions are involved:

$$M^{n+} + qX^{p-} \rightleftharpoons [MX_q]^{n-pq} \quad (11)$$

where q is the coordination number, then the Nernst equation is modified:

$$E = E^o - \frac{RT}{nF}\ln K_f + \frac{RT}{nF}\ln\left[a_{MX_q}^{n-pq}/a_{Xp-}^q\right] \quad (12)$$

K_f is the stability constant of the complex ion. Since K_f will be quite large for very stable complexes, the potential can shift substantially negatively.

trolling factor in electrodeposition processes. (In more rigorous treatments, the term overvoltage is restricted to the excess potential required for a single reaction (usually irreversible) to proceed at a specified electrode whereas the term polarization is more general and includes all reactions at the electrode.)

A minimum energy which the reactants must possess is a requisite for any chemical reaction to occur. For an electrochemical reaction to proceed, an overvoltage is required to overcome the potential barrier at the electrode/solution interface; this is called the activation overvoltage or polarization. It is the overvoltage required for the charge-transfer reaction itself. Cathodic activation overvoltage shifts the energy level of the ions in the inner electrical double layer nearer to the potential barrier, so that more ions can cross it in a given time, producing a deposit on the surface. Activation overvoltage also exists at the anode but in an opposite direction.

Changes in the ion concentrations at the electrodes are major contributions to polarization. Figure 10.3 depicts the increased metallic ionic concentration at the anode and the decreased concentration at the cathode as a result of the dissolution and deposition processes. This results in corresponding changes in the equilibrium potentials per the Nernst equation since it changes the value of log C_E/C_S, C_E being the ionic concentration at the electrode and C_S the concentration in the bulk of the solution [see equation (9)]. This effect due to the concentration changes is called concentration polarization.

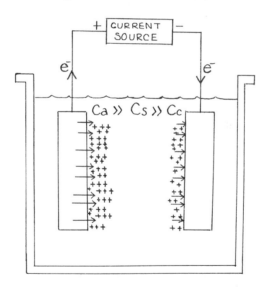

Figure 10.3: Concentration polarization

Increased anode concentration polarization ultimately results in the evolution of oxygen which reacts with the electrode to produce oxide insulating films increasing the ohmic resistance. The oxygen may also react with various solution constituents such as organic compounds or cyanides, thereby consuming them and/or converting them into other compounds which may be detrimental to the electrodeposition process. In some processes, such as anodizing of aluminum or where insoluble anodes are involved such as in chromium plating, anode polarization is desirable.

Cathodic concentration polarization may result in the evolution of hydrogen as the competing reaction. The pH of the cathode film increases and hydrates or hydroxides may precipitate and be occluded in the deposit. The codeposition of hydrogen may result in brittleness of the deposit and, by migration and diffusion in the substrate, result in hydrogen embrittlement.

Hydrogen overvoltage which is the polarization for the specific reaction discharging hydrogen at a specified electrode surface involves at least two steps:

$$2H^+ + 2e \rightarrow 2H_{adsorbed} \rightarrow H_2 \text{ (gas molecule)} \tag{16}$$

The latter step is usually the slower, rate-determining step, and a higher potential is required to discharge the gas.

The factors influencing hydrogen or oxygen overvoltage include:

      a. Electrolyte composition
      b. Type of metal electrode
      c. Nature of electrode surface
      d. Current density
      e. Temperature

Agitation and increased operating temperature of the solution help minimize concentration polarization, permitting higher current densities and faster plating rates. In certain processes, significant cathodic polarization at the higher current density areas produces a substantial decrease in current efficiency, resulting in more uniform deposition over the electrode surface. This phenomenon is called good or high throwing power or, more properly, macrothrowing power since it refers to the thickness distribution over the gross surface with large-scale irregularities. Examples of desirable cathodic polarization are cyanide plating solutions, especially zinc, and the so-called high-throw acid copper plating solutions employed for through-hole plating of printed circuit boards.

An example of "negative" macrothrowing power is chromium plating. This peculiar characteristic is due to the fact that, within limits, the cathode current efficiency increases with increasing current density, thereby greatly exaggerating the non-uniformity of the deposits.

The concept of microthrowing power is discussed later.

ELECTRODEPOSITION

Mechanism of Deposition

Metal deposition differs from other electrochemical processes in that a new solid phase is produced. This dynamic process complicates and introduces new factors in elucidation of the mechanisms involved in the discharge of ions at the electrode surfaces. Factors determining deposition processes include:

    1. The electrical double layer (~ 10 Å thick) and adsorption of ions at the surface—some 2-3 Å away. At any electrode immersed in an electrolyte, a double layer of charges is set up in the metal and the solution ions adjacent to the surface. At solid electrode surfaces, which are usually heterogeneous, the character and constitution of this double layer may exhibit local variations, resulting in variations in the kinetics of the deposition process. This could affect the electrocrystallization processes involved in the overall growth process.

2. The energy and geometry of solvated ions—especially those involving complex ions. All metal ions are associated with either the solvent (water) or complexed with other solution constituents either electrostatically or by coordinated covalent bonding. Desolvation energy is required in transferring the metal ion out of solution to the growing crystal lattice.

3. Polarization effects (see previous section). A symposium[2] on Electrode Processes was held by the Faraday Society in 1947. The excellent papers presented pioneered the concepts upon which the modern concepts of the deposition mechanism are based. Schaeffer and King[3] compiled a chronological annotated bibliography on polarization covering the period 1875-1951.

Thus, the condition of the metal surface to be plated is a basic determining factor in the kinetics of the deposition process and the morphology and properties of the final deposit. The presence of other inorganic ions and organic additives in the double layer or adsorbed on to the surface can greatly modify the electrocrystallization and growth process (Figure 10.4).

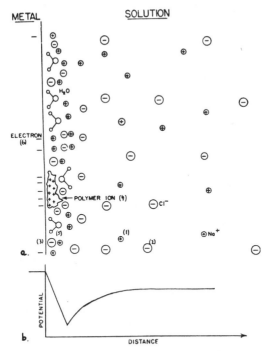

Figure 10.4: (a) the distribution of ions and dipoles in the electrical double-layer. (1) cations, (2) anions, (3) specifically adsorbed anions, (4) adsorbed additives, (5) adsorbed water dipoles, (6) electrons; (b) the potential as a function of distance in the double-layer [corresponding to (a)], measured from the metal surface. (Conway and O'M. Bockris[4])

Based on these considerations, several deposition mechanisms have been proposed.[4-9] The basic or essential steps as shown in Figure 10.5 include:

1. The aquo- or complexed metal ion is transferred or deposited as an adion (still partially bound) to a surface site. Such sites include the plane surface, edges, corners, crevices or holes with the plane surface providing the primary sites.

2. The adion diffuses across the surface until it meets a growing edge or step where further dehydration occurs.

3. Continued transfer or diffusion steps may occur into a kink or vacancy or coordinate with other adions, accompanied by more dehydration until it is finally fully coordinated with other ions (and electrons) and becomes part of the metal being incorporated into the lattice.

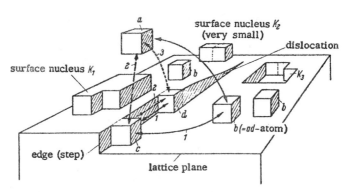

Figure 10.5: Diagram of the crystallization process according to the theory of Kossel and Stranski. Different atom positions: (a) another phase (gas phase, melt, electrolyte), (b) in the lattice plane [ad-atom(ad-ion)], (c) edge (step) site, (d) growth (kink) site. (From *Electrochemical Kinetics* by K.J. Vetter, Academic Press, 1976.[6] Reprinted with permission)

Deposition of metal ions results in depletion in the solution adjacent to the surface. These ions must be replenished if the deposition process is to continue. This replenishment or mass transport of the ions can be accomplished in three ways:

1. *Ionic migration* is least effective. The mobility of the metal ion is very low, its migration rate being dependent on the current and the transport number which is usually less than 0.5. When other salts, conducting salts, are added, these conduct most of the current, reducing the metal ion migration approaching zero. In the case of complex ions where the total charge is negative (complexed as anions), the migration is actually in the reverse direction.

2. *Convection* is the most effective, involving substantial movement of the solution. This is accomplished by mechanical stirring, circulation or air agitation of the solution or moving the electrodes (parts) through the solution. Any one or combination may be employed.

3. *Diffusion* is the effective mechanism for ionic migration in the vicinity of the electrode surface itself where convection becomes negligible. The region near the electrode surface where the concentration of the ions differs from that of the bulk of solution is called the diffusion or boundary layer. It is defined somewhat arbitrarily as the region where the concentrations differ by 1% or more.[10] The diffusion layer is much thicker than the electrical double layer (approximately 15,000 to 200,000 times thicker, depending on agitation). The diffusion rate may be given as:

$$R = D (C_S - C_E)/dN \qquad (17)$$

where R = diffusion rate (moles cm^{-2} S^{-1})
 D = diffusion constant (cm^2 S^{-1})
 C_S = solution concentration (bulk concentration)
 C_E = concentration at the electrode
 dN = the diffusion layer thickness (also called the Nernst thickness)

The diffusion rate increases as dN decreases. On flat, smooth electrode areas, the diffusion layer is fairly uniform. At rough surfaces or irregularities which have a roughness profile with dimensions about equal to the diffusion layer thickness, the diffusion layer cannot follow the surface profile, being thinner at the "micropeaks" than in the "microvalleys." The deposit may be thicker at the peaks than in the valleys, a condition described as poor microthrow. A reverse condition may also exist resulting in good microthrow or leveling. Figure 10.6 represents the three types of microthrowing power. The plating solution composition—especially organic additives—greatly influences the character of microthrowing power and brightening.

Parameters

The parameters generally controlling the composition, structure and properties of the deposit are shown in Figure 10.7. These are briefly reviewed.

Basic Electrolyte Composition: This includes the compounds supplying the metal ions (to be deposited) and the supporting ions. The basic functions of the supporting ions or compounds are to stabilize the electrolyte, to improve solution conductivity, to prevent excessive polarization and passivation (especially anodic), and to provide compatibility to the desired plating conditions. Use of conducting salts or supporting ions reduces the current shared by the metallic ions or complexes, making agitation a more significant factor.

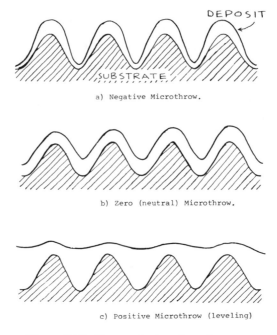

a) Negative Microthrow.

b) Zero (neutral) Microthrow.

c) Positive Microthrow (leveling)

Figure 10.6: Types of microthrowing power

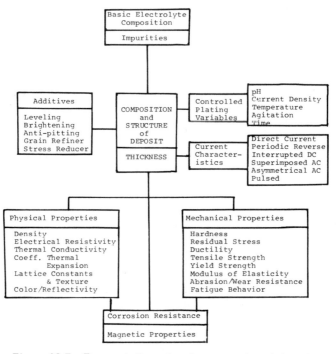

Figure 10.7: Factors influencing the properties of deposits.

Additives: Additives, commonly called addition agents (A.A.), are frequently added to plating solutions to alter desirably the character of the deposit. Read[11] discussed the effects of A.A. on the physical and mechanical properties of deposits—intentional or accidental. They are usually organic or colloidal in nature although some are soluble inorganic compounds. When additives produce a specific effect, they are descriptively called brighteners, levelers, grain-refiners, stress-relievers, anti-pitters, etc.

Profound effects are produced with small concentrations, ranging from a few mg/ℓ to a few percent. In general, the effective concentration range is of the order of 10^{-4} to 10^{-2} M. The mechanisms by which these effects are achieved are not clear in spite of a considerable amount of research and published literature (including a voluminous patent literature, since most commercial additives are proprietary). However, the additive must be adsorbed or included in the deposit in order to exert its effect, and thus appears related to its role in the diffusion layer. Kardos[12] reviews comprehensively the "diffusional" theory of leveling, which, experimentally derived, provides a scientific basis to explain the phenomenon.

To date, no generally acceptable mechanism has been devised to explain satisfactorily the brightening action of addition agents. Brightness, of course, is related to the absence of roughness on a very small scale. The diffusion-controlled leveling theory may be involved for rough surfaces but becomes inapplicable on smooth surfaces or on the sustained growth of bright deposits. Kardos[12] recognized this limitation and for the latter favored the "selective adsorption" of inhibitor on certain growth sites without being diffusion-controlled. The "selective adsorption" of brightening agents on active sites (lattice kinks, crystal projections, growth steps) or random adsorption to suppress crystallographic differences in the deposit represent other proposals of the mechanism, but these are still highly conjectural. Other reviews[13-16], with extensive references, discuss the problems associated with elucidating a brightening mechanism. It seems likely that the "trial and error" method of selecting brightening agents will remain the really effective approach—at least for the near future.

While neither leveling nor brightening may be considered "properties" of the deposit, they reduce the grain size and can greatly influence the physical and mechanical properties due to inclusion of these additives or their decomposition products—especially sulfur and/or carbon—in the deposit. The corrosion characteristics of these deposits are also affected, usually adversely.

The functions of the other types of addition agents are evident. In many instances, the same addition agent performs several of these functions or acts synergistically with other solution constituents or other addition agents.

Controlled Plating Variables: The influence and effects of the operating variables are somewhat dependent upon the solution composition. They are also interdependent. All exert an influence on the structure and properties of the deposit. They are not always predictable, and establishment of optimum ranges is usually determined empirically.

The use of ultrasonic energy agitation in electroplating solutions, i.e., its effect on the polarization, the diffusion layer and properties of deposits has received considerable interest since the 1950's. Rich[17] determined that low frequency vibrations (16-30 kHz) produced more uniform results, and Roll[18] obtained best results in the frequency range 20-50 kHz with intensity (power) range at 0.3-0.5 watt cm^{-2}. However, Hickman[19] found that results based only on reported frequencies and intensities provide an inadequate description and suggested the use of the limiting current method with characterization of the ultrasonic agitation intensity in terms of diffusion layer thickness. The considerable research by Russian investigators is reported by Kapustin and Trofimov.[20] Walker and Walker[21] reviewed the effects of ultrasonic agitation on properties of deposits and noted that the conflicting results reported in the literature may be due to the differences in frequency, intensity and methods of application. Forbes and Ricks[22] were able to reduce the number of preparatory steps required to silver-plate aluminum bus bars from 11 to 4, using ultrasonic agitation in key operations. In this connection, ultrasonic agitation has been widely employed in degreasing, cleaning and pickling pre-plating operations.

Some of the advantages attributed to the use of ultrasonics in electrodeposition are:

1. Higher permissible current densities resulting in higher rates of deposition.
2. Suppression of hydrogen evolution in favor of metal deposition, i.e., a shift in limiting current density.
3. Improved adhesion.
4. Reduced porosity.
5. Reduced stress.
6. Increased brightness.
7. Increased hardness (especially in chromium deposits).

The influence of ultrasonic agitation on grain size appears to be the most important factor, controlling most of the other property changes. However, no specific effects or trends can be attributed to ultrasonics.[21]

Impurities: It is practically impossbile to maintain a plating solution free of impurities. Common or potential sources include:

1. Chemicals used for make-up and maintenance.
2. Impure anodes.
3. Improperly cleaned anodes, anode bags and filters.
4. Rubber or plastic linings and hoses.
5. Rack coatings or maskants.
6. Decomposition of addition agents.
7. Improper rinsing and drag-in of solution from the previous step.
8. Accumulated dissolution of parts during plating.
9. Corrosion of electrical bus bars suspended above the solution.
10. Improper or insufficient cleaning of parts prior to plating.
11. Fall-in of air-borne dirt and oil particles.
12. Chemicals in water used for volume replenishment (e.g., hard water).
13. Generally poor housekeeping.

Particles suspended in the solution may become attached to the surface, resulting in rough, nodular deposits, or leave pits if they fall off; either result produces adverse effects on the integrity and corrosion resistance of the deposit. [A notable exception is the dispersion of controlled particles to be included into the deposit (q.v.).]

Organic impurities generally contribute to pitting, poor covering power, poor adhesion, and harder, more brittle, darker deposits.

Metallic impurities contribute to pitting, poor throwing power, poor adhesion, lower cathode efficiency, stress and cracking, brittleness, burning and off-color deposits. These cations may co-deposit or become entrapped in the deposit, altering its structure and properties. Furthermore, the distribution of the impurity in the deposit may be current density dependent—usually more concentrated in the low C.D. areas. ("Dummying," i.e., removal of impurities from solution by electrolysis, is a common practice in certain plating solutions—especially nickel—at C.D.'s between 0.2 and 0.5 amps/dm^2.) The effects and removal of metallic impurities (copper, zinc and iron) in nickel deposition were studied in detail by D.T. Ewing, et al.[23]

Current Characteristics: All plating processes—with very few exceptions—require unidirectional or direct current (dc). The common current sources are motor generators or rectifiers which convert alternating current (ac) to dc, with the latter almost completely supplanting the former. At present, silicon rectifiers are the most widely used.

Depending on the number of rectifying elements, the type of ac (single or three phase), and the circuitry, the output wave form can be half wave or (usually) full wave with varying percentages of ripple, ranging from 48% to less than 4%. In most plating processes, especially from complex ion type solutions, ripple may not be too significant. However, it can be a significant factor in some plating operations, notably chromium where the ripple should be low (5-10%), since higher ripple may co-deposit excessive oxides and adversely affect the deposit's structure and result in dull deposits.

Figure 10.8 represents some of the modulated current forms employed in attempts to reduce the magnitude and effects of polarization and to alter the structure and properties of deposits.

Ac Superimposed on Dc

Periodic Reverse (PR)   Interrupted Current (IC)   Asymmetrical Ac

Square Wave   Sine Wave   Modified Sine Wave   Modified Skip Sine Wave

Types of Pulsed Wave Forms

Figure 10.8: Modulated current forms.

Superimposed ac on dc, the earliest approach, has not had extensive application. Zentner[24] employed this technique to raise the coercive force (Hc) and decrease remanence (B_R) in cobalt-nickel alloys developed for hard magnetics.

Periodic reverse, or PR[25], under the proper conditions, produces dense fine-grain, striated, leveled and bright deposits. It has its greatest effect and applications on deposits from cyanide solutions, notably copper, permitting smooth, heavy deposits. [Copper deposits produced with dc from cyanide solutions generally become nodular when thicknesses exceed 0.075-0.1 mm (3-4 mils).] A typical PR cycle is 15 seconds plating and 3 seconds deplating; the longer the deplating (reversal) cycle, the smoother is the deposit.

The extended plating time or increased current density required by PR to deposit a given thickness led to the use of interrupted dc employing similar cycles. The interrupted segment's function is to permit the diffusion layer to be replenished.

Asymmetric ac can be considered a variation of PR. In an interesting application, Rehrig[25] used high frequencies (500 Hz), and very high current densities [cathodic C.D. ~8250 A/m^2-11000 A/m^2 (750-1000 A/ft^2) and anodic C.D. at 25% of cathodic C.D.] for high speed spot plating (gold) on lead frames to obtain good bonding properties. In contrast, dc current densities in excess of 3300 A/m^2 (300 A/ft^2) produced deposits with poor bonding characteristics. A minimum anodic C.D. of 2200 A/m^2 (200 A/ft^2) was required to remove any co-deposited metallic impurities, and the degree of removal was proportional to the anodic C.D. The hardness of the deposit decreased and the bond pull strength increased as anodic C.D. increased.

Considerable work is presently being done with pulse plating in electronic plating applications.* Pulse plating may be defined as on/off dc as is interrupted dc mentioned above. The primary differences are that the plating (on) pulses are of very short duration, generally 10-15 milliseconds, and the relaxation (off) time is approximately ten times longer, and much higher current densities are used. Each influences the properties of the deposit, and the optimum conditions to produce the desired quality and properties of deposits from a particular solution must be determined. The pulse wave forms shown in Figure 10.8 are evaluated by Avila.[27] Avila and Brown[28] detail circuitry and power supply requirements. They indicated that the off time is critical since it is based on the requirement of the diffusion layer returning to equilibrium.

Cheh and associates[29] in a research project (No. 35) sponsored by the American Electroplaters' Society (AES) confirmed other (independent) reports that the cathode current efficiency (CCE) in acid copper plating dropped from ~100% to a 93.7%-80.4% range due to pulse plating, shorter pulses (0.5 ms) resulting in lower CCE than longer ones (2-10 ms). They hypothesized that this may be due to a disproportionation mechanism:

$$Cu^{++} + e^- \rightarrow Cu^+ \quad (18)$$

$$Cu^+ + e^- \rightarrow Cu^0 \quad (19)$$

where eq. (18) is faster than eq. (19) during the first interval of the pulse. Thus, the cuprous ion (Cu^+) accumulates and during the relaxation period disproportionates:

$$2Cu^+ \rightleftarrows Cu^0 + Cu^{++} \quad (20)$$

---

*In 1979, the Proceedings of an "International Pulse Plating Symposium" was published by the American Electroplaters' Society.

The throwing power of copper, as measured with a Haring Cell, was somewhat reduced by pulsed plating while that of gold from a citrate-phosphate solution was improved with the improvement diminishing rapidly at increasing peak current densities, becoming practically negligible. Using a rotating-disk electrode, they found a slight improvement in the microthrowing power of the gold solution due to pulsing; however, the C.D.'s employed and especially the agitation had stronger effects.

Reid[30] found that pulsed plating in cobalt-hardened gold deposits virtually eliminated polymer formation under low C.D. ($\sim$5 ma/cm^2) and high off-to-on ratios (100 ms/10 ms). Other effects in the properties noted were:

   a) Improved ductility without any significant decrease in hardness.

   b) Increased density—even in the presence of polymer—from 17.1 g/cm^3 for DC plating (1 ma/cm^2) to 19.2 g/cm^3.

   c) Significant reduction in electrical resistance for cobalt-hardened gold, from 14 $\mu\Omega$-cm to 6 $\mu\Omega$-cm, but an insignificant reduction in pure gold deposits, from 3 $\mu\Omega$-cm to 2.4 $\mu\Omega$-cm.

Effects of pulsed plating on the deposit compositions and properties of gold and gold alloys are reviewed by Raub and Knödler.[31] They show increased alloying element content (Ni or Co) as a function of off-time and a decrease in carbon content. The tensile stresses are reduced in alloy deposits, while the hardness is about 10% higher than that of comparable dc plated alloys. The gas content (H_2, N_2, and O_2) of pulsed plated deposits is also substantially reduced.

Some of the advantages claimed for pulse plating are:

1. Faster plating rates due to increased permissible current densities.
2. Denser deposits (less porosity).
3. Higher purity of deposits, less tendency for impurities to deposit.
4. Smoother, finer-grained deposits.
5. Reduced need or elimination of addition agents.
6. Less hydrogen evolution, providing sharper, finer lines at masking interfaces and possibly less hydrogen embrittlement.
7. Decreased stress in deposits.
8. Increased Ni or Co contents in alloy-hardened gold deposits with less polymer formation.

The price of pulse units depends on the circuitry involved. Smaller capacity units up to 100 amps differ slightly in price, while units larger than 100 amps increase substantially in price.

Some of the effects and advantages attributed to pulse plating and other modulated current forms are very similar to those for ultrasonic agitation. Both "techniques" attempt to reduce polarization effects by increasing diffusion in the boundary layer. In many instances, pulsed current has an effect similar to organic addition agents—especially as related to reduced grain size.

PROCESSING TECHNIQUES

The preparation of metal surfaces for plating involves the modification or replacement of interfering films to provide a surface upon which deposits can be produced with satisfactory adhesion. The type and composition of the soils present as well as the composition and metallurgical condition of the substrate determine the "preparation cycle" and the materials used. The operations involved are designed to accomplish these objectives:

1. Clean the surface.
2. Pickle or "condition" the surface.
3. Etch or "activate" the surface.
4. "Stabilize" the surface. Strike.

In some cycles, several objectives are combined in the same operation. Rinsing steps follow each treatment step.

Examining these steps separately:

1. Cleaning. There are two functions to the cleaning steps: (i) Removal of bulk soils (oils, grease, dirt). This may involve mechanical operations such as wet or dry blasting with abrasive media, brushing or scrubbing or chemical cleaning with solvents (degreasing) or emulsions. (ii) Removal of last "trace" residues. Usually chemical soak (or spray) and electrochemical cleaners are employed. These can affect the substrate and therefore should be compatible to it. Such cleaners may contain alkaline chemicals, surfactants, emulsifying or dispersing agents, water softeners, inhibitors, and chelating agents. Acidic formulated cleaners are also available.

2. Pickling or "conditioning." These are acid dips which neutralize and solubilize the residual alkaline films and "micro-etch" the surface. The common acid dips are either sulfuric acid (~5-15% v/v) or hydrochloric acid (~5% to full strength) and are satisfactory for most alloys. Where undesirable reactions or effects may occur, the acid dip should be formulated to be compatible with the substrate composition.

3. Etching or "activating." Undesirable (from the plating viewpoint) metallurgical micro-constituents are removed or rendered non-interfering; e.g., silicides in aluminum alloys or nickel or chromium in stainless steels or super alloys. Or these steps remove or reduce oxides or other passive conditions prevalent to some surfaces.

High nickel and/or chromium containing alloys usually have tenacious oxide or passive films which must be destroyed with strong acids or anodic etching in strong acids. Solutions containing 15-25% v/v or more sulfuric acid are usually employed at low current densities, 2.2 A/dm^2-5.5 A/dm^2 (20-50 A/ft^2) if metal removal is desired or at high current densities, 10-30 A/dm^2 (100-300 A/ft^2) if smut removal or oxide alteration is desired. Both current density ranges may be employed to maximize adhesion of thick deposits.

In special cases, activation may be accomplished by cathodic treatment in acid or alkaline (cyanide) solutions. Hydrogen is deposited at the surface to reduce superficial oxide films. Solution contamination must be avoided or minimized since such contamination—especially heavy metal ions—may be codeposited as smut.

4. Stabilizing. Very active materials—alloys of aluminum, magnesium or titanium—tend to oxidize or adsorb gases readily, even during rinsing and transfer. These continue to interfere with adhesion of deposits. Therefore, a necessary step involving an immersion deposit of zinc or tin, electroless coating, or

modified porous oxides is required to make the surface receptive to an adherent electrodeposit.

The electrodeposition of thin coatings from specially formulated solutions called strikes can also be considered a stabilizing step, since it provides a new, homogeneous, virgin surface upon which subsequent deposits are plated. These strike solutions and plating conditions are designed to be highly inefficient electrochemically. The considerable hydrogen gas evolution assists any final cleaning, reduction of oxides and activation of the surface while the thin deposit covers surface defects and remaining soils (smut).

The most widely used strike is the cyanide copper strike. The pH and "free" cyanide content are varied depending on the alloy being plated. A typical formulation range is:

> Copper cyanide, CuCN — 15-25 g/ℓ (2-3.5 oz/gal)
> Sodium cyanide, NaCN — 22-40 g/ℓ (3-5.3 oz/gal)
> "free cyanide," NaCN — 2.5-11 g/ℓ (0.3-1.5 oz/gal)
> Sodium carbonate, Na_2CO_3 — 15-60 g/ℓ (2-8 oz/gal)
>
> pH — 10.5-13.0
> Temperature — RT or slightly elevated, 38-45°C
> C.D. — 0.55-1.1 A/dm^2 (5-10 A/ft^2)
> — 1.1-2.2 A/dm^2 (10-20 A/ft^2)

The lower pH's and free cyanide are used for sensitive metals such as aluminum or zinc alloys; the higher pH is used for steels. The cleaning or activating ability is increased with increasing pH, free cyanide, C.D. and temperature. The deposit thickness is approximately 0.25 μm-0.50 μm (0.01-0.02 mils); thicker deposits— 1.25-2.5 μm (0.05-0.1 mils) are applied at slightly higher temperatures.

The second most common strike is the Woods nickel strike or its modifications. This strike is effective (and preferred) on high nickel or chromium containing alloys. A typical formuation is:

> Nickel Chloride, $NiCl_2 \cdot 6H_2O$ — 240 g/ℓ (32 oz/gal)
> Hydrochloric Acid (conc.) — 125 ml/ℓ (16 fl. oz/gal)
>
> Temperature — R.T. (20-30°C)
> C.D. — 5-20 A/dm^2 (50-200 A/ft^2)
> Time — 0.5-3 minutes

Silver or gold strikes are used prior to plating thicker deposits of these metals. Either a gold strike or the strike of the particular precious metal is used prior to plating the specific metal. These are generally formulated similar to the plating solution except that they contain approximately one-tenth the metal ion concentration. The strikes may be applied directly to the substrate or, more commonly, on the copper or nickel strikes discussed. The use of these strikes minimizes the possible contamination of precious metal plating solutions.

Unusual strikes are sometimes employed in special procedures. For example, a chromium strike appears to be most effective for plating on molybdenum alloys or an acid copper or electroless nickel strike on titanium alloys followed by a diffusion treatment to obtain adhesion of subsequent deposits.

Properly designed preparation cycles and the establishment of a stable receptive surface are prime requisites for good quality deposits. However, the condition and integrity (or lack of it) of the surface prior to plating also affect the quality of the deposit; this is becoming more evident as quality and functional requirements of electrodeposits are increased.

Some plating processes require post-plating treatments. To improve the corrosion resistance of zinc or cadmium deposits or the tarnish resistance of silver, chromate conversion coatings are applied by chemical or electrochemical treatments; these gel-like films also improve adhesion of paint films.

Since most preparation and plating processes generate hydrogen which can be occluded and can migrate into the substrate, possibly causing hydrogen embrittlement, stressed articles or high-strength materials are usually given a relief bake in air at 190°C (350-400°F) for 3-24 hours within 3-4 hrs. after plating.

Procedures for the preparation of difficult-to-plate substrates have been prepared as "Standard Recommended Practices" by ASTM. These are listed in Appendix A. The "Standards" reference the literature upon which they are based. Included in the Appendix is a discussion of preparation of less common metals.

SELECTION OF DEPOSITS

Individual Metals

Only 19 or so of all the known individual (single) metals are presently of practical interest in aqueous electrodeposition. Of these, only 10 have been reduced to large scale commercial practice. These are indicated in Table 10.1 with the most widely used ones underlined. Holt[32] reviews the electrodeposition of "uncommon" elements from aqueous, organic and fused salt media.

Alloy deposition, electroless deposition and deposition with dispersed particles ("inclusion plating") extend the practical use of aqueous coating systems considerably. These are discussed separately.

In order to make a proper selection of a deposited coating, one must be cognizant of the fact that these coatings can vary widely in structure and physical and chemical properties, depending on the electrolyte composition and operating conditions as discussed above. For example, the hardness of as-plated chromium deposits can be varied from 350 to 1100 DPN and nickel from about 150 to 650 DPN. The corrosion protection afforded by a coating depends upon its electrochemical relationship to the substrate, its thickness, continuity (porosity) and the environment as well as its overall quality. The important factors to be considered in the selection of a deposit are the purpose of the deposit and the use (function) of the finished article. Other factors which must be considered are the size, shape, and expected useful life of the article and the costs and environment involved.

Table 10.2 comprises a list of various engineering functions of deposited coatings and the deposits usually employed. Table 10.3 gives "representative" hardness values for various deposits in relation to some common materials and Hardness Scales. Spencer[33] discusses selection factors for coatings, their properties and characteristics and uses.

Alloy Deposition

Alloy deposition extends the availability and applicability of coatings from aqueous solutions. It is an area of increasing research and development, although most of the systems have not attained commercial acceptance. An extensive literature has developed. Brenner's 2-volume comprehensive, definitive monograph[34] details compositions, operating conditions, structures and properties of the deposits, covering developments up to 1960. A Russian monograph[35] details their extensive research in this area. Brenner[36] updated the state of the art to 1964. Krohn and Bohn[37] reviewed the literature to 1973 with a count of more than 200 binary alloys; Figure 10.9 summarizes the binary alloy combinations reported to June, 1972. Over 1000 abstracts on alloy deposition were reported in *Chemical Abstracts* between 1964 and 1972.

Deposition from Aqueous Solutions: An Overview 405

Table 10.1: Periodic Table

Enclosed area contains "metallic" elements which have been deposited from aqueous solutions.
— Elements most commonly plated commercially
- - - Elements less frequently plated commercially
· · · · Elements infrequently plated commercially

(¹) Also known as Columbium, Cb

Table 10.2: Selection of Deposits

Primary Function of Coating	Most Widely Employed Coating	Representative Application
Corrosion resistance	Zn, Cd	Sacrificial coatings, fasteners, hardware fittings
	Sn	Food containers
	Ni, Cr	Food processing equipment (wear resistance required)
Decorative	Cu/Ni/Cr composite, Brass (Cu-Zn)	Household appliances, automotive trim
	Ag, Au, Rh	Jewelry
Dielectrics	Anodized oxide coatings of Al & Ti, Ta	Condensors, capacitors, coatings
Electroforms	Ni, Cu, Fe, (Cr), Co, composites	Radar "plumbing," screens, bellows. containers, molds
High temp. oxidation resistance Diffusion barrier	Cr, Rh, Pd, Pt, Au, Ni	Air and space craft Electronic devices
Maskant	Cu, Sn, bronze	Selective carburizing & nitriding
	Sn, Pb-Sn	Etch resists
	Au, Rh, Sn-Ni	
Reflectors	Ag, Rh, Cr	Visible light reflectors
	Au	Infra-red reflectors
Salvage	Cu, Ni, Cr, Fe	Mismachined, worn parts
Soldering, Bonding	Pb, Sn, Sn-Pb Cu, Ag, Au, Sn-Ni, Cd, Ni	Containers, printed circuit and other electronic assemblies, chassis
Wear resistance	Ni, Cr, E-Ni hard anodizing	Air & space craft, hydraulics
	Rh, Au, Au alloys	Electronic contacts

Table 10.3: Approximate Comparison of Hardness Scales, Substrates and Deposits

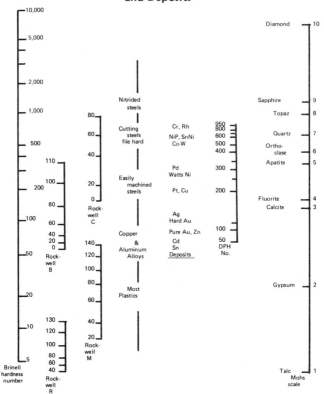

(modified and based on *Metal Progress,* p. 131, September 1959)

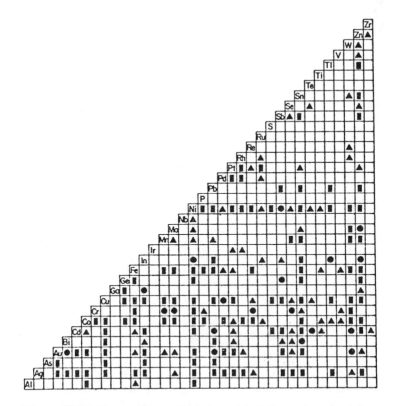

Figure 10.9: Binary alloys which have been electrodeposited from aqueous solution: ■ indicates alloys reported up to 1960, ● indicates alloys electrodeposited for the first time between 1961 and 1964, and ▲ indicates alloys reported since 1964. (Krohn and Bohn[37])

The most widely used plated alloys are:

Cu-Zn: brasses ranging from red brass to white brass, primarily decorative and for rubber bonding.

Cu-Sn: bronzes, decorative, antiquing and as corrosion resistant undercoats substituting for a copper strike.

Pb-Sn: compositions ranging from 5% Sn to 65% Sn. Applications include bearings, corrosion resistant coatings, solderable coatings and etch-resists in electronic assemblies.

Au-Co and Au-Ni: hardened gold alloy deposits used for electronic contacts and wearing surfaces.

Sn-Ni: for corrosion resistance and solderability and etch-resists.

Ni-Fe: as substitute for nickel plating (decorative), soft magnetics (permalloy).

Ni-P: deposited either electrolytically or (more prevalently) electrolessly for its hardness, wearability and corrosion resistance.

Co-Ni: for decorative plating, magnetic applications electroforming (molds for plastics).

Co-P: for hard magnetics, sometimes as ternary alloys containing Ni, Fe, Zn, W, Mo, etc.

The electrodeposition of tungsten alloys[38-41] of Fe, Ni and especially Co is commercially feasible but has remained largely experimental although their properties should be of sufficient interest to engineering applications. While the "as-deposited" hardness is lower than chromium or Ni-P, these alloys can be precipitation hardened. One drawback is the high optimum temperature (600°C) for the Co-W alloys, which can be detrimental to the substrate. The deposits retain hot hardness similar to the Stellites.

Ternary alloys of Fe-Cr-Ni have been produced almost as stainless steel coatings[42,44]; other studies[10,37,45] include reviews. Machu[46] investigated the problems with anodes—especially oxidation to higher valence states and the use of insoluble anodes, alone and in combination with soluble anodes. Other recent work with ternary alloys has been with gold alloys to increase hardness; Au-Ag-Sb alloys[47] reportedly showed wear resistances 25-33 times greater than pure Au.

Methods other than co-deposition have been developed to produce alloy coatings. These include diffusion, dispersion of particles in deposits (electrophoretic phenomena), and mechanical plating.

Diffusion Coatings: These processes involve the deposition of coatings sequentially—similar to the composite: Cu under Ni under Cr for decorative finishes—followed by a thermal diffusion treatment. Such techniques have been applied to improve the adhesion of deposits to difficult-to-plate substrates (diffusion bonding). They have not, however, been extensively applied to producing alloy coatings by deposition possibly due to temperature requirements and the formation of intermediate diffusion zones with undesirable properties (brittleness, etc.).

A proprietary alloy of Ni-Zn called "Corronizing"[48] was used commercially as an improved corrosion resistant coating. (Subsequently, co-deposited Ni-Zn alloys were developed.) The substitution of Cd for Zn by Moeller and Snell[49] produced a corrosion preventive coating for jet engine parts permitting the use of low alloy steels operating at temperatures up to 535°C (1000°F). The coating consisted of 5-10.2 μm (0.2-0.4 mils) Ni plus 2.54-5 μm (0.1-0.2 mil) Cd diffused at 332°C (630°F) [M.P. of Cd = 321°C (611°F)]. The satisfactory function of this diffused alloy coating is dependent on the quality and characteristics of the Ni component.[50]

Composite coatings of Co-W alloy and Cr diffused in air and in a carburizing atmosphere are shown in Figure 10.10 to illustrate the potential of producing unique alloy coatings by controlled heat treatments.

Dispersion Coatings: One of the common problems in electroplating is roughness of the deposit, the primary cause of which is the presence and suspension of discrete particles in the solution with subsequent entrapment in the deposit. Continuous or periodic filtration is part of the operation for many types of plating solutions to overcome this problem. Thus, it is not difficult to include foreign material into a deposit. However, the purposeful addition of a second, dispersed phase of controlled particle size into a plating solution, said particles to be incorporated into the deposit to change or extend the properties of the

Deposition from Aqueous Solutions: An Overview   409

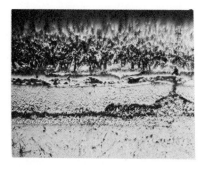

Figure 10.10: Diffused Co-W/Cr/Co-W composite coatings. (a) H.T. in air, 1680°F, 10 hrs (X 500) (unetched). (b) H.T. in carburizing atmosphere, 1680°F, 10 hrs. (X 500 unetched). (c) H.T. in carburizing atmosphere, 1680°F, 10 hrs (X 500) (etchant: hot Murakami.)

deposit, is referred to variously as: dispersion, inclusion, occlusion, composite or electrophoretic plating, deposition or coating.

The requirements are simple:

1. The particles must be insoluble (or only slightly soluble) in the solution.
2. The particles must be compatible with the solution, i.e., not produce any detrimental effects.
3. The particles must be dispersed either "naturally" (as colloidal size particles) or mechanically (stirring, agitation) in order to contact physically the surface being coated.
4. The particle size is usually in the colloidal range (~0.005 to 0.2 μm) or slightly larger, usually less than 0.5-1.0 μm although there are exceptions for certain applications.

The possibilities are numerous. Satin nickel deposits[51] were developed to reduce glare on automotive trim, also providing improved corrosion characteristics. Kilgore[52] describes various applications including: (a) non-galling Ni deposits containing 1000 mesh silicon carbide for pistons and cylinder walls on internal combustion engines, (b) inclusion of Cr in Ni deposits producing nichrome by subsequent heat-treatment, (c) 120 grit diamond dust in nickel to produce permanent abrasive grinding wheels. The hardness and wearability of Cd deposits from acid baths were improved by inclusion of corundum or boron carbide particles.[53]

An important, desired result of dispersion plating is the improved strength, hardness, creep and other properties of the deposit, including the retention of strength after thermal treatments. Sautter[54] reported increased yield strength from 8 kg/mm^2 (11,375 psi) for pure Ni deposits to 35 kg/mm^2 (50,000 psi) for dispersed alloys containing 3.5-6.0 volume percent (v/o) Al_2O_3; the particle size ranged from 0.01-0.04 μm to 0.3 μm and the plating parameters other than agitation had little or no effect. Electroformed lead and lead alloys were strengthened only by additions of TiO_2 (0.01-0.03 μm) although Al_2O_3, $BaSO_4$, Pb_3O_4 and W additions were also studied,[55] indicating the possibility of specificity with respect to the dispersoid. Greco and Baldauf[56] found 2-15% of Al_2O_3 to be the effective range for dispersion-hardening of Ni deposits from a sulfamate bath. The increase in hardness appeared to be linear to the square root of the volume fraction of dispersoid with Al_2O_3 showing a higher slope than TiO_2. The deposits contained 3 times (v/o) more TiO_2 than Al_2O_3 at the same solution concentrations and plating conditions; the particle size averaged 0.074 μm Al_2O_3 (0.013-0.339 μm range) and 0.2 μm TiO_2 (0.037-0.313 μm range).

Table 10.4 indicates the variations in mechanical properties of particle-dispersed nickel alloys due to the dispersoid material and the plating solution composition.

Electrophoretic Deposits: Electrophoresis is the term used to describe the migration, by virtue of the electric charge on their surfaces, of colloidal or near-colloidal particles in a suspending medium when a potential is supplied. This migration is analogous to ionic migration through a solution. The electrical double layer of charges discussed above is involved. The process has been applied to the deposition of a variety of materials including metal powders, oxides, cermets and other particles to metal substrates. Usually the particles ranging in size between 0.5 and 45 μm are suspended in a non-conducting (or poor con-

Deposition from Aqueous Solutions: An Overview   411

Table 10.4: Mechanical Property Data for Nickel-Particle Composites

Dispersed Material[a]	Plating Bath	Tensile Strength kg/sq mm	Tensile Strength psi	Yield Strength kg/sq mm	Yield Strength psi	Hardness, kg/sq mm	Modulus of Elasticity kg/sq mm	Modulus of Elasticity psi × 10⁶	Reference
Alumina, 1.0%	Sulfamate (50 C)	77[b]	110,000[b]	52.5[b]	75,000[b]	270 to 320[b]	—	—	1
Alumina, 1.7%	Watts (50 C)	45[e]	64,000[e]	—	—	—	—	—	2
Alumina, 2%	Watts (50 C)	21[d]	30,000[d]	—	—	514[e]	—	—	4
Alumina, 2.5%	Chloride (50 C)	—	—	35[f]	50,000[f]	350	—	—	3
Alumina, 4.5%	Watts (50 C)	68	97,000	45	64,000	380	22,000[g]	31[g]	5
Alumina, 6.0%	Watts (50 C)	75	107,000	56	80,000	400	22,000[g]	31[g]	5
Alumina, 9.4%	Watts (50 C)	70	100,000	45	64,000	—	—	—	5
Silicon carbide, 3.2%[h]	Sulfamate	174	249,000	—	—	—	—	—	6
Silicon carbide, 3.6%[h]	Sulfamate	202 to 229	288,000 to 327,000	—	—	—	31,500	44	6
Silicon carbide, 5.6%	Watts (50 C)	85	121,000	60	86,000	410	21,000[g]	30[g]	5
Silicon carbide, 9.6%	Watts (50 C)	74	105,000	55	79,000	450	21,000[g]	30[g]	5
Titanium dioxide, 5%	Sulfamate (50 C)	81[b]	115,000[b]	55[b]	78,000[b]	300 to 420[b]	—	—	1

(a) Weight percent.
(b) The effects of heat treating are summarized in Table 14.2.
(c) Tensile strength was 32 kg/sq mm (46,800 psi) after heat treating at 1100 C. The 100- to 500- angstrom particles formed 2500 angstrom clusters during heat treating.
(d) Tensile strength was 2.3 kg/sq mm (3,300 psi) at 1095 C. Elongation at 1095 C was 34%.
(e) Hardness decreased to 205 kg/sq mm at 260 C and 74 kg/sq mm at 425 C.
(f) Yield strength for alloy deposit after heating to 750 C. The yield strength of nickel containing no alumina, which was deposited in a similar solution, was only 8 kg/sq mm (11,400 psi) after a similar heat treatment.
(g) The modulus of the matrix nickel was 18,000 kg/sq mm (26,000,000 psi).
(h) Whiskers about 1 μm in diameter were incorporated in the deposit.

From *The Properties of Electrodeposited Metals* by W.H. Safranek, Elsevier Publishing Co., 1974.[80] Reprinted with permission ducting) medium and a high potential (50-1000 v) is applied to the electrodes. High rates of deposition are obtained and coating thicknesses can be varied by controlling voltage, electrode spacing, suspension concentration and time. The coating is air dried and baked to remove the solvent medium. The coating is nonadherent and must be processed further by compression and/or sintering or by subsequent electrodeposition to bond it to the substrate. Electrophoretic deposition has been applied to produce Ni, Ni-Cr, Ni-Cr-Fe coatings to base metals as well as inclusion of such dispersoids as molybdenum disulfide or silicon carbide.[57] Ortner[58] applied electrophoretic deposition of TaC-Fe-Ni coatings onto graphite, sintered at 2300°C (4170°F) for the protection of rocket nozzle inserts and oxidation resistant coatings for refractory alloys.

A recent development involves mechanical coating of soft metals (Cd, Zn), and alloys by peening powders or flakes onto a substrate with glass beads during a tumbling operation. "Alloys" of Cd-Sn deposited in this manner exceeded 2000 hrs. in salt fog corrosion tests. The process is proprietary to the 3M Company.

Alloy deposits, however produced, offer certain advantages over single metal deposits:

1. Increased corrosion resistance due to greater density and finer grain structure.
2. Combination of properties of the individual constituents.
3. New properties, unlike the individual constitutents.
4. "Tailor-made" properties by proper selection of the constituents.

The disadvantages include the greater control required, the difficulty of reproducing the alloy composition, the greater attention to the anode systems used and their effects on the solution constituents and complexes.

SELECTED SPECIAL PROCESSES

Electroless Deposition

Electroless plating processes differ from electroplating processes in that no external current source is required. Metal coatings are produced by chemical reduction with the necessary electrons supplied by a reducing agent (R.A.) present in the solution:

$$M^{n+} + ne^- \text{ (supplied by R.A.)} \xrightarrow[\text{surface}]{\text{catalytic}} M^0 \text{ (+ reaction products)} \quad (21)$$

The uniqueness of the process is that the reduction is catalyzed by certain metals immersed in the solution and proceeds in a controlled manner on the substrate's surface. The deposit itself continues to catalyze the reduction reaction so that the deposition process becomes self-sustaining or autocatalytic. These features permit the deposition of relatively thick deposits. Thus the process is differentiated from other types of chemical reduction: (a) simple immersion or displacement reactions in which deposition ceases when equilibrium between the coating and the solution is established (e.g. copper immersion on steel from copper sulfate solutions) and b) homogeneous reduction where deposition occurs over all surfaces in contact with the solution (e.g. silvering-mirroring).

To prevent spontaneous reduction (decomposition), other chemicals are present; these are generally organic complexing agents and buffering agents. Other additives provide special functions as in electroplating solutions: additional stabilizers, brighteners, stress relievers.

The reducing agents most widely used are:

>Sodium hypophosphite (for Ni, Co)
>
>Sodium borohydride (for Ni, Au)
>
>Dimethylamineborane (or other substituted amine boranes) (for Ni, Co, Au, Cu, Ag)
>
>Hydrazine (for Ni, Au, Pd)
>
>Formaldehyde (for Cu)

The process was reported by Brenner and Riddell[59] in 1946 for nickel and cobalt coatings and has enjoyed very active interest since, resulting in extension* to electroless plating of copper, gold, palladium, platinum, silver and a variety of alloys involving one or more of these metals. Comprehensive reviews[60-66] with extensive bibliographies cover the considerable technology, solution composition and operating conditions, and literature (including patent) which have accumulated. Representative solution formulations are given in Appendix B.

Nickel deposits produced with hypophosphite or the boron-containing reducing agents are alloys containing the elements: P or B. They are very fine polycrystalline supersaturated solid solutions or amorphous metastable alloys[67-69] with hardness ranging approximately 500-650 VPN and can be precipitation hardened, being converted to crystalline nickel and nickel phosphide (Ni_3P) or boride (Ni_3B). Maximum hardness ranging from 900-1100 VPN is obtained at 400°C (750°F) for 1 hour. The effects of heat treatment at various times and temperature on the hardness of electroless Ni-P have been extensively investigated.[67,70-72] Johnson and Ogburn[73] supplement more fully previous work, showing the influence of phosphorus contents and the specific heat treatments on the range of hardness obtained (Figure 10.11). Higgs[74] investigated the effects of heat treatments on the hardness and structure of the deposits reporting the presence of several Ni_xP_y compounds present other than the usually reported Ni_3P. Schwartz and Mallory[75] found differences in the increasing ferromagnetism of alloys from various solutions as a result of heat treatments.

The phosphorus content** of the deposit increases as the hypophosphite concentration increases and the pH decreases in the solution. The complexing agents in the solution influence deposition rate[76] (along with pH) and may also have an effect on the as-plated deposit; Mallory[77] related differences in salt fog corrosion tests to this factor. It appears then that the properties of the deposit may vary considerably depending on the phosphorus content which, in turn, is determined by the solution used and its operating pH. Graham, et al.[68] observed abrupt changes in structure, strength and ductility of deposits at a phosphorus content of about 7 w/o with both strength and ductility increasing with increasing phosphorus content. They also observed that the lamellar banded

---

*These are commercially available. Other electroless processes for iron, chromium, cadmium, tin have been reported but either not confirmed or commercially applied. Undoubtedly, new developments will continue to be reported.

**The boron content in Ni-B systems is generally similar.

structure was 10 times broader (5 μm ≡ 0.2 mils) in deposits from alkaline solutions than in acid solution deposits (0.5 μm ≡ 0.02 mils).

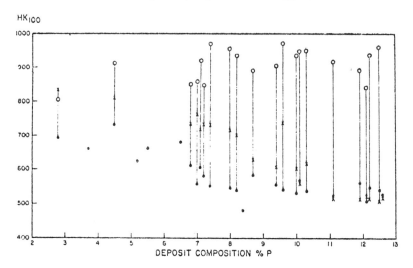

Figure 10.11: Hardness of Ni-P alloy: ● as-plated, x after 8 h at 200°C, o after ½ h at 400°C (Johnson and Ogburn[73])

Parker and Shah[78] determined that the stress in electroless Ni-P alloys varies from tensile to compressive as the phosphorus content of the deposit decreases. They also observed variations in stress depending on the thermal expansion coefficient of the substrate. However, increased thickness reduced the stress on most substrates. Baldwin and Such[79] indicated that zero stress can be obtained by adjustment of solution pH and that any desired value between 11.25 kg/mm^2 (16,000 psi) (tensile) and 5.6 kg/mm^2 (8,000 psi) (compressive) is achievable; maximum ductility was obtained with a 5.5 w/o P alloy from a solution at pH 5.6 ± 0.2. The least wear of hardened electroless Ni-P vs. quenched annealed steel was obtained with deposits containing 8-12 w/o P and the maximum and minimum values of average friction coefficient are 0.43 and 0.57, respectively, compared to 0.63-0.64 for pure nickel.[72]

Thus, it is evident that the compositions, structures and properties of electroless deposits can vary widely and are dependent on many factors. Safranek[80] reviewed those for electroless nickel and cobalt, and Okinaka[81] those for electroless gold. Saubestre[82] studied various reducing agents for electroless copper, concluding that formaldehyde was the most suitable. He also studied the effects of inhibitors or stabilizers to extend the useful life of the solution.[83]

The costs of the complexing and reducing agents used in electroless plating solutions make them non-competitive with electroplating processes. The application of electroless plating is usually based on one or more of the following advantages over electroplating:

> 1. Deposits are very uniform without excessive build-up on corners or projections or insufficient thickness in recessed areas. Internal surfaces are also evenly coated. The uniformity is limited only by the ability

of the solution to contact the surface and be replenished at the surface.

2. Deposits are usually less porous and more corrosion resistant than electroplated deposits (of equal thickness).
3. Almost any metallic or non-metallic, non-conducting surfaces, including polymers (plastics), ceramics, glasses can be plated. Those materials which are not catalytic (to the reaction) can be made catalytic by suitable sensitizing and nucleation treatments (see plating on plastics).
4. Electrical contacts are not required.
5. The deposits have unique chemical, mechanical, physical and magnetic properties.

The disadvantages compared to electroplating include:

1. Solution instability
2. More expensive
3. Slower deposition rates
4. Frequent replacement of tanks or liners
5. Greater and more frequent control for reproducible deposits.

Electroforming

Electroforming is defined[84] as the "production or reproduction of articles by electrodeposition upon a mandrel or mold that is subsequently separated from the deposit." (Occasionally the mandrel may remain in whole or in part as an integral functional part of the electroform.) The mandrels used are classified as permanent or expendable. The choice, composition, design considerations, preparation cycles, and methods of removal of mandrels are probably the most vital aspects of electroforming.[84-87] Various types of mandrels are given in Table 10.5.

Since the electrodeposits, called electroforms, are used as separate structures, they are usually substantially thicker than plated coatings. The fixturing or tooling of the mandrel and the anode positioning are quite critical. These determine the current distribution and resulting thicknesses of the deposit. A wide range of current densities produces changes in the structure, concentration of impurities and properties of the deposit which, in view of the function as an electroform, now are of paramount interest.

Braddock and Harris[90] reported increases in carbon content of nickel deposits from 0.004 w/o to 0.008 w/o at low C.D. ($21.5 \text{ A/m}^2 \equiv 2 \text{ A/ft}^2$) and sulfur contents from 0.0002 to 0.0014 when the C.D. was reduced from "normal" ($537 \text{ A/m}^2 \equiv 50 \text{ A/ft}^2$) to very low C.D. ($21.5 \text{ A/m}^2 \equiv 2 \text{ A/ft}^2$). Dini et al.[88,89] discussed the effects of variations in carbon and sulfur contents of nickel electroforms from sulfamate solutions; these are shown in Figures 10.12-10.14. Sulfur in nickel deposits causes embrittlement and cracking, limiting high temperature applications ($370°C \equiv 700°F$ max.).

Table 10.5: Comparison of Matrix Materials

Type	Material	Advantages	Disadvantages
Permanent	Carbon steel	Availability, low cost	Attacked by some plating solutions, such as acid copper and hot ferrous chloride.
	Carbon steel, chromium or silver plated	Improved hardness and/or corrosion resistance. Coating may be stripped and renewed.	Chromium coatings may be pitted by hot chloride type baths.
	Stainless steel	Inert to most plating solutions.	Costly. Soft surface of non-hardenable types is easily scratched.
	Inconel	Natural oxide film prevents adhesion of most deposits.	
	Invar, Kovar	Low temperature coefficient of expansion facilitates removal from electroform. Non-adherent.	Costly, poor machinability.
	Brass, Ni, Cr, Ag plated	Good machinability, low cost.	Surface easily scratched.
	Glass, Quartz	Close tolerance, high finish.	Costly, fragile and requires a conductive coating.
	Wood, plaster, plastic, etc.	Low cost, moldable. Flexible types can be withdrawn from undercuts.	Large tolerances. Requires a conductive coating and/or sealing.
Soluble	Aluminum	Good machinability. Good finish. Close tolerances can be held in complex non-withdrawable shapes. Soluble in sodium hydroxide.	Costly. Surface easily scratched.
	Zinc and zinc base alloys	Can be die-cast.	Acid stripping solution more likely to attack electroform than caustic solution used for dissolving aluminum.
	Plastics	Moldable. Low cost. Fairly close tolerances.	Cannot be used in hot plating baths. May swell in some baths. Requires conductive coating.
Fusible	Low melting alloys (Pb-Sn-Bi types)	Can be cast at low cost.	Difficult to remove from electroform completely.
	Waxes	Can be cast or molded at low cost.	Easily scratched. May deform by creep. Requires a conductive coating.

From Spencer.[96]

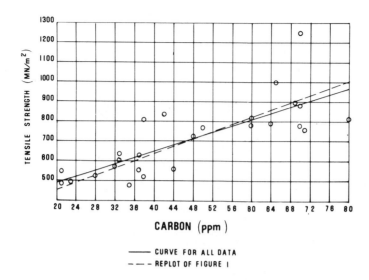

Figure 10.12: Influence of carbon on tensile strength: full curve, all data. (Dini and Johnson[88])

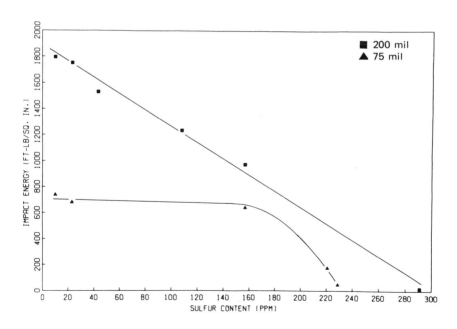

Figure 10.13: Influence of sulfur content on impact strength of electroformed nickel (Dini, Johnson and Saxton[89])

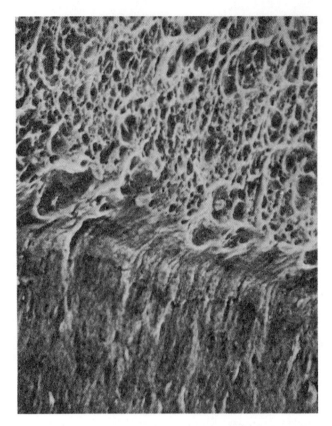

Ductile fracture
(~ 100 ppm sulfur)

Brittle fracture
(~ 200 ppm sulfur)

Figure 10.14: Fracture surface of part with sulfur content varying from 88 to 210 ppm (X 1000) (Dini, Johnson and Saxton[89])

Nickel, copper and iron are the most widely used electroforming deposits. Knowledge of the solution compositions, operating conditions and resulting structures and properties of deposits makes it possible to specify a given solution and the desired results. These are tabulated by DiBari.[85]

Electroforming is very costly and is a very slow method for producing parts. It finds application when:

1. Producing parts by mechanical or other means is unusually difficult or costly.
2. Extremely close dimensions and tolerances must be held, especially on internal dimensions or surfaces with irregular contours.
3. Very fine reproduction of surface details is required.
4. Thin walls are required.
5. Unusual physical, chemical and/or mechanical properties are required in the part.

Anodizing

Anodizing is an electrochemical process in which the part is made the anodic (positive) electrode in a suitable electrolyte. Sufficiently high voltage is deliberately applied to establish the desired polarization to deposit oxygen at the surface (O_2 overvoltage). The metal surfaces or ions react with the oxygen to produce adherent, oxide coatings, distinguishing the process from electrobrightening or electropolishing processes.

Industrial anodizing processes are confined mainly to aluminum and to a much lesser extent to magnesium and titanium alloys. Anodized tantalum is used in capacitors. Anodic coating applications include:

1. Protection – corrosion, wear and abrasion resistance.
2. Decorative – clear coatings on polished or brightened surfaces, dyed (color) coatings.
3. Base for subsequent paint or organic coating.
4. Base for plating on aluminum.
5. Special – based on some specific property or the coating, e.g., thermal barrier films, (refractory films), electrolytic condensers (dielectric films).

Anodizing of aluminum has been investigated intensively. Wernick and Pinner[94] definitively discuss the various processes and the nature and properties of the oxide coatings produced. Specification MIL-A-8625(C), used for both military and non-military applications, describes the most widely used processes and the expected requirements and tests for quality coatings. Three types of anodized coatings are called out:

Type I – from chromic acid solutions

Type II – from sulfuric acid solutions

Type III – from cold sulfuric acid processes (plus additives), producing thicker deposits (0.5-5.0 mils), primarily for wear and abrasion resistance. (Table 10.6 presents the most widely used processes in the U.S.A.)

Types I and II are usually sealed with a 5% (w/v) sodium dichromate solution (Class 1) or after absorption of a dye (Class 2) with a nickel (or cobalt) acetate solution. Typical processing cycles are illustrated in Figure 10.15.

The advantages and limitations of these three types of anodizing processes are analyzed in Appendix C.

Spooner[97] emphasized the importance of sealing methods, operating conditions, water quality and the detrimental effects of contaminants on the quality of the sealed coating, especially corrosion resistance. The suggested maximum levels for contaminants are:

Sulfate ($SO_4^=$)	250 ppm
Chloride (Cl^-)	100 ppm
Silicates ($SiO_3^=$)	10 ppm
Phosphates ($PO_4^\equiv$)	5 ppm
Fluorides (F^-)	5 ppm

Table 10.6: Aluminum Hard Anodizing Processes

Conditions	Alumilite	Martin	Hardas	Sanford
Solution composition	12 w/o sulfuric acid + 1 w/o oxalic acid	15 w/o sulfuric acid, saturated with CO_2	Sulfuric, oxalic acids	Sulfuric, organic acids
Temperature, °C	48-52	25-32	32	0-15
C.D., A/ft²	36	25-30	100-300	12-15
Voltage*	10-60 or higher	10-75	DC or DC/AC in various proportions	15-150
Film growth rate (mils)	1 mil/hr	1 mil/40 min.	1 mil/5-10 min.	1 mil/10-20 min.
Alloy limitations	4% Cu 7% Si 7-9% Cu+Si	5% Cu	—	—

*At a film thickness of approximately 2 mils, voltage requirement is approximately 40-45 volts.

Notes: Typical Properties of Hard Anodize Coatings[94]
Hardness: usually ranges between 350-450 DPH (35-55 Rc).
Abrasion resistance (Tabor): 30,000-40,000 cycles/μm.
Porosity: 5-15%
Heat resistance: to approximately 750F (400C)
Break-through voltage: 7-10 v/μm

From Sweet.[96]

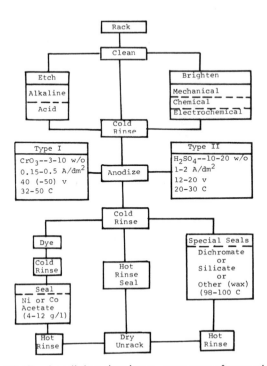

Figure 10.15: Anodizing aluminum—sequence of operations.

ASTM Specification B 580-73 designates seven types of anodizing:

Type	Description	Minimum Thickness μm	Mils
A	Hard coat	50	2.0
B	Architectural, Class I	17.5	0.7
C	Architectural, Class II	10	0.4
D	Automotive – exterior	7.5	0.3
E	Interior – Moderate abrasion	5.0	0.2
F	Interior – Limited abrasion	2.5	0.1
G	Chromic acid	1.2	0.05

The chemical composition of unsealed sulfuric acid anodized films is approximately:

> 80% aluminum oxide
> 18% aluminum sulfate
> 2% water*
> + traces of alloying elements

The coatings can probably be considered as approximating $2Al_2O_3 \cdot H_2O$ and after sealing convert to $Al_2O_3 \cdot H_2O$ with accompanying increased volume, providing enhanced corrosion resistance. Sealing reduces the hardness of the coating as much as 40%.

The oxide coating consists of two different structures: an inner (non-porous) barrier or dense structure and an outer, thicker, porous cell-like hexagonal structure.[98] The barrier layer is approximately 250 Å thick and constitutes about 1-2% of the total film. The pore diameter may range from 100 Å to 300 Å, depending on the electrolyte, operating temperature and voltage. The porosity of the coating is very high; a coating which exhibits 15% porosity contains approximately 400 x 10^9 pores/in^2.

Since the oxide film is a growth film—at the expense of the aluminum substrate (and not simply an add-on-film as in electrodeposition)—the dimensional changes depend on the equilibrium set up between film growth and the dissolving action of the electrolyte. For Type I and II films, it may be assumed that the dimensional increase per surface is about one-third the actual thickness of the film. For Type III, Hard Anodized coatings, the dimensional increase per surface is about one-half the actual oxide thickness. Thus, stripping and re-anodizing would require approximately twice the original film thickness to meet the same dimensional requirements. This could present serious problems in salvaging rejected parts.

The wear resistance of Hard Anodized coatings may vary significantly with coating thickness and alloy composition. George and Powers[99] proposed a more concentrated modified Alumilite (Alcoa's Hard Anodizing Process) solution which appeared to provide improved wear characteristics for difficult-to-coat alloys.

Some of the trends of the effects of operating conditions on the properties of the coatings are summarized in Table 10.7. The following observations are noted:

> 1. Recesses of parts receive lower current densities (at least initially) resulting in softer coatings.

---

*The water content may vary between 1-6%, being probably entrapped.

2. Conversely, projecting surfaces, especially sharp corners, receive higher current densities which produce harder coatings, resulting in cracking.
3. Cracking can occur at either concave or convex corners due to stresses.
4. Coatings which grow laterally as the dielectric film spreads are softer than coatings formed rapidly.
5. Properties of the coatings are influenced by the geometry of the parts as well as the alloying constituents or the electrolyte and its operating conditions.

Table 10.7: Effect of Operating Conditions on Anodic Film Characteristics

	Limiting Film Thickness	Hardness*	Corrosion Resistance	Porosity	Adhesion/ Dye Absorption
Temperature increased	↓	↓	→	↑	↑
Current density increased	↑	↑	→	↓	↓
Anodizing time increased	↓	↓	↑	↓	↑
Acid concentration increased	↓	↓	→	↑	↑
Use of less agressive electrolyte	↑	↑	→	↓	↓
Alloy homogeneity increased	↑	↑	↓	↓	↓

↑ = increases, ↓ = decreases, → = passes through a maximum

*Hardness of sealed coatings is approximately 60% of unsealed coatings. Sealing time also affects hardness, inversely; increased sealing time results in decreased hardness.

Note: Effects on hardness and dye absorption ability of coating are usually in opposite directions.

Voltage requirements increase for all above conditions.

From Wernick and Pinner.[100]

Additions of certain organic acids* to sulfuric acid anodizing solutions produce integral colored anodized coatings, ranging from light bronze or gold to black. These have been used in architectural applications.[101]

Recent developments[102] involve application of pulsed currents at the high voltages required. These claim superior coatings on more alloys in shorter times.

The anodizing of magnesium alloys has not found extensive use, possibly because it is somewhat more difficult than anodizing aluminum. Magnesium

---

*Sulfosalicylic acid, Kaiser Aluminum & Chemical Company, U.S. Patent 3,031,387 (April 24, 1962) and Sulfophthalic acid, Aluminum Company of America, U.S. Patent 3,277,639 (June 4, 1966).

oxide (MgO) is more water-soluble and considerably softer than aluminum oxide (Al_2O_3). The anodizing processes are similar and sealing is also required. The primary purpose is as a preparatory coating for painting or for corrosion and abrasion resistance. The older processes are referred to as Dow 12, Dow 14, and Manodyzing; these are AC or DC low voltage processes. The "newer" processes are fluoride-containing solutions and include Dow 17, Cr 22 and HAE (Hardcoat). These are high voltage (80—up to 320 V) processes. Cr 22 and HAE require AC. Solution formulations and operating conditions can be found in the referenced Handbooks (e.g.,[65,85]).

Plating on Plastics

Commercial plating on plastics became feasible with the development of electroless plating processes—especially the low temperature electroless nickel and copper processes. Large scale, high production automatic decorative (Cu/Ni/Cr) plating on plastics is increasing on automotive trim, houseware and other articles. The technology for manufacturing printed circuit boards (PCB) is another development of electroless plating. This discussion is limited to these developments.

The plastics most widely plated today for decorative applications are (in decreasing order and increasing difficulty): Acrylonitrile-butadiene-styrene (ABS), polyphenylene-butadiene-styrene (Noryl), polysulfones, polypropylenes, and nylons. It should be noted that these plastics are "filled," i.e., they contain mineral fillers, additives, modifiers, or are co-polymers or mixtures of co-polymers. ABS is a mixture of acrylonitrile-styrene and butadiene-styrene; polysulfones generally contain ABS; nylons are mineral-filled. In the etching step one or more of the components is selectively etched, providing a non-uniformly roughened surface for improved mechanical bonding of the deposits (with possible chemical bonding).

Preparatory steps[103-5] for decorative plating (may) include:

1. Deglazing—mechanically or chemically
2. Etch
3. Activation (1 or 2 steps)
4. Electroless Deposit (Cu or Ni)
5. Racking
6. Bright acid copper plate
7. Acid dip
8. Bright nickel plate or dual nickel plate
9. Chromium plate, microcracked preferred

(Note: Rinses are critical between various steps.)

The etch step is a critical one. Usually chromic acid, either supersaturated*, or mixed with sulfuric acid or with sulfuric-phosphoric acids[105] is employed. An interesting alternative etching technique involves the use of plasmas.[106]

The adhesion of the deposits is also related to the activation of the surface. The most widely used system is the stannous chloride ($SnCl_2$)/palladium chloride ($PdCl_2$) 1 or 2 step treatment based on the redox reaction:

---

*L. Kadison, U.S. Patent 3,668,130 (June 6, 1972), (assigned to Crown City Plating Company).

$$Sn^{2+} + Pd^{2+} \rightarrow Sn^{4+} + Pd^0 \qquad (22)$$

The 2-step activation involves first adsorption of $SnCl_2$ on the etched surface, followed by the reduction in a solution of $PdCl_2$. The 1-step or mixed catalyst systems includes both components and is considered either a complexed chloride of Sn and Pd or a colloidal mixture. There is considerable controversy regarding the nature of the system.[107]

Perrins[105] determined that the adhesion of electroless nickel and electroless copper (on polypropylene co-polymers) was dependent on the amount of palladium present. Low palladium gave high copper adhesion and low nickel adhesion. High palladium initially gave low adhesion to both which improved, peaking with 3-5 week ageing. Accelerated ageing at 70°C for 15 hours gave improvement of over 70% of control values. (Improved adhesion by heating is also found with other plastics.) An oxidation mechanism at the deposit/polymer interface is suggested as being responsible for increased adhesion.

Selective plating of plastics[108] can be accomplished by applying an organic stop-off which remains on the surface as a finish coat. Deposition is prevented on the stop-off film by use of a chromating treatment after etching.

Plating Printed Circuit Boards

The printed circuit board (PCB), also called printed wiring board, has made rapid advances since its development in the late 1930's. It is a pre-determined electrical conducting design or path on a non-conductive base whose primary function is to carry an electronic impulse or signal. The non-conductive base or board can be made of a wide variety of materials including wood, masonite, or resins such as epoxy, epoxy-glass, phenolics (flame-retardant or paper-reinforced), polybutadienes, polyimides, and ceramics. Presently, the most popular materials seem to be phenolics, epoxies (and glass), and polyimides.

The types of PC boards fabricated today include:

1. Print and etch*
2. Plate and etch
3. Plated-through-hole (PTH)
   a. Panel plate
   b. Pattern plate
4. Multi-layered (MLB)**
5. Additive circuits
6. Integrated circuits
7. Flexible circuits

The pre-plating preparation steps involve alkaline cleaning, acid etching as do other plating cycles. Additional steps such as abrasive cleaning or honing to remove smeared polymer in the drilled holes and "etch-back" (of polymer) to expose the intermediate layers of copper in MLB's (Figure 10.16) are required.

---

*Print and etch involves no plating. A (photo) resist is applied exposing unwanted copper (on a copper-clad board) which is etched away. Holes are drilled and eyelets inserted for connecting circuitry.

**This is similar to plated-through hole except 2 or more PCB's are bonded together using an epoxy/glass pre-preg. Inter-connections are made by drilling holes after laminating layers.

Deposition from Aqueous Solutions: An Overview 425

Also required for through-hole plating is the $SnCl_2/PdCl_2$ activation treatment discussed above. After activation, electroless copper is deposited over the exposed outer circuits and through the hole. This is followed by electrodeposited copper.

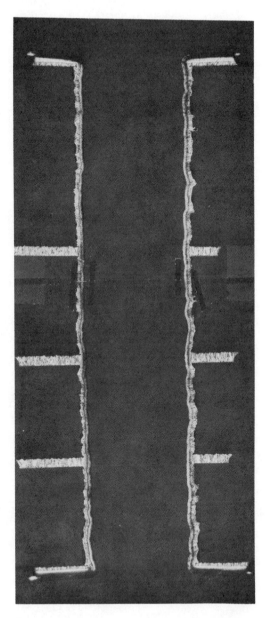

Figure 10.16: Through-hole tin-lead (solder) plate on multi-layer printed circuit board. (X50) (courtesy of B.F. Rothschild.)

Halva[109] and Smith[110] describe in detail fabrication and plating operations. A trouble-shooting chart[111] and manual[112] identify possible sources of trouble and their rectification or suggested cures.

Rothschild and Kilgore[113] discuss the problems of plate distribution (throwing power) in MLB's and relate T.P. to the ratio of surface to minimum hole thickness (S/H) and the ratio of total board thickness (hole length) to hole diameter (B/D). They also discuss fabrication and plating steps, the choice of deposits (Au, Sn-Pb, Sn-Ni), solderability and diffusion and/or migration problems.

Copper plating is used for through-hole plating. Acid sulfate and acid fluoborate plating solutions possessing high throwing power have been deveoped. These are low-metal ion, high acid concentration formulations (to promote desirable polarization at high current density surfaces) with grain refiners to eliminate columnar structures which may develop cleavage planes at corners, resulting in cracking. The pyrophosphate copper solution is the other type of solution employed. (Cyanide copper formulations damage the board due to the high alkalinity and cyanide content.) Which solution is the preferred plating solution is a moot question and invokes considerable controversy among the "practitioners of the art."

A high-throw Sn-Pb, solder plate, has also been developed for through-hole plating.[114]

Other electrodeposited coatings used on PCB's include: Sn-Pb, Sn-Ni, Sn, Ni, Au with various functions as etch resists and to provide solderability, corrosion resistance and wear resistance.

STRUCTURES AND PROPERTIES OF DEPOSITS

The structure and properties of a deposit are related to the deposition parameters and deposit thickness (Figure 10.7). Changes in these parameters may produce significant differences in a deposit, making generalizations difficult, if not misleading. Some investigators have omitted indicating important plating parameters or deposit thickness or testing conditions when reporting property measurements, making these data at least suspect. Further, extrapolation from a narrow set of conditions and data could also be misleading. Figures 10.12, 10.13 and 10.17 emphasize the influence of impurities and thickness on properties of nickel from sulfamate solutions. They also illustrate the importance of taking measurements in a thickness range related to the intended application of the deposit. There appears to be a certain degree of specificity, yet trends have been established that correlate structure with deposition parameters and properties with structure. A few examples and data for copper, nickel and chromium deposits are used in an attempt to illustrate these points.

Since approximately 1947-8, the American Electroplaters' Society (AES) has initiated and supported research programs at various institutions on structure and properties of electrodeposits.* These studies and other published data (about 1300 references) have been "compiled and systematized" into a single source book by Safranek.[80] Most of the data presented here are based on these sources.

The structures of electrodeposits are classified as:

Columnar	Fine-grained (usually equiaxed)
Fibrous	Banded (or striated or lamellar)

*The current AES Research Project is Project #38, "Property-Structure Relationships," at Stevens Institute.

Deposition from Aqueous Solutions: An Overview

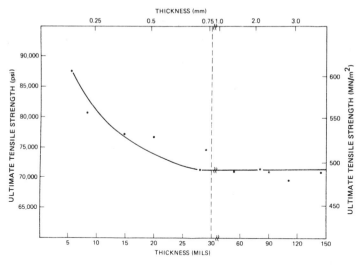

Influence of Thickness on Tensile Strength For Flat Electrodeposited Nickel Specimens

INFLUENCE OF THICKNESS OF SULFAMATE NICKEL DEPOSITS ON TENSILE AND DUCTILITY PROPERTIES

Thickness		Yield Strength[c,d]		Ultimate Tensile Strength		Elongation	RA[g]
(mm)	(mils)	(MN/m^2)	(psi)	(MN/m^2)	(psi)	(%)	(%)
0.14[a]	5.5	302	43 700	588	87 400	-	-
0.21[a]	8.3	265	38 400	556	80 700	-	-
0.38[a]	15	-	-	532	77 100	10.9[e]	90.8
0.51[a]	20	-	-	530	76 800	12.1[e]	93.9
0.74[a]	29	274	39 600	515	74 700	12.6[e]	94.4
1.4[a]	54	267	38 700	488	70 900	-	-
2.3[a]	91	229	33 200	490	71 000	31.8[e]	90.8
3.7[a]	144	270	37 600	489	70 900	29.2[e]	88.4
3.2[b]	125	274	39 600	524	75 900	22.5[f]	89.7

[a] Flat tensile specimens 242 mm (9.5 in.) long with a reduced section of 64 mm (2.5 in.).

[b] Round tensile bars with a reduced section 19 mm (0.75 in.) long and 3.2 mm (0.125 in.) in diameter, machined from a plate 6.4 mm (0.250 in.) thick.

[c] All values for yield, tensile strength, elongation, and RA are the average of at least two specimens. Strain rate was 1.1×10^{-3} s^{-1}.

[d] Determined by 0.2 percent offset.

[e] In 44.5 mm (1.75 in.).

[f] In 12.7 mm (0.5 in.).

[g] Determined by measuring the reduced area with the aid of a microscope at 10X or greater magnification.

Figure 10.17: Influence of thickness on mechanical properties of deposits. (Johnson, Dini and Stoltz[115])

Columnar structures are characteristic of deposits from solutions (especially acid solutions) containing no additives, high metal ion concentration solutions with high deposition rates or from low metal ion concentration solutions at low deposition rates. They usually exhibit lower tensile strength, percent elongation and hardness than other structures; they are generally more ductile. Such deposits are usually of highest purity (high density) and low electrical resistivity.

Fibrous structures represent a grain refinement of columnar structure. Stress relieving additives (such as saccharin or coumarin) promote such refinement as do high deposition rates. These may be considered intermediate in properties between columnar and fine-grained structures.

Fine-grained deposits are usually obtained from complex-ion solutions (such as cyanide) or with certain addition agents. These deposits are less pure, less dense and exhibit higher electrical resistivities due to presence of foreign material.

Banded structures are characteristic of bright deposits (as a result of brightening addition agents—usually S-containing organic compounds which result in small amounts of S and C in the deposit) and some alloy deposits. The use of plating current modifications (P.R., IC, pulse) favors the conversion of "normal" structure from a solution to a banded structure. These deposits generally possess higher tensile strength, hardness and internal stress and decreased ductility than the other structures.

Grain structure and size can be correlated to properties. Grain size can vary widely—from 100-50,000 Å; the grain size of fine-grained or banded deposits is usually between 100 and 1000 Å. Read[116] observed that frequently the grain size is much larger than indicated by etched specimens—the metallographic procedures usually employed—and that X-ray techniques are reliable, especially for large grain sizes.

Some metals (notably Cu, Ni, Co and Au) can be deposited in all four types of grain structure, depending on the solution composition and plating conditions. This is shown in Figure 10.18 for copper deposits. Typical properties of these deposits are given in Table 10.8.

Zentner, Brenner and Jennings[117] (AES Res. Project No. 9) studied the property-structure relationships of nickel plating to the operating variables. The effect of temperature, current density, pH and chloride content on deposit structure are shown in Figures 10.19, 10.20, 10.21 and 10.22. The trends appear to be:

1. Grain structure changed from fine-grain to coarse grain as temperature increased.

2. Significant structural changes occurred at low and high current densities. Typical columnar structure is obtained between 2 and 25 A/dm^2 (20-250 A/ft^2) in Watts-type solutions. The structural changes at low C.D. may be explained by the increased sulfur and carbon contents of the deposit as shown in Table 10.10. Thus, low C.D. produced a banded structure similar to bright nickel deposits.

3. There is essentially no structural change in Watts-type deposits in the pH range 1-5. At pH's above 5 there is a distinct change from columnar to fibrous or fine-grained which is probably due to inclusion of basic material [Ni(OH)$_2$?].

4. Deposits from Watts solutions produced the coarsest, columnar deposits. Increasing the chloride content of the solution results in finer-grained deposits. All-sulfate (no chloride) solution showed a somewhat finer columnar structure than a Watts deposit, with some evidence of a banded structure.

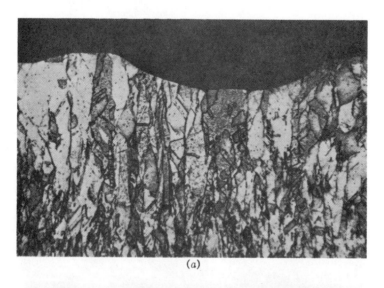

(a)

(b)

Figure 10.18: Structure of copper deposits. (X500). (etchant: ferric chloride). (From Modern Electroplating, 3rd Edition by F.A. Lowenheim, Ed, John Wiley and Sons 1973.[10] Reprinted with permission) Structures are typical for: a) Acid sulfate (no A.A); b) Acid sulfate with A.A. (gelatin + phenosulfonic acid); c) Acid sulfate with brighteners or Pyrophosphate solution; d) Cyanide solution with PR.

430   Deposition Technologies for Films and Coatings

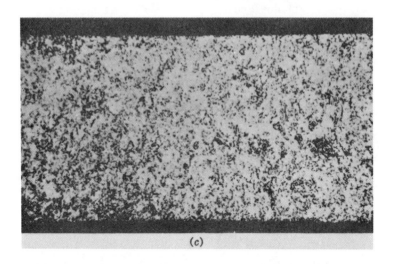

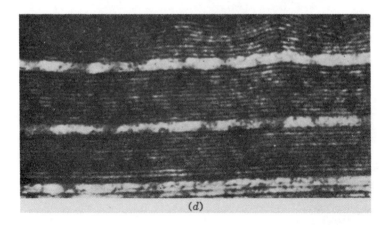

Figure 10.18: (continued)

A good correlation was found to exist between structure and properties as shown in Figure 10.23. Typical values of the mechanical properties of nickel deposited from various engineering electroplating solutions are given in Table 10.9.

Table 10.8: Comparison of Structure and Properties of Copper Deposited at 4 amp/sq dm in Several Different Copper Solutions

Plating Solution	Structure	Modulus of Elasticity kg/sq mm	Modulus of Elasticity 10⁶ psi	Tensile Strength kg/sq mm	Tensile Strength kpsi	Yield Strength kg/sq mm	Yield Strength kpsi	Elongation, percent (2 in.)	Density, g/cu cm	Electrical Resistivity, microhm-cm	Hardness, KHN$_{100}$
Amine[a]	Fine-grained	—	—	42	60	31	44	4	8.906	2.10	169[b]
Cyanide Low Cu conc. and low temperature[c]	Columnar	11,000	16	31	44	—	—	7	8.900	1.80	144
Low Cu conc. and High temperature[d]	Fine-grained	11,000	16	55	78	—	—	6	8.912	1.79	131
High Cu conc. and high temperature[e]	Ditto	—	—	26	37	—	—	22	8.919	1.80	114
Ditto with 2 g/l KSCN[f]	"	12,000	17	60	86	40	57	10	8.917	2.02	144
Fluoborate[g]	Fibrous	8,500	12	26	37	11	16	31	8.926	1.73	56
Pyrophosphate[h]	Fine-grained	12,000	17	28	40	15	22	33	8.926	1.74	92
Sulfate[i]	Coarse, columnar	10,000	14	21	30	6	9	24	8.922	1.72	53
Sulfate with 10⁻⁴ g/l Se[j]	Fibrous	11,000	16	35	50	20	28	20	8.928	1.75	108
Sulfate with 3.5 g/l triisopropanol-amine[k]	Fine-grained	10,500	15	50	71	30	43	7	8.913	1.89	144

[a] Solution at 55 C containing 100 g/l CuSO$_4$·5H$_2$O, 20 g/l (NH$_4$)$_2$SO$_4$, 4 ml/l NH$_4$OH and 80 ml/l ethylene diamine.
[b] Hardness of copper deposited in a similar bath at a slightly lower temperature was 140 to 150 kg/sq mm. (70)
[c] Solution at 40 C containing 40 g/l CuCN, 6 g/l free NaCN and 30 g/l Na$_2$CO$_3$ with a pH of 11.7 to 12.8.
[d] Solution at 80 C containing 40 g/l CuCN, 10 g/l free NaCN and 30 g/l Na$_2$CO$_3$ with a pH 11.8 to 12.8.
[e] Solution at 80 C containing 75 g/l CuCN, 10 g/l free KCN and 40 g/l KOH with a pH of 13.6.
[f] Copper deposited at 6 amp/sq dm.
[g] Solution at 30 C containing 177 g/l Cu(BF$_4$)$_2$, 12 g/l HBF$_4$, and 12 g/l H$_3$BO$_3$.
[h] Solution a] 50 C containing 90 g/l Cu$_2$P$_2$O$_7$·3H$_2$O, 350 g/l K$_4$P$_2$O$_7$, 80 g/l K$_2$P$_2$O$_7$, 15 ml/l KNO$_3$ and 2 ml/l NH$_3$ with a pH of 8.5.
[i] Solution at 30 C containing 187 g/l CuSO$_4$·5H$_2$O, and 39 g/l H$_2$SO$_4$. In comparison with data for these deposits, commercial electrolytic copper (produced with addition agents) exhibited a tensile strength of 28 to 34 kg/sq mm (40,000 to 48,000 psi). (71)
[j] Solution at 30 C containing 187 g/l CuSO$_4$·5H$_2$O, 74 g/l H$_2$SO$_4$ and 10⁻³ g/l Se (2 amp/sq dm).
[k] Solution at 30 C containing 187 g/l CuSO$_4$·5H$_2$O, 74 g/l H$_2$SO$_4$, and 3.5 g/l triisopropanolamine (5 amp/sq dm).

From *The Properties of Electrodeposited Metals* by W.H. Safranek, Elsevier Publishing Co., 1974.[80] Reprinted with permission.

Table 10.9: Nickel Baths for Heavy Plating

Type	Ingredients[a]	Concentration, g/l	pH (electrometric)	Temperature, °C	Cathode Current Density,[b] A/dm²	Hardness, Vickers	Tensile Strength, kg/cm²	Tensile Strength, psi	Elongation, %	Residual Stress, kg/cm²	Residual Stress, psi
Watts	Nickel sulfate, NiSO₄·6H₂O Nickel chloride, NiCl₂·6H₂O Boric acid, H₃BO₃	330 45 38	1.5–4.5	45–65	2.5–10	140–160	3850	55,000	30	1260	18,000
Hard	Nickel sulfate, NiSO₄·6H₂O Ammonium chloride, NH₄Cl Boric acid, H₃BO₃	180 25 30	5.6–5.9	43–60	2–10	350–500	10,500	150,000	5–8	3080	44,000
Chloride	Nickel chloride, NiCl₂·6H₂O Boric acid, H₃BO₃	300 38	2.0	50–70	2.5–10	230–260	7000	100,000	20	2800–3500	40,000–50,000
Chloride-Sulfate	Nickel sulfate, NiSO₄·6H₂O Nickel chloride, NiCl₂·6H₂O Boric acid, H₃BO₃	200 175 40	1.5–2.0	45	2.5–10						
Chloride-Acetate	Nickel chloride, NiCl₂·6H₂O Nickel acetate, Ni(C₂H₃O₂)₂·4H₂O	135 105	4.5–4.9	30–50	2–10	350	14,000	200,000	10		
Nickel-Cobalt	Nickel sulfate, NiSO₄·6H₂O Nickel chloride, NiCl₂·6H₂O Boric acid, H₃BO₃ Ammonium sulfate, (NH₄)₂SO₄ Nickel formate, Ni(CHO₂)₂·2H₂O Cobalt sulfate, CoSO₄·7H₂O	240 22.5 30 1.5 15 2.6	4.7	40	5	450–500				1050–1190	15,000–17,000
Fluoborate	Nickel (as fluoborate) Free fluoboric acid, HBF₄ Free boric acid, H₃BO₃	75 3.7–37.5 (colorimetric) 30	2.0–3.5	40–80	4–10	183	5250	75,000	15–30	1120	16,000
Sulfamate	Nickel sulfamate, Ni(NH₂SO₃)₂ Boric acid, H₃BO₃	450 30	3.0–5.0	40–60	2–30	250–350	6300	90,000	20–30	35	500
Sulfamate-Chloride	Nickel sulfamate, Ni(NH₂SO₃)₂ Nickel chloride, NiCl₂·6H₂O Boric acid, H₃BO₃	300 6 30	3.5–4.2	28–60	2–25	190	7560	108,000	15–20	105	1500

[a] An antipitting agent is normally used in these baths. [b] Higher current densities can be used with increasing rate of agitation.

From *Modern Electroplating*, 3rd Edition, by F.A. Lowenheim, Ed., John Wiley and Sons 1973.[10] Reprinted with permission.

Deposition from Aqueous Solutions: An Overview

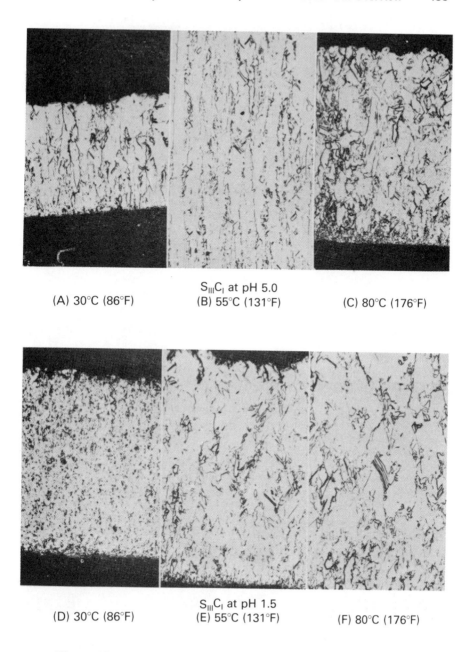

$S_{III}C_I$ at pH 5.0
(A) 30°C (86°F)   (B) 55°C (131°F)   (C) 80°C (176°F)

$S_{III}C_I$ at pH 1.5
(D) 30°C (86°F)   (E) 55°C (131°F)   (F) 80°C (176°F)

Figure 10.19: Effect of temperature of the plating bath on the structure of nickel deposited at 5 amp/dm² (46 amp/ft²). Cross section × 250. Etchant: glacial acetic and nitric acid. (Zentner, Brenner and Jennings[117])

434 Deposition Technologies for Films and Coatings

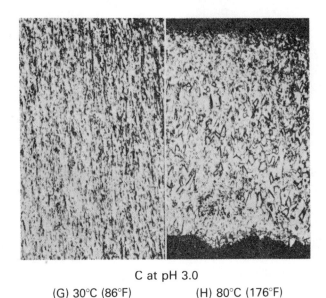

C at pH 3.0
(G) 30°C (86°F)   (H) 80°C (176°F)

Figure 10.19: (continued)

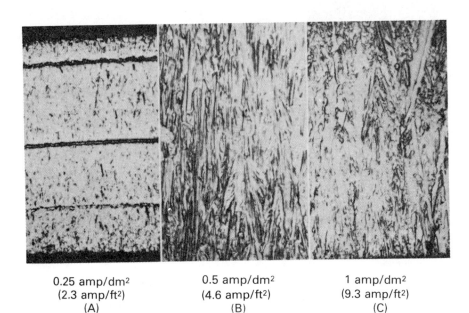

0.25 amp/dm²	0.5 amp/dm²	1 amp/dm²
(2.3 amp/ft²)	(4.6 amp/ft²)	(9.3 amp/ft²)
(A)	(B)	(C)

Figure 10.20: Effect of current density on the structure of nickel deposited from the $S_{III}C_I$ bath at 55°C (131°F), and a pH of 3.0. Cross section x 250. Etchant: glacial acetic and nitric acid. (Zentner, Brenner and Jennings[117])

2 amp/dm² (19 amp/ft²) (D)

5 amp/dm² (46 amp/ft²) (E)

10 amp/dm² (93 amp/ft²) (F)

20 amp/dm² (186 amp/ft²) (G)

25 amp/dm² (230 amp/ft²) (H)

50 amp/dm² (460 amp/ft²) (I)

Figure 10.20: (continued)

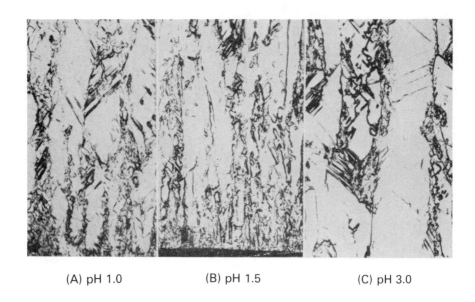

(A) pH 1.0 (B) pH 1.5 (C) pH 3.0

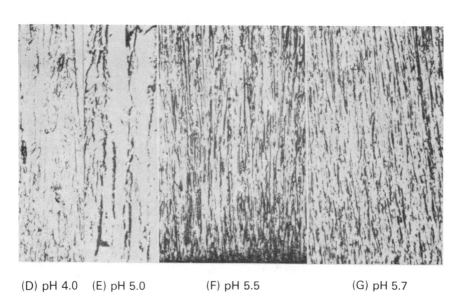

(D) pH 4.0 (E) pH 5.0 (F) pH 5.5 (G) pH 5.7

Figure 10.21: Effect of the pH of the plating bath on the structure of nickel deposited from the $S_{111}C_1$ bath at 5 amp/dm^2 (46 amp/ft^2) and 55°C (131°F). Cross section x 250. Etchant: glacial acetic and nitric acid. (Zentner, Brenner and Jennings[117])

Deposition from Aqueous Solutions: An Overview   437

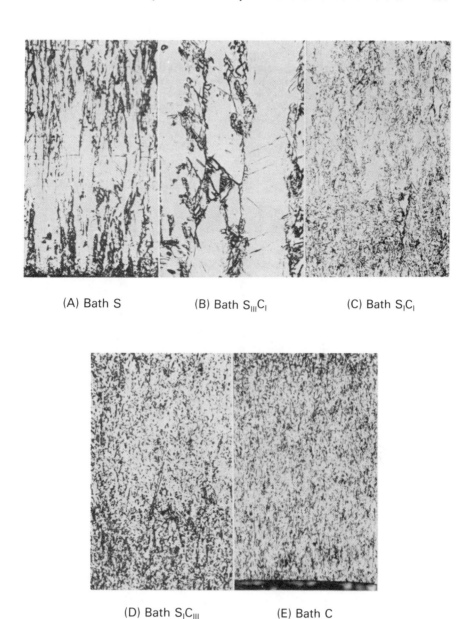

(A) Bath S   (B) Bath S$_{III}$C$_I$ (C) Bath S$_I$C$_I$

(D) Bath S$_I$C$_{III}$ (E) Bath C

Figure 10.22: Effect of increasing chloride content of the bath on the structure of nickel deposited at 55°C (131°F), 5 amp/dm^2 (46 amp/ft^2), and a pH of 3.0. Cross section x 250. Etchant: glacial acetic acid and nitric acid. (Zentner, Brenner and Jennings[117])

Table 10.10: Results of Elemental Analyses of Nickel Electrodeposits

Type of Nickel	Carbon	Copper	Iron	Sulphur
		(Weight Percent)		
As deposited nickel produced at 21.5 A/m²	0.008	0.046	0.04	0.0014
Nickel produced at 21.5 A/m² and annealed in mass spectrometer experiments	0.007	0.054	0.02	0.0012
As-deposited nickel produced at 537 A/m²	0.004	0.058	0.03	0.0002
Nickel produced at 537 A/m² and annealed in mass spectrometer experiments	0.003	–	0.03	0.0004
Nickel anode material used for both low and high C.D. deposits	0.009	–	–	–

From Braddock and Harris.[90]

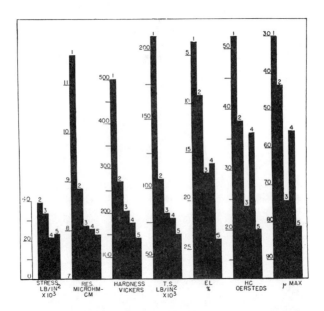

Figure 10.23: Range and trend of physical properties of nickel deposited from 5 different types of baths, each point is the average of the properties of 5 or more deposits obtained under various conditions of plating. 1 equals a, b, c, d, e, f, OB, and OBT baths. Bright nickel. 2 equals S_IC_{III}, C, Ac, C (-4N) baths. Chloride nickel. 3 equals S_IC_I bath. 4 equals S and oS baths. 5 equals $S_{III}C_I$, $oS_{III}C_I$, $S_{III}C_I$(-1N), NH_4, Na, and F baths. Watts nickel. (Zentner, Brenner and Jennings[117])

Properties of chromium deposited under a wide variety of plating conditions and solution compositions were extensively covered by Brenner, Burkhead and Jennings.[118] The deposits—especially the bright deposits—are very fine-grained, as small as 10 Å on the basis of X-ray data. They concluded that the oxide content of the deposit had far greater influence on the properties than crystal orientation or structure. Increased plating temperature from 10°C to 100°C caused reduction of oxygen content from ~1 w/o to ~ 0.1 w/o. The hydrogen content of the deposit also decreases with increasing plating temperatures.

The hardness of chromium is probably its most important engineering property. The oxygen content of the deposit is one of the most important factors affecting its hardness. Above 0.12 w/o O_2, the hardness ranges between 850-1000 KHN (Knoop Hardness Number) and when below 0.12 w/o O_2, the hardness ranges from 625 to 325 KHN. However, it was noted that hardness values may fluctuate as much as 200 points KHN for the same oxygen content. It also appears that bright deposits are hardest. The hardness of chromium deposits, therefore, is probably the result of oxide inclusion, small grain size and internal stress.

The hardness of the substrate along with that of the deposit is an important factor in the application for improved wear resistance of various tools (Table 10.11). In other wear applications the coefficient of friction is a factor; Table 10.12 gives values for various combinations.

Table 10.11: Recommended Basis Metal Hardness and Chromium-Plate Thickness for Various Applications

Application	Recommended Rockwell C Hardness of Basis Metal	Thickness of Chromium, μm
Drills	62–64	1.3–13
Reamers	62–64	2.5–13
Burnishing bars	60–62	13–75
Drawing plugs or mandrels	60–62	38–205
Drawing dies	62 inside, 45 outside	13–205
Plastic molds	55–60	5–50
Gages	48–58	2.5–38
Pump shafts	55–62	13–75
Rolls and drums	–	6–305
Hydraulic rams	–	13–100
Printing plates (engraved steel)	–	5–13

From *Modern Electroplating*, 3rd Edition by F.A. Lowenheim, Ed., John Wiley and Sons 1973.[10] Reprinted with permission.

Table 10.12: Coefficient of Friction for Various Metal Combinations

Metal	Static Coefficient	Sliding Coefficient
Chromium-plated steel on chromium-plated steel	0.14	0.12
Chromium-plated steel on babbitt	0.15	0.13
Chromium-plated steel on steel	0.17	0.16
Steel on babbit	0.25	0.20
Babbitt on babbitt	0.54	0.19
Steel on steel	0.30	0.20
Bright chromium plate on cast iron		0.06
Bright chromium plate on bronze		0.05
Bright chromium plate on babbitt		0.08
Hardened steel on cast iron		0.22
Hardened steel on bronze		0.11
Hardened steel on babbitt		0.19

From *Modern Electroplating*, 3rd Edition by F.A. Lowenheim, Ed., John Wiley and Sons 1973.[10] Reprinted with permission.

Different etching techniques reveal interesting structural characteristics in chromium deposits.[119] In fact, no single etchant reveals all possible features and it is advisable to use several techniques. Structures which have been observed include: fibrous texture, banded or striations associated with the crack pattern (and not found in crack-free deposits), bands delineating changes in plating variables (C.D. and temperature) during deposition.

The internal stress, negative coefficient of thermal expansion (initial shrinkage) and the effect on fatigue strength of the substrate are properties (besides hardness) of interest in engineering applications. These are adequately covered in references already cited.[10,80,118] The reported stress values for chromium deposits cover a very broad range, from highly tensile to compressive in microcracked deposits (>1000 cracks/linear inch). It is influenced by the solution composition and concentration, C.D., temperature, deposit thickness and probably other factors. The high tensile stress and resulting cracking lower the fatigue limits* of substrates (primarily steel, but possibly also aluminum and titanium).

No ductility was found for chromium deposits from aqueous solutions.

In general, the physical properties of electrodeposits approach those of metallurgical wrought metals as the purity increases. Observations regarding the physical properties are:

1. The density is related to pores, voids and impurities in the deposit. Corrosion and high temperature characteristics can be significantly affected by low density.

2. The coefficient of thermal expansion is also affected by impurities in the deposit. Thermal properties are not too well established for electrodeposits. Most deposits expand with thermal cycling, notable exceptions being chromium and cobalt-tungsten alloys. Deposits which expand appreciably develop voids on thermal cycling and could not be considered for high temperature service since they would exhibit decreased corrosion and oxidation resistance.

3. Electrical resistivity is quite sensitive to the presence of small concentrations of impurities. Most deposits, therefore, exhibit higher values than wrought counterparts. Impurities such as oxides, sulfides, hydrates or inclusions tend to concentrate at grain boundaries—especially after a thermal treatment or annealing.

With respect to mechanical properties, the relationship of hardness to strength is not always similar to wrought metals where a constant relationship exists. Although the generalization that the strength of a deposit increases with hardness and ductility varies inversely with strength and hardness holds in many cases, the exceptions are too numerous to make it reliable. Other observations regarding mechanical properties of deposits are:

1. Hardness (or microhardness) of the deposit is the most widely measured property (probably due to the ease of measurement). It may also be the most abused. The literature is replete with inconsistencies and contradictions. This

---

*The higher the stress (in tension) of the deposit, the greater the reduction in fatigue strength.

may be due, in part, to methods of measurements, differences in deposit thickness, plating solution differences, quality of deposit, inadequacy in reporting data (neglect to indicate load applied*), type and condition of substrate, converting from one scale to another, and other factors.

To obtain reasonably reliable microhardness measurements:

1. The deposit thickness should be at least 10X the depth of the indent. For the same load, the depth of a Knoop indent is approximately $\frac{1}{7}$ that of a Vicker's indent.
2. The distance of the indent (to the substrate interface) should be at least ½ the diagonal of the indent (the short diagonal for the Knoop indent) to minimize the "anvil effect."
3. When taking multiple measurements on the same specimen, use a transverse track with the distances between indents as in (2).

Vickers microhardness measurements are less sensitive to errors arising from elastic properties than are Knoop measurements and result in less serious errors as loads are increased.

It appears that too much value is sometimes placed on hardness measurements. The assumed relationship between hardness and strength was discussed above. The same may be said to some degree for the correlation of hardness to wear resistance. The excellent wear resistant characteristics of chromium deposits are related to the low coefficient of friction (Table 10.12) as much as to hardness. The wear resistant characteristics of electroless nickel alloys is related to the presence of phosphorus (or boron) as well as to the hardness.

Despite these comments, hardness measurements are useful in evaluating deposits and predicting their usefulness. They are especially useful in evaluating alloy deposits since changes in hardness reflect (possibly) changes in structure or composition of the alloy deposit.

It is not unexpected that the hardness values of deposits (of the same metal) vary greatly (Figure 10.24). Noteworthy are the great ranges reported for chromium and iron deposits and the ability of some alloy deposits to undergo precipitation hardening.

2. In many instances, the tensile strengths of deposits exceed those of annealed metallurgical counterparts (Table 10.13). The primary reason is the finer-grain structure of electrodeposits. Coarser grained or columnar structures may exhibit lower strengths.

3. The ductility of electrodeposits may equal metallurgical counterparts but is usually lower in the as-plated condition.

4. The tensile strength and ductility and internal stress of the deposit are interrelated in determining the degree of resistance to stress corrosion cracking when deformation may be involved or anticipated.

5. Internal or residual stress in electrodeposits may be either tensile (usually) or compressive.

The mechanism by which stress arises is not completely understood, but undoubtedly a distorted lattice structure is involved. If the atoms are deposited closer together than in a normal lattice spacing, the tendency is for the atoms to

---

*Hardness values should be reported with designated loads, e.g., VHN_{100}, or KHN_{25}, where 100 and 25 (as subscripts) represent the load in grams. Loads less than 25 grams are subject to serious errors and are undesirable due to poor reproducibility.

push further apart, pulling on the substrate and resulting in tensile stresses. Conversely, if the deposited atoms are farther apart than they should be (in a normal lattice spacing), they tend to pull closer together, resulting in a compressive stress on the substrate.

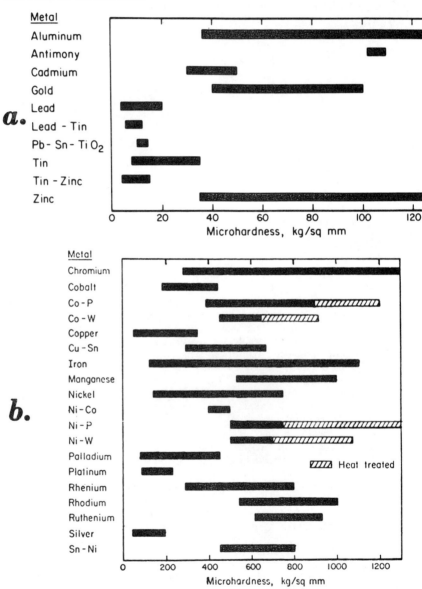

Figure 10.24: Microhardness Ranges a) Microhardness ranges for relatively soft electrodeposited metals and alloys b) Microhardness ranges for relatively hard electrodeposited metals and alloys (From *The Properties of Electrodeposited Metals* by W.H. Safranek, Elsevier Publishing Co., 1974.[80] Reprinted with permission)

Table 10.13: Strength and Ductility Data for Electrodeposited Metals

Metal	Plating Bath	Minimum Tensile Strength kg/sq mm	Minimum Tensile Strength psi	Maximum Tensile Strength kg/sq mm	Maximum Tensile Strength psi	Elongation, percent	Wrought Metal[a] Tensile Strength kg/sq mm	Wrought Metal[a] Tensile Strength psi	Wrought Metal[a] Elongation, percent
Aluminum	Anhydrous chloride-hydride-ether	7.7	11,000	21.7	31,000	2 to 26	9.15	13,000	35
Cadmium	Cyanide	—	—	7.0	10,000	—	7.2	10,300	50
Chromium	Chromic acid	10	14,000	56	80,000	<0.1	8.4	12,000	0
Cobalt	Sulfate-chloride	54	76,500	121	172,000	<1	26	37,000	—
Copper	Cyanide, fluoborate, or sulfate	18	25,000	66	93,000	3 to 35	35	50,000	45
Gold	Cyanide and cyanide citrate	12.7	18,000	21.2	30,000	22 to 45	13.5	19,000	45
Iron	Chloride, sulfate, or sulfamate	33	47,000	109	155,000	2 to 50	29	41,000	47
Lead	Fluoborate	1.4	2,000	1.6	2,250	50 to 53	1.85–2.1	2,650–3,000	42 to 50
Lead-TiO$_2$ composites	Fluoborate	3.2	4,500	4.1	5,800	7 to 15	3.2[b]	4,500[b]	18[b]
Nickel[c]	Watts, and other type baths	35	50,000	107	152,000	5 to 35	32.2	46,000	30
Silver	Cyanide	24	34,000	34	48,000	12 to 19	16–19	23,000–27,000	50 to 55
Zinc	Sulfate	4.9	7,000	11.2	16,000	1 to 51	9.1	13,000	32

(a) Annealed, worked metal.
(b) Powder compact prepared with 1.5% PbO particles.
(c) Data do not include values for nickel containing >0.005% sulfur.

From *The Properties of Electrodeposited Metals* by W.H. Safranek, Elsevier Publishing Co. 1974.[80] Reprinted with permission.

Stress measurements are subject to variations in testing procedures and conditions and are generally not reproducible. However, a particular stress measuring instrument or technique is useful in controlling a plating solution and its operating conditions as well as predicting the quality of the deposit within bounds experimentally established.

In general, thin deposits exhibit higher stresses than thicker deposits produced from the same solution.

6. Magnetic properties of deposits are usually restricted to ferromagnetism and characterized by B-H hysteresis loops, where H is the applied field and B the induced magnetic flux density. Magnetic materials are classified as soft or hard, depending on the value of the coercive force, Hc, which is the magnitude of H when B = O, i.e., the force required to cause random orientation to the domains.

If Hc is small, the magnetic material is considered "soft." These are generally materials which are mechanically soft, i.e., they have a low yield strength. Permalloy (55 Fe - 45 Ni) is such an alloy. If Hc is large, usually >200 oe, the material is considered a hard magnetic material, useful in fast switching computer memory components. Alloys of Co with P and other constituents are usually of this type.

The saturation flux density (B_S) is a physical property determined by the chemical composition of the material. The remanent flux density or retentivity (B_R) and Hc are structure-sensitive properties. The composition, microstructure (grain size and orientation and defects), stress, thickness and impurities of the deposit affect these properties.

If the lattice of the substrate and deposit are similar, the substrate structure can be extended into the deposit. This is called epitaxial growth. High rates of deposition, the presence of addition agents and impurities tend to break down epitaxy. If the lattices differ, then the initial epitaxial growth shifts toward the structure of the deposit. The thickness of the epitaxial/transition zone may vary from 0 to >5 μm before the deposition variables control growth. Also, certain crystal faces grow more rapidly than others, resulting in grain orientation. These factors may be significant for thin film applications such as semi-conductor or magnetic applications.

REFERENCES

(1) C. Wagner, *J. Electrochem. Soc., 101* (4), 181 (1954); A.N. Frumkin and G.M. Florianovich, *Doklady Akad, Nauk SSSR, 86*, 907 (1951).
(2) *Electrode Processes*, Discussions of The Faraday Society, No. 1, 1947, Butterworths (1961).
(3) R.A. Schaeffer and W. King, *Plating, 39*, 487, 627, 769 (1952). (American Electroplaters' Soc. Res. Proj. No. 8)
(4) B.E. Conway and J. O'M. Bockris, *ibid., 46* (4), 371 (April 1959).
(5) A. Damjanovic, *ibid., 52* (10), 1017 (Oct. 1965).
(6) K.J. Vetter, *Electrochemical Kinetics*, p. 282 ff., Academic Press (1967).
(7) H. Fischer, *Plating, 56* (11), 1229 (Nov. 1969).
(8) H. Fischer, *Electrodep. and Surf. Treatment, 1*, 239 (Jan. 1973).
(9) S. Nageswar, *ibid., 3*, 417 (Sept./Nov. 1975).
(10) *Modern Electroplating*, F.A. Lowenheim, Ed., 3rd edition, p. 17 ff., John Wiley and Sons (1973). (Sponsored by the Electrochemical Society.)
(11) H.J. Read, *Plating, 49* (6), 602 (1962).
(12) O. Kardos, *ibid., 61* (1), 61 (1974); (2), 129 (1974); (3), 229 (1974); (4), 316 (1974).

(13) *Theory and Practice of Bright Electroplating*, Yu Matulis, et. al., Editors, Proceedings of All-Union Conference, Dec., 1962, Akademiya of Sciences of the Lithuanian SSR. Translated from Russian by the Israel Scientific Translations, Jerusalem, 1965. TT 65-50000, U.S. Dept. of Commerce.
(14) A.T. Vagramyan and Z.A. Solov'eva, *Technology of Electrodeposition*, pp. 151-175, Robert Draper, Ltd. (1961).
(15) E. Raub and K. Müller, *Fundamentals of Metal Deposition*, pp. 105 - 135, Elsevier Publ. Co. (1967).
(16) K. Bato, *Electrodep. and Surf. Treatment, 3*, (2), 77 (March 1975).
(17) S.R. Rich, *Plating, 42* (11), 1407 (Nov. 1955).
(18) A. Roll, *Metal Finishing, 55* (9), 55 (Sept. 1957).
(19) R.G. Hickman, *Plating, 52*, (5), 407 (May 1965).
(20) A.P. Kapustin and A.N. Trofimov, *Electrocrystallization of Metals in an Ultrasonic Field*. Translated from the Russian by the Israel Scientific Translation, 1970, TT-70-50036, U.S. Dept. of Commerce.
(21) C.T. Walker and R. Walker, *Electrodep. and Surf. Treatment, 1* (6), 457 (July 1973).
(22) C.A. Forbes and H.E. Ricks, *Plating, 49* (2), 279 (1962).
(23) D.T. Ewing and Associates, *ibid.*, 36, 1137 (1949); *37*, 1157 (1950); *39*, 1033 (1952); *39*, 1343 (1952). (Sponsored by the American Electroplaters' Society, Research Project No. 5.).
(24) V. Zentner, *Proc. American Electroplaters' Soc., 47*, 166 (1960).
(25) G.W. Jernstedt, *ibid., 36*, 63 (1949); *ibid., 37*, 151 (1950).
(26) D.L. Rehrig, paper presented at American Electroplaters' Society 65th Annual Technical Conference, Washington, D.C., June, 1978.
(27) A.J. Avila, *Insulation/Circuits*, (Jan. 1977).
(28) A.J. Avila and M.J. Brown, *Plating, 57* (11), 1105 (1970).
(29) H.Y. Cheh, H.B. Linford and C.C. Wan, *Plating & Surf. Finish., 64* (5), 66 (1977). H.Y. Cheh, P.C. Andricacos, and H.B. Linford, *ibid., 64* (7), 42 (1977). H.Y. Cheh, P.C. Andricacos, and H.B. Linford, *ibid., 64* (9), 44 (1977).
(30) F.H. Reid, *Metalloberfläche, 30* (10), 453 (1976).
(31) Ch J. Raub and A. Knödler, *Gold Bulletin, 10* (2), 38 (April 1977).
(32) M.L. Holt, *Modern Electroplating*, F.A. Lowenheim, Ed., 3rd edition, pp. 461 - 485, John Wiley & Sons (1973).
(33) L.F. Spencer, *Metal Finish., 69* (10), 69 (Oct. 1971).
(34) A. Brenner, *Electrodeposition of Alloys, Principles and Practice*, Academic Press (1963).
(35) V.A. Averkin, Ed., *Electrodeposition of Alloys*, (1961). Translated from Russian by the Israel Program for Scientific Translations, 1964, OTS 64-11015, U.S. Dept. of Commerce.
(36) A. Brenner, *Plating, 52* (12), 1249 (1965).
(37) A. Krohn and C.W. Bohn, *Electrodep. & Surf. Treatment, 1* (3), 199 (Jan. 1973); *Plating, 58* (3), 237 (Mar. 1971).
(38) A. Brenner, P.S. Burkhead, and E. Seigmiller, *J. Research, Natl. Bureau of Standards, 39*, 351 (1947).
(39) W.E. Clark and M.L. Holt, *J. Electrochem. Soc., 94*, 244 (1948).
(40) T.P. Hoar and I.A. Brickley, *Trans. Inst. Metal Finish., 32*, 186 (1955).
(41) T.F. Frantsevich and A.I. Zayats, *Zhur. Priklad Khim, 31*, 234 (1958). Eng. translation, p. 224; *Ukrain. Khim Zhur., 24*, 585 (1958).
(42) W. Machu and El Ghandour, *Werkstoffe u. Korrosion, 11* (7), 420 and *11* (8), 481 (1960).
(43) W. Machu and M. Fathi, U.S. Patent 3,093,556 (June 11, 1963).
(44) L. Domnikov, *Metal Finish., 62* (3), 61 (March 1964).
(45) C.V. Chisholm and R.J.G. Carnegie, *Plating, 59* (1), 28 (1972).
(46) W. Machu, *Metalloberfläche, 30* (10), 460 (1976).
(47) L. Domnikov, *Metal Finish., 68* (12), 54 (1970).
(48) U.S. Patent 2,315,740, also G. Black, *Metal Finish., 44*, 207 (1946).
(49) R.W. Moeller and W.A. Snell, *Proc. American Electroplat. Soc., 42* 189 (1955).
(50) R.W. Moeller and W.A. Snell, *ibid., 43*, 230 (1956).
(51) T.W. Tomaszewski, R.J. Clauss, and H. Brown, *ibid., 50*, 169 (1963).

(52) C.R. Kilgore, *Products Finish.*, 34 (May 1963).
(53) R.S. Sayfullin and R.A. Safina, *Zashchita Metal. (USSR), 3* (2), 215 (1967). See also *ASM Rev. of Met. Lit., 24* (7), 99 (July 1967).
(54) F.K. Sautter, *J. Electrochem. Soc., 110*, 557 (1963).
(55) H.J. Weisner, W.P. Frey, R.R. Vanderwoort, and E.L. Raymond, *Plating, 57* (4), 358, 362 (April 1970).
(56) V.P. Greco and W. Baldauf, *ibid., 55* (3), 250 (March 1968).
(57) J.J. Syne, H.N. Barr, W.D. Fletcher, and H.G. Scheible, *ibid., 42* (10), 1255 (Oct. 1955).
(58) M. Ortner, *ibid., 51* (9), 885 (Sept. 1964).
(59) A Brenner and G. Riddell, *J. Research Natl. Bur. of Standards, 39*, (Nov. 1947), (Res. Paper R.P. 1835); *Proc. Amer. Electroplat. Soc., 33*, 23 (1946) and *34*, 156 (1947).
(60) A. Brenner, *Metal Finish., 52* (11), 68 (Nov. 1954); *52* (12), 61 (Dec. 1954).
(61) ASTM Special Technical Publication No. 265, "Symposium on Electroless Nickel Plating," American Soc. for Testing & Materials, Phila., Pa.
(62) K.M. Gorbunova and A.A. Nikiforova, *Physiochemical Principles of Nickel Plating* Translated from Russian by Israel Program for Translation, OTS 63-11003, U.S. Dept. of Commerce.
(63) E.B. Saubestre, *Metal Finish., 60* (6), 67; (7), 49; (8), 45; (9), 59 (1962).
(64) G. Gawrilov, *Metalloberfläche, 25* (4), 118 (1971); *25* (8), 277 (1971); *26* (4), 139 (1972).
(65) G. Gutzeit, E.B. Saubestre, and D.R. Turner, *Electroplating Engineering Handbook*, 3rd ed., A.K. Graham, Ed., pp. 486 - 502, Reinhold Publ. Co.
(66) F. Pearlstein, *Modern Electroplating*, 3rd ed., F.A. Lowenheim, Ed., Chapter 31, John Wiley & Sons (1974).
(67) A.W. Goldenstein, W. Rostocker, F. Schossberger, and G. Gutzeit, *J. Electrochem. Soc., 112*, 104 (1957).
(68) A.H. Graham, R.W. Lindsay, and H.J. Read, *ibid., 112*, 401 (1965).
(69) (a) J.P. Morton and M. Schlessinger, *ibid., 115*, 16 (1968). (b) S.L. Chow, N.E. Hedgecock, M. Schlessinger, and J. Resek, *ibid., 119*, 1614 (1970).
(70) K.T. Ziehlke, W.S. Dritt, and C.H. Mahoney, *Metal Progress, 77*, 84 (1960).
(71) W.G. Lee, *Plating, 47*, 288 (1960).
(72) J.P. Randin and H.E. Hintermann, *ibid., 54*, 523 (1967).
(73) C.E. Johnson and F. Ogburn, *Surf. Technol., 4* (2), 161 (March 1976).
(74) C.E. Higgs, *ibid., 2* (3), 315 (1973/74).
(75) M. Schwartz and G.O. Mallory, *J. Electrochem. Soc., 123* (5), 606 (May 1967).
(76) C.H. deMinjer and A. Brenner, *Plating, 44* (12), 1297 (1957).
(77) G.O. Mallory, *ibid., 61* (11), 1005 (1974).
(78) K. Parker and H. Shah, *ibid., 58* (3), 230 (March 1971).
(79) C. Baldwin and T.E. Such, *Trans. Inst. of Metal Finish., 46*, 73 (1968).
(80) W.H. Safranek, *The Properties of Electrodeposited Metals and Alloys*, Elsevier Publ. Co. (1974). (Sponsored by American Electroplaters' Soc.).
(81) Y. Okinaka, *Gold Plating Technology*, F.H. Reid and W. Goldie, Editors, Chapter 11, Electrochemical Publications, Ltd. (1974).
(82) E.B. Saubestre, *Proc. American Electroplat. Soc., 46*, 264 (1959).
(83) E.B. Saubestre, *Plating, 59* (6), 563 (June 1972).
(84) ASTM, Standard B431-65, "Recommended Practice for Processing of Mandrels for Electroforming." Also see, *Plating, 51* (11), 1075 (Nov. 1964).
(85) G.A. DiBari, "Electroforming." p. 435, *64th Metal Finish. Guidebook and Directory* (1978), Metals & Plastics Publ., Inc.
(86) L.F. Spencer, *Met. Finish., 57* (5), 48 (May 1959).
(87) P. Spiro, *Electroforming*, 2nd ed., International Publ. Services, N.Y. (1971).
(88) J.W. Dini and H.R. Johnson, *Surf. Technol., 4* (3), 217 (May 1976).
(89) J.W. Dini, H.R. Johnson, and H.J. Saxton, *Electrodep. & Surf. Treatments, 2* (2), 165 (1973/74).
(90) D.M. Braddock and S.J. Harris, *ibid., 2* (2), 123 (1973/74).
(91) J.C. Withers and E.F. Abrams, *Plating, 55* (6), 605 (June 1968).
(92) V.P. Greco, W.A. Wallace, and J.N.L. Cesaro, *ibid., 56* (3), 262 (March 1969).
(93) W.A. Wallace and V.P. Greco, *ibid., 57* (4), 342 (April 1970).

(94) S. Wernick and R. Pinner, *The Surface Treatment & Finishing of Aluminum*, 3rd ed., 1 vol. (1964); 4th ed., 2 vols. (1972), Robert Draper, Ltd.
(95) S. Wernick, *Met. Finish.*, 53 (6), 92 (1955).
(96) A.W. Sweet, *Plating, 44* (11), 1191 (Nov. 1957).
(97) R.C. Spooner, Paper No. AN-10, Aluminum Finishing Seminar, Detroit, Mich. (1968), sponsored by Aluminum Assoc.
(98) F. Keller, M.S. Hunter, and D.L. Robinson, *J. Electrochem. Soc., 100* (9), 411 (1953).
(99) D.J. George and J.H. Powers, *Plating, 56* (11), 1240 (1969).
(100) S. Wernick and R. Pinner, *Met. Finish., 53* (11), (1955).
(101) E.L. Coulston, Paper No. AN-6, Aluminum Finishing Seminar, Detroit, Michigan (1968), sponsored by Aluminum Assoc.
(102) J.L. Woods, U.S. Patent 3,857,766 (Dec. 31, 1974); F.S. Newman, J.T. Hartman, and F.A. Dedona, U.S. Patent 3,983,014 (Sept. 28, 1976); M. Kondo and T. Shizuoka (Japan) U.S. Patent 3,996,125 (Dec. 7, 1976).
(103) E.B. Saubestre, L.J. Durney, and E.B. Washburn, *Met. Finish., 62* (11), 52 (1964).
(104) E.B. Saubestre, *Trans. Inst. Met. Finish., 47*, 228 (1969).
(105) L.E. Perrins, *ibid., 50*, 38 (1972).
(106) C.L. Courduvelis, "Applications of Plasmas in the Electroplating of Plastics," paper presented at 65th Annual Technical Conf., American Electroplaters' Soc. (1978), preprint, American Electroplaters' Soc., 1201 Louisiana Ave., Winter Park, Fl. 32789.
(107) C.R. Shipley, U.S. Patent 3,011,920; E. Matijevic, *Plating, 63* (11), 1051 (1974); R.L. Cohen and K.W. West, *J. Electrochem. Soc., 120* (4), 502 (1973); *Plating, 63* (5), 52 (May 1974), (colloid hypothesis). R.J. Zeblinsky, U.S. Patent 3,672,938; A. Rantell and A. Holtzman, *Trans. Inst. Met. Finish., 51*, 62 (1973) and *Plating, 63* (11), 1052, 1054 (1974). (complex hypothesis).
(108) J.J. Martin, *Plating, 58* (9), 888 (1971).
(109) C.J. Halva, "Plating and Finishing of Printed Circuit Boards." American Electroplat. Soc. Illustrated Lecture #41. American Electroplaters' Soc.
(110) C.M. Smith, *Plating, 56* (4) (April 1969).
(111) B.F. Rothschild, M.E. Farmer, and T.W. Brewer, *ibid., 49* (12), 1269 (Dec. 1962).
(112) M.W. Jawitz, *Insulation/Circuits*, p. 5 (April 1976).
(113) B.F. Rothschild and L.C. Kilgore, "Electroplating: Cornerstone of Multilayer Board Fabrication," presented at Western Regional Technical Session, American Electropalters' Soc., March, 1966. (Available as pre-print from Autonetics Div., Rockwell International, Paper X6-362/3111.)
(114) B.F. Rothschild and D. Sanders, *Plating, 56* (12), 1363 (Dec. 1969).
(115) H.R. Johnson, J.W. Dini, and R.E. Stoltz, "On the Mechanical Properties of Sulfamate Nickel Electrodeposits." Presented at 65th Annual Technical Conference, American Electroplaters' Soc., June, 1978. (Pre-print).
(116) H.J. Read, *Plating, 49* (6), 602 (1962).
(117) V. Zentner, A. Brenner, and C.W. Jennings, *ibid., 39*, 865-894, 899-927, 933 (1952).
(118) A. Brenner, P. Burkhead, and C. Jennings, *J. Research Natl. Bureau of Standards, 40*, 31 (Jan. 1948) (R.P. 1854).
(119) M.H. Jones, M.G. Kenez, and J. Saiddington, *Plating, 52* (1), 39 (1965).

SUPPLEMENTARY REFERENCES

Journals

Electrochim. Acta.
*Electronic Packaging and Production**, Milton S. Kiner, Publ., 222 W. Adams, Chicago, Il. 60606.
*Gold Bulletin**, International Gold Corp. Ltd., P.O. Box 61809, Marshalltown 2107, South Africa.
*Product Finish.**, Gardner Publ. Co., 600 Main St., Cincinnati, Ohio 45202.
Trans. Faraday Soc.

---

*Trade journals (complimentary on controlled circulation).

Books

AES, *Symposium on Electroforming,* American Electroplat. Soc. Winter Park, Fl. (1967).
A.E.S., *Symposium on Plating in the Electronics Industry*, (Proceedings), 1st-1966, 2d-1969, 3d-1971, 4th-1974, 5th-1975, 6th-1977, 7th-1979, 8th-1981.
AES, *Illustrated Lecture Series* (Slides and text, 41 subjects available.)
ASM, *Metals Handbook*, vol. 2: "Heat Treating, Cleaning and Finishing," American Soc. for Metals, Metals Park, Ohio.
ASTM, *Anodizing Aluminum,* STP 388, American Soc. for Testing and Materials, Philadelphia, Pa. (1965).
ASTM, *Electroforming-Applications, Uses and Properties of Electroformed Metals, (1962).*
ASTM, *Hydrogen Embrittlement Testing,* STP 543 (1974).
H. Bennington and R. Draper, *Tables and Operating Data for Electroplaters,* R. Draper Ltd., Teddington, Eng.
I.M. Bernstein and A.W. Thompson, (Ed.), *Hydrogen in Metals,* Amer. Soc. for Metals (1974).
W. Blum and G.B. Hogaboom, *Principles of Electroplating and Electroforming,* 3d ed., McGraw-Hill Publ. Co., N.Y. (1949).
A.F. Bogenschutz, *Surface Technology and Electroplating in the Electronics Industry,* Porticullis Press, Ltd., London, Eng. (1974).
J. O'M. Bockris and A.K.N. Reddy, *Modern Electrochemistry,* (2 vols.), Plenum Press, N.Y. (1970).
R. Brugger, *Nickel Plating,* International Publ. Services (Porticullis), N.Y. (1970).
R.M. Burns and W.W. Bradley, *Protective Coatings for Metals,* 3d ed., Reinhold Publ. Corp., N.Y. (1967) (ACS Monograph series).
Cobalt Monograph, (prepared by staff, Batelle Memorial Inst.) ed. by Centre D'Information du Cobalt, Brussels, Belgium (1960).
C.F. Coombs, Jr. (Ed.), *Printed Circuits Handbook,* McGraw-Hill Book Co. (1967).
G. Dubpernell, *Electrodeposition of Chromium,* Pergamon Press, N.Y. (1977).
J. Fischer and D.E. Weiner, *Precious Metal Plating,* R. Draper, Ltd., Teddington, Eng. (1964).
W. Goldie, *Metallic Coating of Plastics,* (2 vols.), Electrochemical Publ., Ltd., Middlesex, Eng. (1968).
A.K. Graham, (Ed.), *Electroplating Engineering Handbook,* 3d. ed., Van Nostrand, Reinhold Co., N.Y. (1971).
J.D. Greenwood, *Hard Chromium Plating,* 2d ed., Int'l. Publ. Serv. (Porticullis), N.Y. (1971).
J.D. Greenwood, *Heavy Deposition,* R. Draper, Ltd., Teddington, Eng. (1970).
N. Hall, (Ed.), *Metal Finishing Guidebook - Directory,* Metals and Plastics Publ., N.J. (1978). (revised annually)
C.A. Hampel, (Ed.), *Encyclopedia of Electrochemistry,* Reinhold Publ. Corp., N.Y. (1964).
G.D.R. Jarrett, C.R. Draper, G. Muller, and D.W. Baudrand, *Plating on Plastics,* 2d. ed., International Publ. Services (Porticullis), N.Y. (1971).
A. Kutzelnegg, *Testing Metallic Coatings,* R. Draper, Ltd., Teddington, Eng. (1963).
F.A. Lowenheim, *Electroplating,* McGraw-Hill Book Co., N.Y. (1978). (Sponsored by American Electroplaters' Soc.).
J.A. Murphy, (Ed.), *Surface Preparation and Finishes for Metals,* McGraw-Hill Book Co., N.Y. (1971).
H. Narcus, *Metallizing of Plastics,* Reinhold Publ. Co., N.Y. (1960).
H.J. Read, (Ed.), *Hydrogen Embrittlement in Metal Finishing,* Reinhold Publ. Co. (1961). (Sponsored by American Electroplaters' Soc.).
F.H. Reid and W. Goldie, *Gold Plating Technology,* Electrochemical Publ., Ltd. (1974).
E. Raub and K. Muller, *Fundamentals of Metal Deposition,* Elsevier Publ. Co., N.Y. (1967).
R. Sard, H. Leidheiser, Jr., and F. Ogburn, (Eds.), *Properties of Electrodeposits, Their Measurements and Significance,* The Electrochemical Soc., Princeton, N.J. (1975).
H.H. Uhlig, *Corrosion and Corrosion Control,* 2d. ed., John Wiley & Sons, N.Y. (1971).
H.H. Uhlig, (Ed.), *Corrosion Handbook,* John Wiley & Sons, N.Y. (1948). (Sponsored by Electrochemical Society).

K.R. Van Horn, (Ed.), *Aluminum,* 3 vols., American Society for Metals, Metals Park, Ohio (1967).

J.M. West, *Electrodeposition and Corrosion Processes,* Van Nostrand Co., N.Y. (1965).

APPENDIX A: PREPARATION OF SUBSTRATES FOR ELECTROPLATING

ASTM Recommended Practices

Data from *Book of ASTM Standards,* Part 9, revised annually. Also approved as American National Standards by the American National Standards Institute.

Number	Title
B 177-68(73)	Rec. Practice for Chromium Plating on Steel for Engineering Use
B 183-72	Rec. Practice for Preparation of Low-Carbon Steel for Electroplating
B 242-54(71)	Rec. Practice for Preparation of High-Carbon Steel for Electroplating
B 253-73	Rec. Practice for Preparation of and Electroplating on Aluminum Alloys by the Zincate Process
B 254-70	Rec. Practice for Preparation of and Electroplating on Stainless Steel
B 281-58(72)	Rec. Practice for Preparation of Copper and Copper-Base Alloys for Electroplating
B 322-68(73)	Rec. Practice for Cleaning Metals Prior to Electroplating
B 343-67(72)	Rec. Practice for Preparation of Nickel for Electroplating with Nickel
B 431-69	Rec. Practice for Processing of Mandrels for Electroplating
B 450-67(72)	Rec. Practice for Engineering Design of Electroformed Articles
B 503-69	Rec. Practice for Use of Copper and Nickel Electroplating Solutions for Electroforming
B 480-68	Rec. Practice for Preparation of Magnesium and Magnesium Alloys for Electroplating
B 481-68(73)	Rec. Practice for Preparation of Titanium and Titanium Alloys for Electroplating
B 482-68(73)	Rec. Practice for Preparation of Tungsten and Tungsten Alloys for Electroplating
B 488-71	Spec. for Electrodeposited Coatings of Gold for Engineering Uses
B 558-72	Rec. Practice for Preparation of Nickel Alloys for Plating
B 580-73	Spec. for Anodic Oxide Coatings on Aluminum

Preparation for electroplating of less common substrates including those used in nuclear, electronic or high temperature alloys of Fe, Co, Ni or Cr usually requires activation treatments* in order to obtain satisfactory adhesion. Other techniques involve diffusion bonding with thermal treatments.

---

*See C. Levy, *Proc. AES, 43,* 219 (1956) for activation for Cr plating and W.W. Sellers and C.B. Sanborn, Ibid., 44, 36 (1957) for Ni and Ni alloys prior to Ni plating for detailed formulations.

Beach and Faust* and Friedman** review procedures for light metals and for high temperature applications including plating on refractory metals—U, Mo, W, Th, Zr, Nb and Si.

For plating Cr on previously plated Cr, the following procedure has been satisfactory:

1. If Cr is oiled (due to grinding), degrease and polish lightly. Then clean in alkaline cleaner by immersion, scrubbing or cathodically.
2. Provide a light etch anodically in alkaline, sulfuric or chromic acid solutions.
3. Immerse in Cr plating solution and allow parts to reach solution temperature.
4. Plate at very low C.D.: 77.5 ma/cm^2 ($\sim$0.5 A/in^2) to deposit only hydrogen to activate the surface, for 0.5-3 minutes approximately.
5. Slowly increase C.D. to 0.5-1.0 A/cm^2 (3-6 A/in^2) for 15-30 seconds to guarantee coverage then reduce to normal plating C.D.: 0.15-0.5 A/cm^2 (1-3 A/in^2).

APPENDIX B: REPRESENTATIVE ELECTROLESS PLATING SOLUTION FORMULATIONS

	............(g/ℓ)............			
Nickel-Phosphorus	Ref. (a)	Ref. (a)	Ref. (a)	Ref. (b)
Nickel sulfate	35	35	30	25
Sodium hypophosphite	10	10	10	25
Sodium hydroxyacetate	10			
Sodium acetate		10		
Sodium citrate			100	
Sodium pyrophosphate				50
Ammonium chloride			50	
pH	4.5-5.5	4.5-5.5	9-9.5	10-10.5
Temp., °C	90-95	90-95	90-95	70-75
w/o P in deposit	7-9	7-9	5-7	4-6
Nickel-Boron	Ref. (c)			
Nickel chloride	30			
Dimethylamine borane	3.5			
Malonic acid	34			
pH (with NH$_4$OH)	5.5			
Temp., °C	77			

(continued)

---

*Modern Electroplating, 3rd ed., Chapter 27, 618. F. Lowenheim, ed., John Wiley & Sons, (1974).
**Plating, 54, 1035 (Sept. 1967).

APPENDIX B: (continued)

.... (g/ℓ) ....

Cobalt-Phosphorus	Ref. (d)		
Cobalt sulfate	24		
Sodium hypophosphite	20		
Sodium citrate	70		
Ammonium sulfate	40		
Sodium laurylsulfate	0.1		
pH	8.5		
Temp., °C	92		
Copper	**Ref. (e)**		
Copper sulfate	29		
Sodium potassium tartrate	142		
Versene T	17		
Sodium hydroxide	42		
Sodium carbonate	25		
Formaldehyde (37%), ml/ℓ	167		
Temp., °C	25		
Palladium	**Ref. (f)**		
Palladium chloride (as ammine complex)	5.4		
EDTA Na_2	33.6		
Ammonium hydroxide	350		
Hydrazine	0.3		
Temp., °C	80		
Gold	**Ref. (g)**	**Ref. (g)**	**Ref. (g)**
Potassium cyanoaurate	5.8	0.86	5.8
Potassium cyanide	13	6.5	1.3
Potassium hydroxide	11.2	11.2	45
Potassium borohydride	21.6	10.8	
Dimethylamine borane			23.6
Temp., °C	75	75	85
Silver	**Ref. (h)**		
Sodium silver cyanide	1.83		
Sodium cyanide	1.0		
Sodium hydroxide	0.75		
Dimethylamine borane	2.0		
(Thiourea)	(0.25)		
Temp., °C	55–65		

References

(a) A. Brenner & G. Riddell, *Op. cit.*[59]
(b) M. Schwartz, *Proc. AES*, 176 (1960).
(c) G.O. Mallory, *Plating*, 58, 319 (1971).
(d) L.D. Ransom & V. Zentner, *J. Electrochem. Soc.*, 111, 1423 (1964).
(e) E.B. Saubestre, *Proc. AES*, 46, 264 (1959).
(f) R.N. Rhoda, *Trans. Inst. Met. Finish.* 36, 82 (1959).
(g) Y. Okinaka, *Plating*, 57, 914 (1970).
(h) F. Pearlstein & R.F. Wightman, *Plating*, 58, 1014 (1971).

Note: Some of the above formulations are protected by U.S. Patents. Their listing here does not imply any rights to infringe.

APPENDIX C: COMPARISON OF ALUMINUM ANODIZING PROCESSES (TYPES I, II AND III)

Advantages of Type I Coatings

1. Corrosion resistance of coatings are as high (if not higher) than Type II coatings.
2. Provide excellent bond for organic coatings.
3. Chromic acid is a corrosion inhibitor and therefore it is not essential to assume (or provide for) complete removal from crevices, joints or recesses due to spot welding, riveting, bolting or blind holes.
4. It has practically no effect on the fatigue strength of the part.
5. Although thinner, less porous, and somewhat opaque due to pick up of chromate ion and alloying constituents, the coating is capable of absorbing dark dyes for Class 2 requirements.
6. It is preferred as maskant for selective Hard Anodize since it is less porous than Type II films and especially for assemblies with joints or recesses.

Limitations of Type I Coatings

1. A smaller increase in abrasion resistance is obtained as compared to Type II coatings due to lower thickness and structure differences.
2. Limited to alloys containing less than 5% copper or 7% silicon.
3. Higher voltage is required with extended time as compared to Type II coatings.
4. Under conditions used for wrought alloys, casting alloys tend to use excessive current and "burning" may occur. In such cases, conditions might require changes to 30-35 volts at 90F with compensating increase in time to obtain adequate coating thickness.
5. Alloys in the annealed condition do not anodize satisfactorily. Heat treatable alloys should be tempered by solution heat treatment and approved aging.
6. Wrought and cast alloys with high alloy content (such as 7075) tend to develop thinner coatings and may behave erratically or poorer in salt spray tests.

Advantages of Type II Coatings

1. Less expensive (compared to Type I coatings) with respect to chemicals involved (and waste treatment thereof), heating and power costs, length of time to obtain required coating.
2. More alloys can be treated.
3. Coatings are harder than Type I coatings.
4. Coatings may be slightly more corrosion resistant *after sealing* than Type I coatings (due to thicker and more porous coating).
5. Clear coating permits dyeing with greater variety of colors.

Limitations of Type II Coatings

1. Cannot be used where possibility of solution entrapment exists, especi-

ally joints, laps or recesses since any sulfuric acid residue may be corrosive.
2. Reduces the fatigue characteristics of the alloy.
3. Difficult to control where small dimensional changes are desired or required since coatings grow faster and are thicker for corrosion resistant requirements as compared to Type I coatings. (Thus, Type I coatings should be considered on close tolerance parts such as threads.)

Characteristics of Hard Anodize Coatings

1. Corrosion resistance is excellent, several thousand hours in salt spray tests have been reported (after proper sealing).
2. Abrasion and wear resistance excellent.
3. Chemical resistance is poor as compared to calcined aluminum oxides; will not resist alkalies or acids as well.
4. Coefficient of thermal expansion is different from that of the aluminum alloys and spalling may result at temperatures above 200°-300°C.
5. Film crazing--As part temperature increases from formation temperature (25°-32°F) to room temperature or the higher sealing temperatures (200°-210°F) or post honing temperatures, the coating may craze or fracture since it is tensively stressed; this phenomenon becomes aggravated as film thickness increases. Sometimes this crazing seems to disappear after aging.
6. "Chalking"—This refers to a white film which sometimes appears on the surface after drying. It is not considered detrimental and is usually not noticed unless (or until) surface is wiped. The mechanism is not understood; it may possibly be a bleed-out phenomenon.

Effect of Alloying Elements on the Hard Coating

1. Thicker coatings are obtained with the purer or higher conducting alloys containing magnesium or zinc:

    | Purer alloys | EC, 1100, 3003 |
    | Al-Mg alloys | 5005, 5050, 5052, 5252 |
    | Al-Mg-Si alloys | 6061, 6063 |
    | Al-Zn alloys | 7075 |

2. Copper-containing alloys produce intermetallic compounds (after HT) which increase the ohmic resistance resulting in thinner coatings. Type III Hard Anodize is restricted to those alloys containing less than 5% Cu.
3. High silicon-containing alloys also produce intermetallic compounds and do not anodize readily. These involve most castings which depend on reduction of the alloy's melting point by the eutectics formed with the silicon (even less than 7% Si). The Si or silicides do not anodize, being "inert" and acting as inclusions, depending on "bridging" for continuity of coating.
4. Since copper and silicon constituents may result in poorer coatings, a total of 7-9% of the combination of these two elements is usually considered as a maximum in an alloy to be hard anodized.
5. The color of the Hard Anodized Coating reflects the alloying constituents.

11

Plasma and Detonation Gun Deposition Techniques and Coating Properties

Robert C. Tucker, Jr.
*Union Carbide Corporation
Coatings Service Department
Indianapolis, Indiana*

INTRODUCTION

Plasma and detonation gun (d-gun) coatings have been used in industry for over twenty years. The d-gun process was developed by Union Carbide Corporation[1] and d-gun coatings are only available through their Coatings Service Department. Plasma deposited coatings (also developed by Union Carbide[2]) are available from a number of coatings service organizations and the equipment is available from several sources for in-house use.

Both are line-of-sight processes in which powder is heated to near or above its melting point and accelerated (by either a detonation wave or plasma gas stream). The powder is directed at a substrate (surface to be coated) and on impact forms a coating consisting of many layers of overlapping thin lenticular particles or splats. Almost any material that can be melted without decomposing can be used to form the coating. The substrate is seldom heated above 150°C. Typical coating thicknesses range from 0.002 to 0.020 inches, but in a few applications may exceed 0.2 inches.

The description of the processes and coatings that follows is divided into three sections: equipment, processes, and coating structure and properties.

EQUIPMENT

In this section plasma torches, d-guns, auxiliary equipment and equipment related coating limitations are discussed. A description of the physics of plasma or detonation generation would be too lengthy to be included here and is unnecessary to an understanding of the utilization of the process.

Plasma Coating Torches

The essential elements of a plasma torch are shown in Figure 11.1. The anode is usually copper and the cathode tungsten. A gas, usually argon or nitrogen or a mixture of these with hydrogen or helium flows around the cathode and through the anode which serves as a constricting nozzle. A direct current arc, usually initiated with a high frequency discharge, is maintained between the electrodes. The arc current and voltage used vary with the anode/cathode design, gas flow, and gas composition. The power used varies from about 5 to 80 kilowatts depending on the type of torch and the operating parameters. In one variant of a coating torch, a partially transferred arc is used; i.e. part of the arc goes to the anode and part to the substrate being coated. This causes substantial heating of the substrate and is used only in special situations. Fully transferred arc surfacing torches will not be discussed here.

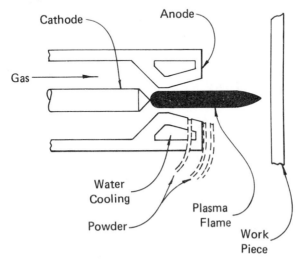

Figure 11.1: Schematic of a plasma coating torch showing alternative powder inlet positions.

The gas plasma generated by the arc consists of free electrons, ionized atoms, and some neutral atoms and undissociated diatomic molecules if nitrogen or hydrogen are used. The specific anode/cathode configuration, gas density, mass flow rate and current·voltage determine the plasma temperature and velocity. Plasma gas velocities with most conventional torches are subsonic, but supersonic velocities can be generated by using converging-diverging nozzles with critical exit angles. The temperature of the plasma may exceed 50,000°F. The enthalpy of the plasma and efficiency of heat transfer to the powder particles can be increased substantially with the inclusion of diatomic gases, such as hydrogen or nitrogen, Figure 11.2.

The velocity that powder achieves in a plasma stream depends on the integrated effect of mass flow rate of the plasma and the distance the powder is carried in the stream. Similarly, the temperature the powder achieves is a function of the integrated effect of the plasma temperature, plasma composition and the transit time in the plasma stream. (Both powder velocity and temperature are functions of other factors as well, such as particle size, powder composition, heat capacity, density, emissivity, etc., to be discussed elsewhere.) It follows, therefore, that the point of entry of the powder into the plasma stream is very important. The ideal location would be in a uniform pattern upstream of the anode throat since this would probably allow the best distribution of the powder in the plasma stream, expose the powder to the highest plasma temperature and provide the longest path or time in transit. Most torch manufacturers, however, have been unable to prevent powder adherence to the entry or throat of the nozzle and excessive superheating using this approach. As a result, powder entry is usually in the diverging portion of the nozzle or just beyond the exit as shown in Figure 11.1. Fairly recent attempts,[3] in high velocity torches, have been made to adjust the point and angle of entry of the powder into the plasma stream for the melting point of the powder. The goal in this case was to heat the powder to close to, but not over, the melting point. In another high velocity torch design[4] in which shock diamonds are generated, the powder is introduced a short distance beyond the exit in a region of rarefaction in the plasma stream. The commercial success of these approaches has yet to be determined.

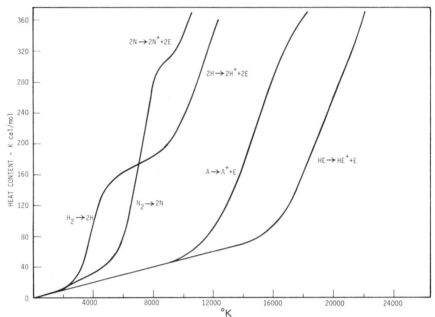

Figure 11.2: Enthalpy of gases commonly used in plasma spraying.

The most important parameters relative to the powder particles at impact on the substrate are their temperature, velocity and extent of reaction with the gaseous environment. The velocity of the powder, as previously mentioned, is a function of the mass flow rate of the plasma, the density, mass and shape of the powder, and the distance the powder travels in the plasma. With most of the conventional commercial torches available up to the mid 70's, velocities varied from about 400 to 1000 ft/s. Higher velocity torches have since become available[3-5] with powder velocities claimed[6] to be in excess of 1800 ft/s (measured by a rotating mirror), but velocities measured with a more sophisticated technique (Doppler laser) for similar torches were 1200 ft/s.[7]

It has often been stated that any material that can be melted without decomposition can be used as a plasma coating. There appears, however, to be two schools of thought on whether or not the powder should be molten on impact. Certainly the combination of particle plasticity or fluidity and velocity must be high enough to allow the particle to flow into a thin, lenticular shape that molds itself to the topology of the substrate or previously deposited material. The extent to which this is achieved determines the density and strength of the coating. With the relatively low velocity torches, reasonably high densities can only be achieved if the particles are substantially molten. As noted previously, the intent of at least some high velocity torch designs is to achieve highly plastic, but not molten, particles.

Care should be exercised in developing the coating parameters not to heat the powder to an excessive temperature. The most obvious hazard is vaporization of all or part of the powder. This is most likely when:

(a) The difference between the melting and boiling point of a single phase powder is too small;

(b) One or more of the components in a multiphase powder has a substantially lower boiling point than the others;

(c) One or more of the components in a mixture of powders has a substantially lower boiling point than the others; or

(d) The powder size distribution is too wide with a single component or not adjusted for heating rates with a mixture.

In addition to vaporization through boiling, there may be some loss of a component in an alloy or compound that has a particularly high vapor pressure. This is generally not a significant problem because transit times are so short.

The temperature in a plasma is high enough to melt (or decompose) any material, given enough time. Comparison of the relative heating rates of powders is not as simple as comparing their melting points, however. Heat transfer in the plasma jet is primarily the result of the recombination of the ions and reassociation of atoms in diatomic gases on the powder particle surfaces and absorption of radiation.[3] The ultimate temperature of the powder particles, therefore, is a function of the catalytic activity of their surface, their emissivity (particularly in the ultraviolet range), their heat capacity (including any heats of phase transformations and heat of melting), their thermal conductivity, and their surface to volume ratio (shape). Many metals, having high absorption in the UV range, high surface activity, and high thermal conductivity, tend to heat much more rapidly than most oxides. Specific tables of heating rates are not available, but tables of the pertinent physical properties can be used as guidelines in selecting appropriate coating parameters.

The extent of reaction of the powder with its gaseous environment during transit depends both on the composition of the plasma gas and the amount of intermixing of the plasma gas with the ambient gas between the nozzle and the substrate. It is generally assumed that argon and helium are inert and no degradation of the powder occurs in the torch when they are used as the only plasma gases. Obviously for this to be true, the gas source must be free of oxygen and other contaminants and the torch and other equipment must be gas tight.

Whether or not hydrogen or nitrogen when used in the plasma gas are effectively inert relative to the powder depends on the composition of the powder. The transit time and temperature of the powder in the plasma determine the extent of reaction and/or solution of the gas in the powder in those cases where the gas is not thermodynamically inert. The use of hydrogen to reduce the amount of oxidation during spraying may be somewhat effective, but may be due as much to a shielding effect (by reacting with oxygen from the air inspirated into the plasma stream) as to actual reduction of oxide formed on metallic powder. On the other hand, oxide powders or oxide films on metallic powders may be decomposed in the plasma spray; e.g., zirconia coatings sprayed with an argon plasma are oxygen deficient and the amount of oxygen in copper can be lowered simply by thermal decomposition of its oxides.

Usually of greater concern than reaction with the plasma gas is the extent of reaction of the powder with oxygen or nitrogen from the air inspirated into the plasma stream after it exits the nozzle. This effect is strongly a function of the type of torch used as illustrated in Figure 11.3. None of these coatings were shielded from the atmosphere, yet the differences in extent of oxidation are dramatic. If coatings with even less oxide than that shown in Figure 11.3(a) are desired, several means of shielding the plasma stream are available. One of the best, and certainly the most adaptable to production, is a patented inert gas shroud that surrounds the effluent with argon.[9] A comparison of the results obtained with this shield compared to those obtained with the same torch that produced the relatively clean microstructure of Figure 11.3(a) is shown in Table 11.1. Note that the oxygen contents of molybdenum, copper and nickel are all lower in the coating than in the starting powder, while that of titanium, a very reactive metal, is only slightly higher. Alternative methods of excluding air include spraying in a vacuum chamber or in an enclosure filled with argon. An extreme example used for coating large parts is an entire room or cubicle filled with argon in which the operators wear life support suits.[10] Nitridation of some materials may also occur, but has not been extensively studied.

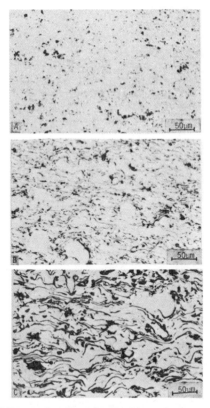

Figure 11.3: Microstructures of aluminum bronze coatings made with three types of standard plasma spray torches illustrating varying degrees of oxidation during deposition. As-polished.

Table 11.1: Oxygen Content of Plasma Deposited Coatings

Coating Material	Starting Powder	Oxygen Content, % Conventional Coating	Coaxial Gas Shielded Coating
Copper	0.126	0.302	0.092
Nickel	0.172	0.456	0.151
Tungsten	0.027	0.274	0.030
Titanium	0.655	>2.0	0.730
Molybdenum	0.419	0.710	0.160

Detonation Gun

The detonation gun, shown schematically in Figure 11.4, consists of a water-cooled barrel about 3 ft long with an inside diameter of about one inch and associated gas and powder metering equipment. In operation a mixture of oxygen and acetylene is fed into the barrel along with a charge of powder. The gas is then ignited and the detonation wave accelerates the powder to about 2400 ft/sec while heating it close to, or above, its melting point. The maximum free burning temperature of oxygen/acetylene mixtures occurs with 45% acetylene and is about 3140°C, but under detonation conditions probably exceeds 4200°C, so most, but not all, materials can be melted. The distance that the powder is entrained in the high velocity gun is much longer than in a plasma device which accounts for the much higher particle velocity. After the powder has exited the barrel, a pulse of nitrogen purges the barrel. The cycle is repeated about four to eight times a second.

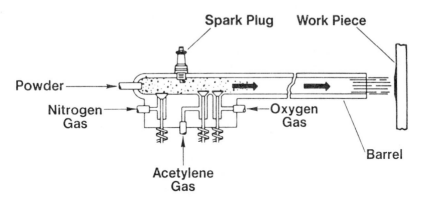

Figure 11.4: Schematic of a detonation gun.

Each pulse of powder results in the deposition of a circle of coating about 25 mm in diameter and a few microns thick. This circle of coating is, of course, composed of many overlapping thin lenticular particles or splats corresponding to the individual powder particles. The total coating is, in turn, produced by many overlapping circles of coatings. This pattern of overlapping is closely controlled to produce a smooth coating and minimize substrate heating and residual stress. Auxiliary CO_2 cooling of the part is frequently used as well.

Because of the gases used in the d-gun, the powder may be exposed to either an oxidizing or carburizing environment, although an essentially inert mixture can be achieved with precise control. Carburizing conditions, in particular, can be used to advantage[11] as will be illustrated elsewhere.

Auxiliary Equipment

In addition to the plasma torch or d-gun itself, gas controls, power supplies, and powder feeders are required. Most of these are supplied with the basic unit. A detailed discussion of their characteristics is not appropriate here, but there are several general criteria that all such equipment should meet. Excellent gas control can be achieved with either rotometers or critical flow orifice columns, but, in either case, attention should be paid to downstream pressures to insure that the control device is capable of accurately measuring flow. All gages, rotometers and orifices used should be periodically calibrated as well. Power supplies should be reasonably ripple-free and again all meters used should be periodically calibrated.

For optimum plasma spraying, powder must be distributed uniformly in the plasma stream at a constant rate. There are a variety of powder dispensers designed to do this including those based on an auger, aspirated flow or fluidized bed.[12,13] Conversely, a pulsed flow of powder is required for a d-gun. Again, however, uniform distribution of the powder in the barrel is important, as is the constancy of the amount of powder in each pulse.

The highest quality of plasma or d-gun coatings can only be achieved with automated or semiautomated torch and part handling. Hand held torches lead to varying stand-off, poor thermal control and nonuniform thickness—all of which result in varying coating properties across the part. The most commonly used method of part and torch motion control utilizes a modified lathe concept with the torch mounted on what would be the tool post and the parts to be coated either rotated as a cylinder or mounted on an annulus plate. Predetermined torch-to-part surface speeds and overlap can then be maintained by varying the rotation and torch speeds. A variety of cam actions can be used to maintain a uniform deposition rate from the center to the outside of an annulus plate. Another method of controlling relative motion, particularly suitable for d-gun because lower surface speeds can be used, is that of traversing and indexing in a raster pattern. Using this technique, very large flat surfaces can be coated. More extensive automation has been developed including part transfer handling and computer control of the torch which tremendously increases the productivity of the equipment.

Equipment Related Coating Limitations

Plasma deposition is a line-of-sight process in which the structure of the coating is a function of the angle of deposition; i.e., the angle between axis of the plasma jet effluent and the surface of the substrate being coated. Normally coatings with the highest density and bond strength are achieved at a 90 degree angle of deposition. The extent of changes in structure is a function of the type of plasma torch and the operating parameters. With some low velocity torches, angles less than 75° may cause significant degradation of properties,[13] with others, angles as low as 60° can be tolerated. This limitation may cause some problems in coating complex parts, particularly those with narrow grooves or sharp angles, and may require several set-ups to adequately coat the different faces or surfaces of a part. The detonation gun, with its higher particle velocity, can usually tolerate a wider deviation from 90° (down to about 45° in many cases).

Another limitation, of course, is the size of the torch and the required stand-off (distance from the nozzle or front face of the torch to the workpiece) when an inside diameter must be coated. The smallest torch available today can apply a metallic coating to a 1¼ inch or a ceramic coating to a 1⅞ inch inside diameter at 90°. Another torch with an effluent at 45° to the torch axis can apply a coating to the inside of a blind cylinder 1⅞ inches in diameter.

The d-gun, of course, can not fit into a cylinder or other cavity. It can be used, however, to coat an inside diameter to a depth about equal to the diameter; i.e., to an angle of deposition of about 45°. While there is some change in microstructure as the angle decreases, the inherently high density and bond strength, as previously mentioned, still allows superior coatings to be deposited at the lower angles.

COATING PROCESS

The total coating process includes specification and procurement of powder, substrate preparation, masking, and finishing, in addition to the coating operation itself. Each of these is discussed in the following subsections.

Powder

Most of the powder used for plasma and d-gun deposition falls between 5 and 60 microns in size. To achieve uniform heating and acceleration of a single component powder, it is advisable to have as narrow a size distribution as possible. The additional cost of sizing is, at least partially, recovered in higher deposition efficiency and better coating quality. The specific powder size range to be used is a function of the torch or d-gun design and the heating characteristics of the powder discussed earlier. Generally speaking fine powders, of course, are accelerated and heated more rapidly in the plasma stream, but they also tend to lose momentum more rapidly when spraying at the longer distances (greater standoffs). They generally result in denser, but more highly stressed coatings. Finer powders also tend to create more torch operating problems and have higher oxide contamination levels.

Good quality control of powder is essential, not only during manufacture, but during storage and handling. Powder specifications and quality control should include, as a minimum, chemical analysis (including interstitials for metallic powders), shape characterization, size distribution, and flowability. A wide variety of equipment is available for analyses, and selection of a specific technique or type of test will vary with the type of powder. It is obvious that the powder should be kept clean and dry, but too little attention paid to this will result in dispensing problems, torch clogging and lumps in the coating.

Substrate Preparation

It seems quite obvious that any part to be coated (substrate) must be clean, yet this step in the total coating process is frequently given too little attention. Not only must all oxide scale or other foreign matter be removed, but all oils, machining lubricants, etc. must be eliminated. It is therefore usually good practice to degrease a part after any descaling, machining or grinding is done. Grit blasting, to be discussed momentarily, should not be relied upon to remove heavy scale, since it may simply embed it in the surface leading subsequently to a weakly bonded area or a site for corrosion.

Virtually all plasma coatings require a roughened substrate surface. Although machining, chemical etching and other techniques are sometimes used, the most frequently used method is grit blasting. The type of grit and grit blasting pressure used should be determined by the composition and heat treat condition of the substrate. For many relatively soft substrates, chilled steel grit is satisfactory. It does not shatter and does not embed excessively in the surface. For harder substrates, alumina or silicon carbide grit has better cutting action. For some applications, special grit may be used to achieve unusually low levels of grit inclusions.[14] Regardless of the method used, the surface roughness should normally exceed 150 micron inches, rms. In addition, the surface topology should be sharply peaked, not smoothly undulating. Excessive grit blasting, on the other hand, can be detrimental due to work hardening, blunting of the peaks and increased grit entrapment.

For d-gun coatings, grit blasting may not be necessary. The unusually high particle velocity in itself results in substantial surface roughening. This is generally true for titanium substrates, for example, but for harder substrates grit blasting is usually used.

Grit blasting, of course, increases the surface area significantly, so whether bonding, discussed elsewhere, is due to a mechanical interlocking, to interdiffusion, surface reaction, or a combination of these, it is advantageous in increasing bond strength. In any case, the coating should be applied as soon after grit blasting as possible to insure an active, clean surface.

Masking

A wide variety of masking techniques are used to limit the deposition to the required area on the part. In most cases masking is less expensive than subsequent removal by grinding. Many types of tapes and oxide-loaded paints or stop-off lacquers are satisfactory for low velocity, long standoff plasma torches. For high velocity, short standoff torches, more substantial masking is required; e.g., glass fiber reinforced high temperature tape, adhesive-backed steel or aluminum foil, or sheet metal masking. For d-gun coatings, metal masking is used most frequently. Efficiently designed masking can significantly reduce the total cost of a coating and deserves careful consideration.

Coating

The torch parameters that must be selected to apply coating of a given powder composition and size distribution include the types of gases to be used and their flow rates, the anode design, the power level to be used, and, for some torches, the point of powder entry. All of these vary with the specific torch model used. The torch manufacturer should be able to provide specific instructions, or at least detailed guidelines.

It is always advisable to coat a quality control specimen to verify the coating deposition rate and coating microstructure before coating any parts. Metallographic examination of this specimen should include, as a minimum, general phase content, the amount of oxidation occurring during deposition, apparent porosity, and microhardness. It is also advisable to check the grit inclusion level and/or amount of substrate surface contamination, but this is only meaningful if the quality control specimen is made of the same material, is in the same heat treat condition, and has undergone the same surface preparation as the parts to be coated.

One of the major advantages of plasma and d-gun coatings is that they may be applied to substrates without significantly heating them above room temperature. As a result, a part can be fabricated and fully heat treated, without changing the substrate microstructure or strength. This also avoids any possibility of distortion or volumetric change during any post coating heat treatment that is common to other coating methods. It is, however, advisable to warm the surface slightly, usually with a pass of the torch without powder flowing, to remove any adsorbed gases from the surface before applying coating. The surface temperature usually does not exceed 125 to 150°C during this warming pass. During coating deposition a substantial amount of heat is transmitted to the part through the plasma gas and the molten powder. To insure uniform coating thickness and minimize residual stress within the coating, it is necessary to carefully control the areal rate of deposition. This can only be accomplished satisfactorily by using automated part and torch handling equipment with the selection of appropriate surface speed, overlap pattern and deposition rate. Cooling air or CO_2 jets may be used as well. Under normal circumstances the part temperature does not exceed about 150°C during coating.

Finishing

For many applications plasma and d-gun coatings can be used as-coated. In fact, in at least one application, a d-gun tungsten carbide-cobalt coating is grit blasted to further roughen the surface for better gripping action. Probably in the majority of applications, however, the coatings are finished before being placed in service. Finishing techniques vary from brush finishing to produce a nodular surface, to machining, grinding, and lapping to produce surfaces with surface roughness down to less than 2×10^{-6} inches, rms. Machining can be used on some metallic coatings, but most coatings are ground with silicon carbide or diamond (diamond is usually preferred for d-gun coatings). The best surface finish that can be obtained is a function not only of the finishing technique, but of the coating composition and the deposition parameters.

Recommendations for the machining, grinding, and lapping techniques to be used with a specific coating can be obtained from coatings service organizations or coating equipment manufacturers. Great care should be exercised in finishing operations to avoid damaging the coating through heat checking, pull-out, or edge chipping. A typical check list[15] for grinding is:

(1) Check diamond wheel specifications.
    (a) Use only 100 concentration.
    (b) Use only resinoid bond.
(2) Make sure your equipment is in good mechanical condition.
    (a) Machine spindle must run true.
    (b) Backup plate must be square to the spindle.
    (c) Gibs and ways must be tight and true.
(3) Balance and true the diamond wheel on its own mount - 0.0002" maximum runout.
(4) Check peripheral wheel speed - 5,000 to 6,500 SFPM.
(5) Use a flood coolant - water plus 1-2% water soluble oil of neutral pH.
    (a) Direct coolant toward point of contact of the wheel and the workpiece.
    (b) Filter the coolant.
(6) Before grinding each part, clean wheel with minimum use of a silicon carbide stick.
(7) Maintain proper infeeds and crossfeeds.
    (a) Do not exceed 0.0005" infeed per pass.
    (b) Do not exceed 0.080" crossfeed per pass on revolution.

(8) Never spark out - stop grinding after last pass.
(9) Maintain a free-cutting wheel by frequent cleaning with a silicon carbide stick.
(10) Clean parts after grinding.
   (a) Rinse in clean water - then dry.
   (b) Apply a neutral pH rust inhibitor to prevent atmospheric corrosion.
(11) Visually compare the part at 50X with a known quality control sample.

Similarly, a typical check-list[15] for lapping is:

(1) Use a hard lap such as GA Meehanite or equivalent.
(2) Use a serrated lap.
(3) Use recommended diamond abrasives - Bureau of Standards Nos. 1, 3, 6, or 9.
(4) Imbed the diamond firmly into the lap.
(5) Use a thin lubricant such as mineral spirits.
(6) Maintain lapping pressures of 20 to 25 psi when possible.
(7) Maintain low lapping speeds of 100–300 SFPM.
(8) Recharge the lap only when lapping time increases 50% or more.
(9) Clean parts after grinding and between changes to different grade diamond laps - use ultrasonic cleaning if possible.
(10) Visually compare the part at 50X with a known quality control.

COATING STRUCTURE AND PROPERTIES

In this section the macro- and microstructure of plasma and d-gun coatings will be discussed as will several important characteristics in coating design, bond strength, residual stress, and density. In the balance of the section the mechanical, wear, thermal and electrical properties of the coatings will be discussed including a few illustrations taken from service experience.

Macrostructure

The surface roughness of most plasma and d-gun coatings is greater than 100×10^{-6} and usually greater than 200×10^{-6} inches rms. Most of the metallic and cermet coatings are a dull grey, but some, sprayed with an argon shroud, may be a fairly bright metallic, light grey. The oxide coatings vary from black to white with the color frequently differing from the powder or a conventional ceramic part of the same composition. This is usually due to some dissociation and/or oxygen deficiency of the coating. Very slight deficiencies, in some cases, can produce substantial color changes. Exposure to air at high temperatures often returns the oxide to stoichiometry and its normal color without any other noticeable changes in the coating.

Microstructure

Both plasma and d-gun coatings consist of many layers of thin lenticular particles, Figure 11.5, the result of the impact of molten or semimolten powder particles. The major microstructural difference between the two types of coatings is that d-gun coatings have a higher density. The impacting particles may split with some small droplets branching out or separating from the central particle. Thus the average splat volume may be smaller than the average starting powder size, and the total surface area much larger in the coating. Typical powder size, and the total surface area much larger in the coating. Typically a splat may be about 5 microns thick and 10 to 50 microns in diameter.

The cooling rate of the impacting particles has been estimated[16] to be 10^4 to 10^{6}°C/s for oxides and 10^6 to 10^{8}°C/s for metals. It is evident, however, that rates may vary significantly with the substrate material and thickness of the coating. As a result of the rapid cooling, some coatings have been found to have no crystallographic structure by x-ray[17] or neutron diffraction.[18] Others may have a thin amorphous layer next to the substrate followed by crystalline layers.[16] Many coatings form columnar grains within the splat in one or two layers perpendicular to the surface of the substrate, Figure 11.6.

In almost all cases where crystalline structure can be determined by x-ray diffraction the peaks are quite broad, indicative of high local residual stresses due to the rapid quenching. Also as a result of the rapid quenching, nonequilibrium phases may be present; e.g., alumina coatings consist of a high volume fraction of gamma in addition to the equilibrium alpha when the particles are only slightly superheated on impact or the substrate has poor thermal conductivity (or in the outer layers of a thick coating). When the particles are highly superheated and impact on a substrate with high thermal conductivity, delta and theta may be formed in addition to gamma, with alpha suppressed. Similar effects may occur in d-gun coatings, as illustrated in Figure 11.7 for a Laves phase coating.

In addition to phase shifts due to the rapid quench, some changes in composition may occur due to selective evaporation of one component in an alloy, to decomposition to a gas, or to reaction with the atmosphere as previously mentioned. If the loss of a component with a high vapor pressure can be predicted, it can obviously be compensated for in the powder manufacture. It must be kept in mind, however, that such a loss will be more rapid from a fine powder than a coarse powder, and it becomes even more imperative to use a very narrow powder particle size distribution to ensure a homogeneous coating composition. The slight decomposition or loss of oxygen has already been noted relative to color changes. Zirconia coatings are an example of this effect with the coatings being a dirty yellow color that changes to white with a simple air oxidation.

The reaction of the powder particles with their local environment in transit, particularly the extent of their oxidation, is very important to the properties of the coatings. The loss of carbon from tungsten carbide plasma coatings through oxidation of WC to form gaseous CO, W_2C and free tungsten has been reported.[19-21] Metallic or cermet coatings may also react with air inspirated into the plasma stream, as previously noted, forming oxide scales on the particles, or dissolving the gases in the molten droplet. The effects on the properties of the coating can be extensive as will be shown later. The extent of these reactions varies greatly with the type of plasma torch used as shown earlier in Figure 11.3. None of these torches used an inert gas shroud and none were made in an inert gas chamber, yet the extent of oxidation is extremely different. Similar effects can be obtained with d-gun coatings, both by reaction with the detonating gas mixture and with air after the powder leaves the barrel. An example of reaction with the gas mixture is the carburization of a Laves phase alloy for added wear resistance, Figure 11.7.

466 Deposition Technologies for Films and Coatings

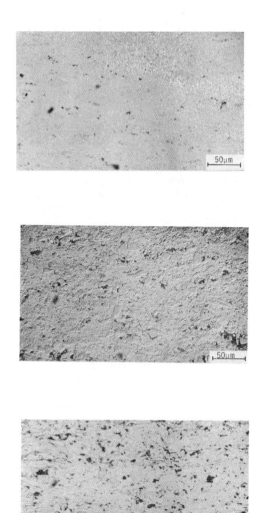

Figure 11.5: Cross section micrographs of d-gun WC-9Co (top), d-gun WC-15Co (center), and plasma WC-12Co (bottom). As-polished. DIC.

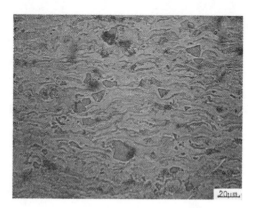

Figure 11.5A: A comparison of the microstructures of cross sections of plasma and detonation gun tungsten carbide-cobalt coatings; ~9 wt/o Co d-gun (top), ~15 wt/o Co d-gun (center), ~12 wt/o Co plasma (bottom). DIC, As-polished.

468 Deposition Technologies for Films and Coatings

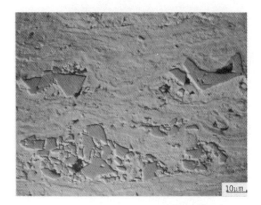

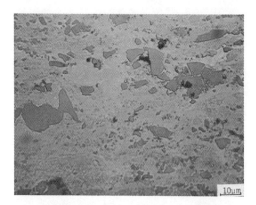

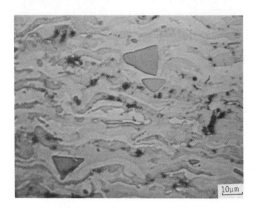

Figure 11.5B: A comparison of the microstructures of cross sections of plasma and detonation gun tungsten carbide-cobalt coatings; ~9 wt/o Co d-gun (top), ~15 wt/o Co d-gun (center), ~12 wt/o Co plasma (bottom). DIC, As-polished.

Figure 11.5C: A comparison of the microstructures of cross sections of plasma and detonation gun tungsten carbide-cobalt coatings; ~9 wt/o Co d-gun (top), ~15 wt/o Co d-gun (center), ~12 wt/o Co plasma (bottom). DIC, Etched in sat. $KMnO_4$–8KOH.

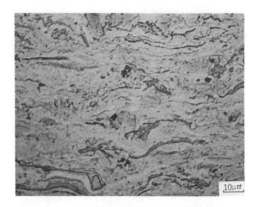

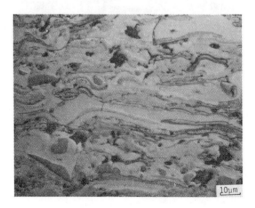

Figure 11.5D: A comparison of the microstructures of cross sections of plasma and detonation gun tungsten carbide-cobalt coatings; ~9 wt/o Co d-gun (top), ~15 wt/o Co d-gun (center), ~12 wt/o Co plasma (bottom). DIC, Etched in sat. $KMnO_4$–8KOH.

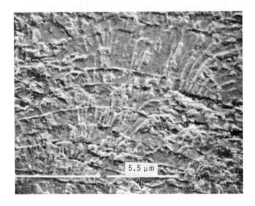

Figure 11.6: Scanning electron micrographs of a fractured plasma deposited tungsten coating above and a cross section of the same coating polished and etched showing the columnar grain structure within the lenticular particles.

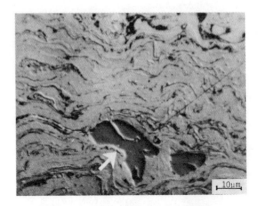

Figure 11.7: Cross section of a Laves phase d-gun coating (LDT-400), as-coated (above) and after 4h at 1080°C in vacuum (below) illustrating the metastability of the as-coated structure. The arrow identifies a carbide formed by reaction with the detonation gases during deposition. Etched, DIC.

Bond Strength

Bond strength is, quite naturally, an important property of a coating. It is most frequently measured in a tensile test (ASTM-633) in which the coating is applied to the face of a one inch diameter round bar and a mating bar is epoxied to it. The limit of the test is the strength of the epoxy, currently about 10,000 to 12,000 psi. Most plasma coatings have bond strengths below this, but most d-gun coatings and a few plasma coatings have strengths that exceed it with the test serving only as a "proof" test. In a few instances where the mating bar was brazed to a d-gun coating, the tensile strengths exceeded 25,000 psi. Some reduction of residual stress with enhancement of bond strength could have occurred during brazing, however.

The ASTM test procedure specifies a coating thickness of at least 0.020 inches. This was established to prevent penetration of the epoxy through porous coatings, such as oxy-acetylene flame spray coatings, and is frequently reduced for use with the denser d-gun and plasma coatings to more closely measure their strength at the thicknesses more typically used in service; e.g., 0.010 to 0.012 inches.

A variety of other tensile bond strength and shear strength tests have been used, but most introduce undue stress risers. An epoxied lap shear test is still used for some quality control purposes. It is unfortunate that more satisfactory shear tests have not been developed, since the coatings are more often loaded in shear than in tension.

The mechanism of bonding of plasma deposited coatings in many respects is still in dispute.[22-24] Mechanical interlocking has been considered the most important mechanism by most investigators.[23] Grit blasting, as previously discussed, provides an ideal surface topology for interlocking and it has been shown in several studies,[25-27] that bond strength increases with increasing surface roughness in both shear and tensile tests, although it may diminish again above $250 - 300 \times 10^{-6}$ inches, rms. Few published reports have taken into account the detrimental effects of surface embrittlement, peak blunting, and grit inclusions[19,28,29] when excessive grit blasting is used.

Other mechanisms have been suggested as contributing to the bond strength including van der Waals forces, interdiffusion or alloying, epitaxy, oxide cementation or spinel formation, and surface reaction. There is some microstructural evidence that when the refractory metals, tungsten or molybdenum (with high melting points and heat capacities), are sprayed on steel or nickel or chromium on aluminum, there is some interdiffusion; i.e., a metallurgical bond is formed.[30] Similar results are reported for the so-called exothermically reacting nickel aluminide coatings.[31] Interdiffusion and the more nebulous surface reaction may depend in part on added surface energy in the substrate due to grit blasting, as evidenced by recrystallization of the surface.[32] The bond strength of ceramic coatings is generally attributed to interlocking, but some degree of spinel formation or similar reaction has been reported for Al_2O_3 on steel[17] and Al_2O_3/TiO_2 on aluminum.[33]

Oxide cementation was cited as important in bonding metals or cermets to metals[34,35] in earlier work, but is not considered desirable in modern practice. In general, oxides on the surface of the substrate or oxidation occurring during spraying[36] decrease bond strength.

Most of the factors that affect the bond strength of plasma deposited coatings also apply to d-gun coatings, but the situation is somewhat more complex. Because of the unusually high velocity of the particles, they are actually driven into the surface of most metallic substrates. As previously noted, some substrates require no grit blasting to achieve adequate bonding, since the coating itself roughens the interface. This embedding/roughening process creates atomically clean interfaces between the coating and substrate over most of the coating area which facilitates chemical bonding and can be likened to the explosive bonding of sheets of metals. This undoubtedly plays a large role in forming the unusually high bond strengths of d-gun coatings.

Residual Stress

Residual stress has already been discussed to some extent, but a few additional remarks may be in order. It occurs as a result of cooling individual powder particles or splats from above their melting point to the temperature of the part. The magnitude of the residual stress is a function of torch parameters, deposition rate, the relative torch to part surface speed, the thermal properties of both the coating and the substrate, and the amount of auxiliary cooling used. It has also been found that the use of finer powders leads to higher residual stresses, but this can generally be controlled by adjusting the coating parameters. If the part temperature is allowed to rise above room temperature, there will be a secondary change in the state of stress of the coating as both the part and the coating cool to room temperature. Residual stress increases linearly with coating thickness above some minimal initial thickness.[37-39] The rate of increase, however, is a function of the parameters of deposition already listed and the coating material.

Residual stress has a significant effect on bond strength, as already noted, and must be considered when the coating is placed in service, since it detracts from the inherent mechanical strength of the coating. For example, coatings are normally in tension as a result of the residual stress, and this stress must be subtracted from the allowable fracture stress calculated from mechanical property tests of free standing specimens. Residual stress is, however, reproducible and can be accounted for with adequate knowledge of the stress and adequate control of the coating parameters.

Density

The density of detonation gun coatings is greater than 95 percent of theoretical, usually greater than 98 or 99 percent. This high density is due, as with other properties, to the unusually high kinetic energy of the particles on impact. Plasma coatings have densities varying from less than 80 to 95 percent of theoretical with some of the high velocity coatings currently being reported with densities greater than 95 percent. The density of a plasma sprayed coating is, of course, a function of the deposition parameters. In addition, it is a function of the powder size, as illustrated in Table 11.2[40] for tungsten carbide-cobalt, with finer powders producing denser coatings. The same effect has been noted in many other systems; e.g., Ni-Cr-Al[41,42] and chromium carbide-nickel aluminum.[26] It has also been shown that oxidation during deposition can decrease coating density as shown by comparing the densities (and tensile strength) of stainless steel and aluminum coatings sprayed in argon with those sprayed in air), Table 11.3.[43] The combined effects of powder size and oxidation during deposition are shown for tungsten, in Table 11.4.[43] Similar results were found for nickel.[26]

Table 11.2: Effect of Powder Size on the Structure of Plasma Deposited Tungsten Carbide

	........Powder Size.........		
Coating Property	Coarse (10-105 μ)	Medium (10-74 μ)	Fine (10-44 μ)
Apparent density, g/cc	10.5	13.0	14.2
Bulk density, g/cc	8.7	11.1	13.0
Percent of theoretical	60	77	89
Porosity	17.1	14.5	8.5
Apparent hardness, Kn500	538	684	741

Table 11.3: Properties of Plasma-Deposited Coatings Sprayed in Argon and Air

Coating	Atmosphere	Density (%)	Tensile Strength (psi)	Elongation (%)	Reduction in Area (%)
Stainless steel	Argon	91	33,900	1	1
	Air	84	19,200	1	1
Aluminum	Argon	86	5,600	1	1
	Air	76	4,000	1	1

Table 11.4: Properties of Plasma-Deposited Tungsten

Powder Size	Deposition Atmosphere	Density (%)	Modulus of Rupture (psi)	Average Grain Diameter (μ)
−200+325 mesh	Argon	90	31,900	3.5
	Argon*	70*	21,700	3.5
	Air	86	17,000	3.0
−400 mesh+10 μ	Argon	91	51,000	2.2
	Air	85	29,000	1.5

*Intentionally produced with low density.

The porosity in plasma and detonation gun coatings is partially interconnected and hence may have a strong influence on the corrosion rates of the coatings in some environments. Some detonation gun coatings have been shown to have sufficiently small pores as to be unimportant in oxidation.[44] This will be discussed more fully in a subsequent section.

Mechanical Properties

The mechanical properties of plasma and d-gun coatings are sensitive to the deposition parameters used, the substrate, cooling, etc. Therefore any general tabulation of properties based only on coating composition would be meaningless. Moreover, most of the data has been generated on specimens coated under ideal conditions of angle and standoff while in many service applications both of these variables may be less than ideal on part or all of the coated area. Nonetheless, a good deal of data has been compiled on a wide range of coatings to serve as very useful guidelines to equipment designers and other users. For purposes of illustration in subsequent discussion and to provide a general feeling of comparison with other types of materials, the mechanical properties of a few specific coatings are listed in Table 11.5.

The mechanical (as well as other) properties of plasma and d-gun coatings are anisotropic because of their splat structure and directional solidification. This anisotropy is probably more pronounced for cermets and metallic coatings with somewhat oxidized splat boundaries than it is for either pure ceramic or pure metallic coatings. An example of this anisotropy is given in Table 11.6.[45] Although most coatings are used with loading normal to the surface, measurement of mechanical properties normal to the surface is particularly difficult because of the limited thickness of most coatings and is seldom done. Properties parallel to the surface are also important, however, particularly if the substrate expands or contracts thermally or under mechanical loading.

The most frequently quoted mechanical property is hardness. The hardness of a detonation gun coating is generally higher than that of a plasma coating of same composition as shown in Table 11.5. This is primarily due to their higher density and greater cohesive strength. For a plasma coating with a given composition, the hardness usually increases with increase in density. Thus, for example, hardness generally increases with the use of a finer powder, as already shown in Table 11.2. Hardness is usually reduced for a given material if the coat is applied in an inert atmosphere as compared to spraying in air, as has been noted for WC-Co,[36] for Mo[46] and for Ti, Nb, and Zr.[47] It should be noted, however, that excessive oxidation, although it may increase the hardness of the coating, will weaken its internal cohesive strength and thus may be detrimental to the coating's performance.

476 Deposition Technologies for Films and Coatings

Table 11.5: Properties of D-Gun and Plasma Coatings

Coating Characteristics	LW-1	LW-1N40	LW-11B	LA-2	LA-6	LN-2	LS-31
Composition, wt %	91 Tungsten carbide 9 Co	85 Tungsten carbide 15 Co	88 Tungsten carbide 12 Co	>99 Al$_2$O$_3$	>99 Al$_2$O$_3$	>99 Ni	25 Cr 10 Ni 7 W Bal Co
Method	D-gun	D-gun	Plasma	D-gun	Plasma	Plasma	Plasma
Diamond pyramid hardness, kg/mm^2 300 g load	1,300	1,050	750	1,100	825	200	350
Tensile bond strength, psi*	>12,000**	>12,000**	10,000	10,000	7,500	10,000	10,000
Modulus of rupture, psi	80,000	100,000	55,000	20,000	20,000	55,000	53,000
Modulus of elasticity, 10^6 psi	31	31	22	14	5.7	14	8
Density, g/cm^3	14.2	13.2	12.5	3.4	3.38	7.5	8.0
Metallographic porosity, vol %	½	1	2	2	3	1	2

*Determined with ASTM method C633-69. Bond strength varies somewhat with the substrate composition, and the values shown are for either steel or aluminum.

**Bond strength of these coatings produced by the Detonation process is too high to be measured by ASTM method C633-69 which is used to measure bond strengths of plasma applied coatings. Special laboratory methods for testing detonation coating bond strengths reveal values in excess of 25,000 psi.

Table 11.6: Mechanical Properties of Cu-2Be* in Compression

	Parallel to Surface	Perpendicular to Surface
Elastic modulus	13×10^6 psi	10×10^6 psi
0.2% yield strength	82×10^3 psi	73×10^3 psi
Ultimate strength**	97×10^3 psi	164×10^3 psi
Strain to fracture**	3.3%	26%

*Union Carbide, UCAR LCU-3.
**Function of specimen geometry.

Hardness is used not only as a guideline for wear resistance, but for the strength of the coating. In both cases it may be quite misleading. The measurements of hardness are usually made on metallographic cross-sections of the surface, even though loading is usually perpendicular to the surface and the hardness in the two directions may be different due to the anisotropic microstructure of the coatings. It should also be kept in mind that hardness measurements made on test specimens may differ from those on actual parts due to differences in angle of deposition and standoff and, in some cases, due to differences in residual stress.

The following is an example of a situation in which hardness, used as a guide to wear resistance, was the initial criterion for coating selection and too little weight given to impact resistance or toughness:

> Virtually all midspan shrouds on gas turbine engine compressor blades have a d-gun tungsten carbide-cobalt coating. In the initial development of this application, the most wear resistant grade of tungsten carbide-cobalt with a hardness of 1300 VPN_{300} and a nine percent cobalt binder was tried. This coating was found to fail, however, not from typical wear, but because of surface fatigue which resulted in spallation of the coating. Success was achieved when a more impact resistant grade of tungsten carbide with a hardness of 1050 VPN_{300} in a 15 percent cobalt binder was tried. The greater "toughness", combined with a still superior wear resistance of this coating makes it the only production approved coating for midspan shrouds today with extremely few exceptions.

The modulus of rupture, elastic modulus and strain-to-fracture in bending plasma and d-gun coatings is measured more often than conventional uniaxial tensile and compressive properties because the former measurements can be made on easily fabricated free-standing rings of coatings as thin as 0.010 inch. On the other hand, it is often difficult to produce thick enough coatings for conventional specimens. This difficulty arises from the thickness limitations of some coatings due to residual stress and the inherent brittleness of the coatings. Even most metallic coatings have a strain-to-failure of less than one percent. Some typical values from ring tests are shown in Table 11.5. It is evident that the d-gun coatings have a higher modulus of rupture than comparable plasma coatings; for example, the tungsten-carbide cobalt coatings. Also note that, as expected, increasing the cobalt content increases the strain-to-fracture (either measured directly or calculated from the ratio of modulus of rupture to the elastic modulus).

An example of the use of this kind of test data is as follows:

A plasma chromium oxide coating was specified on the interior surface of an aluminum hydraulic cylinder in an aircraft landing gear, because of its earlier success on another landing gear and cyclic pressure bench testing on prototype cylinders without pistons. Even though visual examination of the bench tested cylinders revealed no irregularities, the coating failed when the complete assembly was placed in service. The cylinder expansion under pressure clearly exceeded the strain-to-failure of the coating and it cracked. The additional stress of the piston caused spalling. Re-examination of the bench tested cylinder revealed microscopic cracks. The designer had not adequately taken into account the difference in elastic moduli between the coating and the substrate and the limited strain-to-failure of the coating.

Evaluation of data from ring tests indicated both aluminum bronze and nickel coatings had adequate strain-to-failure. Subsequent tests verified this as well as the fact that they had sufficient wear resistance. The aluminum bronze was chosen for service, and, even though it does not have the wear resistance of chromium oxide, this design compromise has proven to be successful in service, providing more than adequate life. Nickel has been used successfully in similar situations as well.

It is obvious, of course, that all the coating process variables and the resulting microstructures strongly affect the mechanical properties of the coating. For example, tungsten coatings made with fine powder have a higher modulus of rupture than those made with coarse powder when both are protected from oxidation by spraying in an inert atmosphere,[43] as shown in Table 11.4. Referring to Tables 11.3 and 11.4, it is apparent also that oxidation during deposition can seriously weaken a coating. In a study of the effect of oxidation on aluminum bronze it was found that even minor oxidation during deposition was detrimental to compressional strength both parallel and perpendicular to the surface, Table 11.7.[45] Additions of discrete oxide particles, on the other hand, not only strengthened the coatings, but added wear resistance (discussed elsewhere).

Table 11.7: Mechanical Properties of Plasma-Deposited Aluminum Bronze (Cu-10Al)

Type of Deposition	Alumina Addition	Volume Percent Al_2O_3	Hardness** DPH_{300}	Compressional Properties*			
				Perpendicular to Surface		Parallel to Surface	
				E (10^6 psi)	YS (10^3 psi)	E (10^6 psi)	YS (10^3 psi)
Standard	No	2.36	246	7.1	58	8.4	62
Oxidizing	No	3.20	200	2.9	47	7.0	42
Standard	Yes	7.26	170	—	—	—	—
Standard	Yes	12.4	202	8.5	73	9.7	89
Standard	Yes	21.0	246	—	—	—	—
Standard	Yes	40.5	186	—	—	—	—
Oxidizing	Yes	10.8	142	7.6	57	10.9	70
Wrought (AMS 4640)	—	—	252	—	—	—	—

*E is the elastic modulus, YS is the 0.2% yield strength.
**Hardness perpendicular to surface.

Before leaving the subject of mechanical properties, it might be well to mention that the properties of the substrate can not be ignored in considering a coating application. One of the first considerations is that the substrate must be able to support the coating without yielding beyond the coating's strain-to-failure. For example:

> A d-gun tungsten carbide-cobalt coating has extended the life of roller guides used in steel mill pickle lines more than ten times. In the first trial of this coating, however, the coating occasionally cracked and spalled because the substrate yielded under the heavy impact of the steel sheet. This deformation exceeded the strain-to-failure of the coating. When a change was made to insure that all substrates had a hardness greater than 55 R_c, no failures were experienced.

Also of concern in some applications is the effect of the coating on the fatigue life of the substrate. Some coatings, particularly d-gun coatings, are so well bonded that a crack generated in the coating may propagate into the substrate under sufficient cyclic stress. The results of a number of studies, especially by airframe and gas turbine engine manufacturers, suggest that so long as the strain-to-failure of the coating is not exceeded, the coating will have no measurable effect on the fatigue strength of the substrate. More work needs to be done, however, before the effects of a specific coating on a given substrate can be predicted without experimental verification. In those cases where stresses are very high and the component is particularly susceptible to fatigue, care should be taken to prevent both direct coating and overspray. For example:

> The mid-span shrouds or stiffeners used on many titanium compressor blades must be coated with a d-gun tungsten carbide-cobalt coating, as previously mentioned. The root area of the mid-spans is extremely sensitive to fatigue and all coating and overspray must be excluded. This is successfully achieved by either very careful masking or directing the coating away from the radii during deposition. No fatigue failures of these blades have ever been attributed to the coating.

Wear and Friction

The major use of plasma and d-gun coatings today is for wear resistance, particularly for adhesive and abrasive wear resistance. Their use in erosive situations is growing steadily as well, particularly for d-gun coatings. No attempt will be made here to tabulate the wear resistance of coatings or conversely to recommend specific coatings for the various types of wear. To do so it would be necessary to assume that all coatings of a given composition are the same (while in fact they are a function of the specific coating device and operating parameters used) and to assume that all wear situations can be fit into a relatively few, well defined categories (which is definitely not the case). The situation is far from hopeless, however, and experienced coatings service engineers or equipment manufacturers can, after analysis of a specific situation, usually make reasonably accurate recommendations of one or two coatings that will solve the problem. Some of the considerations that are involved are listed in Table 11.8.

Although no specific recommendations will be made here, a few general comments may be in order. Hardness is a useful first approximation to abrasive and adhesive wear resistance so long as materials of the same type and general composition are compared. For example, a d-gun WC-9Co coating is harder and more wear resistant than d-gun WC-15Co which, in turn is more wear resistant than a plasma WC-13Co coating, Table 11.9. (An example, compressor midspans, has already been cited of an application where hardness and wear resistance had to be tempered with toughness for success, however.) Hardness can be misleading, however, when comparing coatings with wrought materials of the same composition. For example:

> Plasma-deposited aluminum bronze or beryllium copper coatings are softer than their wrought counterparts. In an adhesive wear test under boundary lubrication conditions simulating many bearing applications, the plasma coatings were far more wear resistant as shown in Figure 11.8.[49]

Table 11.8: Considerations in Coating Selection for Wear Resistance

I. Wear System
   A. Adhesive or Abrasive
      1. Type of relative motion - unidirectional, oscillating, impact
      2. Surface speed - velocity and frequency if cyclic
      3. Load or impact energy
      4. Abrasive particles or wear debris - trapped or removed, size and composition
      5. Conformability requirements
      6. Embedability requirements
   B. Erosive
      1. Gas or solid particle
      2. Media - gas or liquid
      3. Gas or particle velocity and angle of impingement
      4. Particle size, shape, mass, and composition.

II. Environment
   A. Temperature - maximum, minimum and rate of change
   B. Media - gas or liquid
   C. Contaminants
   D. Corrosive characteristics - chemical, galvanic
   E. Lubricant

III. Mating Material
   A. Composition
   B. Heat treatment condition
   C. Hardness
   D. Surface roughness and topology

IV. Substrate Material
   A. Composition
   B. Heat treatment condition
   C. Dimensional changes after coating
      1. During assembly due to press fit, shrink fit, etc.
      2. In service due to thermal expansion/contraction or mechanical loading

V. Coating Requirements
   A. Cost limitations
   B. Required life, time or maximum wear
   C. Compositional limitations
   D. Thickness limitations
   E. Coefficient of friction requirement
   F. Surface finish
   G. Geometric constraints
   H. Overspray limitations

Table 11.9: Wear Tests

Material	Dry Rubbing Wear Rate (10^{-6} in/1000 ft of sliding)	LFW-1* (10^{-6} cm^3)
D-gun tungsten carbide-cobalt	35	10
Plasma tungsten carbide-cobalt	80	23
52100 steel (wrought)	2,000	—
Hard chrome electroplate	3,600	44

*450 lb/load in hydraulic fluid vs steel for 5,400 rev.

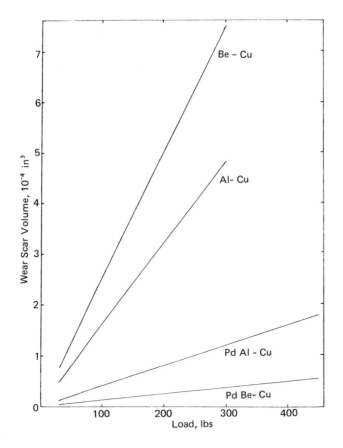

Figure 11.8: Alpha block-on-ring wear test of aluminum bronze, wrought (Al-Cu) and plasma sprayed (PD Al-Cu), and beryllium copper, wrought (Be-Cu) and plasma sprayed (PD Be-Cu) versus SAE 4640 steel (R_c 60) in hydraulic fluid at 65 ft/min for 1950 ft.

There are also situations in the comparison of coatings where hardness can be misleading, particularly in adhesive wear with coatings of somewhat different compositions. For example:

As shown in Table 11.7, the addition of an oxide dispersion to an aluminum bronze coating slightly reduces its hardness, yet in an adhesive wear test its wear resistance is increased significantly as shown in Figure 11.9.[48] Similar results were obtained with carbide additions. It should be noted that this increase in wear resistance should not affect the conformability and embedability of the basic aluminum bronze. It is also important to note again that an intentional oxide dispersion deposited under conditions that do not significantly oxidize the metal matrix is far superior to a coating heavily oxidized during deposition, both in wear resistance and mechanical properties.

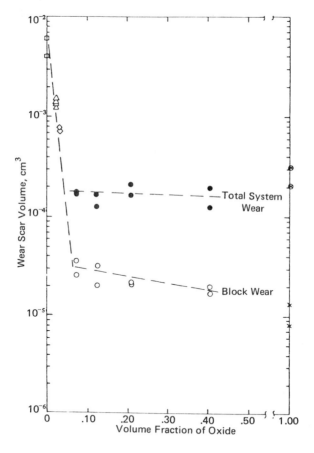

Figure 11.9: Alpha block-on-ring wear test of aluminum bronze with Al_2O_3 addition (block) versus SAE 4640 steel (R_c 60) in hydraulic fluid at 65 ft/min for 1950 ft/min under a 180 lb load. (□) wrought alloy; (△) standard plasma; (◇) oxidizing plasma (●, ○) alumina additions to plasma (⊗, X) plasma pure alumina.

Physical and chemical compatibility of the mating surfaces are, of course, important in selecting a coating. Laboratory testing can be an excellent guide in this aspect of selection so long as the other considerations (temperature, load, lubricant, etc) are similar to those in service. A few examples of satisfactory and unsatisfactory combinations are shown in Table 11.10.[45]

Table 11.10: Mating Materials Selection for Dry Rubbing Wear

Materials	Rating	Wear*	Coefficient of Friction
D-gun WC-Co** vs GA Meehanite	Excellent	33	0.08 at room temperature 0.11 at 400°F
440C stainless	Good	35	0.34 at room temperature 0.25 at 1000°F
Inconel X	Poor	562	0.53 at room temperature 0.42 at 1000°F
D-gun WC-Co	Good	39	0.46 at room temperature 0.33 at 1000°F
D-gun Al_2O_3 vs Haynes 25	Excellent	16	0.25 at room temperature 0.17 at 1400°F
Hastelloy C	Good	35	0.32 at room temperature 0.10 at 1400°F
D-gun Al_2O_3	Poor	245	0.24 at room temperature 0.27 at 1400°F

*Total system wear, 10^{-6} in/1000 ft.
**Rotating member.

In some applications both surfaces can be coated. Combinations of very hard coatings such as chromium oxide provide excellent self-mating characteristics when no conformability or embedability of either surface is required. When these are required, a combination such as plasma-deposited aluminum bronze with an oxide dispersion versus several types of hard d-gun coatings offer both mechanical compliance and greatly increased wear resistance.

The erosion resistance of conventional plasma deposited coatings is not very high, probably because of their porosity and relatively low cohesive strength. D-gun and perhaps some high velocity plasma coatings, on the other hand, have shown exceptional erosion resistance in some applications. For example:

> Some compressor blades of gas turbine engines suffer from severe particle erosion. Tests have shown that d-gun coatings of tungsten, titanium carbide-nickel and tungsten carbide-cobalt coatings tremendously increase the lives of the blades. The coatings are only applied to the outer portion of the blades where the erosion problem is most severe to reduce their cost and avoid any potential fatigue effects on the blades.

Coatings can be used to adjust the frictional characteristics of a system whether or not wear is a problem. Thus, for example, coatings might be used to reduce power losses through frictional heating.

Reduction in frictional heating can also extend the life of mating organic materials. For example:

> A manufacturer of bearing and sealing systems for the power trains of large ships was experiencing difficulties with the forward seal for the tailshaft. This seal is located well inside the ship and is not cooled by the outside water. The seal is

formed by mating a rotating shaft liner with stationary rubber seals. The liner, made of a special alloy, was sufficiently wear-resistant without a coating; however, the heat generated by friction caused a rapid deterioration of the rubber seals. The problem was solved by (a) incorporation of a cooling device for the oil in the system and (b) reducing the friction at the seal by the adoption of a specially finished plasma-deposited chrome oxide coating on the rotating shaft liner. The special finish minimizes contact with the mating rubber material while still maintaining the necessary seal. The rubber seal now operates at a lower temperature, and its life is significantly extended. Needless to say, there is no sacrifice in the wear life of the liner with the addition of the coating.

Occasionally it is necessary to prevent self-welding between essentially static components and ensure that the static coefficient of friction will be low enough to prevent equipment start-up failure. For example:

The sodium-cooled breeder reactor requires that both of these criteria be met by the load pads on the fuel ducts. Uncoated stainless steel, stripped of its oxide film by the sodium, is self-welding. Extensive testing[14] has shown that the best solution to the problem is a d-gun coating of chromium carbide-nichrome.

It is evident from the preceding that the surface finish of a coated surface is extremely important. The smoothest finish that can be obtained on a given coating is a function of its composition and the method of deposition. However, it should be borne in mind that the lowest coefficients of friction are not always obtained with the smoothest surface. A nodular brush finish, for example, provides the best frictional behavior in liquid sodium.

In other applications specific, intermediate-range coefficients of friction are used. For example:

Textile machinery components, such as snick plates, tension gates, and draw rolls are in contact with fast moving fibers being processed and are, of course, subject to high rates of wear. Hard plasma coatings are used to correct this wear. Equally important, however, the coating finish must provide rather precise intermediate frictional properties to hold the fiber in constant tension. The tension control is essential in order to prevent slack or breakage of the fiber.

High friction forces are required in many types of drive mechanism. Coatings can often meet this requirement and provide longer life than most other solutions. For example:

Many large rollers in sheet steel production rely on high surface friction to move the steel sheet through the line without slippage. Others require this gripping action in order to tightly wrap the steel sheet into non-telescoping coils. Experience has shown that a d-gun tungsten carbide-cobalt coating, used as-coated or slightly roughened by controlled grit blasting, resists wear, grooving, and gouging 6 to 40 times longer than the previously used hardened steels or chrome plate. An additional benefit of the coating is its resistance to the transfer of material from the steel sheet to the roll, which is a rather common problem with other materials.

Thermal

The thermal properties of coatings are important both during their formation and in elevated temperature applications. The effects of thermal contraction from their freezing point during coating formation have already been mentioned. Conversely, the relative thermal expansion of the coating and substrate if heating occurs during service is important. As a rough estimate of the strain that may be placed on a coating as a result of such heating, handbook values of coefficients of thermal expansion may be used. Care must be taken to insure that the values used are those for the phases actually present in the coating. Because of their lamellar microstructure, the thermal conductivity of coatings is lower than that of solid, fully dense materials of the same composition. Their absorption characteristics may be very different as well because of their surface topology and, in some cases, slight shifts in composition (already mentioned for some oxides).

One of the most common uses of coatings for their thermal properties is as oxide thermal barriers:

> For purposes of discussion, thermal barriers may be divided into two categories—relatively thin ones, less than about 0.020 inches thick, and thicker ones, up to about 0.25 inches thick. The thinner thermal barriers have been used for years on gas turbine engine combustion chambers and are currently being considered for turbine blade and vane airfoil surfaces, thrust reversers, diesel and combustion engine piston heads and valves, and many other applications. In addition to a low thermal conductivity, these coatings must be resistant to corrosion, thermal shock, gas erosion and, sometimes, particle erosion. They usually consist of a metallic undercoat such as nickel-chromium, nickel-aluminum, or an MCrAlY alloy (where M is Ni, Co or Fe) and an outer layer of an oxide, usually zirconia or magnesium zirconate. Occasionally one or more intermediate layers of mixtures of metal and oxide or a continuous gradation from pure metal to pure oxide is used. This approach improves thermal shock resistance, but if the temperature in service at the first zone of mixed metal and oxide is too high the metal will rapidly oxidize (since the oxide layer is permeable to air) and cause spallation of the outer portion of the coating.[52] This same thing will happen to the metallic undercoating in a two layer coating if it has inadequate oxidation resistance because of its composition or being too porous.
>
> In thick thermal barrier systems, it is necessary to use an essentially continuous gradation from metal to oxide or multiple layers with increasing oxide content to have adequate thermal shock resistance. These thick thermal barriers are being investigated on outer air seals in advanced gas turbine engines. In this case the already complex task of increasing thermal shock resistance without sacrificing oxidation resistance and erosion resistance is complicated by the need for abradability.

Corrosion Properties

Obviously the use of a coating in a corrosive environment requires that the coating itself resist the corrodant, but it should be kept in mind that the corrosion resistance of a wrought, cast or sintered composition may change when plasma or d-gun deposited. For example, alpha alumina is very corrosion resistant, but plasma sprayed alumina is a mixture of phases all of which are not corrosion resistant.

All plasma and d-gun coatings have varying degrees of interconnected porosity that allow attack of the substrate in corrosive environments. In most cases at temperatures up to about 350°F this problem can be overcome by the proper selection and application of a sealant. For example:

> Bronze shaft sleeves running in centrifugal pumps handling saturated brine in a chlorine processing plant were being rapidly worn beyond tolerance. They run against asbestos-filled Teflon with no lubrication at 250°F. The solution was a machineable metallic plasma undercoat to restore the sleeves to size followed by plasma-deposited chromium oxide coating that was sealed with epoxy to prevent substrate and undercoat corrosion. The coating was ground to a 4 to 6 x 10^{-6} inch rms, surface. The coated sleeves not only salvage worn parts, but outlast the original sleeves several times and reduce downtime.

Galvanic corrosion can occur in some environments, most commonly salt water, when an improper selection of coating composition is made. For example:

> An aluminum bronze coating on an aluminum substrate creates a galvanic cell in the presence of an electrolyte. In aircraft landing-gear bearings with this coating/substrate combination, galvanic corrosion of the substrate was observed when the hydraulic fluid became contaminated with salt water in transoceanic service. The problem was solved by sealing the coating in one case and avoiding the problem in another by selecting a modified aluminum coating with an electrostatic potential virtually identical to the substrate.

For corrosion resistance at elevated temperatures, plasma deposited coatings must be sealed by sintering, sometimes combined with mechanical surface treatment. For example:

> Some gas turbine blades and vanes, depending on the type of fuel and operating environments, are subject to hot corrosion. The best solution to this problem at the present time is an MCrAlY (where M is Ni, Co, and/or Fe) type of coating. To date these have been applied by physical vapor deposition, an expensive method with some elemental limitations. Plasma deposition would offer significant economic advantages and have no elemental limitations. To be effective, however, the coatings must be deposited without oxidation and then sealed to prevent rapid internal oxidation of the coating and oxidation of the substrate. Methods have been developed to achieve this using inert gas shrouding during deposition and post-

coating heat treatment to effectively sinter the coating. Alternative methods using deposition in a vacuum or inert chamber and a partially transferred arc are also being developed. In this case, since the substrates are superalloys, the coating heat treatment can be combined with or precede the alloy heat treatment and not interfere with the structural properties of the component.

D-gun coatings, because of their high density, often do not need to be sealed with a high temperature sintering to prevent internal oxidation or oxidation of the substrate. For example:

> For many years d-gun coatings of chromium carbide-nickel chromium have been used on the shroud edges and lacing wire of gas turbine engines to prevent fretting and impact wear. More recently two new families of cobalt base alloys with oxide additions have been developed[44] to provide better performance for more advanced engines. These coatings do not require heat treatment to prevent internal oxidation. (Heat treatment is used to further improve the already superior bond strength in particularly severe impact situations.)

Electrical Characteristics

The microstructure of metallic coatings has an effect on the electrical conductivity similar to that on the thermal conductivity. Thus the resistance is higher than that for wrought alloys of the same composition, and it is somewhat higher perpendicular to the surface than parallel. The conductivity of coatings deposited with an inert gas shroud or in an inert gas chamber with very little oxidation during deposition is much higher than conventional coatings, since the conductivity is particularly sensitive to oxide films in the splat boundaries.

Coatings are used as both conductors and insulators. The use of oxides as an insulator is fairly obvious, but the flexibility that this type of coating offers the designer is often overlooked. For example:

> Aluminum oxide coatings applied to the tips of pliers, screwdrivers, and diagonal cutters for electrical insulation are especially useful in work on confined electrical circuit installations. The coating guards against short-circuiting which would otherwise be possible during accidental contact with adjacent terminals.

> In a steel mill ferrostan tin line, where steel sheet is tin-plated, wringer rolls are used to remove water from the stock. These rolls are rubber-coated for electrical insulation. One such roll, coated with an insulative plasma-deposited aluminum oxide instead of rubber, is in service on an experimental basis. The aluminum oxide coating resists the wear and grooving which in the rubber coated rolls eventually allow arcing and subsequent "arc burns" on the sheet steel.

Coatings are usually used as conductors when the application simultaneously requires wear resistance and/or corrosion resistance. For example:

> Also operating in steel mill ferrostan tin lines are rollers designed to conduct electricity to the sheet stock during plating.

Typically, the conductive surface of the roller has been clad copper. Recent steel mill experience has shown a plasma-deposited tungsten coating to be a better material selection. The conductivity of the tungsten coating is more than adequate, and it is far more resistant than copper to wear, grooving and gouging.

REFERENCES

(1) R.M. Poorman, H.B. Sargent, and H. Lamprey, "Method and Apparatus Utilizing Detonation Waves for Spraying and Other Purposes", U.S. Patent 2,714,553, August 2, 1955.
(2) R.M. Gage, O.H. Nestor, and D.M. Yenni, "Collimated Electric Arc Powder Deposition Process", U.S. Patent 3,016,447, January 9, 1962.
(3) E. Muehlberger, "Coating Heat Softened Particles in a Plasma Stream of Mach 1 to Mach 2 Velocity", U.S. Patent 3,914,573, Oct. 21, 1975.
(4) A.J. Fabel and H.S. Ingham, "Plasma Flame-Spraying Process Employing Supersonic Gaseous Streams", U.S. Patent 3,958,097, May 18, 1976.
(5) E. Muehlberger and R. Kremith, "New Sonic and Supersonic 80 kW Plasma Spray Systems", presented at Ninth Airlines Plating Forum, Montreal, Canada, 1973.
(6) "Selected Coating Properties - The 7M High Energy Plasma System", Metco, Inc., 1975.
(7) F.J. Wallace, "High-Energy Plasma-Sprayed Tungsten Carbide Cobalt Development for Turbine Applications", presented at 14th Annual Airline Plating Forum", Tulsa, Okla., 25–27 April 1978.
(8) H. Meyer, "Fusion of Powder in a Plasma Jet", *Ber. Dtsch. Keram. Ger. 39* (H2) (1963) 115–124.
(9) J.E. Jackson, "Method for Shielding a Gas Effluent", U.S. Patent 3,470,347, 1969.
(10) H. Kayser, "Spraying Under an Argon Atmosphere", *Thin Solid Films, 39* (1976) 243–250.
(11) M.O. Price, T.A. Wolfla and R.C. Tucker, Jr., "Some Comparative Properties of Laves- and Carbide-Strengthened Coatings Deposited by Plasma or Detonation Gun", *Thin Solid Films, 45* (1977) 309–319.
(12) A.J. Fabel, "Powder Feed Device for Flame Spray Guns", U.S. Patent 3,976,332, 24 August 1976.
(13) R.F. Smart and J.A. Catherall, *Plasma Spraying,* Mills and Boon, Ltd., London, 1972.
(14) T.A. Wolfla and R.N. Johnson, Refractory Metal Carbide Coatings for LMFBR Application - A Systems Approach", *J. Vac. Sci. Technol. 12* (1975) 777–783.
(15) "Finishing - UCAR Metal and Ceramic Coatings", Union Carbide Corp.
(16) V. Wilms and H. Herman, "Plasma Spraying of Al_2O_3 and Al_2O_3-Y_2O_3", *Thin Solid Films, 39* (1976) 251–262.
(17) T.A. Taylor, unpublished data.
(18) H. Alperin and T.A. Taylor, unpublished data.
(19) M.A. Levinstein, A. Eisenlohr, and B.E. Kramer, "Properties of Plasma-Sprayed Materials", *Weld. J.; Weld. Res. Suppl. 40,* 8s (1961).
(20) M. Levy, G.N. Sklover, and D.J. Sellers, "Adhesion and Thermal Properties of Refractory Coating-Metal Substrate Systems", U.S. Army Materials Research Agency, AMRA TR 66-01, 1968.
(21) W. Milewski, "Sonic Phenomena Occurring During Plasma Spraying WC+Co Compositions", presented at the 7th International Metal Spraying Conference, London, 1973.
(22) R.C. Tucker, Jr., "Structure/Property Relationships in Deposits Produced by Plasma Spray and Detonation Gun Techniques", *J. Vac. Sci. Technol. 11* (1974) 725–734.
(23) H.A. Matting and H.D. Steffens, *Metall.* 17(6), 583 (1963); 17(9), 905 (1963).
(24) L.H. VanVlack, "The Metal-Ceramic Boundary", presented at the 1964 Metals/Materials Congress, Philadelphia, Penn., Technical Report No. P (10-1-64).

(25) S.J. Grisaffe, "Analysis of Shear Bond Strength of Plasma-Sprayed Alumina Coatings on Stainless Steel", NASA Technical Note, NASA TN D-3113, 1965.
(26) Union Carbide Corp., unpublished data.
(27) H. Marchandise, "The Plasma Torch and its Applications", European Atomic Energy Community, EUR 2439.f, 1965.
(28) T.A. Wolfla, unpublished data.
(29) D.H. Leeds, "Some Observation on the Interface Between Plasma-Sprayed Tungsten and 1020 Steel", Defense Documentation Center, AD-803286, 1966.
(30) S. Kitahara and A. Hasui, "A Study of the Bonding Mechanism of Sprayed Coatings", *J. Vac. Sci. Technol. 11* (1974) 747-754.
(31) F.N. Longo, *Weld. J. 45* (2), 66s (1966).
(32) H.A. Matting and H.D. Steffens, *Metall. 17*(12), 1213 (1963).
(33) G. Durmann and F.N. Longo, *Ceram. Bull. 48*(2), 221 (1969).
(34) H.S. Ingham, Jr., "Flame-Sprayed Coatings", in Composite Engineering Laminates, edited by A.G.H. Deitz (MIT Press, Cambridge, 1966).
(35) H.S. Ingham and A.P. Shepard, Metco Flame Spray Handbook (Metco, Inc., Westbury, N.Y., 1965).
(36) M. Okada and H. Maruo, *Brit. Weld. J., 15,* 371 (1968).
(37) G.E. Poquette, Linde Division, Union Carbide Corporation, private communication.
(38) S. Yu. Sharivker, *Poroshk. Metall. 54*(6), 70 (1967).
(39) C.W. Marynowski, F.A. Halden, and E.P. Farley, *Electrochem. Technol.* 3(3-4), 109 (1965).
(40) M. Donovan, *Brit. Weld. J. 13,* 490 (1966).
(41) R.C. Tucker, Jr., Linde Division, Union Carbide Corporation, private communication.
(42) D.M. Yenni, Linde Division, Union Carbide Corporation, private communication.
(43) D.R. Mash and I. MacP. Brown, *Met. Eng. Quarterly 18* (1964).
(44) T.A. Wolfla and R.C. Tucker, Jr., "High Temperature Wear Resistant Coatings", presented at International Conference on Metallurgical Coatings", 3-7 April 1978, San Francisco, Cal.
(45) R.C. Tucker, Jr. and T.N. Bishop, "The Utilization of Plasma and Detonation Gun Coatings in Design", presented to AIME Symposium on Interaction of Design and Materials II, 1973.
(46) V.P. Elyutin et al. *Svar. Proizvod, 6,* 72 (1969).
(47) K.N. Muller, "Structure and Properties of Arc-Sprayed Titanium Coatings", presented at the 7th International Metal Spraying Conference, 1973.
(48) R.C. Tucker, Jr., "Wear Characteristics of Modified Plasma-Deposited Aluminum Bronze", presented to American Society for Testing and Materials Symposium on Erosion, Wear and Interfaces with Corrosion, 1973.
(49) R.C. Tucker, Jr. and P.W. Traub, "Wear Behavior of Wrought and Plasma-Deposited Aluminum Bronze and Beryllium Copper", presented to the Metallurgical Society of AIME, 1971.
(50) R.C. Tucker, Jr., T.A. Taylor, M.H. Weatherly, "Plasma Deposited MCrAlY Airfoil and Zirconia/MCrAlY Thermal Barrier Coatings", presented at Third Conference on Gas Turbine Materials in a Marine Environment", 20 to 23 September 1976, Bath, England.

General Reading

D.A. Gerdeman and N.L. Hecht, *Arc Plasma Technology in Materials Science,* Springer-Verlag, New York, Wien, 1972.
R.F. Smart and J.A. Catherall, *Plasma Spraying,* Mills and Boon Limited, London, 1972.

12

Organic Polymer Coatings

John G. Fish

*Texas Instruments, Inc.
Dallas, Texas*

INTRODUCTION

The other discussions on coatings accompanying this topic have centered on metallic or refractory materials that are deposited on a surface by transport of atoms or ions across a transport medium. This medium varies from a hard vacuum in the case of sputtering, to aqueous solutions in the case of electroplating. In most of the cases, the activating force for transport and deposition relies on a significant energy source such as an electron gun, plasma field, electrical potential, etc., and a concomitant assortment of rather expensive hardware. Other aspects of the discussion center on the methods used to characterize surfaces, morphology, and composition of both the substrate and the coating, and how these variables affect the performance of the system in their intended application environments.

In this chapter, an overview will be given of some of the same variables as applied to organic and polymeric coatings.

HISTORY AND ART OF ORGANIC COATINGS

Coatings have been used for decoration and protection for many centuries. Early man used natural pigments such as minerals, chalk, and charcoal for their artistic manifestations. Ancient Egyptian paintings have been found to contain organic binders to help keep the pigment intact. These binders were made from available animal and vegetable sources such as egg albumen, gelatin, beeswax, gum arabic, and casein. Water proofing has been recorded twice in the ancient Biblical writings. The first as a pitch covering the wooden sides of Noah's Ark and the second as the use of bitumen and pitch to render seaworthy the basket carrying young Moses in the shallow waters of the Nile. Many artists through the centuries experimented in developing better binders for their works of art. By the sixteenth century oil base paints had become quite common.

The use of coatings and paints for protection of architectural buildings and industrial machinery for many years relied on formulas of individual craftsmen or home concoctions. The long term protection afforded was negligible, requiring frequent repainting. The modern coatings industry has developed many new synthetic binders for pigments and have evaluated the degradation mechanisms due to sunlight, moisture, fungus, heat, and air. Additives to impact specific resistance to the above mentioned antagonists is a large business. The resultant coatings can be tailored to be durable and long lasting in a variety of conditions and applications.

Today, the coatings industry has surpassed the five billion dollar mark in annual sales. Of those sales, about half are for industrial finishes and the rest for architectural paints. Traditionally such coatings were made from dilute solutions of polymers in organic solvents. Since 1966, when the Los Angeles Rule 66 was enacted, there has been a shift toward the use of less solvent. The shift has further been compounded by the recent OPEC pricing pressures. The shift from the petroleum solvents has resulted in the further development of water-borne coatings, non-aqueous dispersions, solventless and high solids coatings, powder methods, electrodeposition, and radiation curing.

INTRODUCTION TO POLYMER CHEMISTRY AND PHYSICS

The science of organic surface coatings involves the study of both the medium and the pigment, the combination of the two in the paint system, the application methods, and finally the properties of the coating that is formed. This study requires the use of organic, inorganic, and physical chemistry as well as many sub-disciplines of physics: such as light scattering, viscoelasticity, surface and interfacial phenomena, etc.

During the evolution of the chemical sciences, several branches of study became evident, i.e. inorganic, physical, analytical, and organic. Organic chemistry was mainly a study of the chemistry of living things or the products therefrom, such as natural fibers, wood, rubber, etc. When it was discovered that some of the natural products, as well as compounds not found in nature, could be synthesized in the laboratory, a new concept of synthetic organic chemistry began. With the advent of petrochemical resources, the organic chemist began to use his synthetic tools to manipulate the structure of compounds to such an extent that even very complicated compounds such as chlorophyl have been made. Even though the chemical study of living matter has been taken over by the biochemist and physiological chemists, the term organic chemistry is still used to cover all the compounds that use carbon as the basic building block. (Other elements that are commonly incorporated into the organic molecules include hydrogen, oxygen, nitrogen, sulfur, halogens, and phosphorus.) However, all of the elements of the periodic table (except noble gases) can be incorporated as part of an organic compound, whether as a metal salt of an organic acid, organometallic complexes and compounds, or as esters of inorganic acids. (To date, the *Chemical Abstracts* has catalogued over a million compounds in their computer, two thirds of which are classified as organic compounds.) The area of polymer chemistry began as a subset of organic chemistry and rightfully remains there. The synthesis of the starting materials and the subsequent reactions required to make the polymer rely on the basic synthetic guidelines of organic chemistry. Polymer science today however requires that not only the synthesis, but also the physical properties, molecular structure, and applications be studied. Statistics show that the majority of graduate chemists and chemical engineers are involved wth some aspect of polymer science in their careers.

Most organic compounds have a fixed structure and molecular weight and can be identified as unique by physical properties such as boiling and melting points, refractive index, and a variety of spectral "fingerprints." A polymer is defined as a high molecular weight compound made up from a small repeating organic unit (poly = many; mer = parts). The magnitude of molecular weight ranges from 1000 to several million (amu) and depending on polymerization conditions, a statistical distribution of molecular weights will be present in a given reaction product. Since a given polymer "batch" contains such a mixture, a definite set of physical and chemical properties can be defined only for the batch. If conditions of preparation change, these properties can change to some extent. The simplest example of the variation of properties with molecular weight is that of the homologous series of compounds and polymers called the aliphatic hydrocarbons—those derived from petroleum. The lower molecular weight species of this group are the gases and liquids that are familiar as heating fuel (such as methane, propane, and butane), up through the gasoline and fuel fractions. Heavier, or higher molecular weight, compounds comprise the oils, waxes, and asphaltic portions of crude oil fractions. Extension of the series to high molecular weights beyond the waxes is best described in terms of the synthetic oligomers and polymers of polyethylene. It is formed by the catalytic reaction of gaseous ethylene to form high molecular weight polymers. A variety of catalysts and reaction systems allow tailoring of properties and molecular weights to yield products that may have relatively low melting characteristics, flexibility, and optical clarity, making them useful for things such as plastic bags. On the other extreme, ultra-high molecular weight polyethylene (UHMWPE) is very hard, tough, and has a very low coefficient of friction that allows it to be used in structural applications in which wear is a problem. From this simple description it can be seen that calling out of the name of a given polymer is not sufficient to identify it accurately.

There are two basic techniques for monomers to form polymers: condensation and addition reactions. In both cases the monomer has to be capable of reacting in at least two sites in order to form a chain of monomer units. The condensation reaction is so named because it usually results in the loss of a small portion of the original starting molecules. For example the reaction of two bifunctional monomers, a diacid and a dialcohol (or diol), results in the formation of ester links and the elimination of water:

$$x\text{HO-R-OH} + x\text{HO-CO-R'-COOH} \rightarrow \text{HO-(R-O-CO-R'-CO-O)}_x\text{H} + (2x-1)\text{H}_2\text{O}$$

This definition is not rigorous since there are many examples where no by-product (such as water) is formed. For the present discussion we will hold it as a generality. The polymer resulting from the condensation reaction will generally have interunit functional groups in the chain, that is, carbon and some other atom or combination of atoms in the chain.

Some typical examples of condensation polymers are shown:

$$\text{Polyester} \quad {\leftarrow}\text{C-C-}\overset{\overset{\text{O}}{\|}}{\text{C}}\text{-O-C-C-O-}\overset{\overset{\text{O}}{\|}}{\text{C}}\text{-C-C}{\rightarrow}_x$$

$$\text{Polyacetal} \quad {\leftarrow}\text{O-}\overset{\overset{\text{R}}{|}}{\text{C}}\text{-O-C-O-}\overset{\overset{\text{R}}{|}}{\text{C}}\text{-O-C-O-}\overset{\overset{\text{R}}{|}}{\text{C}}{\rightarrow}_x$$

$$\text{Polyamide} \quad {\leftarrow}\text{C-C-NH-}\overset{\overset{\text{O}}{\|}}{\text{C}}\text{-C-C-NH-}\overset{\overset{\text{O}}{\|}}{\text{C}}{\rightarrow}_x$$

Polyurethane

$$\{C-C-O-\overset{O}{\overset{\|}{C}}-NH-C-C\}_x$$

Polyurea

$$\{C-C-NH-\overset{O}{\overset{\|}{C}}-NH-C-C\}_x$$

The parentheses indicate only one segment of the repeating units that make up the polymer chain, and hydrogens attached to carbon are omitted for clarity. One condensation polymer with no carbon in the chain is the silicone:

Silicone

$$\{Si(CH_3)_2-O-Si(CH_3)_2-O-Si(CH_3)_2-O\}_x$$

Addition polymers are those that result from the reaction of an unsaturated monomer with an initiator that begins a chain reaction at an activated site to start the growing polymer chain. The active sites can be anions, cations, or free radicals as well as certain coordination complexes. For brevity we will describe only a simple free radical reaction of a vinyl monomer. The polymerization requires three steps: initiation, propagation, and termination.

A free radical initiator (1) is broken down to generate free radicals (designated $R\cdot$), which then react with monomer (2) to form a new radical species (3):

(1) $\rightarrow$ 2R$\cdot$

$$R\cdot + CH_2=CHX \rightarrow R-CH_2-\overset{H}{\underset{X}{C}}\cdot \xrightarrow{n(CH_2=CHX)} R-CH_2-CHX(CH_2-CHX)\cdot_n$$

(2)   (3)   (4)

This new radical reacts with more monomer to propagate a growing chain (4). The chain is terminated by any number of reactions that can deactivate the growing radicals.

POLYMER CHARACTERISTICS

Thermoplastic and Thermosetting Polymers

The polymers just described are generally called thermoplastic, i.e., they soften and melt at elevated temperatures. If during polymerization or subsequent processing, a significant amount of interaction between polymer chains creates cross-links, the polymer becomes insoluble and will not melt below its thermal decomposition point. Such polymers are loosely called thermosetting, or cross-linked polymers. This is an important concept when we will later discuss the "curing" of coatings.

Molecular Weights of Polymers

As mentioned earlier, there is a statistical distribution of molecular weights of the molecules in a polymer. This distribution can be varied by altering the polymerization processing and blending procedures.

The distribution curve is typically that shown below:

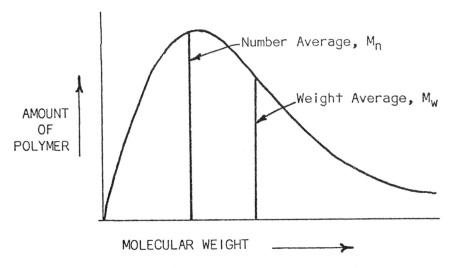

In the distribution, the polymer chemist is concerned with two types of molecular weight values: M_n is the number average and M_w is the weight average. Number average molecular weights are determined by solution techniques that measure colligative properties, such as membrane and vapor pressure osmometry, ebulliometry, and cryoscopy. These techniques, in effect, count the number of polymer molecules per unit weight of polymer.

The weight average molecular weight M_w is measured by techniques that are dependent on the molecular mass or size. These include light scattering, viscometry, and ultracentrifugation. The values of M_n and M_w can be described mathematically, but it is beyond the scope of this paper to delve into it. However, the relationship of these two values is important: M_w is always greater or equal to M_n, and the ratio M_w/M_n is called the polydispersity. The polydispersity of the system can greatly effect the physical properties of the polymer during processing and application. For example, if the ratio is close to one, that means that the distribution of the molecular weight is very narrow and the melting point will be very distinct. In a higher ratio polymer, the melting would cover a broader range of temperatures. Other properties are similarly affected.

The analytical techniques required to measure molecular weights have been very time consuming, but the advent of gel permeation chromatography (GPC) has led to rapid determination of both M_n and M_w within a couple of hours, making it useful as a quality control measurement for production of polymers and coatings. (GPC separates the polymers by molecular size in a high pressure column filled with a packing containing pores of controlled dimensions. Output of the instrument is essentially a trace of the molecular weight distribution shown earlier. Calculations of M_w/M_n and dispersity are made from the output.)

Solubility of Polymers

The solubility characteristics of a polymer are very important for solution coatings. They determine the ability to make a good coating solution, and are relevant to solvent resistance of the applied coating.

The dissolution of a polymer occurs in two stages: 1) diffusion of the solvent to form a swollen gel followed by 2) disintegration of the gel to form a true solution. Two types of effects prevail: kinetic and thermodynamic. In order to obtain rapid swelling, a good penetrating type solvent is needed but, to get true solution, a good thermodynamic solvent is needed. Often a single solvent can be found to give an acceptable balance of the two, but solvent blends are more often used. If a polymer is crosslinked (loosely defined earlier as thermoset) the polymer will not be soluble in any solvent. It will swell however, to an extent depending on the degree of crosslinking and the inherent solvent-polymer interaction. This process of swelling is taken advantage of in the removal or stripping of paint or varnish. The bond to the substrate is weakened and the mechanical properties of the swollen polymer allow it to be removed easily from the surface. Factors other than crosslinking also affect solubility. Highly crystalline polymers are difficult to dissolve unless the temperature is raised to near the crystalline melting point.

The most important factor in solution formation is the thermodynamic effect. In essence, it is the concept of "like dissolves like." For example the very polar solvent, water, will dissolve polar molecules such as ethanol, acetic acid, etc. but will not dissolve non-polar molecules such as hexane, chlorinated solvents, or esters. The miscibility of two solvents (in the absence of hydrogen bonding) is possible if the free energy of mixing is negative:

$$\Delta G = \Delta H - T \Delta S$$

Hildebrand (1950) showed that the heat of mixing is:

$$H = v_1 v_2 (\delta_1 - \delta_2)$$

where v is the volume fraction of solvents and δ is the solubility parameter of the two solvents, 1 and 2. The solubility parameter is derived, and can experimentally be measured, from the cohesive energy density expression:

$$\delta = \frac{H_v}{MW/d} = \frac{H_v}{V}$$

where H_v is the latent heat of evaporation and V is the molar volume. Since polymers are not volatile, the value of δ has been calculated from structural groups in the polymer chain (Burrell 1966; Hoy 1970). Burrell showed that, as a first approximation, if the difference in δ between the solvent and polymer is $(\delta_1 - \delta_2) \leq 1.7\text{-}2.0$ the polymer will be soluble. This assumes that hydrogen bonding is not an important factor in the solvent and/or polymer (i.e. no alcohol, amine or acid protons are present). Burrell (1966) later added a hydrogen bonding index and Crowley (1966) discussed a third parameter, the dielectric constant.

The most convenient combination of parameters is solubility parameter (δ) and hydrogen bonding index (γ). Figures 12.1a and 12.1b show a chart of the solvent maps with typical solvents, and Figure 12.2 shows the solubility characteristics of a specific polymer. The polymer is soluble in any solvent that has δ and γ values within the closed loop, and insoluble outside the loop. Solvent combinations can be made such that:

496   Deposition Technologies for Films and Coatings

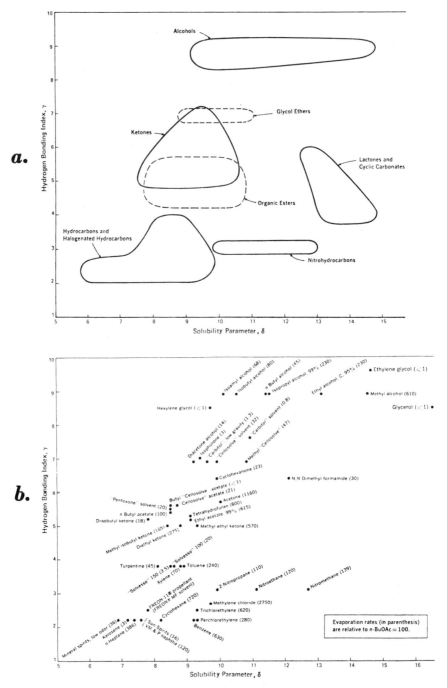

Figure 12.1: Solvent maps for typical solvents.
(a) Parameter locations for major solvent groups.
(b) Parameter locations and relative evaporation rates for typical solvents.

$$\delta_{mix} = v_1\delta_1 + v_2\delta_2 .....v_i\delta_i$$

$$\gamma_{mix} = v_1\gamma_1 + v_2\gamma_2 .....v_i\gamma_i$$

where v_i is the volume fraction, δ_i the solubility parameter and γ_i the hydrogen bonding index of the respective solvents. In coatings applications, the cost and relative volatility of each component is also of prime concern. Experimentally, the polymer solubility maps are determined by mixing 5% polymer with the solvents of varying δ and γ values.

Obviously this concept can also be used to prescribe or negate the use of certain polymer coating types in applications where exposure to solvents is expected.

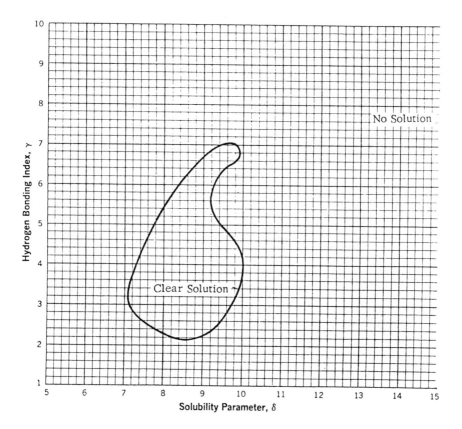

Figure 12.2: Solubility characteristics of a specific polymer.

Solubility/Molecular Weight Considerations

For application of a polymer to a surface from solution, it is necessary for the viscosity to be amenable to the method, and the ratio of solvent to polymer needs to be low enough to minimize excessive shrinking and to afford complete coverage. As discussed earlier, one method of molecular weight determination

M_w was to use viscosity. Briefly stated, for a given weight concentration of polymer in solution, the viscosity will increase with the molecular weight of the polymer. If one has a solution with a satisfactory viscosity but a low solids content it is imperative that a lower molecular weight polymer be used if the solids content is to be increased. Several approaches to obtain high molecular weight coatings from low viscosity solution include crosslinking or polymerization after drying, and use of suspensions of high molecular weight solids that, on solvent evaporation, coalesce (with or without heat) to form the film. It is obvious that the incorporation of pigments, fillers and other additives dramatically affect the flow properties of the solution. These topics will be discussed later.

TYPICAL POLYMERS USED FOR COATINGS

Virtually every polymer that can be made to form a continuous film on a surface has been used for coatings.

Natural Products

Materials in this category include shellac derived from the larvae of lac beetles, vegetable oils derived from seeds and nuts of certain plants, and marine oils from a few species of fish. Their drying to form a film depends on solvent evaporation and varying degrees of air oxidation and subsequent crosslinking. Typical oils are linseed, cottonseed, soybean, and cod. These are typically very slow drying, produce soft films and lack uniformity of raw material.

Oleoresinous coatings are made by reacting the above mentioned oils with natural or synthetic hard resins to produce a harder more durable coating. Also, the incorporation of other low molecular weight synthetic monomers such as maleic anhydride or styrene helps improve the hardness. These are called treated, synthetic or styrenated oils.

Alkyd resins are made from polyhydric alcohols (such as glycerine and pentaerythritol) with polyacids (such as phthalic or maleic acids) and one of the natural oils or fatty acids. On curing these polymers form a tough, tightly crosslinked coating.

Cellulose derivatives are widely used to form lacquers; coatings that only require the evaporation of solvent to form the final film. Nitrocellulose, cellulose acetates, propionates, butyrates, and ethylcellulose are most widely used. Acrylics, polyvinyl chloride, polyvinyl butyral, styrene-butadiene and many other synthetic polymers can be used in lacquers.

Many resins require either catalytic or thermal mechanisms to go from low molecular weight to tough, highly crosslinked films. Epoxy, phenolics, polyesters, polyurethanes, and silicones are typical examples.

MECHANISM OF FINAL FILM FORMATION

In this section we will discuss the chemical or physical processes associated with the drying and curing of films.

The physical requirements of a coating system requires first that, as applied, the solution or molten polymer thoroughly wets the substrate being coated. The surface tension of the fluid and of the substrate need to be compatible for wetting, and is sensitive to the presence of oils, dirt or other contaminants. Very often the substrate has to be treated to afford a more wettable surface. This will be discussed under the section entitled Surface Preparation.

The flow properties of the fluid are variable depending on the method of application and type of coating. Ordinary varnishes and lacquer have Newtonian viscosity characteristics in that the viscosity is independent of shear rates. In this case, as the coating is applied it will level to a smooth finish as the solvent evaporates. If too rapid solvent evaporation occurs the film will skin over, causing subsequent solvent evaporation from under the skin to be retarded. The result is that the finish will take on a wrinkled or "orange-peel" texture. The appropriate blending of solvents with both fast and slow evaporation rates is used. Also, too rapid solvent evaporation, coupled with humid conditions, results in formation of blush or whitening of the coating. Leveling agents and fillers added to the solution also modify its drying performance. Non-Newtonian solutions, or more correctly, suspensions, occur when the suspension causes either dilatency or thixotropy. A dilatent suspension is quite fluid until stressed, at which time it becomes resistant to motion. Thixotropy is the opposite, in that the suspension is thick and gelatinous but when stirred or agitated it flows readily. Many of the no-drip household latex paints are thixotropic. When applied, it is necessary for the wet film to level rather quickly before the thickening occurs. Graphically, the three important types of viscosity behavior are shown below:

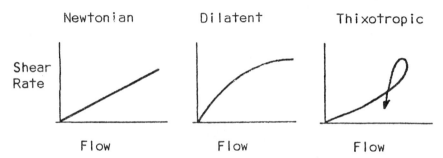

Latex Paints rely on another unique property to affect formation of the film. The binder polymer is not in solution, but in suspension as tiny particles on the order of 1 to 5 μm diameter. The original monomer is mixed with water and surfactants to form an emulsion of tiny monomer droplets surrounded by the surfactant molecules. When a polymerization catalyst is added to the emulsion, it migrates into the droplet and polymerization takes place. (In this way a given polymer can also be lightly crosslinked to make it insoluble after film formation.) The same surfactants used for polymerization stabilize the particles for further blending with pigments, other additives and packaging. On application and evaporation of the water, the polymer particles begin to touch, coalesce, then collapse into a continuous film. Since a degree of crosslinking is incorporated at the polymerization stage, it is unnecessary for additional curing to take place. The types of polymers often used for latex paints are butadiene-styrene, and acrylics.

Chemical curing of polymer films is quite varied. Therefore we shall simply discuss the more typical examples of several classes of reactions.

Varnishes and the oleoresin type coatings contain "drying oils" or vegetable oil esters that contain unsaturated fatty acid residues. These oils, when exposed to air are oxidized at the unsaturated part of the molecule and ultimately crosslink at those sites. Small amounts of "driers" are added to the paint to accelerate the oxidation. These are generally salts of cobalt, lead, zinc, manganese or calcium in the form of oleates, resinates, or naphthenates. As a rule, only these natural oil coatings rely on oxygen from the air for the cure.

Another set of coating polymers utilizes moisture from the air to be part of the crosslinking or polymerization mechanism. Some silicones and urethanes fall into this class. Polyurethanes are made from low molecular weight glycols (1) or polyols and diisocyanates (2) which react to form polymeric urethanes with excess isocyanate groups on the end of the chains (3).

$$nHO-R-OH + (n+1)O{=}C{=}N-R'-N{=}C-O \rightarrow$$

(1)   (2)

$$O{=}C{=}N-(R'-NH-CO-O-R-O-CO-NH-)_n R'-N{=}C{=}O$$

(3)

After application, exposure to ambient moisture causes the isocyanate ends to react to form urea crosslinks:

$$2{\sim}\!\!\sim\!\! N{=}C{=}O + H_2O \rightarrow {\sim}\!\!\sim\!\! NH-CO-NH{\sim}\!\!\sim$$

Likewise, room temperature vulcanizing (RTV) silicones are made of relatively low molecular weight polymers with reactive end and side groups, and often contain a multifunctional crosslinking siloxane. The basic polymer is made up of a silicon-oxygen backbone with methyl and/or phenyl groups attached to each silicon atom.

$$X-(Si-O-)_n-X \quad (X = \text{halide, acetate, or alkoxy})$$

At the ends, or interspersed along the chain, are hydrolyzable groups that react to form silanols.

$$\sim\!\! Si-O-CH_3 + H_2O \rightarrow Si-OH + \text{methanol}$$

$$\sim\!\! Si-O-CO-CH_3 + H_2O \rightarrow Si-OH + \text{acetic acid}$$

The silanol groups then react with a crosslinker such as methyl triacetoxy, or methyl trimethoxy silane to form the crosslink:

$$3\sim\!\! Si-OH + CH_3Si(OCH_3)_3 \rightarrow \sim\!\! Si-O-\underset{\underset{\sim\!\! Si\sim}{\overset{|}{O}}}{\overset{CH_3}{\underset{|}{Si}}}-O-Si\sim$$

The reaction generally has a catalyst to accelerate or initiate the reaction. (There are other silicone crosslinking schemes that do not require moisture.)

Two part polymer or coating systems require the addition of catalyst to the base resin to cause curing. The polyester used for the manufacture of glass reinforced articles or gel coatings for other substrates is one such example. The

Organic Polymer Coatings

base resin is a polyester resin made typically from ethylene glycol and maleic acid anhydride, an unsaturated compound.

maleic anhydride + glycol → H–(O–CO–CH=CH–CO–O–CH$_2$–CH$_2$)$_n$–OH

This polymer is dissolved in styrene monomer which in itself contains a double bond. Addition of a peroxide catalyst propagates polymerization of the styrene. The unsaturated sites in the polyester also are copolymerized into the totally crosslinked network, (here shown very hypothetically):

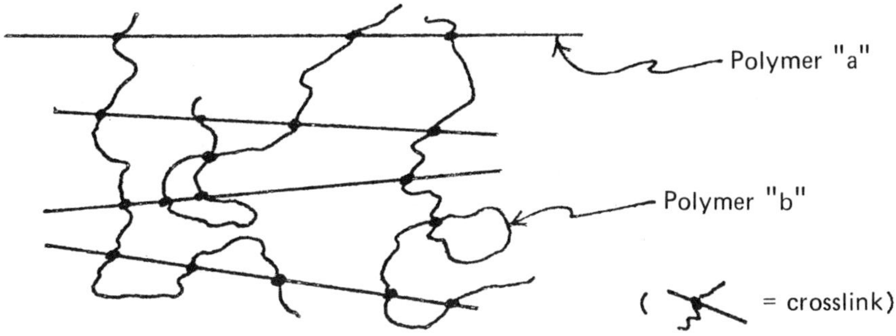

Polymer "a"

Polymer "b"

(✕ = crosslink)

Epoxy resins can likewise be cured by addition of a catalyst and a hardener: Typically, an amine, amide, or acidic compound is added to a low molecular weight epoxy prepolymer, shown below:

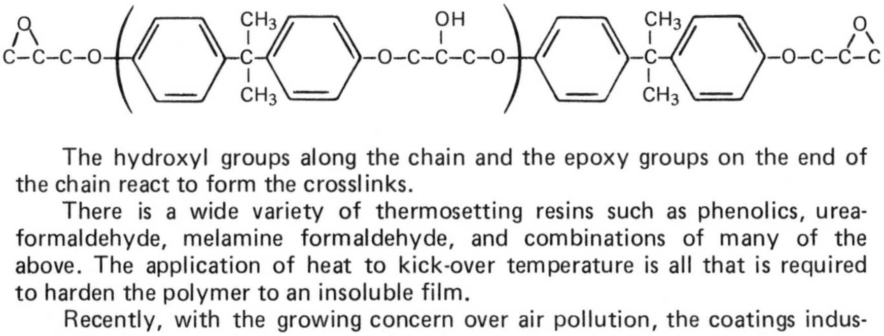

The hydroxyl groups along the chain and the epoxy groups on the end of the chain react to form the crosslinks.

There is a wide variety of thermosetting resins such as phenolics, urea-formaldehyde, melamine formaldehyde, and combinations of many of the above. The application of heat to kick-over temperature is all that is required to harden the polymer to an insoluble film.

Recently, with the growing concern over air pollution, the coatings industry is turning toward the use of solventless coatings that can be cured by either ultraviolet light or by high energy electrons from a generator. The coating is made up of reactive groups, mostly acrylate esters of a polyfunctional alcohol such as pentaerythritol or propylene glycol:

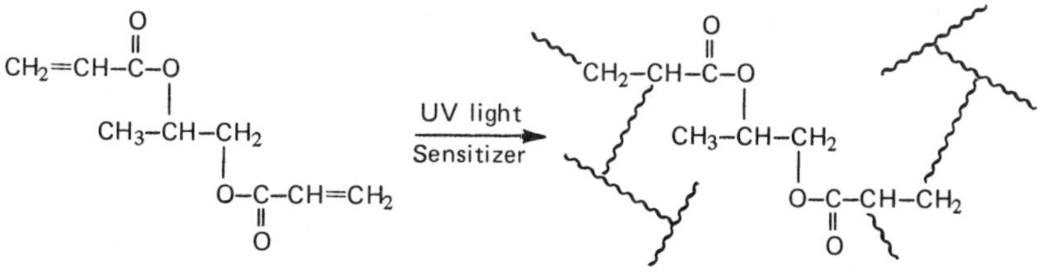

$$CH_2=CH-\underset{\underset{O}{\|}}{C}-O-\underset{\underset{\underset{O}{\|}}{O-C-CH=CH_2}}{CH_3-CH-CH_2} \quad \xrightarrow[\text{Sensitizer}]{\text{UV light}} \quad \text{Polymer network}$$

Propylene glycol diacrylate · · · · · · · · · · · · · · · Polymer network

Any combination of these reactive monomers can be combined to give the appropriate viscosity for the particular application method. A photosensitizer is incorporated into the mixture which, when exposed to the U.V. radiation, generates free radicals to initiate crosslinking. The highly reactive acrylic groups can be cured in a matter of seconds to a tough highly crosslinked coating. There has also been extensive use of acrylic derivatives of epoxy and urethane polymers that give impoved properties. Some of the large applications of radiation curing include finishes on wood products, no-wax vinyl floor covering top coats, and the printing of aluminum beverage cans.

COATING FUNCTIONS

There are many reasons for applying coatings to substrates. In this section we will discuss the particular result desired and the type of coating system or additive required to accomplish the result. Many of the components of a system serve multiple roles and these will be pointed out where appropriate.

Decoration

The applications of coatings may be used to enhance the substrate appearance in one of several ways. A clear varnish with or without a staining agent, certainly enhances the natural grain of a piece of hardwood. Not all substrates, however, have such interesting patterns and beg to be hidden by our desire for aesthetics. Therefore opacifiers are added to the binder. The ability of a white pigment to hide a dark background depends on its ability to reflect and refract light back to the observer before it can penetrate through to the substrate. This is a function of particle size, concentration of pigment in binder, and the difference in refractive index of the pigment and binder. The particle size is optimum when it is of dimensions on the same order as the wavelength of light being reflected. For visible light in the 0.3-0.6 μm range, the particle size will be best at 0.3-0.6 μm. The refractivity difference allows the particle to act as a small irregular prism that statistically reflects most of the light back toward the observer, providing that enough thickness of pigment particles are present in the binder. Ideally, the concentration of pigment should be high enough to allow an average distance between particles to be on the order of three times the particle size. The most effective hiding pigment is titanium dioxide which has a refractive index of 2.71 while 1.3-1.5 is typical for many organic binders. Hiding power can also be accomplished by the use of microvoids in the coating. With tiny bubbles of the submicron size, the refractive index of air is sufficiently different from the polymer that reflection of light occurs. These microvoids can be incorporated into the binder as glass or polymer microballoons filled with air. Several other techniques such as incorporation of blowing agents, thermally decomposable components, and solvent extraction or precipitation schemes have been developed. (The topic was very thoroughly discussed in the symposium on Microvoid Coatings sponsored by the Organic Coatings and Plastics Division of the American Chemical Society, August 27, 1973. The papers were also published in *Ind. & Eng. Chem., Prod. Res. Develop.*, V. 13, #1, 1974.)

Other fillers and colored pigments are used to add reinforcement to the binder, affect surface gloss, and of course, add color. Colored pigments are generally insoluble organic dyes, metal complexes of organic compounds, or inorganic compounds. Special effects can be created by the use of metallic flakes with color imparted by soluble organic dyes. Pearlescent coatings are prepared by incorporated flake pigments that are made up of many thin layers of material with differing refractive indices or with minute mica flakes. These refractive and reflective layers cause interference colors to appear, much like mother of pearl. Zinc oxide is often added to the pigment composition because of its ability to absorb and dissipate the effect of ultraviolet light. Carbon black and aluminum powder effectively block the U.V. light. With the advent of interest in solar energy applications, the paint chemist has a significant chore in finding a coating that will effectively absorb all the thermal energy while still being able to maintain an integral coating after years of exposure to U.V. and high temperatures.

Preservation of Substrate

Coatings can afford either passive or active control in the protection of the surface that it covers. Passive coatings simply cover the surface and act as a barrier to corrosion, rotting, weathering, etc. Active coatings have chemical agents or additives that specifically retard the deteriorating mechanism. The incorporation of zinc chromate as a pigment for corrosion protection on steels is a good example. An important part of corrosion on assemblies composed of more than one metal is that, if an external conductive path such as salt water is available, a galvanic couple occurs. Depending on the metals, the pH of the moisture, and the type of ionic contaminant, gross corrosion may occur. (For example, zinc in contact with iron will dissolve and corrode preferentially.) A good protective coating will prevent the external conductive path and hence the corrosion. Wood surfaces are protected in several ways by coatings. The solvent carries the binder and pigment into surface pores and cracks to seal out moisture and add strength to the surface. Biocides added to the paint help retard the growth of mildew and other fungi.

A variety of coatings are available to protect surfaces from chemical agents such as acids, alkali, salts, and solvents.

Self-Preservation of Paint Surfaces

The coating of any surface simply exposes a new surface to the environment. Therefore additives to the coating are used to protect its function. For outdoor applications, U.V. stabilizers or absorbers are used. In addition to the effect of zinc oxide mentioned earlier there are a number of organic components that absorb the energy and dissipate it in a harmless fashion. Self cleaning paints contain pigments and binders that age in a controlled manner, such that traces of the surface degrade and wash off with age. Tin and lead salts have been incorporated into paints used on marine surfaces to prevent fouling from algae and barnacles. The metal salts act as biocides or inhibitors that prevent the growth on the surface.

Lubrication and Wear Resistance

Fluorocarbon polymers are known for their very slippery surfaces. Parts fabricated from solid fluorocarbons are used as bearing surfaces in applications varying from micro-valve seats to bridge supports. The most common fluorocarbon is tetrafluoroethylene (TFE). One of the disadvantages of the material is that it is both insoluble and infusible, so it cannot be applied to surfaces in a conventional manner. When the TFE is synthesized, it is in the form of very tiny particles. Fabrication into parts requires that the powder be molded or sintered at high temperature and pressure. Coatings of TFE are made from a fluid suspension of the particles, followed by drying and sintering. The coating is quite soft so modifications have been made by codeposition with harder, thermally stable polymers such as polyphenylene sulfide. These coatings have a characteristic brown color due to the sulfide polymer.

Nylon, acetal, and ultrahigh molecular weight polyethylene (UHMWPE) are other polymers that have inherently low friction. They can be incorporated into a coating as would a pigment, melt extruded onto a surface, powder coated, or laminated.

Temporary Coatings

The protection of surfaces during processing can be accomplished by the use of strippable, temporary coatings. Typical applications are for highly polished or decoratively textured surfaces, or for the working edges of precision cutting tools. Any polymer that is soluble can be used, but the use of solvent to remove it can be cumbersome. Peelable coatings can be formulated from any number of polymers such as polyvinyl acetate or polyvinyl chloride. Plastisols offer a convenient form of coating. The plastisol is a suspension of polymer particles in a low viscosity fluid that will not dissolve the polymer at room temperature. When heated, the polymer dissolves in the fluid to form a very viscous solution. Parts dipped into the melt will retain a thick coating of the flexible polymer when withdrawn and cooled. The original solvent is non volatile and is retained in the polymer to act as a flexibilizer and as a surface lubricant to prevent tight adherence.

Antistatic Coatings

Most polymers are highly dielectric, having restivities in excess of 10^{14} ohm-cm. Therefore, friction of any type is likely to create some degree of static electricity on the surface. A thin coating on grounded metal parts is not likely to present any problems, but plastic substrates in electronic instrumentation can create havoc with the equipment functions. Antistatic coatings can be made in several ways. The most obvious is to incorporate metallic aluminum or silver powder into the coating as a pigment. Conductive carbon can also be used. These pigments have to be loaded in heavily enough to have enough intimate particle contact to form a conductive path. Another approach is to incorporate organic components into the polymer backbone that can ionize to some extent to afford a slight amount of conductivity. Carboxylic acid or amine groups have been used. These polymers generally rely on their inherent hygroscopicity to absorb moisture from the air. The equilibrium dissociation of the water to form H^+ and OH^- ions is enhanced by the acid or base groups on the polymer, and it is these ions that enhance the surface conductivity that allows a static charge to dissipate.

Water Repellence

Almost no assurance of hermetically sealing a surface from water vapor can be realized by using polymeric coatings. Polytetrafluoroethylene and vinylidene chloride are the best barriers with a water vapor transmission rate (WVTR) of 4.8 (g)(mil)/(m^2)(24 hr). Polymethyl methacrylate has a value of 550 and cellulose acetate is 1200, for comparison.

An important point about moisture resistance is that water vapor and liquid water behave very differently. Silicones are touted as being good waterproofing materials for brick, masonry, textiles, leather, etc. They do, however, have a very high MVTR. On porous surfaces that have been treated with silicones, the pores become hydrophobic. Any water droplets that collect on the surface will bridge over the pores instead of wetting and wicking into them. A common technique for evaluating the hydrophilicity of a surface is to measure the contact angle that is made between the surface and a small drop of water placed on it. A low contact angle means that the surface is wetted by the water, while conversely, a high contact angle means it is hydrophobic.

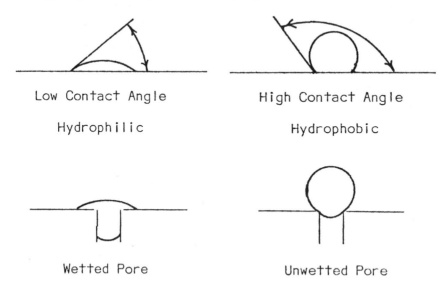

Hydrophilic Coatings

Only one particular application of a water wettable coating will be discussed here. The fogging of windows, mirrors, etc. is the result of condensation of moisture droplets on the cool surface. On a hydrophobic surface, the condensate is in the form of tiny droplets that refract and reflect light to give a diffuse appearance. If a water wettable coating is present, the contact angle of the droplet will approach zero and a continuous layer of water will be present. By nature these coatings are somewhat soluble, necessitating periodic reapplication.

Electrical Properties

Many coatings are used in electrical and electronic applications that requires them to have good insulation resistance, dielectric strength, dielectric constant, dissipation factor, and physical continuity. Specific applications include: coatings on wire, coating of a completely assembled printed circuit board, encapsulation of active semiconductor devices, potting of high voltage power supplies, etc. Each of the applications have different requirements for toughness, electrical properties, and perfection. Many also need to be repairable, that is, they can be cut away, a component changed, and a new layer of coating material applied. The dielectric strength of a coating is a measure of its ability to isolate a high voltage from discharging to ground through the coating. The application of coatings to sensitive microelectronic structure and printed circuit boards requires careful control of a number of factors. Ionic contamination on the surface of the substrate, or present in the coating material, can cause a conductive path to occur when a humid environment is encountered. It was pointed out in the section under water repellance that, even the polymer with the best moisture barrier properties does have a finite rate of penetration. Therefore, if ionic contaminants are present, the moisture that permeates the coating will create a conductive path, causing corrosion, electrochemical migration and ultimately short circuiting. In the applications of polymers to protect electronic circuits it is important to remember that the coatings are not able to guarantee a hermetic seal toward moisture. They do protect the circuit from dirt, fingerprints and other external influences.

SURFACE PREPARATION

Cleanliness

Any coating problem requires that the surface be prepared to afford the best combination of qualities desired on the finished product. Of prime importance is cleanliness. The presence of oils, dirt, ionic residues, or loosely-bound corrosion products will result in a weak bond and sites for corrosion, degradation, and blistering. Many fabricated metal parts have shop oils or mill scale present that must be removed. The sequence of cleaning is important in that the oily residues need to be removed first since they will prevent the wetting of other soils by an aqueous cleaning agent. Vapor or solvent degreasing or alkaline detergent cleaners are useful for removing these oils. Good rinsing and drying are necessary to remove the alkaline or other ionic salts. A visible measure of the cleanliness of a surface is the lack of "water-breaks" after the surface is rinsed. A water-break-free surface means that the water runs off the surface in sheets without breaking into droplets or rivulets.

Pickling or acid etching of the surface removes rust and mill scale and also slightly roughens the surface to give better adhesion. Some corrosion products are not necessary to remove if it is thin and tightly adherent to the surface.

Texture

Surface roughness has utility in promoting adhesion in that it presents a network of interpenetrating grooves and channels into which the coating may flow and become locked in place.

Primer Technology

A given combination of substrate and coating material may not have a good inherent adhesion to each other. A primer coat is one that bonds tightly to the substrate and in turn allows the topcoat to bond well to the primer. In many cases the primer may also have other functions such as the incorporation of zinc chromate pigment for corrosion protection. Some primers are designed to form direct chemical bonds to the substrate and/or the topcoat. One example is the silane primers used on glass surfaces. Typical applications are for the glass fiber used in fiber-glass-reinforced polyester or epoxy resins. If the polymers do not bond well to the glass, the effectiveness of their mutual reinforcement is lost. The silane primers are designed to react with the glass surface that contain surface silanol groups (Si–OH). When the silane reacts with the silanols, a strong chemical (glass-like) bond network is formed:

$$(X)\ HO-\underset{\underset{OH}{|}}{\overset{\overset{R}{|}}{Si}}-OH$$

$$+$$

(silane reacting with hydroxyl groups on glass substrate to form $-(O-\underset{|}{\overset{R}{Si}}-O-\underset{|}{\overset{R}{Si}})_{x/2}$ bonded to glass substrate)

The other end of the silane molecule contains organic functional groups (R) that are tailored to react with the particular resin used, i.e. vinyl groups to co-react with polyesters, and glycidyl or amine groups to co-react with epoxies.

Conversion Coatings

Chemical conversion coatings are adherent surface layers of low solubility oxide, phosphate or chromate compounds produced by the reaction of suitable reagents with the metallic surface. The reaction results in the conversion of a small amount of metal to a crystalline or amorphous metal compound. Phosphate conversion of ferrous metals is accomplished in dilute chemical solutions containing phosphoric acid and other chemicals. Variations of the chemical compositions and processing conditions results in variations in the surface texture, color, crystallinity, and tightness of the bond to the metal. With no metal ions in the phosphate solution, the resulting coating is an iron phosphate. Addition of zinc or manganese to the solution results in a co-deposition of zinc or manganese phosphates along with the iron. The phosphates, in addition to promoting adhesion of coatings, offer a significant degree of corrosion protection. There are several companies that sell tried and true proprietary formulations to the industry.

Chromate conversion coatings are used to protect the surfaces of zinc, cadmium, aluminum, and magnesium. This coating is formed as a result of immersion in chromic acid, dichromates, and other salts. The attack on the metal and subsequent reactions of the chromium compounds results in the formation of a tightly bonded chrome complex. As with the phosphates, these coatings afford inherent corrosion protection as well as a bonding interlayer for organic coatings. The coating has a characteristic color of chrome-yellow or -orange, although some colorless formulations have been developed for use under clear lacquers where the metallic color of the metal is desired. Again, several companies have expertise in applications of their proprietary formulations.

Anodization is a process used mainly for aluminum and magnesium surfaces. This process requires that the part be used as an anode while immersed in an acid such as phosphoric, sulfuric, chromic, or oxalic. A hard aluminum or magnesium oxide is formed that affords good corrosion and abrasion resistance, and an adhesion base for coatings. The oxide is also very receptive to certain types of organic dyes, and is often used as the final finish.

Corona discharge is a process that, while not a conversion coating, is a treatment that modifies the surface of plastic substrates. A polymer surface is passed through a corona discharge in the presence of air or some reactive gas. The discharge activates the surface molecules and the oxygen so that the resultant surface is highly oxidized. The surface is then more receptive to coatings, including vacuum metallization, electroless plating, and organic coatings.

DEPOSITION METHODS

Now that we are thoroughly familiar with all the variabilities of coatings and substrates, we can add a new dimension of variability by looking at methods of application. We can only briefly mention each technique and point out some key advantages or disadvantages. We will also include some processes that would better be called printing, but since they are essentially organic material transfer processes, they will be discussed.

Each of the processes to be described is usually best adapted to a particular type of substrate. For instance, some objects to be coated are discrete objects with complex surfaces (i.e. furniture, automotive, or appliance parts), while others may be a continuous strip of "web" such as cloth, paper, metal sheet, or plastic film. A given process, such as dip coating, may be used for both the discrete and strip materials, but the method of parts handling will be significantly different. Properties of the substrate may also dictate the selection process. For example, electrostatic or electrophoretic coatings require that the substrate be electrically conductive.

Mechanical Processes

Brush, pad, and roller coating are all well known to all who have been involved in home improvement or maintenance. In some industrial applications, these may be the method of choice. Automated roller coating of both sides of strip metal is very fast and gives uniform coatings. Automation on substrates of greater complexity than strips is not very effective. Offset printing is a variation of roller coating where the pattern or color to be coated is selectively transferred to certain portions of the roll and subsequently transferred to the substrate. Several examples of mechanical processes are shown below:

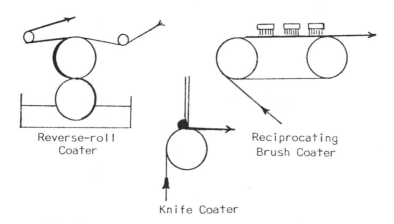

Reverse-roll Coater

Knife Coater

Reciprocating Brush Coater

Dip Coating

Dip coating is useful for parts with complex shapes and inaccessible recesses. After immersion, the parts are withdrawn and the excess coating material removed by centrifugation or simple gravity runoff (with occasional rotation to assure removal from recesses). Dip coating of roll stock is very easily accomplished as shown in the diagram.

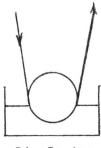

Dip Coater

Spray Coating

Several types of spray applications are available; compressed air, airless, hot spray, and electrostatic. All the spray techniques atomize the fluid into tiny droplets and propel them to substrate. There is usually a significant waste of material due to overspray. Electrostatic spraying is effective in reducing these losses when painting metal. The part to be painted must be conductive and connected to ground. The electrostatic sprayer imparts a positive charge on each paint droplet. The difference in charge and its mutual attraction causes about 90% of the paint to flow to the substrate. To some extent, the paint will bend around corners and will effectively coat sharp corners because of the high electric fields present there. However, it will not reach deep recesses.

Screen Printing

Screen printing is often used on flat, rigid surfaces that require a patterned coating. The applicator is a tightly stretched fine wire or fiber screen that is coated with a patterned emulsion. The pattern is often generated by a photographic process. After patterning, the mesh of the screen is open to allow passage of ink. The remainder of the screen has closed pores. The coating material (ink) is spread across the top side of the horizontal screen, the screen pressed onto the surface to be printed, and a squeegee passed over the screen to force the ink through the screen onto the surface to be printed. The process can be automated, offers very good pattern resolution, and can apply fairly thick films of materials that cannot be applied by other techniques.

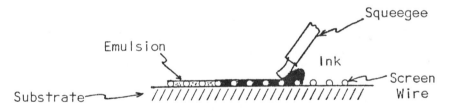

Lamination

Two or more layers of materials can be bonded together with the application of pressure. Adhesion can be achieved by the presence of an adhesive layer between the layers or by partially melting one or both layers to affect coalescence of the interfaces.

Melt Extrusion

Molten polymer is forced through a slot of a calibrated thickness and is deposited on the substrate. Since the polymer is a fluid, it is capable of bonding intimately to the surface as it cools. The substrate can be in the form of a flat sheet that would simply pass under the free falling polymer melt. The thickness of the coating is determined by the die opening and the relative rate of motion of the substrate and polymer flow.

Flow Coating

Flow coating is similar to melt extrusion, the difference being that the coating material is generally a solution of polymer rather than a melt. Finishing of the operation requires solvent evaporation and curing. A modification of flow coating is called curtain coating. One of the principle features of this process is that there is a continuous "curtain" of polymer solution flowing from a slot die into a trough. The material collected in the trough is filtered and recycled back into the reservoir. A conveyer belt is arranged to pass a flat substrate under the curtain at a controlled speed.

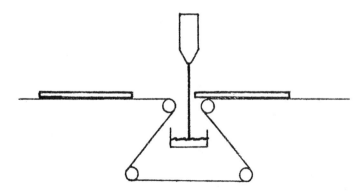

Roll Coating

There are numerous processes that are variations of roll coating and are only applicable to flat surface, most often rolled stock. Offset printing is the most obvious example of such a process. The essential feature of the process is that fluid (ink or polymer solution) is coated onto a transfer roll that ultimately contacts the substrate. The many variations in the process encompass the methods of coating the transfer roller and assuring that the coating is uniform and of the correct thickness. In the case of printing a pattern, the transfer roll is especially treated to allow ink to be present only in the areas needed for pattern (i.e. text letters).

Graft Polymerization

This is a chemical process that grows a polymer directly onto a surface.

The surface is activated by an initiator that chemically bonds to it. In the presence of monomer in gas, liquid, or solution form, the activated sites begin to grow polymer whiskers, ultimately forming a bonded film. Different initiators are used for glass, metal, wood textile, plastic, leather, paper, etc. The activation of the surface can also be accomplished with a glow discharge or R.F. coupled plasma.

Fluid Bed Coating

Solid polymer powder can be levitated in a chamber by the use of air passing through a perforated plate in the bottom of the chamber. So levitated, the particles have a much lower bulk density than the quiescent powder, and the "fluidized" powder actually appears to be gently boiling. Coating is accomplished by preheating a solid substrate to a temperature higher than the melting point of the polymer powder and immersing it into the chamber. As the powder strikes the hot substrate, it melts and coalesces into a continuous film on the surface. As the part cools, the polymer solidifies. If the polymer is reactive, such as an epoxy, it may need additional oven curing to complete the crosslinking. It is obvious that this method can only be used on materials that can be heated to temperature typically greater than $90°C$. Several distinct advantages of fluid bed coating include (a) relatively thick coatings can be applied, even over sharp edges, (b) it is a solventless operation that has cost and environmental advantages, and (c) polymers such as polyethylene, nylons, and fluoropolymers that cannot be applied from solutions can be used for coatings.

Electrophoretic Deposition

Electroplating of metals requires a conductive substrate held at a negative potential and immersed in a bath containing metal ions. The positive metal ions migrate and deposit on the substrate thus building up a metallic layer. Electrophoretic polymer coating is similar. The polymer is in the form of charged colloidal particles suspended in an aqueous medium. The electric field in the solution caused the colloid to migrate to the substrate which is held at a potential charge opposite to that of the colloid. As the particles migrate to the substrate, their charge is neutralized, causing them to deposit and coalesce into a continuous coating.

Spin-On Coatings

The manufacture of semiconductors requires the use of photo-sensitive polymers to impart a pattern onto a silicon slice in a process called photolithography. The substrate configuration presents a unique problem that is most effectively solved by the spin-on technique. Since silicon substrates used for semiconductors are circular with a diameter of 1-5 inches, a small turntable was developed that holds one slice (or wafer) at a time. The polymer solution is metered onto the slice and the turntable accelerated very rapidly to several thousand RPM. A very uniform coating ranging in thickness from 500 to 20,000 Å can be formed in this way.

BIBLIOGRAPHY

Introduction to Polymer Chemistry and Physics

(1) F.W. Billmeyer, *Textbook of Polymer Science,* Wiley Interscience, 1971.

(2) H.S. Kaufman and J.J. Falcetta, *Introduction to Polymer Science and Technology,* Wiley, 1977.
(3) J.M. McKelvey, *Polymer Processing,* Wiley 1971.
(4) R.W. Lenz, *Organic Chemistry of Synthetic High Polymers,* Interscience Publishers, 1971.

Polymer Characteristics

(1) L. Mandelkern, *Crystallinity of Polymers,* McGraw Hill, 1964.
(2) J.H. Hildebrand, J.M. Prausnitz, and R.L. Scott, *Regular and Related Solutions,* Van Nostrand Reinhold Co.
(3) Chemical Society Special Publication No. 20, *Molecular Relaxation Processes,* Academic Press, 1966.

Polymer Formulations

(1) Z. Tadmor and C. Gogos, *Principles of Polymer Processing,* Wiley Interscience, 1979.
(2) G.D. Parfitt, *Dispersion of Powders in Liquids: With Special Reference to Pigments,* Wiley, 1973.
(3) J.A. Manson and L.H. Sperling, *Polymer Blends and Composites,* Plenum Press, 1976.
(4) B. Ranby, J.F. Rabek, *Photodegradation, Photo-Oxidation, and Photostabilization of Polymers,* Wiley 1975.
(5) H.H.G. Jellinek, *Aspects of Degradation of Polymers,* Elsevier, 1978.

Film Formation

(1) R.M. Burns and W.W. Bradley, *Protective Coatings for Metals,* American Chemical Society Monograph Series No. 163, Reinhold, 1967.
(2) Solomon, *The Organic Chemistry of Film Formers,* –.
(3) R.B. Seymour, *Hot Organic Coatings,* Reinhold, 1959.

Coating Functions

(1) J.J. Licari, *Plastic Coatings for Electronics,* McGraw Hill, 1970.
(2) W. Goldie, *Metallic Coating of Plastics,* Electrochemical Publications Ltd., 1968.
(3) J. Crank and G.S. Park, *Diffusion in Polymers,* Academic Press, 1968.
(4) N. Lakshminarayanaiah, *Transport Phenomena in Membranes,* Academic Press, 1969.

Surface Preparation

(1) N.M. Bikales, *Adhesion and Bonding,* Wiley Interscience, 1971.
(2) M.J. Schick, Ed., *Polymers at Interfaces,* Journal of Polymer Science, Part C, Polymer Symposia, Interscience Publishers, 1971.
(3) L.H. Lee, *Adhesion Science and Technology,* Polymer Science and Technology, Volume 9A and 9B, Plenum Press.
(4) J.A. Mock, *A Guide to Conversion Coatings,* Materials Engineering, December 1969, pp. 52-58.
(5) T. Lyman, Ed., *Metals Handbook,* Volume 2, Heat Treating, Cleaning, and Finishing, American Society for Metals, 8th Edition.

Characterization and Testing of Organic Coatings.

(1) N.M. Bikales, Ed., *Characterization of Polymers,* Wiley Interscience, 1971.
(2) J. Haslam and H.A. Willis, *Identification and Analysis of Plastics,* D. Van Nostrand, 1965.
(3) H. Lee and K. Neville, *Handbook of Epoxy Resins,* McGraw Hill, 1967.
(4) J.V. Dawkins, *Developments in Polymer Characterization,* Applied Science Publishers, Ltd., 1978.
(5) D.T. Clark and W.J. Feast, *Polymer Surfaces,* Wiley, 1978.

Trade Journals

Journal of Water-Borne Coatings
Radiation Curing
High Solids Coatings
Journal of Radiation Curing
Materials Digest
Powder Coatings
    Published By: Technology Marketing Corp., Stamford, Conn.
Adhesives Age, Palmerton Publishing Co., New York
Modern Plastics, McGraw-Hill, New York
Plastics World, Cahners Publishing Co., Boston, MA
Plastics Compounding, Industry Media, Inc., Denver, CO
Journal of Coatings Technology, Federation of Societies for Coatings Technology, Philadelphia
Modern Paint and Coatings, Abby Mann, Ed., New York
Journal of Applied Polymer Science, Wiley, New York
Oil, Paint, and Drug Reporter, Chemical Pricing Patterns, New York
Resin Review, Rohm & Haas, Philadelphia
Plastics Technology, Bill Brothers, New York
Plastics Engineering, Society of Plastics Engineers, Greenwich, Conn.
Urethane Abstracts, Technomic Publishing Co., Westport, Conn.

13

Deposition Techniques and Microelectronic Applications

P.B. Ghate
*Texas Instruments Incorporated
Dallas, Texas*

INTRODUCTION

Microelectronics covers a broad spectrum of devices from discrete diodes and transistors to the most complex integrated circuits. Metal and dielectric thin film requirements are so diverse that no single film deposition process or equipment can satisfy all the requirements. Each process needs to be evaluated in terms of the device requirements.[1-8]

In this chapter, the discussion will be confined to basic concepts common to a wide variety of microelectronics applications. Physical principles and fabrication techniques of thin film resistors and capacitors will be discussed. The evolution trend in microelectronics is set mostly by continued improvements in silicon integrated circuits (ICs). For all microelectronic devices, there is one common requirement, namely to establish an electrical contact to the external world through some metal wire, interconnect and so on. Elementary concepts related to the formation of rectifying and ohmic contacts to silicon are discussed. Use of aluminum and platinum silicide contacts is reviewed. In ICs, metal film interconnections provide the necessary connections between the various devices on the chip (die). Also the minimum requirements a metal system must satisfy are reviewed. Large scale integrated (LSI) circuits require multilevel interconnections. A two level metallization scheme (AL/insulator/AL) and its application are presented. It has been observed that electromigration-induced failures in film interconnections adversely affect IC reliability. A brief discussion on electromigration in thin films is presented. All devices need to be appropriately packaged so that they can be connected to other components on a circuit board, a chassis and so on. A brief discussion on the bonding and assembly technologies is also included. Finally, metal and dielectric film deposition processes must be compatible with the device fabrication processes and must meet performance requirements. The cost of processing will be the ultimate determining factor in microelectronics or in any other applications.

THIN FILM RESISTORS

Resistor Characteristics

Resistor is a basic component in electronic industry.[1,4] The resistor value R is given by

(1) $\quad R = \rho \times (\ell/A)$

where ρ = resistivity of the material (Ω-cm), ℓ = length of the current path and A is the cross-sectional area. According to Ohm's Law, Resistance = Voltage/Current. Resistance is one of those unique physical parameters that varies over a wide range ($\sim 10^{25}$). Resistivity of metals at low temperature may be on the order of 10^{-9} to 10^{-10} Ω-cm whereas resistivity of dielectric films can be as high as 10^{14} to 10^{15} Ω-cm.

Resistance of a thin film resistor length ℓ, and cross-sectional area A (width W x thickness t) of a material characterized by resistivity ρ, is given by

(2) $\quad R = \rho \times \dfrac{\ell}{A} = \rho \times \dfrac{\ell}{W \times t} = \dfrac{\rho}{t} \times \dfrac{\ell}{W}$

Here it is convenient to rewrite equation (2) as

(3) $\quad R = R_s \times \dfrac{\ell}{W} = R_s \times N$

where $R_s = (\rho/t)$ is sheet resistance of the film, and N is the number of squares along the length of the resistor. R_s has the dimension of ohms and it is customarily referred to in units of ohms per square written as $\Omega/\square$. A schematic of a straightline thin film resistor of four-squares is shown in Figure 13.1.

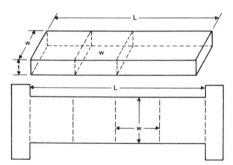

Figure 13.1: Thin film resistor.

A resistor of desired value R, in principle, can be fabricated from a given film of sheet resistance R_s simply by selecting an appropriate number of squares along a current path of the resistor. In practice, however, the available range of sheet resistivity of the films and the allocated area for the resistors on the substrate do not permit straight line resistors. In order to accommodate large numbers of squares, resistor designs use meandering patterns. These conductor paths have L-shaped bends in them and appropriate correction factors need to be applied.[9] For L shaped resistor shown in Figures 13.2a and 13.2b, correction factors are 0.559 and 0.469, respectively. A meandering resistor, shown in Figure 13.3 patterned in area A_r will have a resistance R, neglecting the correction factor,

(4) $\quad R = R_s \dfrac{A_R}{W_\ell^2 (1 + W_s/W_\ell)}$

where W_ℓ and W_s are the width and spacing of the resistor pattern. For $W_s = W_\ell = W$, $R = R_s \times A_r/2W^2$.

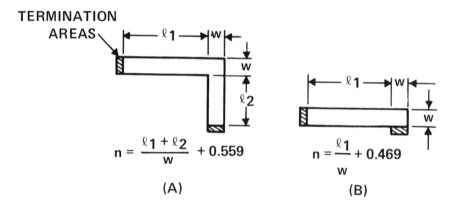

Figure 13.2: L-shaped resistor. (From *Thin Film Technology* by R.W. Berry, P.M. Hall and M.T. Harris. Copyright © 1968, Reprinted by permission of Bell Laboratories.)

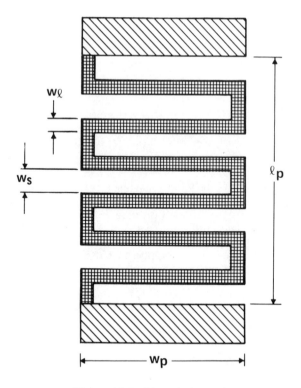

Figure 13.3: Meandering resistor.

The power dissipated in a resistor depends on the operating conditions. For a specified resistor with permissible power dissipation P, the minimum resistor Area (A_R) is given by

(5) $(A_R)_{min} = P/Q_{max}$

where Q_{max} is the maximum power density dissipated. Q_{max} depends on the resistor material, the substrate, thermal impedance of the package and ambience.

Resistance of a thin film resistor is temperature dependent and the temperature coefficient of resistance (TCR) is given by

$$TCR = \alpha = \frac{1}{R(T_1)} \cdot \frac{R(T_2) - R(T_1)}{T_2 - T_1} = \frac{1}{R}\left(\frac{\partial R}{\partial T}\right)$$

The TCR is generally expressed in units of ppm/°C for a given temperature range. Typical TCRs of thin film resistors lie in the range of ±200 ppm/°C.

Voltage coefficient of resistance (VCR) is defined as

$$VCR = \frac{1}{R(V_1)} \frac{R(V_2) - R(V_1)}{V_2 - V_1}$$

where $R(V_2)$ and $R(V_1)$ are the resistance values of the resistor at applied voltages V_1 and V_2. Typical VCR values lie in 0 to 50 ppm/V range.

For film resistors, another parameter of interest is noise. Noise measurements are carried out over a specified frequency range in the operating temperature and voltage ranges. It is a comparative measurement. The coefficient of noise is expressed in decibels as: 10 x log (noise power of sample/noise power of standard). Typical values are –30 to +30 dB.

Conduction in Thin Films

Electrical conduction in metals is due to electrons, while electrical resistivity results from the scattering of these electrons by lattice vibrations, point defects, dislocations, surfaces and so on. In a perfect lattice, at absolute zero, i.e., 0°K, all the atoms are quiescent and the electrons pass through the lattice without any scattering, and the resistivity of this lattice is almost zero. The atoms in a metal at finite temperature are vibrating, and the electrons are scattered by the lattice vibrations (phonons). Electron-phonon scattering is the dominant scattering mechanism giving rise to the resistivity of the metal. The resistivity "ρ" of a metal can be expressed as follows:

$$\rho = \frac{h}{2q^2}\left[\frac{3}{\pi n^2}\right]^{1/3}\left(\frac{1}{\lambda}\right)$$

where h is Planck's constant, q is the electronic charge, n is the density of conduction electrons, λ = electron mean free path = $v_F \tau$ where τ is the relaxation time and v_F velocity of colliding electrons (velocity at the Fermi surface). At room temperature, the electron mean free path for most metals is on the order of 30 nm; however, it is possible to obtain crystals of copper so pure that their conductivity σ (= reciprocal of resistivity, i.e., $1/\rho$) at liquid helium temperature (4°K) is nearly 10^5 times that at room temperature corresponding to the electron mean free path on the order of 0-3 cm.

According to Debye's theory of specific heats,[10] a crystal is assumed to possess a wide spectrum of lattice vibrational modes with a fixed upper limit v_{max}

and the Debye temperature θ is given by the expression $\theta = h\nu_{max}/k$, where h and k are Planck's constant and Boltzman's constant, respectively. At low temperatures, $(T \ll \theta)$, resistivity varies linearly with T. For many metals, the Debye temperature happens to be close to or below room temperature ($\approx 27°C$); hence the temperature variation of resistance above room temperature is approximately linear.

The presence of a defect (e.g., a vacancy, an impurity atom, a dislocation, grain boundary surface, etc.) in a solid disturbs the periodicity of the lattice and this defect serves as a scattering center for the electrons and contributes to the resistivity of the solids. The defect resistivity, for all practical purposes, is temperature independent. Hence, the resistivity of a metal containing defects (e.g., impurity atoms) may be written in the form

$$\rho = \rho_L(T) + \rho_I$$

where ρ_L and ρ_I are the resistivities caused by lattice vibrations and impurity atoms, respectively. This relationship is referred to as *Matthiessen's rule*. At low temperatures, $\rho_L(T)$ tends to zero and the resistivity of a metal is dominated by ρ_I, and it is referred to as the residual resistivity.

For thin films of thicknesses (size) comparable to the mean free path of electrons, scattering of electrons by the physical surface of the films must also be considered in the resistivity analysis. In the framework of Fuchs-Sondheimer theory,[11,12]

$$\sigma = \sigma_\infty [1 - \frac{3}{8k}(1 - p)] \quad (k \gg 1)$$

$$\sigma = \sigma_\infty \frac{3}{4} k (1 + 2p)(\ln\frac{1}{k} + 0.423) \quad (k, p \ll 1)$$

where σ = conductivity of the film - reciprocal of the resistivity = ρ^{-1}, σ_∞ = conductivity of a bulk sample (size $\gg \lambda$), λ is the mean free path of the electrons, $k = d/\lambda$ where d is the thickness of film, and p is a parameter that describes the fraction of electrons elastically scattered at the surface and $(1 - p)$ fraction of electrons is diffusely scattered.

Similar reasoning has been applied to analyze the temperature coefficient of resistance (α) of thin films. The result that one obtains is

$$\alpha = \alpha_0 [1 - \frac{3}{8k}(1 - p)] \quad (k \gg 1)$$

$$\alpha = \alpha_0 \frac{1}{\ln(1/k) + 0.423} \quad (k \ll 1)$$

In practical applications, film thicknesses exceed the mean free path of electrons and the so-called thick film approximation is considered to be appropriate. The important point to notice here is that the coefficient of resistance is positive for continuous metallic films of uniform thickness.

For vacuum deposited films, thin films are anything but continuous during their early stages of growth. As the atoms condense on the substrate, small islands are formed and with increasing amount of deposit these islands connect with each other. Eventually when all the islands are connected and empty regions between the islands are filled, a continuous film is formed. Thin discontinuous films with island structure display a high resistivity and a *negative* temperature coefficient of resistance. Electron conduction in discontinuous films is discussed in the framework of an activated process.[13] Activation energy is required to transfer charge from one island to another at some finite distance. Charge transport mechanism based on tunneling between the allowed states of the neighboring islands has also been considered.

Deposition Techniques and Microelectronic Applications

"Films that are partially continuous have their resistance governed by both an activated process and the normal scattering mechanism discussed earlier." Structure of thin films is governed by the cleanliness of the substrate and also by the vacuum ambience. Films that normally grow into continuous films with characteristic positive TCRs may start showing negative TCRs symptomatic of the discontinuous island structure films. If the grain boundaries of continuous thin films are oxidized, the film will start behaving like an island structure film with negative TCRs. Some of the stabilization bakes employed to lower the TCRs of films can be rationalized in the framework of island structure films. Detailed understanding of the conduction mechanism in discontinuous island structure thin films giving rise to negative TCRs is still lacking.

Resistor Design

Design Considerations: The design of thin film resistors starts with the consideration of several factors such as (i) value of the resistor and the desired tolerance, (ii) TCR and power rating, (iii) film material and method of preparation, (iv) substrate, (v) sheet resistivity control and stability of films, (vi) operating environment, (vii) patterning techniques for line width and spacing control, (viii) method used for resistance adjustment, (ix) process compatibility with other components on the same substrate, if any, and (x) cost. No attempt will be made here to present a detailed discussion of all these factors; however, a brief discussion on the geometric design of the resistors and the commonly used resistor materials will be presented in the following paragraphs.

Geometric Design: Geometric design of the resistor has to comprehend the variation in sheet resistance (R_s) of routinely deposited films, spatial variation of R_s on the substrate, change in R_s due to processing, stabilization and patterning techniques to control line width and spacing. If the value of the resistor and sheet resistance of the film material permit, then a straight line resistor with contact pads at the two ends is highly desirable. High value resistors, in spite of the use of high sheet resistivity films, may have to be designed in a meandering pattern. The design must make provision to trim the resistor (i.e., adjust the value) to a prescribed value. Normally, a resistor is designed so that initial value of the processsed resistor R_I prior to trimming is smaller than the final desired value R_F. A trimming operation proceeds to increase the value of this resistor either by (a) reducing film thickness, e.g., selective anodization of TaN films, or (b) reducing width of the conductor path with a laser beam, or (c) removing the shunt resistance with a laser beam. Some basic resistor designs that facilitate laser trimming are shown in Figure 13.4. The final value of the trimmed resistor R_F is within the specified tolerance limit of the resistor value R.

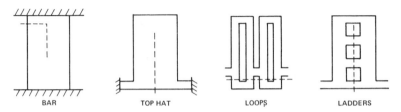

Figure 13.4: Resistor designs for laser trimming. (Glaser/Subak-Sharpe, *Integrated Circuit Engineering.* Copyright © 1977, Bell Laboratories and B. Subak-Sharpe. Published by Addison-Wesley, Reading, Mass. Reprinted with permission of the Publisher.)

Resistor Materials

A wide variety of materials have been employed for thin film resistor fabrication. High sheet resistance for a reasonable film thickness, low TCR and reproducibility of a process are the primary considerations. Resistivities of pure metals lie in the range of 1 to 20 $\mu\Omega$-cm and the TCRs fall in the 1,000 to 4,000 ppm/°C range in the 0° to 100°C temperature interval. Resistivities of vacuum deposited metal films are higher than the bulk values because of their inherent large grain boundary surfaces which contribute to the scattering of conduction electrons. Impurities embedded in the films during film growth also increase the resistivity. Alloy films have higher resistivities than those of the pure constituents. Resistor materials can be classified in three main categories: (a) single metal system (b) alloy films and (c) cermet films. Resistivities of the films are not only sensitive to their source composition and mode of preparation but they also depend on the quality of the substrates such as roughness and waviness. A perfectly smooth surface is desirable to promote the growth of defect free thin films. Since resisitivities of thin films depend on their microstructure (grain size, island structure and so on), the resistivities of films deposited on rough substrates are generally higher than those on smooth substrates. Polished 99.94% alumina substrates (ultra-fine grains) with surface roughness on the order of 25 nm are preferred for resistor fabrication. Other substrates that find use are oxidized silicon substrates, polished sapphire, Corning code 7059 glass and soda lime glass. Alumina substrates are commonly used.

Single Metal System—Example: Tantalum. Tantalum has been widely used as a resistor material. Tantalum films are vacuum deposited either by electron-beam evaporation or by sputtering and the latter method is preferred in high volume production. Tantalum is a reactive material and the sputtered films are generally contaminated by the constituents of the vacuum ambience. Tantalum films can exist in at least three forms: (a) BCC alpha-phase – similar to that of the bulk, (b) tetragonal beta-phase, and (c) a mixture of alpha and beta-phases. Formation of a particular form is strongly dependent on the constituents of background gases in the vacuum chamber. Nitrogen and oxygen both exert profound influence on the resistivity and TCR of these films. Increase in partial pressure of nitrogen from 2.6×10^{-4} Pa ($\approx 2 \times 10^{-6}$ torr) to 1×10^{-3} Pa change the TCR from a negative value to a positive value. (See Figure 13.5.) Incorporation of oxygen in the films and/or oxidation of the grain boundaries of the films will cause the TCR to decrease very rapidly to a negative value on the order of -500 ppm/°C. Tantalum films are normally subjected to stabilization bakes in air ambience. For example, when a stabilizing heat treatment of 1 to 2 hours at 200°C is applied, sheet resistivity of these films increases somewhat but the TCR drops to values in the 0 to -300 ppm/°C range. Detailed understanding of the effects of stabilization bakes is still the subject of active research. However, the negative TCR can be reconciled in terms of the oxidation of grain boundaries or connecting links of the island structure. Tantalum films of resistivities in the range of 100 to 200 $\mu\Omega$-cm (and sheet resistances 30 to 300 $\Omega/\square$) are used to fabricate resistors with ±100 ppm/°C TCR values.[14,15]

Alloy Films—Example: Nichrome. Nichrome alloy films have been successfully employed for thin film resistors.[15] An alloy composition in the vicinity of 80% nickel and 20% chromium is generally used. Other compositions have also been used on some occasions. Since the vapor pressures of Ni and Cr differ considerably, composition control of Ni and Cr in films is achieved by flash evaporation of the source material in the form of a wire; thermal evaporation from a molten source of suitably chosen composition has been employed to achieve the desired composition in films; the problem of composition control of Nichrome films has also been solved by sputter deposition from a cathode of desired Ni and Cr composition. Since chrome readily reacts with oxygen, resistivity of Nichrome films is dependent not only on alloy composition of the films but also on the evaporation conditions. The TCRs of sputtered films are found to be much less variable than those of evaporated films. Nichrome films of resistivities 100 to 130 $\mu\Omega$-cm and sheet resistances 10 to 400 $\Omega/\square$ have been used to fabricate resistors with ±100 ppm/°C TCR. (See Figure 13.6.)

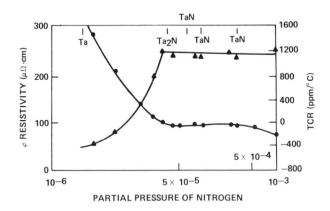

Figure 13.5: Effect of N_2 partial pressure in the sputtering chamber on the resistivity and TCR of Ta films. (Maissel). (From *Handbook of Thin Film Technology.* Copyright © 1970, McGraw-Hill. Used with permission of McGraw-Hill Book Co.)

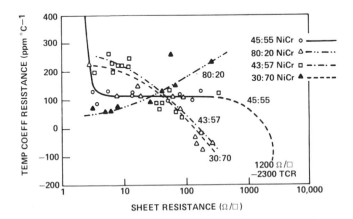

Figure 13.6: TCR of nichrome films as a function of sheet resistance for several different compositions. (Maissel). (From *Handbook of Thin Technology.* Copyright © 1970, McGraw-Hill. Used with permission of McGraw-Hill Book Co.)

Cermets—Example: Cr-SiO. A solid mixture of metal-insulators such as Cr-SiO, Au-SiO, Au-Ta$_2$O$_5$ is referred to as a cermet; the Cr-SiO cermet films have been successfully employed.[16] The resistivities of cermets are generally higher (10^2-10^4 $\mu\Omega$-cm) than those of alloy films and they are stable over a fairly large temperature range (0° to 500°C). The resistivities and TCRs are composition dependent, and thermal treatments affect the final resistor characteristics. In Figure 13.7, resistivity vs composition is given for films as deposited at 200°C as well as after annealing treatments at 400°, 500° and 600°C for 1 hour in argon.

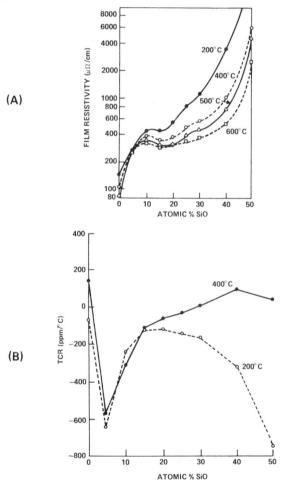

Figure 13.7: (A) Resistivity of Cr-SiO films as a function of composition and thermal history and (B) TCR of Cr-SiO films as a function of composition, both as deposited at 200°C and after heat treatment at 400°C. (Maissel). (From *Handbook of Thin Film Technology*. Copyright © 1970, McGraw-Hill. Used with permission of McGraw-Hill Book Co.)

Resistor Fabrication: The following flow chart summarizes the salient features of resistor design and fabrication.

THIN FILM CAPACITOR

Capacitor

A capacitor is the basic component for storing electric charge; $Q = CV$. (Q = charge, C = capacity, V = voltage.) Thin film capacitors usually have a parallel plate configuration.[3,5] Schematic cross sections of parallel plate capacitors are shown in Figure 13.8, and the basic structure essentially consists of a bottom electrode, an insulator film and a top electrode. The capacitance C of such a structure is given by

$$C = k\epsilon_o \frac{A}{d}$$

where d = thickness of film, k = dielectric constant or the relative permittivity of the insulator material, ϵ_o is the permittivity of free space (= 8.85×10^{-12} farad/meter), A is the effective area of the insulator film that has both upper and lower electrode plates and d is the insulator film thickness. Thin film capacitors are not ideal charge storage elements. Dielectric (insulator) films always have a finite parallel resistance R, and the impedance Z of a capacitor is obtained from a vector diagram and is given by

$$Z = R - \frac{j}{\omega C}$$

where ω is 2π times the frequency of the applied voltage, j is the imaginary number $\sqrt{-1}$. Furthermore $\tan \delta_o$ is a measure of the energy absorbed in the dielectric. We note that

$$\tan \delta_o = \frac{1}{R\omega C}$$

and it is referred to as the dielectric loss tangent. Resistance R is fairly large for insulator films (typical resistivity range 10^{10}–10^{14} Ω-cm), and at low frequencies $\tan \delta_o$ is fairly small and is almost independent of frequencies; for example, a typical value of $\tan \delta_o \approx 0.005$ for tantalum oxide film capacitors.

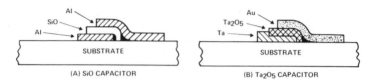

Figure 13.8: Thin film capacitors.

Since the resistance of the electrodes and leads contribute to the energy loss of a capacitor, it is customary to express the total dissipation factor of a thin film capacitor as

$$\tan \delta = \tan \delta_o + \omega C R e$$

where $\tan \delta_o$ is the frequency independent part, ω is the angular frequency and Re is the electrode resistance. At high frequencies ($\approx$1 MHz), the electrode resistance is a major contributor to the energy loss. For example, at 1 MHz, an electrode resistance of 1 ohm will essentially double the loss factor of a 500 pF (picofarad = 10^{-12} F) tantalum oxide capacitor.

The dielectric film thickness d, the maximum dielectric strength E_B (or the breakdown electric field strength) and the breakdown voltage V_B are related by

$$V_B = d \times E_B$$

The rated voltage V_R is some fraction K (K $\cong$ 0.1 to 0.5) of the breakdown voltage V_B; thus

$$V_R = KV_B = K \times d \times E_B$$

Another parameter of interest is the charge storage factor (CSF) for a given material, and it is given by the product of the rated voltage V_R and the capacitance density C/A; charge storage = $V_R \times$ C/A serves as a figure of merit in the evaluation of dielectric materials. (For example, CSF $\cong$ 3 microcoulomb/cm^2 for tantalum oxide, k = 25; and CSF $\cong$ 0.5 microcoulomb/cm^2 for SiO_2 oxide, k = 3.8.)

Dielectric Films

A number of insulator materials have been investigated for thin film capacitor fabrication; some notable films commonly used are silicon monoxide, silicon dioxide and tantalum oxide.[17] In practice, deposited dielectric films thinner than 500 to 700 Å have fairly high pinhole density and the yields (number of good units produced/total number of potential units) are poor. Pinholes lead to resistive shorts between electrodes and thereby increase the leakage currents.

Thick dielectric films (e.g., d > 20,000 Å) also are undesirable because of the high internal stress levels in these films. High compressive stresses in films pose adhesion problems, as it is not uncommon to observe the films peel off from the substrates; on the other hand large tensile stresses are relieved by crazing. Once the dielectric, electrode and substrate materials combination is chosen for a thin film capacitor fabrication, then dielectric film deposition conditions and thickness must be optimized; for example, the optimum thickness range for tantalum oxide films on glazed ceramic substrates is 250 to 350 nm.

Silicon Monoxide Films: Evaporated silicon monoxide films have been extensively studied for applications as dielectric films in thin film capacitors, insulation layers between different levels in integrated circuits, protective coatings and so on. Silicon monoxide films are deposited by thermal evaporation (heated crucible or electron beam).

Film properties such as composition, internal stresses, are sensitive to O_2 partial pressure in the vacuum system, total background pressure during film deposition, substrate temperature and film deposition rate. Low compressive stresses are preferred. SiO is a metastable phase and it is susceptible to oxidation on exposure to O_2 and water vapor; capacitors made with film composition near SiO have high leakage currents, high dissipation factors between 0.05 to 0.1 and large values of temperature coefficient of capacitance (TCC) on the order of 1,000 ppm/°C. Improved capacitor performance, such as low leakage currents, low dissipation factors and low TCC, is realized for SiO films heat treated in O_2 ambience. Dielectric constant (K) and conducitivity (δ) of these silicon oxide (Si_xY_y) films depend on silicon to oxygen ratio and parameters K and δ vary from 6 to 3.8 and 10^{-10} to 10^{-16} Ω^{-1}-cm^{-1} respectively as film composition changes from SiO to SiO_2. It is observed that silicon oxide films closer to $SiO_{1.5}$ rather than SiO are well suited for capacitor applications. Thin film capacitors are generally fabricated on smooth ceramic substrates with aluminum electrodes. Since aluminum is susceptible to corrosion under high humidity conditions some protective coating is applied. Silicon oxide film capacitors in the 10 to 1,000 pF range requiring an area well below 0.05 cm^2 (capacitance density $\approx$0.03 to 0.003 microfarad/cm^2) find usage in microwave circuits.

Silicon Dioxide Films: Silicon dioxide films are extensively used as diffusion masks, insulating layers between levels, protective contatings and so on in integrated circuits. Thermally grown oxides of varying thicknesses are used as diffusion masks in IC fabrication. MOS (metal oxide-semiconductor) technology routinely uses thermally grown SiO_2 films on the order of 1000 Å in MOS IC production. Several other methods have been employed for SiO_2 film deposition; (a) chemical vapor deposition: e.g., oxidation of silane $SiH_4 + 2O_2 \rightarrow SiO_2 + 2H_2O$; (b) plasma glow discharge; (c) sputter deposition from a quartz cathode; (d) reactive sputtering of silicon in O_2 ambience, (c) electron beam evaporation of quartz and so on. Physical and electrical properties of these deposited films strongly depend on the methods of preparation. Film composition and dielectric properties of reactively sputtered films depend strongly on the O_2 partial pressure during deposition whereas the sputtered quartz films are less sensitive to the sputtering environment. In the last few years, SiO_2 films deposited by plasma assisted chemical vapor depositon method are widely used as interlevel insulator and protective coatings; these oxides are referred to as plasma SiO_2.

Though deposited oxides have a high pinhole density, deposition methods carried out in clean environments are capable of yielding high quality insulating layers (e.g., plasma SiO_2) compatible with large scale integrated (LSI) circuit processing; e.g., LSI circuits with 25,000 cross overs 10 x 10 μm^2 each (metal/insulator/metal) are feasible. For some applications, where low defect (or pinhole) density insulator films are desired, two layers of films are utilized. Since the nucleation sites of these defects are distributed randomly in the film, defects in the second layer are also randomly distributed and hence there are extremely few defects in the composite films.

Thermally grown SiO_2 films are far superior than any of the deposited films. Thermal oxides are characterized by a dielectric constant k = 3.8, breakdown field strength $\sim 4 \times 10^6$ V/cm, and a conductivity of 10^{-14}-10^{-16} ohm^{-1}-cm^{-1}. Metallic ion contaminants, such as Na, diffuse readily in SiO_2 films subjected to an electric field. Great care is exercised to eliminate Na ion contamination of these oxides during oxide growth and metal electrode fabrication for metal oxide semiconductor integrated circuit (MOS IC) applications.

Silicon Nitride Films: Silicon nitride films are generally deposited by three methods: (a) chemical vapor deposition by silane-ammonia reaction at an elevated temperature $\approx 900°$-$1000°C$, (b) in a plasma glow discharge and (c) reactive sputtering of silicon in N_2 ambience. Silicon nitride films prepared by the first method, referred to as high temp. nitride have an almost ideal silicon-nitrogen ratio and can be symbolized as Si_3N_4. These films have a dielectric constant of 9.4 and conductivity of 10^{-13}-10^{-14}-ohm^{-1}-cm^{-1}. Since the deposition temperature is high, these films are not well suited for capacitor fabrication; however, these films are widely used as diffusion masks and insulating layers in IC fabrication. Silicon nitride films are excellent diffusion barriers for Na ion migration in oxides. In the last 5 to 6 years, interest in silicon nitride films deposited in a plasma reactor has been growing. These films are deposited in the 1-10 pascal (10-100 millitorr) pressure range; the silicon nitrogen ratio in these films depends on gas flow conditions in the reactor, plasma conditions, substrate temperature and so on; since the substrates are held at a temperature in the $50°$ to $350°C$ range, these films are referred to as low temp. nitride. In general, internal stresses in these films are tensile in nature, and these films are prone to cracking when subjected to high temperature (>$350°C$) excursion. Deposition conditions and film thicknesses are optimized to minimize the effects of internal stresses. Dielectric constant, conductivity and etch rates are controlled to specifed limits, and these films are used as protective coatings on semiconductor devices.

Finally, the composition of reactively sputtered silicon nitride films varies as a function of the depsoition parameters. Reactively sputtered films on the order of 2000 to 3500 Å are employed for protective coatings; however, these films are far from ideal for thin film capacitor applications.

Tantalum Film Capacitors

Tantalum oxide film capacitors have found widespread use in thin film circuits because of their process compatibility to fabricate thin film resistors and capacitors on the same substrate. Tantalum oxide has a fairly high dielectric constant (~25), and high quality defect-free films are readily formed by anodic oxidation of tantalum films. Because of their technological significance, tantalum films have been extensively studied and results have been summarized in a recent book entitled "Tantalum Thin Films".[114] Ta films are vacuum deposited either by electron beam evaporation or by sputtering, and the latter method is generally preferred for high volume production. Ta films are found to exist in at least threee phases: (a) BCC - alpha-phase, similar to that of the bulk material, (b) tetragonal - beta-phase and (c) a mixture of alpha and beta-phases. The particular phase in which the film deposit has grown is strongly dependent on the vacuum ambience and film deposition conditions. Tantalum oxide films can also be deposited by reactive sputtering from a tantalum cathode in O_2 ambience. However, tantalum oxide films formed by anodic oxidation of Ta are preferred for thin film capacitor fabrication. Anodization is the electrochemical oxidation of a metal in an electrolyte cell. Several electrolytes such as HNO_3, H_2SO_4, H_3PO_4, citric acid and acetic acid can be used for Ta_2O_5 formation. The formation of tantalum oxide as a capacitor dielectric is carried out at constant current density until a preselected voltage is reached. As illustrated in Figure 13.9, during this period, the oxide thickness increases at a constant rate and the potential also increases. As the preselected voltage is reached, anodization is continued at constant voltage until the ion current decays to a low level on the order of 1 to 2% of its original value.

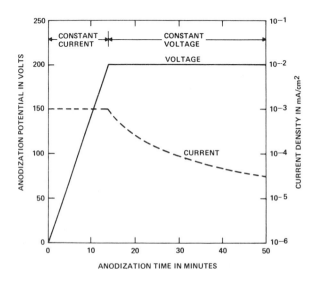

Figure 13.9: Constant current-constant voltage anodization steps for tantalum oxide formation. (Gerstenberg). (From *Handbook of Thin Film Technology,* Copyright © 1970, McGraw-Hill. Used with permission of McGraw-Hill Book Co.)

The quality of the anodically formed Ta_2O_5 films depends mainly on the cleanliness and purity of the Ta films. Ta_2O_5 films are amorphous in nature, pinhole free and display a uniform breakdown strength. The conduction characteristics of capacitors formed by anodic films generally depend on the polarity of the applied voltage. Capacitors formed on pure Ta films for example have breakdown voltages which are about 80 to 90% of the anodization voltage when the Ta electrode is positive (anodic polarization), while upon cathodic polarization this value is only 5 to 10% of the anodization voltage. Addition of impurities such as N_2, O_2, C or Mo to Ta films results in capacitors with almost symmetric conduction characteristics.

Thin film Ta_2O_5 film capacitor fabrication process begins with the selection of a glazed smooth ceramic substrate. The substrate is cleaned and beta-Ta films on the order of 4000 Å thick are sputter-deposited. Bottom electrode area is patterned. Capacitor area is defined and part of the film is anodically oxidized in a dilute aqueous electrolyte. Top electrode material such as Ti/Pt/Au or NiCr/Au film layer is deposited and the electrode is patterned. Capacitance per unit area of the anodic film depends on the formation voltage and film thickness, assuming a growth constant of 17 Å/V while consuming 6.3 Å of Ta per volt. Thus the capacitance density C/A of anodic oxide formed at 130 volts is approximately given by $(k\epsilon_o/d) = 25 \times 886 \times 10^{-14}/17 \times 130 \times 10^{-8} \approx 10^{-7}$ $\mu F/cm^2$. The loss tangent (the dissipation factor) varies between 0.002 and 0.01 for frequencies between 0.1 to 10 kHz and the capacitance changes by about 1% in this frequency range. The temperature coefficient of capacitance (TCC) is about 200 ppm/°C for 0 to 100°C temperature range. Capacitors are considered acceptable if their leakage current is well below $1\mu A\mu F^{-1}$ at about 60% of the anodization voltage.[17] Normally, a predetermined value of leakage current is used as a criterion for the quality of tantalum capacitors.

CONTACTS

Contact formation to devices is an essential process step in device fabrication.[4,5,18,19] In general, three types of contact formation are encountered: (a) metal-metal contacts, e.g., resistor bond pads, (b) metal-insulator, e.g., capacitors, (c) metal-semiconductor contacts – transistors, integrated circuits, Schottky rectifiers and so on. In general, the current transport across an interface of dissimilar materials is asymmetric; i.e., the current is transported more easily in one direction than another. This phenomenon was observed by K.F. Braun as early as in 1877.[20] Figure 13.10 shows a current-voltage characteristic for a contact between a metal wire and lead sulfide crystal. The direction of the rectification was such that the current flowed easily when the metal was positive with respect to the semiconductor; very little current flowed if the metal was negative. Contacts displaying such current voltage characteristics are referred to as rectifying contacts. Even though current transport across interfaces of dissimilar materials has been extensively studied by a large number of investigators and over the years considerable understanding has been gained, still this subject continues to be an active field of research.

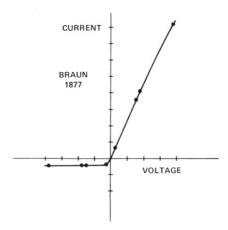

Figure 13.10: Current voltage characteristics of a metal wire contact to lead sulfide crystal. Taken from K.F. Braun (1877).

In this section, barrier heights of metal-semiconductor contacts are briefly reviewed in the framework of the Schottky model. Expressions for current voltage characteristics of M–S contacts are presented. Finally, fabrication processes of forming M–S contacts and ohmic contacts to silicon substrates are discussed.

The Schottky Model

When a metal makes an intimate contact with a semiconductor, the Fermi levels in the two materials must be coincident at thermal equilibrium and as a result a potential barrier is formed between the two. (For the present discussion, silicon is the representative semiconductor material.) According to the Schottky model, the barrier height ϕ_B is given by the difference between the metallic work function ϕ_m and the electron affinity χ_s of the semiconductor

$$\phi_B = \phi_m - \chi_s$$

The energy band diagram of a metal n-type silicon is shown in Figure 13.11. Here ϕ_B represents the barrier to the movement of electron from the n-silicon into the metal. If the junction is forward biased, with an external source V_B, so that the semiconductor is negative with respect to the metal, the forward barrier is reduced to $\phi_B - V_B$, and the current increases; for a reverse bias the barrier increases to $\phi_B + V_B$, and the current decreases. Energy band diagrams for metal-semiconductor junctions are shown in Figure 13.12. Similarly, the barrier height for a metal-p-type semiconductor ϕ_{Bp} is given by the expression

$$\phi_{Bp} = [E_g/q - (\phi_m - \chi)]$$

where E_g is the energy gap and q is the electronic charge.

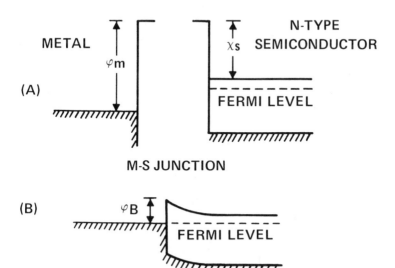

Figure 13.11: Schottky barrier formation: (A) non-equilibrium band structure at metallic and n-type semiconducting surfaces and (B) metal-semiconductor junction at thermal equilibrium.

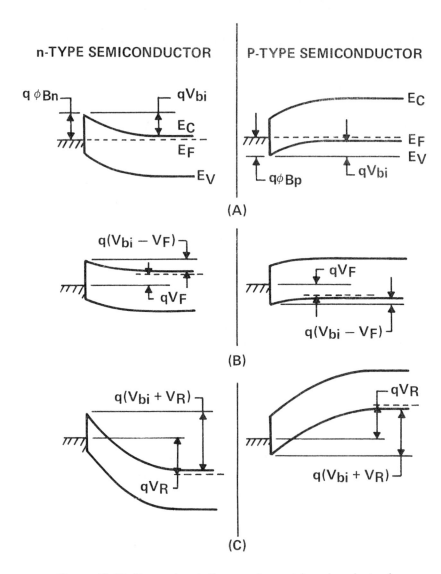

Figure 13.12: Energy band diagrams for metal-semiconductor junctions: (A) thermal equilibrium, (B) forward bias and (C) reverse bias. (Glaser/Subak-Sharpe, *Integrated Circuit Engineering.* Copyright © 1977, Bell Laboratories and B. Subak-Sharpe. Published by Addison-Wesley, Reading Mass. Reprinted with permission of the Publisher.)

Current transport in metal-semiconductor barriers is mainly due to majority carriers in contrast to p-n junctions where the minority carriers are responsible. Expression for the current voltage characteristic developed in the framework of Thermionic Emission (TE) model is as follows:

$$J = A^*T^2 \exp\left(\frac{-q\phi_B}{kT}\right)\exp\left(\frac{q(V + \Delta\phi)}{kT}\right) \text{ for } V \gg \frac{3kT}{q}$$

$$\approx J_{ST}\exp\left(\frac{qV}{nkT}\right)$$

where A^* = effective Richardson constant = 120 amp/cm^2/°K
ϕ_B = barrier height
$\Delta\phi_B$ = lowering of the barrier height due to image forces
V = bias voltage
n = parameter in the range of 1 to 1.3
J_{ST} = saturation current for a reverse bias

$$= A^*T^2 \exp\left(\frac{-q\phi_B}{kT}\right)$$

The TE model satisfacorily explains the current-voltage characteristics for a lightly doped semiconductor ($<10^{18}$ cm^{-3}) in contact with a metal. The specific contact resistance R_c (expressed in ohms-cm^2) for a Schottky barrier diode (SBD) is found to be temperature dependent as can be seen from the following relationship.

$$R_c = \frac{\partial V}{\partial J} \quad J \to 0$$

$$\cong \frac{k}{qA^*T}\exp\left(\frac{q\phi_B}{kT}\right)$$

For a disc contact of radius r, the spreading resistance R_s is given by

$$R_s = \frac{\rho}{4r}$$

where ρ is the resistivity of the semiconductor. Hence for small contacts, the voltage drop across a SBD is not only governed by R_c but also by R_s.

In many semiconductor device applications, metal contacts are made to semiconductors whose doping density lies in the range of 10^{18} to 10^{20} cm^{-3}. As the doping density starts increasing, the depletion width W starts decreasing as $1/\sqrt{\text{doping density}}$. For highly doped semiconductors (10^{20} cm^{-3}) the depletion width W drops to a range of 1 to 3 nm. For metal-semiconductor contacts with such small depletion widths, the current transport across the interface is dominated by quantum mechanical tunneling through the barrier rather than by thermionic emission. For intermediate doping density, the current voltage characteristics are developed in the framework of a thermionic field emission (TFE). It is sufficient to note here that once the current transport is dominated by tunneling the specific contact resistance is primarily governed by the shape (width and height) of the barrier and is almost independent of temperature.

Experimentally measured barrier heights for different metals on n-type silicon are presented in Table 13.1. Current-voltage characteristics of forward biased Schottky diodes are presented in Figure 13.13.

Table 13.1: Typical Schottky Barrier Heights for Metal-n-Type Silicon

Metal	ϕ_{BN} (eV)
Al	0.65
Au	0.79–0.80
Mo	0.57–0.59
Ni	0.66
Pd	0.75
Pt	0.80–0.87
W	0.65–0.67

Note: $\phi_{BN} + \phi_{BP} = E_g = 1.1$ eV for silicon.

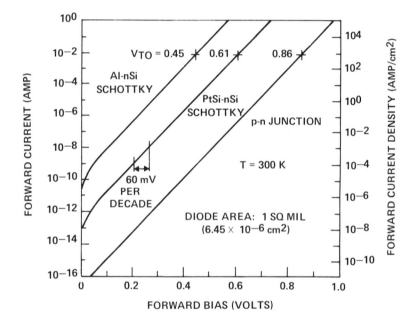

Figure 13.13: Current voltage characteristics for Schottky diodes. (Glaser/Subak-Sharpe, *Integrated Circuit Engineering*. Copyright © 1977, Bell Laboratories and B. Subak-Sharpe. Published by Addison-Wesley, Reading Mass. Reprinted with permission of the Publisher.)

The Schottky barrier diode (SBD) is a majority carrier device with inherent fast response in contrast to a p-n junction where the current is transported by minority carriers and the response time is longer. Because of the fast response of the SBD, it is used to shunt excess base current through the diode as a majority carrier current in a clamped transistor[21] (Figure 13.14). The measured saturation time can be reduced to about 10% of that of the original transistor, and it is on the order of a few nanoseconds. SBDs are widely used in bipolar ICs to improve circuit performance.

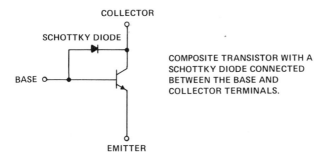

Figure 13.14: Schottky clamped transistor.

Ohmic Contacts

For most applications, a low impedance ohmic contact is desirable. The current-voltage characteristic of an ohmic contact is linear. In general, ohmic contacts to silicon are achieved by establishing metal contacts to heavily doped n+ (or p+) regions. In such a metal-semiconductor contact, the depletion width in heavily doped semiconductor region is small, and the current transport takes place by a tunneling mechanism.

Specific contact resistance "R_c" is defined as the reciprocal of the derivative of current density with respect to voltage and is expressed in units of $\Omega\text{-cm}^2$. It is an important figure of merit for charge transport characteristics across a metal-semiconductor contact. For a given metal-silicon system, contact resistance depends on surface concentration. Typical R_c data are shown in Figure 13.15. Under simplifying assumptions (neglecting current crowding, perimeter effects and so on) contact resistance r_c for a contact of area A is given by $r_c = R_c/A$.

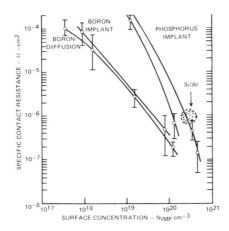

Figure 13.15: Specific contact resistance data for platinum silicide contacts. (J.C. Blair, C.R. Fuller and P.B. Ghate.)

Fabrication Methods: Formation of ohmic contacts to devices consists of joining two dissimilar materials. This interface is then expected to offer almost zero resistance to the passage of current and provide linear current voltage characteristics. In practice, so long as the contact resistance is small relative to that of the device and the voltage drop across the contact does not affect significantly the current-voltage characteristics of the device, that contact will be considered as ohmic. Whatever may be the metal chosen for establishing the ohmic contacts, so long as it is compatible with the total device processing and reliability requirements, that metal system will be an acceptable solution for ohmic contact formation. A large number of pure metals and alloys have been investigated for various applications. Several processes have been employed to establish an intimate contact between the metal and the device region, e.g., alloying, plating, evaporation, bonding, and so on but it is the cost effective solution that finds application in practice.

It is one of the pre-requisites in contact formation that the adhesion (or wetting) between the materials to be joined is good and reproducible. Clean surfaces are highly desirable and several techniques such as ultrasonic cleaning, chemical etching, plasma glow discharges, sputter etching, ion milling, and so on have been employed for surface cleaning. Surface preparation must be considered an integral part of the contact formation process.

In most device fabrications, more than one type of process is used for forming ohmic contacts at different process steps. Consider the fabrication of silicon ICs: (a) an *evaporation process* is employed to form ohmic contacts between two metal layers, e.g., NiCr resistors and Al conductors; this process is also used to form metal-semiconductor contacts, e.g., Al or PtSi contacts to emitter, base and collectors; (b) an *alloying* process is used to attach Si chips (dies) to the package; the backside of a chip has an evaporated layer of Au; the package (lead frame or TO-5 header, etc.) is gold plated, and a preform (an Au-Ge alloy) is placed between the chip and the package. When this sandwich structure is heated above the eutectic temperature, the preform wets the surfaces of the chip and the package and an ohmic contact is formed by alloying; (c) thermocompression or ultrasonic *bonding* techniques are used to attach Au or Al wire (25 μm in diameter) to the aluminum bond pads of the IC; (d) a *plating* process is also employed to increase the thickness of Au interconnects; here metal-metal contact is established; in certain cases electroless nickel is employed to establish ohmic contacts to semiconductors.

In general, forming ohmic contacts between two metal systems is relatively easy so long as their work function differences are small, and clean surfaces are used. The system is heated to an appropriate temperature to promote intermixing of atoms at the interface. Thin oxide layers or foreign materials usually interfere in contact formation. If interatomic diffusion of these metals leads to the formation of an intermetallic compound layer, the characteristics of this contact are greatly influenced by this compound formation.

In semiconductor technology, metal films are vacuum deposited by several methods such as electron beam evaporation, induction heated source, RF sputtering, DC magnetron sputtering onto clean semiconductor surfaces. The metallized substrates, after appropriate patterning steps as needed, are heated to a temperature slightly below the metal-silicon eutectic temperature to form ohmic contacts. During this annealing step, normally referred to as "contact sintering," a thin layer of metal silicide or silicon rich metal compound is formed due to interdiffusion. The metal-silicon interface is no longer sharp. So long as this diffused layer has a metal-like behavior and that the depletion width in the semiconductor at the contact remains thin enough for the electrons to tunnel through, the primary objective of forming the ohmic contacts is achieved.

Aluminum to Silicon Contacts: Among the large number of metals investigated for achieving contacts to Si, Al is the most widely used one, because it forms low resistance contacts to both p and n type Si and also meets the interconnection requirements. Aluminum films are vacuum deposited by several techniques including (i) filament evaporation, (ii) flash evaporation, (iii) induction heating, (iv) electron beam heating, (v) sputtering, etc. Electron beam, induction heating and DC magnetron sputtering deposition technologies are widely used because of their ability to deposit high purity Al films to within 5-10% of the bulk resistivity ($\rho = 2.7$ $\mu\Omega$-cm). After the appropriate interconnection patterns are delineated on the substrates, Si/Al contacts are effected by annealing the system at some temperature in the 400°-550°C range (for a finite duration in an inert ambient, N_2 or Ar, forming Al+Si alloy at the interface. This step is referred to as contact sintering. In principle, the contact resistance of the Si-Al interface is a function of the silicon surface (surface concentration) and the metal; however, high resistive Si/Al contacts or poor contacts are not uncommon in IC manufacturing. Poor metallurgical reaction at the Si-Al interface either due to surface contamination or due to poor metal deposition technique is considered to be the primary cause of this problem. Experiments have shown that so long as the Al film depositions are carried out in a hard vacuum on the order of 1×10^{-6} torr or below, contact resistance is not significantly affected by vacuum ambience. Silicon surface preparation prior to presenting the slices for Al film deposition plays an important role in contact formation. Chemically cleaned Si surface in a dilute HF solution followed by a deionized water will normally have a natural SiO_2 layer 1-3 nm thick prior to metal deposition because the surface is exposed to air for an interval ranging from a few minutes to a couple of hours. Aluminum reduces the silicon dioxide by a diffusion controlled reaction. A sintering operation of short duration at elevated temperature (e.g., 10 minutes at 500°C) reduces this oxide uniformity at the contact window leading to uniform Al+Si formation. With a lower temperature and longer duration sintering, Al penetrates deeply into the silicon at discontinuities at the oxide layer instead of uniformity reducing the SiO_2 layer to permit uniform Al-Si formation. (See Figure 13.16.)

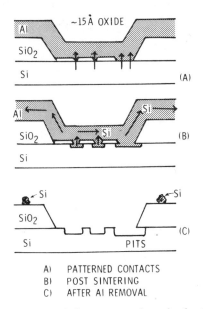

A) PATTERNED CONTACTS
B) POST SINTERING
C) AFTER Al REMOVAL

Figure 13.16: Formation of alloy penetration pits for Si/Al contacts.

Silicon from the contact windows diffuse into Al films along the grain boundaries to satisfy the solid solubility of Si in Al at sintering temperature and precipitates on cooling. Aluminum diffusion in silicon substrate is orders of magnitude slower. The Al film simply falls into the eroded regions. This alloy penetration (or erosion depth) is a function of sintering temperature, time, contact size, and dimensions of Al leads. Scanning electron micrographs of pits formed in (111) and (100) oriented silicon substrates are shown in Figure 13.17.

Figure 13.17: SEMs of pits formed in contacts on (111) and (100) silicon substrate.

A wide Al lead covering smaller contact windows can lead to deep alloy penetration and junction shorting. For devices with shallow emitter-base junction (<1 μm) alloy penetration can cause junction shorting (≈ 1 wt % Si). Alloy penetration can be impeded by the use of Al+Si films for interconnections.

For most devices with emitter-base junction >1 μm the processing of Si/Al ohmic contacts is not a serious problem. If the Si/Al contact is subjected to different heat treatments, the dissolution of silicon from the contact windows into aluminum at elevated temperatures and the precipitation of Si from Al films at the Si/Al interface and at the Al grain boundaries on the cooling cycle, cause changes in electrical behavior of the contacts. Changes in the barrier height of the Si/Al Schottky diode are found to be dependent on heat treatments.

Platinum Silicide Contacts: Platinum silicide contacts are widely used for achieving rectifying (e.g., Schottky contacts) and low resistance ohmic contacts in the large scale integrated (LSI) circuits on silicon substrates.[8] Platinum films of appropriate thickness are vacuum deposited either by evaporation or by sputtering onto chemically cleaned silicon substrates. Sometimes the silicon substrates are in-situ sputter etched prior to Pt deposition. The Pt metallized substrates are sintered at temperatures ranging from 400° to 600°C in an inert ambient such as N_2. Platinum silicide layers are formed by the interdiffusion of Si and atoms. In the initial growth phase of this layer, the region closer to Si substrate is platinum rich and a Pt_2Si compound formation is observed. With the lapse of time, the Pt layer is all converted into a PtSi layer. The PtSi layer thickness varies not only as a function of Pt thickness but depends on the Si surface treatment prior to Pt deposition and on the sintering ambience. If the native oxide on the Si surface is not removed either by dipping the substrate in HF solution or by in-situ sputter etching, Si diffusion into Pt films is impeded by this oxide layer, thinner nonstoichiometric PtSi films being formed. The stoichiometry of the PtSi films is also affected if the films are formed in an oxidizing ambience (e.g., air) rather than in an inert gas (N_2, argon). The formation of PtSi has been studied by several workers. Considerable understanding of the effects of sintering ambience on PtSi formation has been gained with the aid of ion scattering, Auger electron spectroscopy (AES) and ion depth profiles. In Figure 13.18, the results for two PtSi layers formed by sintering in inert (N_2) and oxidizing (air) ambients are presented. Depth profiles for Pt, Si, O, N and C were developed using peak-to-peak heights of the Auger transition and sputtering times. In order to compare the Pt and O concentration vs depth into these PtSi layers, the ratios of the peak to peak heights of Pt:Si and O:Si were developed, and these ratios are presented in this figure. It can be readily seen that oxygen is present throughout the PtSi layer when it is formed in an oxidizing (air) ambience. The presence of SiO_2 in PtSi layers has been found to produce a broad distribution of contact resistance as shown in Figure 13.19, and device yields are adversely affected.

When the PtSi is formed in an inert ambient, and the hot slices are pulled into air, Si atoms on the PtSi surface react to form SiO_2 (or PtSi + SiO_2). It is this oxide layer that protects the PtSi from dissolution during subsequent platinum stripping in aqua regia. The specific contact resistance of Si/PtSi interface is on the order of $(0.5-10) \times 10^{-6}$ Ω-cm^2 for emitter (phosphorus doped) or base (boron doped) contacts with surface concentrations $(4-6) \times 10^{20}$ cm^{-3} and $(3-4) \times 10^{18}$ cm^{-3} respectively. The barrier height of the n- PtSi Schottky diodes lies in the range of 0.8 to 0.87 eV.

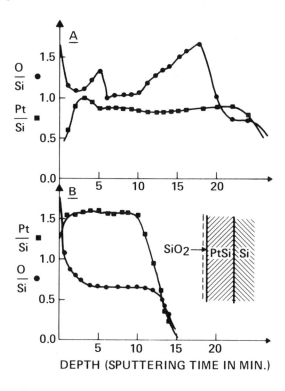

Figure 13.18: The auger peak-to-peak height ratios vs. the sputtering time for PtSi samples annealed in (A) air and (B) N_2.

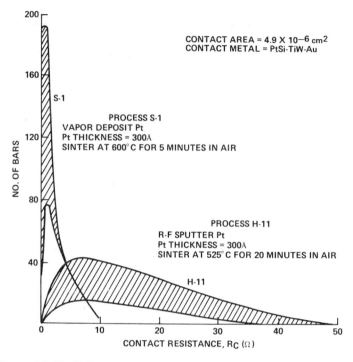

Figure 13.19: PtSi contact resistance distribution for two different processes.

INTERCONNECTIONS FOR INTEGRATED CIRCUITS

Single Level Interconnections

Any metallization system for IC interconnections must satisfy certain minimum requirements.[8,22] It should (i) make good ohmic contacts to both p and n type silicon, (ii) be amenable to practical methods of deposition to provide low resistive interconnection paths ($\rho <$ 3 to 10 $\mu\Omega$-cm), (iii) be adherent to both silicon and silicon dioxide, (iv) be patternable in chemicals not reactive with photoresist, silicon and silicon dioxide, (v) be compatible with bonding techniques, and (vi) be reliable under normal operating conditions. A large number of single, bi- and tri-metal combinations have been explored for interconnections; e.g.,

Single Metal:	Al, Cr, W, etc.
Bi-Metal:	Cr-Ag, Cr-Au, Mo-Au, Ti:W-Au, etc.
Tri-Metal:	Ti-Pt-Au, Ti-Pd-Au, Cr-Ag-Au, etc.

For reasons already stated, Al is the most widely used metal system; however, bi- and tri-metal sandwiches with Au as the main conductor are also used for interconnections. Au films have poor adhesion to SiO_2. Hence in such metallization systems, refractory metals such as Mo, Ti:W, or Ti-Pt provide necessary adhesion to SiO_2 and act as barrier layers between silicon and gold films. PtSi is usually necessary for reproducible metal-silicon contacts.

One of the areas of major concern in interconnections is metal coverage on Si-SiO_2 steps at contact windows.[23] Step coverage depends on the oxide profiles to be covered, the deposition equipment and the deposition parameters. Oxide profiles with nearly vertical or reverse slopes are major causes of discontinuity in metal leads after either the metal deposition or metal patterning. A 20-100% thinning of the metal film is observed at these oxide steps. Furthermore, minute cracks are formed in the metal leads at the Si-SiO_2 steps. Formation of such cracks is found to be directly related to the substrate-source configuration in the film deposition system and also to the profile of the oxide step. This is due to shadowing of the oxide steps by the straight line trajectories of the evaporated atoms. As a result, some region in the oxide window adjacent to the oxide step does not receive metal atoms at particular slice positions in the evaporator. An example of a growth crack in Al film interconnect at the oxide step is shown in Figure 13.20.

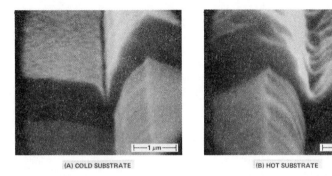

(A) COLD SUBSTRATE   (B) HOT SUBSTRATE

Figure 13.20: Aluminum metal coverage on SiO_2 steps of an IC.

Use of hot substrates (150°-350°C) increases the mobility of the Al atoms and the severity of the crack is reduced. Many techniques including novel slice rotation and appropriate biasing the substrates during sputter deposition of the films have been employed to solve metal coverage problem. Model calculations have been attempted to determine the metal coverage on oxide steps with surface sources. A cosine distribution is assumed. Results of these calculations for metal thickness on oxide steps metallized in (a) a standard canted domed planetary system with a concentric source and (b) horizontal rotating substrate over a circular planar magnetron are shown in Figure 13.21 and they indicate that a thinning factor of 50% or greater should be expected for steps exceeding 60° angle.[24] However, the use of a gently sloped oxide profile (~60°) appears to be the best pragmatic solution to this step coverage problem.

Use of diffusion barriers in interconnections technology is not new.[25,26] It needs to be emphasized that the effectiveness of a diffusion barrier can only be measured under prescribed conditions. Use of Al films for forming contacts and interconnections on shallow (< 1 μm) junction devices poses a potential hazard of junction shorting due to Al alloy penetration in the contacts. Al film conductors are separated from the contacts by a barrier layer such as Ti:W. Here PtSi contacts are used. Other examples of a barrier are Pt and Pd films interposed between Ti and Au films. Finally the choice of a barrier layer is dictated by the operating conditions of the device and its metallization requirements.

540 Deposition Technologies for Films and Coatings

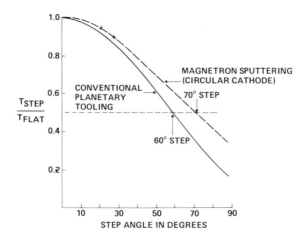

Figure 13.21: Modeling results of metal coverage on oxide steps.

Two Level Interconnections

Improvements in design and fabrication of complex medium scale integrated (MSI) and large scale integrated (LSI) circuits require use of multilevel (at least two) interconnections to achieve higher packing density, speeds and a reduced bar size.[27,29] Impact of two level interconnections on the design and performance of an MSI circuit is shown in Figure 13.22.

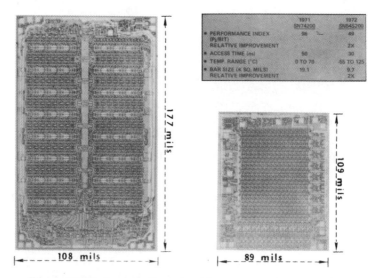

Figure 13.22: Impact of two level interconnections on an IC.

Some metal-insulator combinations employed to achieve multilevel interconnections are as follows: (a) Al/SiO_2/Al, (b) Al/Al_2O_3/Al, (c) Al/polyimide/Al, (d) Mo-Au-Mo/SiO_2/Mo-Mu, and (e) Ti:W-Au-Ti:W/SiO_2/Ti:W-Au. Aluminum and gold films provide the desired low resistive interconnections, and SiO_2, Al_2O_3 or polyimide provide the necessary insulation between layers.

The primary components of a two-level interconnection system are (a) *crossovers* – a second level lead crossing the first level lead separated by an insulator (SiO_2) layer, and (b) *vias* – locations for level to level contact through a hole in the insulator. Topography of the oxide layer covering first level leads, (e.g., re-entrant folds developed over vertical edges of first level leads) has been a major factor affecting integrity of crossovers because of poor metal coverage on these oxide steps.

In the fabrication of etched two-level $Al/SiO_2/Al$ interconnections, desirability of sloped first level leads to generate smooth oxide contours for second level metal coverage has been well recognized. Hillock formation on the first level leads has also been a major concern in fabrication of interlevel short free crossovers. As the aluminum metallized substrate is heated to a temperature in the $100°-180°C$ range for curing photoresist for lead patterning step and/or the patterned substrate is heated to a temperature in the $450°$ to $525°C$ range for Si/Al contact formation, Al film is subjected to a compressive stress due to the difference in coefficients of thermal expansion of Al and silicon substrate. The compressive stress in Al film is relieved by hillock formation. The height of the hillocks can equal metal thickness. Hillocks formed prior to insulation deposition are difficult to cover and those formed subsequent to insulation deposition may crack the insulation layer producing interlevel shorts. Poor photoresist coverage over hillocks can result in insulator removal near the hillocks during via etching, providing a shorting path between levels at crossovers (as shown in Figure 13.23). Various techniques including high substrate temperature deposition to achieve large grain films, alternate metal overcoat, and adding grain growth inhibiting impurities such as Cu, Ni, Mg, Mn, etc., have been employed to suppress hillock formation. Our experience shows that addition of approximately 2 atomic % Cu to Al films adequately suppresses hillock formation for two-level interconnections on ICs.

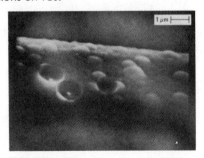

Figure 13.23: Hillock formation on Al film surfaces. Holes etched in interlevel oxide due to resist problems resulting from hillocks in first-level leads.

Thickness of first level metal leads is dictated by current density and metal coverage requirements on oxide steps at contact windows. A 1 to 1.2 μm thick Al film is used for first level leads. Insulator films considerably thicker than the first level leads have been suggested to assure insulator integrity between levels. Thicker insulator films pose internal stress cracking problems. This problem can be eliminated by use of thinner Al+Cu film conductors without loss of reliability. Sloped first level leads, realized by optimizing photoresist and etching operations provide the desired insulator profiles for second level coverage on steps.

Several dielectrics such as sputtered quartz, chemically vapor deposited

(CVD) silicon dioxide, anodic oxide, silicon nitride, polyimide and a combination of sputtered and CVD oxides have been explored for multilevel insulators. Dielectric selection for multilevels is governed by several factors such as pinhole density, coverage at metal edges, internal stresses, via etching, reliability, etc. For most applications CVD SiO_2 provides satisfactory insulation for multilevel interconnections.

Etching of vias in insulator is a critical step. Proper process controls must be established to achieve a reproducible process. An Al to Al interface resistance on the order of 10^{-7} Ω-cm^2 is observed in clean vias. Incomplete removal of oxide in vias produces very high interface resistance ($\sim 10^{-2}$ Ω-cm^2) resulting in nonfunctional integrated circuits. Sputter etching of vias prior to second level metal deposition appears to be one of the solutions to overcome the high resistance via problem. Continuity of the second level leads is no longer a problem with smooth oxide contours over edges of sloped first level leads.

A metallization process scheme useful for a bipolar LSI can be as follows:

$$PtSi/Ti:W-Al+Cu/SiO_2/Al$$

where PtSi forms ohmic and Schottky contacts
Ti:W acts as a barrier layer
Al+Cu acts as a first level conductor
SiO_2 serves the purpose of an interlevel insulator
and Al acts as a second level conductor.

In Figure 13.24, scanning electron micrographs of vias and crossovers for an Al+Cu/SiO_2/Al system are presented.

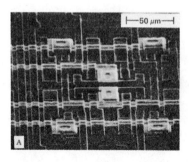
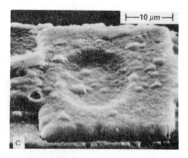
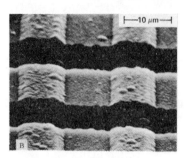
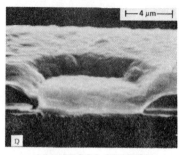

CROSSOVER VIA CROSS SECTION

Figure 13.24: Two level interconnections on an IC: (A) circuit, (B) crossover, (C) via and (D) via cross section.

Reliability and Electromigration

Reliability studies on microelectronic circuits have led to identification of "electromigration induced failures in thin film conductor" as one of the failure modes.[30] Mass migration under the influence of a direct current is called electromigration. Atomic flux (or vacancy flux) is related to current density in the framework of a model proposed by Hungtington, and is given by

$$j_i = (N_i D/kT) \, Z^*e\rho j$$

where j_i = flux of metal ions, N_i = local density of ions, $D = D_0 \exp(-Q/kT)$ = diffusion coefficient, Z^*e = effective charge on the ion, j = current density and ρ = film resistivity. The above expression, though developed for a bulk specimen, is assumed to be valid for films where grain boundary diffusion dominates ion migration. A nonvanishing divergence of the atomic flux (vacancy flux in the opposite direction) leads to void formation. Temperature gradients and structural inhomogeneities (an abrupt change in grain size, a triple point of grain boundaries, a precipitate) are primary causes of nonvanishing divergencies of vacancy flux that lead to void formation. These voids grow to develop open lead failures.

Expressions have been developed for the mean time to failure (MTF) of film conductors subjected to high current stress.

$$MTF = A j^{-n} \exp(Q/kT)$$

where A = parameter depending on sample geometry and physical characteristics of film and substrate, j = current density, n = exponent, Q = activation energy, k = Boltzmann's constant and T = average temperature of the conductor. Since the atom transport in thin films is governed by grain boundary diffusion, Q is related to the activation energy for grain boundary diffusion. For Al film conductors, reported values of Q and n lie in the range of 0.6 ± 0.2 eV and 1 to 3 respectively.

One of the primary objectives of life test studies is to be able to use the data for predicting the time to failure of an IC interconnection. A first glance at the expression for the atomic flux under an impressed current density suggests that MTF should be inversely proportional to the current density (n = 1), when the atomic flux divergence is not dominated by temperature gradients. For accelerated testing at high current densities, temperature gradients strongly influence the current dependence and high values of n are observed. Use of n = 2 has been proposed by Black[31] to extrapolate MTF values for low operating temperatures and current densities. Such a procedure results in very optimistic estimates. For IC interconnections, the exponent n appears to be in the range of 1.0 to 1.5 and the predicted MTFs based on n = 1 can be treated as conservative estimates.

Since d'Heurle and coworkers at IBM reported an improvement in MTF by a factor of 70 by copper doping aluminum films, there has been considerable interest in electromigration in aluminum alloy films. It has been confirmed by other workers that MTF does increase with copper content and peaks out around 8 wt % copper in Al films. However, the addition of copper does not appear to have a dramatic effect on the activation energy which lies in the range of 0.6 ± 0.2 eV.

Our studies on Ti:W/Al+Cu (1.6 wt % Cu) film conductors, with Ti:W as an underlying barrier layer,[32] have shown that

$$Q_{Al+Cu} = 0.71 \pm 0.03 \text{ eV},$$

and this value is consistent with other reported data.

A search for reliability improvement continues and it has led to the introduction of Al-alloy films such as Al+Cu, Al+Si, Al+Cu+Si and so on. In Figure 13.25 mean time failure data on Al, Al+Cu and Al+Cu+Si film conductors (0.8 μm thick, 6 μm wide and 380 μm long) subjected to 1×10^6 A/cm^2 in the 150° to 215°C temperature range are presented. Also, our earlier data on Ti:W/Al and Ti:W/Al+Cu film conductors (170 nm/800 nm thick, 9 μm wide and 1.14 mm long) are presented. Efforts will continue to optimize film microstructure and composition to minimize electromigration induced hazard for IC interconnections.

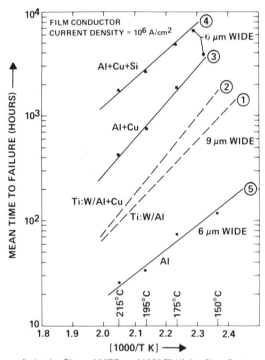

Arrhenius Plots of MTFs vs (1000/T K) for Glass Passivated 9-μm Wide (1) Ti:W/Al (IN-S) and (2) Ti:W/Al-Cu (IN-S) Film Conductors Tested at 1×10^6 A/cm^2; Recent Data on 6-μm Wide (3) M-S Al-Cu, (4) M-S Al-Cu-Si; and (5) M-S Al Film Conductors Tested at 1×10^6 A/cm^2

Figure 13.25: MTF vs current density for Al film conductors.

BONDING AND ASSEMBLY

Two major bonding techniques are employed for the attachment of a wire to a chip and to the leads of a package: (a) thermocompression TC bonding and (b) ultrasonic bonding UB.[5,34,35] In thermocompression bonding, a gold wire, ~25 μm (0.1 mil) in diameter, is attached to Al bond pads on the chip. First a ball is formed on the end of the Au wire using a hydrogen flame, and then it is welded to the Al bond pad using a capillary tool. Welding conditions such as substrate temperature and pressure are optimized to achieve an excellent bond strength. This process is tolerant to some over-deformation of the bond. This wire is then led off in a desired direction to make a stitch (or wedge) bond to one of the leads of the package. For most commercial applications, the TC bonds satisfy the less stringent reliability specifications.

In IC devices with Au wires attached to Al bond pads, Al and Au are in intimate contact, and intermetallic compounds are formed as a result of diffusion controlled reaction. Since the diffusion of Al in Au is faster than that of Au in Al, voids are formed on Al pads. The phenomenon of void formation due to the difference in interdiffusion rates of the components of a couple is referred to as Kirkendall effect. As a result of this intermetallic compound formation, the bond strength decreases over a period of time. Formation of the intermetallic compounds and voids in Al pads are readily seen in samples subjected to accelerated tests. One of the intermetallic compounds is purple in color and the observed discoloration at the bond pads is referred to as purple plague.

In order to eliminate the loss of bond strength and the problems associated with purple plague, three major solutions have been proposed: (a) change the Al interconnect metallization on the chip to Au metallization and establish Au to Au bonds either by thermocompression bonding or ultrasonic bonding, (b) use Al wire and establish Al to Al bond by ultrasonic bonding techniques, or (c) use the Al metallization on the slice and then proceed to provide Au bond pads with use of an interposed barrier layer between Al and Au and then establish Au to Au bonds, e.g., beam leads.

In ultrasonic bonding, an Al wire is attached to Al pad by use of ultrasonic energy. The strength of the bond is determined by how well the ultrasonic energy is coupled to the wire and the pad. Even though purple plague problem can be eliminated by this method, the total reliability of the system depends on the strength of Al wire. Since Al wire softens due to high temperature ($>150°C$) exposure, 1% Si doped Al wire is used. Ultrasonic bonding has received considerable attention over the last 8-10 years, but it has not become the industry standard.

Major advances are being witnessed in the chip bonding technology.[36-40] The bonding of Au wires to Al pads, one step at a time is being upgraded by automated wire bonding techniques to improve productivity. However, competitive cost pressures have led to the development of "tape carrier bonding technology." Here gold or copper bumps are formed on a passivated slice at the bond pad locations. A barrier metal such as Ti, Pt, Cr, Cr-Ag, etc. is deposited on the Al pads of conventionally metallized ICs and gold or copper bumps (≈ 25 μm in height) are formed.

Chips on these silices are separated and carried on an adhesive tape together with a tin-plated copper lead for fluxless contact soldering by a gang lead bonder in a single step. This technique promises high throughput and will be an increasingly important packaging method as labor costs continue to rise.

OTHER APPLICATIONS

In microelectronics, there are numerous applications of film deposition techniques in the fabrication of a wide variety of devices. So far, the discussion has been confined to thin film resistors and capacitors, contacts and film interconnections in silicon integrated circuits, and bonding and assembly of devices. Though it is almost impossible to cover a wide variety of applications, a few of the other interesting applications (it is merely a personal preference) will be mentioned here.

High quality photomasks[3] are essential in device fabrications. Chrome masks used for processing LSI circuits on 75 mm diameter slices demand a defect density of less than 1 defect/cm^2 on 100 x 100 mm^2 plate. Same thing can be said about see-through masks made with semitransparent dielectric films. Device yields strongly depend on the quality of photomasks.

Magnetic bubble memory devices of 1 Megabit complexity are in development stage.[41] Fabrication process is almost an extension of two-level interconnection process, e.g., substrate/epi-layer/SiO$_2$/Al conductor/insulator SiO$_2$/Permalloy films/protective layer/bond pad metallization. A number of metal and dielectric film depositions are involved.

There is considerable development activity in the area of Josephson tunneling devices.[42,43] These devices use superconducting metal layers and operate at low temperatures. A vertical section of a test vehicle structure described by IBM researchers has the following sequence of film layers: oxidized silicon substrate/Nb/Nb oxide/Au/Pb/In/Pb/Au/Pb/SiO/In/Pb/Au. As these test structures are thermally cycled between room temperature and 4.2°K, hillock growth in Pb films poses a reliability problem. In and Au layers are added to suppress hillock growth. These multilayered structures pose some interesting problems in process control.

FUTURE TRENDS

The increasing demand for high performance (speed) and high density (large number of components) LSI circuits is being met by new device development, circuit designs and by continuous shrinkage of geometrical dimensions. The current microelectronic device fabrication processes use optical patterning techniques, and these techniques are reaching their physical limits as linewidths and spacings are approaching the wavelength of light that is used for photoresist exposure. Practical limits in today's device processing appear to be 2.5 μm (0.1 mil) lines and spaces. The trend in increasing the device packing density by reduction of linewidths and spacings will continue through developments in electron beam technology and x-ray lithography for patterning submicron geometries. Complex device structures will demand stable multilayered films. Stability of these structures will require a better understanding of interdiffusion in thin films. Film deposition techniques will have to be carried out under controlled conditions. Future deposition equipments will be under computer control. In the final analysis, the process engineers will have to teach these computers to monitor and control the relevant film deposition parameters.

REFERENCES

Books

(1) R.W. Berry, P.M. Hall and M.T. Harris, *Thin Film Technology,* D. Van Nostrand, Princeton, N.J. (1968).
(2) K.L. Chopra, *Thin Film Phenomena,* McGraw-Hill, New York (1969).
(3) L.I. Maissel and R. Glang, *Handbook of Thin Film Technology,* McGraw-Hill, New York (1970).
(4) Z.H. Meiksin, *Thin and Thick Films for Hybrid Microelectronics,* Lexington Books, Lexington, Mass. (1976).
(5) A.B. Glaser and G.E. Subak-Sharpe, *Integrated Circuit Engineering,* Addition-Wesley, Reading, Mass. (1977).
(6) J.M. Poate, K.N. Tu and J.W. Mayer, editors, *Thin Films—Interdiffusion and Reactions,* John Wiley & Sons, New York (1978).

Articles

(7) J.A. Amick, G.L. Schnable, and J.L. Vossen, *J. Vac. Sci. Technol. 14,* 1053 (1977).
(8) P.B. Ghate, J.C. Blair and C.R. Fuller, *Thin Solid Films 45,* 69 (1977).
(9) P.M. Hall, *Thin Solid Films 1,* 277 (1967).
(10) C. Kittel, *Introduction to Solid State Physics,* John Wiley & Sons, New York (1967), Chapters 6 and 7.
(11) K. Fuchs, *Proc. Cambridge Phil Soc. 34,* 100 (1938).
(12) E.H. Sondheimer, *Adv. Physics 1,* 1 (1952).
(13) C.A. Neugebauer in reference 3, chapter 5.
(14) W.D. Westwood and N. Waterhouse; *Tantalum Thin Films* Academic Press N.Y. (1975).
(15) W.J. Ostrander, *IEEE Trans, PHP 9,* 155 (1973).
(16) L.I. Maissel in reference 3, chapter 18.
(17) D. Gerstenberg in reference 3, chapter 19.
(18) S. Sze, *Physics of Semiconductor Devices,* Wiley, New York (1969).
(19) M.P. Lepselter and J.M. Andrews in *Ohmic Contacts to Semiconductors,* B. Schwartz (ed.), Electrochemical Society, New York (1969), p. 159. Also see H. Sello, p. 277.
(20) K.F. Braun, *Ann. Phys. Pogg. 153,* 556 (1877) cited by C.A. Mead in *Ohmic Contacts to Semiconductors,* B. Schwartz (ed.), Electrochemical Society, New York (1969) p. 8.
(21) K. Tada and J.L.R. Laraya, *Proc. IEEE. 55,* 2064 (1967).
(22) J.A. Cunningham, *Solid State Electronics 8,* 735 (1965).
(23) I.A. Blech, D.B. Fraser and S.E. Haszko, *J. Vac. Sci. Technol. 15,* 13 (1978).
(24) C.R. Fuller and P.B. Ghate, *Thin Solid Films 64,* 25 (1979).
(25) M.P. Lepselter, *Bell System Tech. J. 45,* 233 (1966).
(26) P.B. Ghate, J.C. Blair, C.R. Fuller and G.E. McGuire, *Thin Solid Films 53,* 117 (1978).
(27) J.A. Cunningham, W.R. Gardner and S.J. Wood in *Ohmic Contacts to Semiconductors,* B. Schwartz (ed.), Electrochemical Society, New York (1969).
(28) A.M. Wilson and P.B. Ghate in *Semiconductor Silicon—1977,* p. 1047, Proc. Vol. 77-2, Electrochemical Society, Princeton, N.J.
(29) P.B. Ghate and C.R. Fuller in *Semiconductor Silicon—1981,* p. 680, Electrochemical Society, Princeton, N.J.
(30) F.M. d'Heurle and P.S. Ho in reference 6, chapter 8.
(31) J.R. Black, *IEEE Trans-Electron Devices 13,* 1391 (1970).
(32) P.B. Ghate and J.C. Blair, *Thin Solid Films 55,* 113 (1978).
(33) P.B. Ghate, *Proc. 19th Int. Rel. Phys. Symp. IEEE,* (1981).
(34) E. Philofsky, *Solid State Electronics 13,* 1391 (1970).
(35) K.I. Johnson, M.H. Scott and D.A. Edson, *Solid State Technology 20,* 50 (1977).
(36) R.G. Oswald, J.M. Montante and W.R. Rodriques de Miranda, *Solid State Technology 21,* No. 3, 39 (1978).
(37) A. Keizer and D. Brown, *ibid 21,* No. 3, 59 (1978).
(38) T.G. O'Neal, *Semiconductor International 4,* No. 3, 43 (1981).
(39) K. Levy, *Solid State Technology 21,* No. 5, 69 (1978).
(40) J.E. Levine and H.C. Schick, *Semiconductor International 3,* No. 5, 105 (1980).
(41) P.K. George and G. Reyling, *Electronics 52,* No. 16, 99 (1979).
(42) S.K. Lahiri and S. Basavaiah, *J. Appl. Phys. 49,* 2880 (1978).
(43) M. Murakami, *J. Appl. Phys. 52,* 1309 (1981).

14

Characterization of Thin Films

Gary E. McGuire
Tektronix, Incorporated
Beaverton, Oregon

The characterization of surfaces and thin films has received considerable attention in recent years. This has resulted in rapid improvements in a broad range of analytical techniques capable of sampling films that may range from a few hundred angstroms to microinches.

The surface sensitive techniques of Auger electron spectroscopy, X-ray photoelectron spectroscopy, ion scattering spectroscopy, and secondary ion mass spectroscopy, and the near surface sensitive techniques of Rutherford backscattering, nuclear reaction analysis and ion-induced X-rays are described in this review. The principles of these techniques are discussed, and they are compared based upon elemental sensitivity, resolution, detection limits, quantification, chemical state determination, depth distribution, lateral resolution, sample requirements, and sample alterations. Advantages and disadvantages are discussed for each technique to aid in determining which technique is most suitable for a given problem.

INTRODUCTION

Coatings are used in a variety of applications to achieve the appropriate physical and chemical properties. Applications include optical coatings for lenses and mirrors, metallic coatings for electrical contacts, refractory coatings to increase hardness and decrease wear, protective coatings for corrosion and oxidation resistance. It is important to establish the relationship between deposition parameters and the structure of coatings. There are many techniques used to evaluate physical properties, however the chemical composition is not always easy to determine, especially if light elements are present. New analytical techniques such as Auger electron spectroscopy (AES),[1,5,10] X-ray photoelectron spectroscopy (XPS),[2,6,10] ion scattering spectroscopy (ISS),[3,11] secondary ion mass spectrometry (SIMS),[4,7,12] Rutherford backscattering (RBS),[8,13,14] nuclear reaction analysis (NRA),[14,15] and ion-induced X-rays (IIX),[9,14-17] are useful in determining the elemental composition and chemical states in thin coatings. Certain aspects of these techniques have been discussed in past reviews.[18,19,20] In this chapter a brief description of the principles of each technique will be given followed by a discussion of the sensitivity, resolution, detection limits and quantification. Chemical state effects, spatial resolution, and sample requirements will also be examined.

One method of outlining methods of surface and thin film analysis is to indicate the method of exciting the sample and the signal that is detected.[21] The sample may be excited by a beam of electrons, ions, photons or neutral particles or non-particle probes such as thermal, electric fields, magnetic fields or surface waves. Excitation of the sample surface results in the emission of some form of energy or particle such as electrons, ions, photons or neutral particles. The combinations of input probes and detected particle or beam gives rise to a large number of analytical techniques. This is summarized below. This chapter is restricted to a few of the analytical techniques capable of sampling surfaces and thin films.

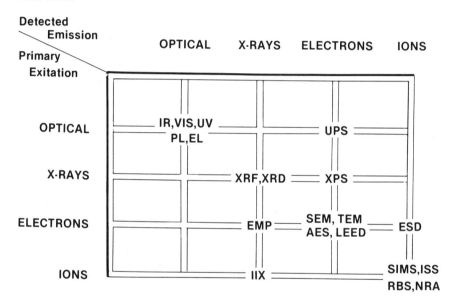

PRINCIPLES

The process of photons in, electrons out is the photoelectric effect. When high energy photons in the X-ray range are used the technique is called X-ray photoelectron spectroscopy (XPS). XPS is used to determine the electronic structure of solid surfaces as well as to chemically identify surface components. This gives rise to a second acronym ESCA, electron spectroscopy for chemical analysis. A flux of X-rays striking a solid will cause ejection of electrons from electron levels with binding energies less than the energy of the incoming photon. The kinetic energy of the photoelectron as shown in Figure 14.1 is given by the energy of the X-ray minus the binding energy of the ejected K-shell electron and minus a work function and wave function relaxation term.[22] Since the set of binding energies are unique for a given element, the photoelectron peaks may be used for elemental identification.

Many photon or X-ray sources may be used but the most frequently used sources are $AlK_{\alpha 1,2}$ and $MgK_{\alpha 1,2}$ radiation. Both offer nearly monochromatic radiation with line widths of less than 1 eV. The Al or Mg X-ray sources actually produce a number of X-ray lines of which the $K_{\alpha 1,2}$ is the most intense. Table

14.1 shows the relative intensities and energies of the X-rays produced. Each of the X-ray energies produces photoelectron transitions which mirror those produced by the $K_{\alpha 1,2}$ radiation. Since the intensity ratio and energy separation is known, the satellite photoelectron spectra may be removed with appropriate software. An alternate method makes use of a crystal monochrometer which allows only a narrow portion of the $K_{\alpha 1,2}$ radiation to strike the sample. Resolution down to 0.4 eV has been achieved by this method.

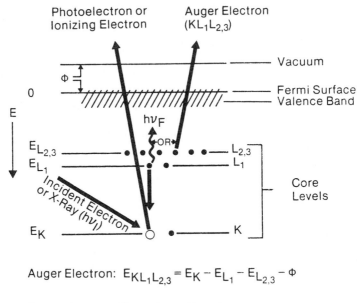

Auger Electron: $E_{KL_1L_{2,3}} = E_K - E_{L_1} - E_{L_{2,3}} - \Phi$

Photoelectron: $E_{PE} = h\nu_I - E_K - \Phi$

X-Ray Fluorescence: $h\nu_F = E_K - E_{L_1}$

Figure 14.1: Energy level diagram depicting the Auger electron and photoelectron effects.

Table 14.1: Characteristic X-Rays Produced from Al and Mg Targets*

X-Ray	Mg Energy, eV	Mg Relative Intensity	Al Energy, eV	Al Relative Intensity
$K_{\alpha 1}$	1253.7	{67 100	1486.7	{67 100
$K_{\alpha 2}$	1253.4	{33	1486.3	{33
$K_{\alpha'}$	1258.2	1.0	1492.3	1.0
$K_{\alpha 3}$	1262.1	9.2	1496.3	7.8
$K_{\alpha 4}$	1263.7	5.1	1498.2	3.3
$K_{\alpha 5}$	1271.0	0.8	1506.5	0.42
$K_{\alpha 6}$	1274.2	0.5	1510.1	0.28
K_β	1302	2	1557	2

*F. Wuilleumier and M.O. Krause, *Phys. Rev. A. 10* 242 (1974)

The kinetic energy of the photoelectron is determined by the electrostatic deflection of the electrons onto a biased electron multiplier. Electrostatic deflection type analyzers come in several configurations including cylindrical mirror, spherical sector and retarding potential. The double focusing cylindrical mirror analyzer first introduced by Palmberg,[23] shown in Figure 14.2, is the most widely used configuration at this time because it combines high transmission with high resolution. The resolution is expressed as a percentage of the kinetic energy of the electron being analyzed. In order to maintain adequate resolution at high electron energies, the electrons are decelerated to a constant kinetic energy before analysis.

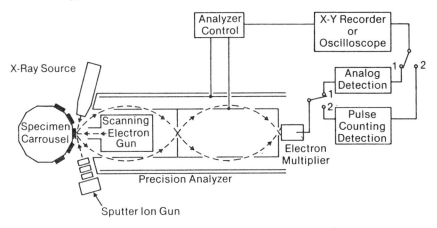

Figure 14.2: Schematic diagram of a Physical Electronics double pass cylindrical mirror analyzer as a photoelectron spectrometer.

Core level ionization, such as shown in Figure 14.1, leaves the atom in an excited state. De-excitation occurs by an upper level electron, L_1 level, decaying to the core hole with the excess energy being transferred to and causing ejection of another electron, $L_{2,3}$ level, which is by definition an Auger electron. The Auger transitions are denoted by the electron levels involved, thus Figure 14.1 illustrates the ejection of a $KL_1L_{2,3}$ electron. The Auger electron, unlike the photoelectron, is independent of the excitation source and leaves the atom with a constant kinetic energy. The kinetic energy is given by the differences in binding energies for the three levels ($E_K - K_{L_1} - E_{L_{2,3}}$) minus a correction term for the work function and electron wave function relaxation. The core level ionization which initiates the Auger process can be caused either by incident X-rays, electrons, or ions.

By convention, AES refers to electron excited AES. Since electron beams are quite versatile the conditions for AES excitation can be quite varied. The acceleration voltage on current spectrometers systems can be varied from 10 KeV down. At optimum conditions the beam voltage should be about 2.5 times the core level to be ionized. Some elements have more than one characteristic series of Auger transitions. For elements with multiple AES transitions or multielement samples, selection of the proper beam voltage will involve some compromise.

AES transitions produced by X-ray excitation may be observed in the XPS spectrum, but the reverse is not the case. AES transitions in XPS spectra contain chemical information that is complimentary to that contained in the XPS transitions. AES spectra taken with a high resolution XPS analyzer show sharp features which are accompanied by a series of transitions at lower kinetic energy due to the various couplings of the core hole with the valence electrons. Photoelectron transitions may be distinguished from Auger features by changing X-ray sources which will produce a shift in the photoelectron kinetic energy.

As with XPS, Auger electron energies are measured by electrostatic deflection. The analyzer typically has lower resolution, 0.5%, and higher transmission (11%) than an XPS analyzer. The electron gun is mounted coaxially inside the analyzer as in Figure 14.3, so that the analyzer and electron gun have the same focal point. Sample alignment simply requires positioning the sample at the source point of the cylindrical mirror analyzer. Electrons ejected from the point of excitation move radially outwards passing through a grid-covered aperture on the inner cylinder.

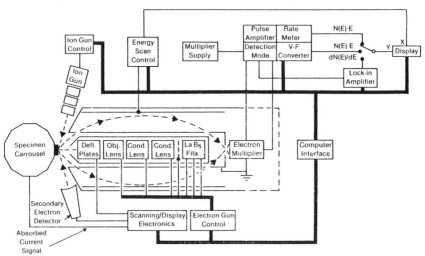

Figure 14.3: Schematic and block diagram of a scanning Auger spectrometer with a coaxial electron gun within a single pass cylindrical mirror analyzer.

While data for XPS has traditionally been taken as the number of electrons with an energy, N(E), versus that energy, E, AES data are normally taken as $dE \cdot N(E)/dE$ versus E spectra with the potential-modulation differentiation[24] scheme first introduced by Harris.[25] The high background caused by inelastically scattered electrons from the primary beam are suppressed by differentiation which in effect gives the slope to the N(E) curve. Since the background of inelastically scattered electrons varies smoothly over the range of interest, the Auger transitions are the primary features that change the slope of the line and that are enhanced by differentiation. Differentiation as introduced by Harris[25] is accomplished by superimposing a small AC voltage on the DC ramp voltage applied to the analyzer and measuring the in-phase signal from the electron multiplier with a lock-in amplifier. An alternate method of background suppression is beam blanking where the modulation voltage is applied to the primary beam in a scanning Auger microprobe.[26] With the advent of scanning Auger systems and improvements in digital signal processing, direct N(E) spectra are taken with signal processing being used to increase sensitivity and for background subtraction. This approach was not used previously because the high beam currents used in static beam AES resulted in a signal which would saturate the electron multiplier.

The low energy ion techniques of ISS and SIMS are shown in Figure 14.4. In ISS, a collimated monoenergetic beam of ions with initial energies of 4 keV or less and of known mass is directed toward a solid surface and the energy of the ions elastically scattered from the surface are measured at a particular angle. The energy loss is determined by the mass of the scattering atom and scattering angle. The energy, E', of the scattered ions of mass M_o with an incident energy, E_o, scattered at an angle θ is

$$(1) \qquad E' = \frac{E_o M_o^2}{(M_o + M_s)^2} \left\{ \cos\theta + \left(\frac{M_s^2}{M_o^2} - \sin^2\theta\right)^{1/2} \right\}^2$$

For a fixed angle of scattering at 90°, the ratio of the energy of the reflected primary to its initial energy is a simple function of the ratio of the mass of the scattering atom to the mass of the bombarding ion as in the expression below:

$$(2) \qquad \frac{E'}{E_o} = \frac{M_s - M_o}{M_o + M_s} \quad \text{provided } M_s > M_o$$

Depending on the energy, less than 1% of the incident ions are scattered as ions. The remainder are neutralized by resonance neutralization or Auger neutralization.[27] Since only ions are detected and the probability of ion neutralization at the sample surface is high, essentially all of the elastically scattered ions are from the outer surface layer of atoms. The scattering cross-section and, hence, the signal intensity increases with increasing atomic number.

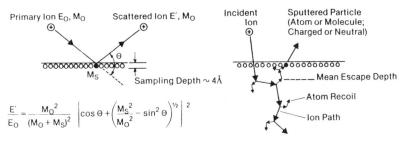

A. Low Energy Ion Scattering B. Secondary Ion Production

Figure 14.4: Schematic illustration of elastic scattering of ions for ion scattering spectroscopy and the energy cascade resulting from ion impact which causes sputtering for secondary ion mass spectroscopy.

The energy of the scattered ions is determined by using an electrostatic spherical sector or cylindrical mirror analyzer. Currently 3M's ISS system shown in Figure 14.5 uses a coaxial ion gun with a cylindrical mirror analyzer in a manner similar to many AES spectrometers. Ion energies between 0 and E_o are measured by sweeping the voltage between the analyzer plates. The spectrum consists of a series of peaks in the scattered ion current at values of E'/E_o determined from Equation 2 by substituting for M_s the masses of the surface atoms.

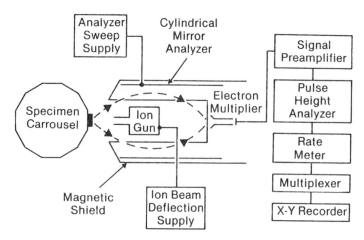

Figure 14.5: Schematic and block diagram of a 3M ion scattering spectrometer with a coaxial ion gun within a single pass cylindrical mirror analyzer.

Some ions suffer inelastic collisions with surface atoms or penetrate a short distance into the lattice where they lose their kinetic energy and create energy cascades which may cause sputtering of material from the surface as charged atoms and molecules or neutrals. The most important parameters influencing the sputtering yield are the energy, mass and incidence angle of the projectile ion. Other factors influencing the sputtering yield include crystal orientation, the electronic and chemical properties of the surface.

By collecting and mass analyzing the sputtered secondary ions (SIMS) the elemental composition of the solid can be determined. Three types of spectrometers are in common usage for SIMS, the magnetic sector ion microprobe shown in Figure 14.6, the ion microscope and the quadrupole mass filter. The energy of the incident primary beam is usually in the range 1-20 keV. High spatial resolution is achieved by the utilization of a duoplasmatron ion source.[28] The availability of quadrupole mass filters has enabled the construction of economical SIMS equipment without high spatial resolution. Positive oxygen ions are used to enhance the yield of positive secondary ions[29] and Cs' primary ions to enhance the yield of negative secondary ions.[30] The secondary ion species are mainly singly charged atomic and molecular ions, however, doubly and triply charged atomic species may be detected due to the high sensitivity of mass spectrometers. When molecules are collected as secondary ions, their cracking patterns may allow identification of the parent molecule.

RBS analysis provides the ability to distinguish the atomic masses of elements and their distribution as a function of depth. RBS is the high energy counterpart of ISS. Ion beams with energies in the range of 100 KeV to 5 MeV are used to bombard a solid sample and then the backscattered ions and neutrals are detected. The most frequently used projectile ions are hydrogen, helium or any other light element. The ions may penetrate long distances into the solid before undergoing an elastic collision. During penetration and escape from the lattice, the primary ion interacts with electrons and loses energy by creation of valence and core level holes, Figure 14.7. Because the energy loss of the ion occurs in a continuous manner, the signals from greater depths will have lower energy. The left edge of a spectrum with a scale of increasing scattered energy from left to right represents scattering from the deepest part of the material.

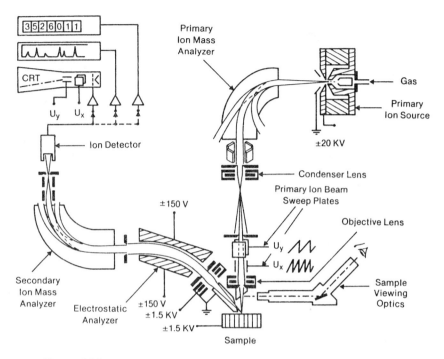

Figure 14.6: Schematic drawing of the Applied Research Laboratories ion microprobe mass analyzer.

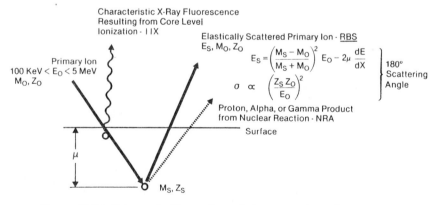

Figure 14.7: Schematic illustration of the generation of X-rays for ion-induced X-rays (IIX), elastic scattering of energetic primary ions in Rutherford backscattering (RBS), and production of reaction products in Nuclear Reaction Analysis (NRA).

The energy difference between higher and lower energy features of a spectrum is directly related to the total thickness of a layer. By knowing the scattering cross-section, which is dependent on the atomic number of the target atom

(Figure 14.8), the total number of atoms per unit area of each type is determined. To obtain quantitative results, the number of counts, scattering geometry, current integration and detector area must be known accurately. For thick targets, a correction should be made for changes in the scattering cross-section with energy.

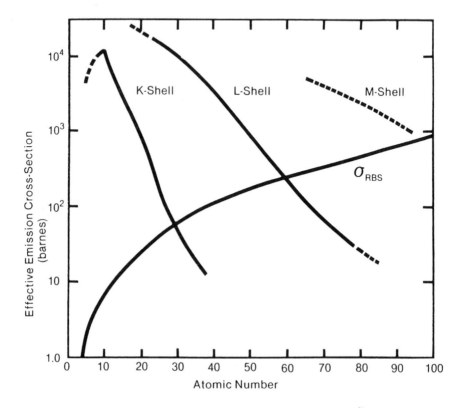

Figure 14.8: Rutherford backscattering cross-section (170° scattering angle) for 1 MeV ^{4}He particles (Ref. 18) and effective X-ray emission cross-section for 5 MeV alpha particles (Ref. 34) versus atomic number.

Energetic particles coming from the target can be detected by surface barrier detectors, electrostatic or magnetic analyzers. Figure 14.9 shows the experimental arrangement for an electrostatic and surface barrier detector. Surface barrier detectors provide large solid angle detection of particles at various energies simultaneously. Surface barrier detectors are insensitive to the charge state of the detected particles and provide good routine analysis capability. The primary advantage of electrostatic or magnetic analyzers over surface barrier detector is resolution at the expense of extended counting time.

If the primary ion has the proper atomic mass and energy it may penetrate a nucleus and cause a reaction in which either a new isotope or new atom is created, hence a nuclear reaction. The nuclear transformation usually results in ejection of a particle with a precise identity and energy which may be used to identify the parent atom. Some nuclear reactions used so far in NRA are listed in Table 14.2.[31] Isotopic tracing is easy since two isotopes of the element result in completely different nuclear reactions. The emitted charged particles are detected with different types of semiconductor detectors depending on the nuclear reaction.

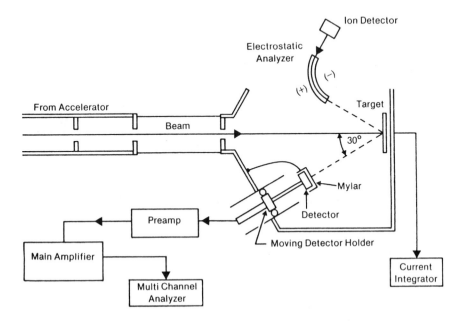

Figure 14.9: Scattering chamber and electronics for energy analysis of scattered particles, nuclear reaction products and ion induced X-rays.

Table 14.2: Useful Nuclear Reactions

Nucleus	Reaction
^{2}H	^{2}H(^{3}He,p)^{4}He
^{3}He	^{3}He(d,p)^{4}He
^{6}Li	^{6}Li(d,α)^{4}He
^{7}Li	^{7}Li(p,α)^{4}He
^{9}Be	^{9}Be(d,α)^{7}Li
^{11}B	^{11}B(p,α)^{8}Be
^{12}C	^{12}C(d,p)^{13}C
^{13}C	^{13}C(d,p)^{14}C
^{14}N	^{14}N(d,α)^{12}C
^{15}N	^{15}N(p,α)^{12}C
^{16}O	^{16}O(d,p)^{17}O
^{18}O	^{18}O(p,α)^{15}N
^{19}F	^{19}F(p,α)^{16}O
^{27}Al	^{27}Al(p,γ)^{28}Si

The experimental arrangement is nearly identical to that of RBS analysis in Figure 14.9 with the exception of a thin mylar film added in front of the detector to stop backscattered particles from the incoming beam. The energy of the beam is chosen so the range of elastically scattered particles from the primary beam is well below the range of alpha particles or protons produced during NRA. The detector is placed at an angle of between 150° and 165° to the axis of the beam since the angular distribution of the emitted particles peaks at backward angles.[32] Depth profiles may be extracted from the spectra similar to RBS but with poorer depth resolution[33] due to energy straggling in the absorber.

Electron holes created by energy loss of the ion while traversing the lattice undergo de-excitation by Auger electron on X-ray emission. While ion excitation does not offer much advantage for Auger emission, it does offer selective X-ray analysis for low atomic number elements. The bremsstrahlung background characteristic of electron beam induced X-ray excitation, is reduced with ion beam induced X-ray production. Figure 14.8[34] shows the effective X-ray emission cross-section as a function of atomic number. The lower atomic number elements show a high cross-section for K-shell X-ray production. X-ray emission for higher atomic number elements may be enhanced by using a high Z primary beam.

The experimental arrangement for IIX again is similar to that shown in Figure 14.9 for RBS and NRA with the exception of the detector. The elemental composition of the sample is determined by analysis of the emergent x-rays using an energy dispersive lithium-drifted silicon detector or wavelength dispersive crystal spectrometer. The energy dispersive detector gives an instantaneous qualitative representation of the sample composition and has a much larger solid angle for detection than the wavelength dispersive detector but lacks the energy resolution of the wavelength dispersive detector.

ELEMENTAL SENSITIVITY AND RESOLUTION

In actual practice XPS can detect all elements with $Z \geqslant 3$ with few interferences. Using $AlK_{\alpha 1,2}$ or $MgK_{\alpha 1,2}$ radiation the K-shell ionization cross-section dominates for $Z \leqslant 10$ but above this atomic number the L-shell cross-section is significant and L photoelectrons may be used for analysis. Figure 14.10 shows the relative sensitivity factors or cross-section as a function of atomic number which shows that higher shell electrons must be used for analysis as the atomic number increases.[35] The relative cross-sections could be changed by varying the energy of the incident radiation. Table 14.3 shows a number of available X-ray sources and their full width at half maximum height. $CuK_{\alpha 1}$ could be used to enhance the ionization cross-section of the deeper core levels but the energy resolution would be degraded. Very soft X-rays may be desirable for studying valence band spectra where high resolution is important. The need for variable excitation sources has generated interest in synchrotron radiation sources for XPS.[36]

Table 14.3: Typical X-Ray Photoelectron Excitation Sources

Radiation	Energy (eV)	Full Width at Half Maximum (eV)
Y Mζ	132.3	0.44
Zr Mζ	151.4	0.84
NaK$_\alpha$	1041.0	0.6
MgK$_\alpha$	1253.6	0.8
AlK$_\alpha$	1486.6	0.9
CuK$_{\alpha 1}$	8047.8	3.0

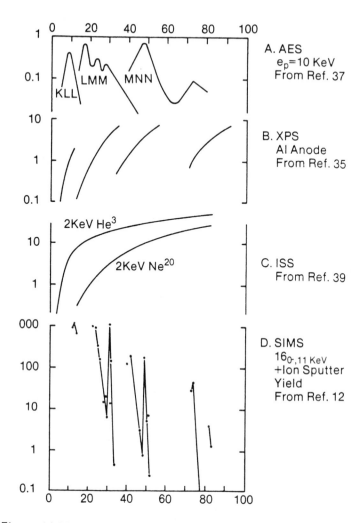

Figure 14.10: Relative elemental sensitivities for Auger electron spectroscopy (AES), X-ray photoelectron spectroscopy (XPS), ion scattering spectroscopy (ISS), and secondary ion mass spectrometry (SIMS) versus atomic number.

XPS is not affected severely by sensitivity effects, or matrix effects. XPS allows for complete analysis in a single determination with an average sensitivity of approximately 0.3% atomic. XPS is at best a semiquantitative analytical technique with an accuracy of no better than 10%.

AES requires three electrons for Auger emission therefore this technique can detect elements with $Z \geqslant 3$. Analogous to XPS, the KLL Auger transitions are initially the most intense, but the LMM transitions increase in intensity with increasing atomic number and a subsequent increase in the MNN transitions. By progressively using the KLL, LMM and MNN series Auger transitions, the elemental sensitivity variation across the periodic table can be held to less than a factor of fifty, as shown in Figure 14.10[37] The rate of core-level ionization may be adjusted by varying the primary electron beam energy so that the relative KLL, LMM and MNN Auger production rates may be altered.

Elemental resolution for AES is excellent since non-overlapping peaks can usually be found. AES has an average sensitivity limit of 0.1% atomic. Peak overlap in multicomponent samples often increases these values. AES has all the inherent problems of XPS in determining ionization cross-sections, electron escape depths plus the additional factors of electron backscattering from the primary beam and competing Auger and X-ray decay processes. These factors prevent AES data from being interpreted with an accuracy of better than 30%. Reference samples are often prepared in order to minimize the influence of analyzer transmission, surface roughness and the other factors just mentioned. Standards can be quite helpful if the surface composition is representative of the bulk. Quite often surface segregation contamination or oxidation complicate the preparation of standards. Ion sputtering may be employed in order to clean the surface but during sputtering other sample changes may result.[38]

The identification of target atoms with ISS is relatively simple and covers a practical range of $Z \geqslant 3$. The mass resolution of ISS is poor since the relationship between the scattering mass and ion energy is non-linear. This makes it difficult to resolve two high Z elements close in mass. Using higher mass primary ion beams will improve resolution but the choice of scattering ions is limited to basically a few inert gas ions.

The ISS sensitivity factor as a function of atomic number as shown in Figure 14.10[39] varies smoothly with a variation of approximately two orders of magnitude. The scattering yield depends on the differential scattering cross-section for the ion-atom collision and on the probability that the scattered particle remains ionized as well as on the number of target atoms. The detection limit for ISS is approximately 0.1% atomic. When standards are used, ISS data may be quantified to within 30% for simple systems. However, since ISS is a surface sensitive technique and of necessity uses an ion beam, sputtering artifacts may complicate any data obtained from standards.

Atomic identification by SIMS covers the widest range of all the techniques, from H to all higher Z elements. Since SIMS determines the sample composition by mass analyzing material sputtered from the surface, element resolution is primarily a function of the mass resolution of the analyzer. Quadrupole mass filters generally have a mass resolution, $\frac{M}{\Delta M}$, of over 300 while magnetic analyzers typically have a mass resolution of 3000. Atomic identification may be complicated by the presence of molecular ions with the same charge to mass ratio. Monomers, dimers, trimers, etc. may be formed but fortunately the abundance of molecular ion species falls off rapidly as the number of particles in combination increases.

SIMS suffers, somewhat unfortunately, from wide fluctuations, $\sim 10^4$, in sensitivity with atomic number. Another way of stating this is that SIMS, fortunately, is very sensitive for certain elements. Since ions sputtered from the sample surface are used for SIMS analysis, the detection limits are functions of the rate at which material is removed and the rate at which secondary ions, as opposed to neutrals, are produced. Both the sputtering rate and ion yield are[38] atomic and chemical matrix functions which produce variations in the SIMS detection limits.[12] For easily ionized materials, the detection limit may be 10^{-4}% atomic when relatively thick samples are available. In instances where "surface standards" with compositions near that of an unknown have been sputtered to a steady state composition, SIMS data have been quantified. With the uncertainties in predicting the effects of matrix changes upon secondary ion yields, quantification of data from an unknown is not easily accomplished.

Back-scattering analysis provides the ability to distinguish the atomic masses of elements and even isotopes. The energy of the elastically scattered particle increases with increasing atomic mass of the target atoms until it approaches E_o. This allows adjacent elements in the periodic table to be distinguished up to about mass 40, at which point the mass resolution of the analyzer, typically 15 keV in 2 MeV, becomes inadequate. Elements with larger mass separations may be distinguished when solid state detectors are used. The ability to distinguish different masses is reduced at low analysis energies in comparison with higher energies.

The relative sensitivity for RBS is largely determined by the scattering cross-section which is shown in Figure 14.8 for 1 MeV ^{4}He. The cross-section is low at low Z but increases significantly with increasing atomic number. For thick targets, a correction should be made to account for the change in the scattering cross-section as it loses energy traversing the film.[40] In addition to having a poor sensitivity for low Z elements because of the scattering cross-section, the signal from the lighter elements appears as a step on top of the background from the RBS spectrum of a heavy element. For a heavy element in a light matrix, the detection limit is much better, approximately 10^{18} atoms/cm^3, because of the higher cross-section for heavy Z elements and the low background of the light matrix. Since the Rutherford scattering cross-sections are known with great accuracy, absolute concentrations can be determined to better than ±5% accuracy without standards.

The basic advantages of nuclear reaction analysis are that it provides background free detection of light elements on a heavy substrate and that it allows isotopic discrimination of the same element. The latter leads directly to isotopic tracer experiments with light isotopes that are not possible using standard radioactive tracer techniques because of the short lifetime of most of the light radioactive nuclei. NRA is not sensitive to high Z elements because of strong coulombic repulsion between the incoming ions and protons in the nucleus. The background-free detection for light elements is due to the high positive Q values which yield emitted particles with energies above that of the beam. The particles backscattered from the substrate may be stopped by an absorber placed in front of the detector. Background interference may arise from other light nuclei. Since the cross-sections for nuclear reactions are well known, absolute elemental concentrations may be calculated to within ±5%, with a detection limit of approximately 10^{15} atoms/cm^2.

Atomic identification by IIX covers nearly the entire periodic table, however, below atomic number six the effective X-ray emission cross-section falls off rapidly. By selecting X-rays originating from progressively K, L and M core levels as the atomic number increases, a variation in sensitivity of ⩽50 can be achieved across the periodic table with excellent elemental resolution. A wavelength dispersive spectrometer covering a range from 1 to 100 Å requires the use of several crystals. Good energy discrimination improves detection sensitivity which is typically 10^{14} atoms/cm^2. Because the ion trajectories are well behaved and the ion/solid interaction well known, this technique is quantitative to within 10%.

CHEMICAL INFORMATION

The unique energy of photoelectrons may be used for elemental identification and chemical state information when changes in the valence band density of states cause small but reproducible shifts in the core level and valence electron binding energies. The relative simplicity of the photoelectron spectra and narrow line widths permit the identification and cataloging of photoelectron binding energies, chemical shifts, for a large number of compounds.[41] Some progress has been made in calculating these shifts from first principles[42] but the technique is still largely empirical where energy shifts are compared to spectra from standards. Multiplet splitting and shake-up transitions, which are due to coupling of the core and valence electrons during excitation, accompany many core-level photoelectron transitions and are another source of chemical information.[43]

Since the energies of Auger electrons are related to the binding energies of the electron levels involved in Auger emission, chemical state changes produce shifts in the Auger electron kinetic energy and changes in the shape of the AES spectrum. While these shifts are often larger for Auger emission than for photoemission, the shifts in Auger electron peaks are more complicated due to the involvement of multiple electron levels. Some Auger chemical shifts have been identified and characterized, but since the resolution of AES analyzers has generally been much lower than for XPS analyzers, many Auger electron shifts have gone unnoticed. Auger electron energies are also more difficult to measure accurately due to surface charges that build up on the sample surface through a combination of the primary electron beam influence and the secondary electron yield.

Auger electron transitions accompany the initial photoionization process and may be observed in a typical photoelectron spectrum. With X-ray excitation the problems of sample charging are minimized so that Auger chemical shifts may easily be characterized with the high resolution photoelectron analyzer. By combining the chemical shift information contained in both the photoelectron and Auger portions of the XPS spectrum the chemical state can be determined more accurately.[44] By measuring the difference in Auger electron peaks and core level photoelectron peaks, the Auger parameter, Wagner[44] has been able to identify different chemical states which had core level (XPS) shifts below the detection limits of the spectrometer.

SIMS is the only other technique currently under discussion which is clearly capable of providing chemical state information. Ion bombardment of surface oxides, carbides, nitrides, etc. produces ionic fragments and molecular clusters when material is sputtered from the surface. Mass analysis of the ionic fragments gives the cracking pattern which may be used to determine the molecular structure of the sample surface. Cracking patterns generated by electron impact or chemical ionization mass spectrometry may be used as guides for SIMS analysis. Standard samples may also be used, however, standards for surface analysis are difficult to prepare.

The main feature of ISS is that each element or isotope of each element produces a single scattering event. A considerable amount of additional information is available in ISS spectra.[45] The yield curve of a typical ISS energy loss spectrum also contains low energy sputtering peaks, multiple scattering peaks, doubly ionized particles and tailing on the low energy side of the backscattered peak. These fine features may be a source of chemical information.

Since RBS and NRA involve interactions between energetic particles and atomic nuclei, there are no chemical effects on the processes. The X-ray energies emitted in IIX are different for different chemical compounds, but the detectors used are usually not sufficiently sensitive to detect these shifts and IIX is considered to be a matrix insensitive technique.

LATERAL RESOLUTION

The maximum amount of information is obtained from a sample by an analytical tool when both spatial and depth resolution are available. The ability to detect sample inhomogeneity in the surface plane is largely controlled by the size of the probing beam. A pictorial representation of the minimum excited areas for the various techniques is shown in Figure 14.11. XPS has the poorest spatial resolution of the techniques under consideration with an imaged area of 2 to 3 mm in diameter, that is controlled by the area imaged by the electron analyzer. The area exposed to X-rays from the excitation source is much larger. On some XPS instruments a set of entrance slits, $\sim$1 x 10 mm, on the electron analyzer determine the area analyzed.

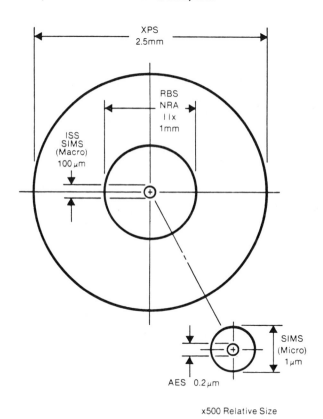

Figure 14.11: Relative sizes of the areas sampled by the various experimental techniques.

High energy ion beams used for RBS, NRA and IIX may be focused to a beam diameter of $\sim$1 mm. Some slight improvements in beam diameter may be achieved but focusing of energetic ion beams is difficult. As the ion beam energy is lowered to the range used for SIMS or ISS, focusing may be achieved more easily. Consequently, the ion beam for ISS and micro-SIMS has a minimum diameter of ~100 μm. As the beam diameter decreases, it becomes possible to raster the ion beam across the surface and generate elemental images with magnifications of up to 25x. Since both ISS and SIMS sputter during analysis, elemental mapping may be complicated by sputtering of material from the bottom of the crater generated by sputtering up onto the crater walls. When an element is removed from the bottom of the crater, the signal persists due to the presence of that element on the crater walls. By blanking the signal from the energy analyzer (ISS) or mass spectrometer (SIMS) while the ion beam is at the outer edges of the scanned area, signals which arise due to sputtering from the crater wall can be avoided.

By using a duoplasmatron ion source the ion microprobe or micro-SIMS can achieve a minimum probe diameter of 1 μm while Auger microprobes use electron beams as small as 0.05 μm. When the primary electrons penetrate the surface, beam spreading occurs as the electrons interact with the sample and are scattered. The back scattered electrons give rise to Auger emission from an area larger than the exciting beam. Since the primary beam does not penetrate over several hundred angstroms into the surface, beam spreading in SIMS is not a significant problem. When the beam diameter decreases, the detection limit for AES or SIMS decreases. The detection limit for AES is near 0.1% atomic for a 10 μm, 3×10^{-9} A/μm^2 beam but decreases to 10% atomic for 0.1 μm, 3×10^{-9} A/μm^2 beam.[46] The detection limit is directly proportional to beam size for a fixed beam current density since the production rate, number of Auger transitions per incident electron, is a constant. Similarly for SIMS, with a current density of 3×10^{-11} A/cm^2 the SIMS detection limit is 5×10^{-4}% atomic for a 50 μm beam, 1×10^{-2}% atomic for a 10 μm beam, and approximately 50% atomic for a 0.1 μm beam.[47] Neither technique offers simultaneous trace element analysis and high spatial resolution when small beam sizes are used.

DEPTH RESOLUTION

The depth resolution may be controlled by the primary beam penetration or the signal escape depth. For AES and XPS the sampling depth is determined by the electron escape depth in solids[18] which is defined as that length where the escape probability has dropped to 37%. The escape probability of electrons decreases exponentially with length from the surface and shows an energy dependence with a minimum of 0.3-0.5 nm at 60-100 eV and an increasing escape depth at lower and higher kinetic energy. Since the sampling depth for AES and XPS is so shallow other means are required to determine the composition as a function of depth. This is accomplished by ion bombarding the surface and measuring the composition as progressive surface layers are removed by sputtering. The depth resolution in sputter profiling is limited first by the electron escape depth but secondary effects due to the ion beam such as surface roughening, selective sputtering, sample reduction, elemental redistribution, etc. also play a role.[49,50]

ISS is the most surface sensitive technique under discussion since it samples primarily the outer atomic layer. The primary beam undergoes a single binary elastic collision with the first monolayer. Multiple collisions or elastic collisions with other than the first atom layer are unlikely to produce a significant signal with the low energy ion beams used in ISS. Ion sputtering for depth profiling is an inherent part of ISS, yet the sputtering rate may be adjusted by varying the mass and energy of the ion beam. In most respects depth profiling with ISS is very similar to depth profiling with AES, XPS, and SIMS with all the limitations of surface roughness, selective sputtering, sample reduction, elemental redistribution, etc.

For SIMS, removal of the surface atoms for detection is necessary for identification of the surface species. During ion bombardment the impinging ions undergo a sequence of binary elastic collisions, during which the energy of the incident ion is transferred to the lattice atoms until the incident ion is implanted or escapes from the solid. As more energy is transferred to the lattice, the mean escape depth of the target atoms increases with secondary ions originating from 1.5 to 2.0 nm below the surface. The sputtering rate can be varied from one atomic layer per hour with a detection limit of 50 ppm up to 100 atomic layers per second with ppb sensitivity.[47] Because of differential sputter rates caused by crystalline orientation and surface contamination, the depth resolution decreases as the film thickness increases. The resolution of a sharp interface is typically no better than 5-10% of the sputtered depth, yet amorphous films have been profiled with a depth resolution of better than 1% due to the absence of crystalline orientation effects on the sputter yield.

RBS and NRA provide the ability to distinguish the atomic masses and their in depth distribution without sputtering because the backscattered ion loses energy to the target as a function of depth. The primary particles with energies from 100 KeV to 2 MeV penetrate up to 0.1-1.0 μm into the target. The depth resolution is limited by the energy resolution of the detector's energy straggling to ±5 nm. Since the energy resolution of the detector is fixed, RBS or NRA at glancing angles may be used to enhance depth resolution.[51]

Besides depth profiling by analyzing the energy loss of nuclear reaction products, many nuclear reaction cross-sections exhibit a resonance with energy which provides depth information. If primary particles are used with an energy equal to the resonance in the cross-section, most of the signal will come from the surface region. As the primary particles penetrate the solid, they lose energy and the reaction cross-section decreases sharply. Profiling is accomplished by sequentially increasing the primary beam energy so that the reaction products originate predominately from deeper and deeper regions in the target. Depth resolutions, which depend on the sharpness of the resonant cross-section, of 20 nm have been observed.

The depth resolution of IIX is limited by both the ion penetration and the X-ray escape depth to a few μm. The ion penetration for excitation may be adjusted by varying the primary ion beam energy, angle or the ion species. Depth resolution may also be enhanced by varying the take-off angle for X-ray detection.

SAMPLE REQUIREMENTS

In general, the samples analyzed are solids, however, liquids and gases may be analyzed by condensation on a solid surface or through the use of a differentially pumped sample chamber. This is necessary since all of the techniques use particles whose mean-free-paths are very short at high pressure. Since contamination can accumulate at a rate of one monolayer per second at 10^{-6} pascals, when the sticking coefficient is unity, the very surface sensitive techniques, XPS, AES, ISS, require a vacuum in the range of 10^{-8} pascals. SIMS does not require a vacuum of 10^{-8} pascals from the standpoint of surface contamination since the surface is removed rapidly during analysis before any buildup can occur. Instead, a good vacuum is required for SIMS to prevent the formation of molecular ions that produce interferences for the sample being analyzed. RBS, NRA and IIX may be accomplished in vacuums of 10^{-3} pascals or better since the depth analyzed is much greater than the few monolayers that might build up on the surface.

Sample size is dictated by the beam size of the technique and by handling and mounting considerations. When samples become too large it is difficult to load the samples and mount them properly in the vacuum chamber. Flat planar samples are easiest to analyze, but surface roughness on the μ scale can influence the data for AES,[52] XPS,[53] ISS,[54] and SIMS. The effect of surface roughness is to decrease the signal intensity sometimes in a non-homogeneous fashion. In RBS, NRA and IIX analysis, surface roughness can appear as a layer thickness variation or can result in multiple scattering.

Powder samples can be analyzed by any of these techniques by pelletizing the sample, mounting it loose in a cup, adhering it to an adhesive surface or pressing it into a soft metal foil. Powders pose the problem of surface roughness occasionally encountered with solid samples, however, limited composition versus depth information can still be obtained.

Since charged particles are used for sample excitation in all these techniques, consideration must be given to maintaining an electrically neutral specimen. For conductors this is not a problem, but for insulators excess charge can accumulate on the sample. A positive charge generally exists when X-ray or ions are used for excitation due to electron emission from the surface and due to the positive charge when ions are used. A positive charge may be neutralized by low energy thermal electrons from a heated filament. Electron bombardment can result in positive or negative charge accumulation depending on whether the secondary electron yield is larger or smaller than the incident beam current. No good intense source of low energy positive ions is available to compensate for negative charge accumulation. Charging can be minimized by reducing the conduction path length with metal masks or grids over the sample, reducing the current density, by defocusing or scanning the beam and by changing the primary beam energy or angle of incidence to increase the secondary electron yield. The charge may be neutralized by simultaneous bombardment of the surface with an ion beam but this results in sputtering.

In order to determine the elemental and chemical composition of a sample in XPS and AES it is important to measure the electron energy accurately. If a stable charge accumulates on the sample, the energy shifted XPS or AES peaks can be referenced against a known peak energy such as C, Au or implanted inert gas from sputtering. If the reference is not in intimate contact with the surface, it may be shifted to a different degree than the sample. Carbon is not an ideal reference since the chemical state of carbon can shift the electron energy by up to 10 eV. If the charge accumulation is not stable, referencing will not be adequate to characterize the sample.

The influx of charged particles may produce changes in the sample. Chemisorbed alkali and halogen molecules are especially susceptible to ionization and subsequent desorption. Organic molecules are quickly reduced by an electron beam to different forms which also desorb or in the case of residual gases reduces them to free carbon on the surface. Although such problems are severe at times, the effects are generally current density and total dose dependent and many of the above effects can be avoided by reducing the analysis time and lowering the current density.

X-ray excitation used for XPS is relatively gentle yet damage will still accumulate due primarily to the accompanying bremsstrahlung after long exposure times. Ion sputtering as a means of sampling as in ISS or SIMS or as a means of profiling as in XPS and AES in combination with X-ray or electron beam exposure will enhance sample reduction.[46] The energetic ion beams used in RBS and IIX are generally considered to be less damaging to samples because analysis can be performed with low fluxes and energy deposition and ionization occur deeper in the solid where desorption and sputtering will not occur. Thus RBS and IIX are generally considered to be non-destructive since changes at the sample surface are minor compared to the depth of analysis. NRA is similar to RBS and IIX in that energy is deposited deeper in the solid. However, because the cross-sections for nuclear reactions are relatively low, much higher fluxes are required with the accompanying danger of sputter damage.

SUMMARY

The characteristics of the analytical technique are summarized in Table 14.4. Each technique has special attributes and disadvantages. Table 14.5 summarizes some types of problems which are best addressed by the various techniques. While this summary is not all-inclusive, it suggests how to choose between the various techniques. For some problems there is not one technique that can give complete analysis. In order to get a better understanding of the sample, several techniques must be used simultaneously or sequentially. Not until one knows the surface and bulk composition and the various treatments a sample has undergone, does he begin to understand the physical and chemical changes that have occurred.

Table 14.4: Summary of Various Characteristics of the Analytical Techniques

Characteristic	AES	XPS	ISS	SIMS	RBS	NRA	IXX
Elemental Analysis							
Sens. Varia.	Good	Good	Good	Poor	Fair	Fair	Good
Resolution	Good	Good	Fair	Good	Fair	Good	Good
Detection Limits	0.1%	0.5%	0.1%	10^{-4}% or higher	10^{-3}% or higher	10^{-2}% or higher	10^{-2}% or higher
Quantification	......with difficulty...... req. standards			very difficult req. standards	........absolute........ no standards		
Chemical State	Yes	Yes	No	Yes	No	No	No
Depth Analysis	............Destructive sputter................				...Non-destructive...		Very difficult
Depth Resolution	..........Atomic layer to ....% of sputter depth..........				10 nm	10 nm	None
Lateral Resolution	200 nm	2 mm	100 μm	100 - 1 μm	........1 mm........		
Sample Alteration	High for Alkali Halogen Organic Insulators	.............Low..............			.........Very Low..........		

Table 14.5: Appropriate Technique for Various Types of Samples and Analytical Requirements

Situation	Appropriate Technique
Minor and major elements at the surface or interface of small samples	AES
Chemical state analysis; analyses of organics, insulators	XPS
Analysis or outer atom layer; analysis of insulators	ISS
Trace elements at surface or interface of medium to small size samples; analysis of insulators; sputter profiling of light elements	SIMS
Depth profiling of higher Z elements and thin films; trace analysis of heavy elements in light matrix; quantitative analysis	RBS
Depth profiling lower Z elements and thin films; trace or minor analysis of light elements; quantitative analysis	NRA
Trace, minor, and major element analysis in thicker samples; quantitative analysis	IIX

REFERENCES

(1) A. Joshi, L.C. Davis, and P.W. Palmberg in: *Methods of Surface Analysis,* Ed. by A.W. Czanderna (Elsevier, New York, 1975), Chapter 5.
(2) W.M. Riggs and M.J. Parker in: *Ibid.,* Chapter 4.
(3) T.M. Buck in: *Ibid.,* Chapter 3.
(4) J.A. McHugh in: *Ibid.,* Chapter 6.
(5) C.C. Chang in: *Characterization of Solid Surfaces,* Ed. by P.F. Kane and G.B. Larrabee (Plenum Press, New York, 1974), Chapter 20.
(6) S.H. Hercules and D.M. Hercules in: *Ibid.,* Chapter 13.
(7) J.J. McCrea in: *Ibid.,* Chapter 21.
(8) W.D. Mackintosh in: *Ibid.,* Chapter 16.
(9) T.B. Pierce in: *Ibid.,* Chapter 17.
(10) T.A. Carlson, *Photoelectron and Auger Spectroscopy,* (Plenum Press, New York, 1976).
(11) G.C. Nelson, *J. Colloid Interface Sci. 55* 289 (1976).
(12) C.A. Andersen and J.R. Hinthorne, *Science 123* 858 (1971).
(13) G. Foti, J.W. Mayer, and E. Rimini in: *Ion Beam Handbook for Material Analysis,* Ed. by J.W. Mayer, and E. Rimini (Academic Press, New York, 1977).
(14) J.A. Borders in: *Contemporary Topics in Analytical and Clinical Chemistry,* D.M. Hercules, et al., Eds. (Plenum Press, New York, 1978).
(15) G. Amsel, J.P. Nadai, E.D'Artemare, D. David, E. Girard, and J. Moulin, *Nucl. Instr. Methods 92* 481 (1971).
(16) R.G. Musket and W. Bauer, *Thin Solid Films 19* 69 (1973).
(17) J.F. Ziegler in: *Ion Beam Surface Layer Analysis,* Ed. by O. Meyer, et. al., (Plenum Press, New York, 1976), Vol. 2, p. 759.
(18) J.W. Mayer and A. Turos, *Thin Solid Films 19* 1 (1973).
(19) C.A. Evans, Jr., *Anal. Chem. 47* 818A (1975).
(20) P.H. Holloway and G.E. McGuire, *Thin Solid Films 53* 3 (1978).
(21) R.E. Honig, *Thin Solid Films 31* 89 (1976).
(22) D.A. Shirley, *Chem. Phys. Lett. 17* 312 (1972).

(23) P.W. Palmberg, *J. Electron Spectrosc.* 5 691 (1974).
(24) J.E. Houston and R.L. Park, *Rev. Sci. Instrumen.* 43 1437 (1972).
(25) L.A. Harris, *J. Appl. Phy.* 39 1419 (1968).
(26) A. Mogami and T. Sekine, Sixth European Congress on Electron Microscopy, TAL Intl. Pub. Co., 1976, Ramat Gan, Israel.
(27) E. Taglauer and W. Heiland, *Appl. Phys.* 9 261 (1976).
(28) H.J. Liebland and R.F. K. Herzog, *J. Appl. Phys.* 34 2893 (1963).
(29) C.A. Anderson and J. Hinthorne, *Science 175* 853 (1972).
(30) C.A. Anderson, *Anal. Chem.* 45 1421 (1973).
(31) W.K. Chu, J.W. Mayer, M.A. Nicolet, T.M. Buck, G. Amsel, and F. Eisen, *Thin Solid Films, 17* 1 (1973).
(32) G. Amsel, *Ann. Phy.* 9 297 (1964).
(33) G. Amsel, G. Beranger, B. de Ge'las and P. Lacombe, *J. Appl. Phys. 39* 2246 (1968).
(34) R.G. Musket, *Research/Development 28(10)* 26 (1977).
(35) C.D. Wagner, *Anal. Chem.* 44 1050 (1972).
(36) K. Codling, *Rep. Progr. Phys. 36* 541 (1973).
(37) L.E. Davis, N.C. McDonald, P.W. Palmberg, G.E. Riach and R.E. Weber, *Handbook of Auger Electron Spectroscopy,* 2nd Edition (Physical Electronics Ind. Inc., Eden Prairie, MN, 1976).
(38) G.K. Wehner in: *Methods of Surface Analysis,* Ed. by A.W. Czanderna (Elsevier, N.Y., 1975), Chapter 1.
(39) A.W. Mullendore, G.C. Nelson, and P.H. Holloway, *Proc. Advanced Tech. Failure Analysis,* Los Angeles (IEEE, 345 E. 47th St., N.Y., N.Y., 1977).
(40) M.H. Bordsky, D. Kaplan and J.F. Ziegler, *Appl. Phys. Lett. 21* 305 (1972).
(41) C.D. Wagner, W.M. Riggs, L.E. Davis, J.F. Moulder and G.E. Muilenberg, *Handbook of X-ray Photoelectron Spectroscopy,* Physical Electronics Ind. Inc., Eden Prairie, Minnesota, 1979.
(42) J. Hedman, G. Johnasson, T. Bergmark, S. Karlsson, I. Lindgren, *ESCA-Atomic Molecular and Solid State Structure Studies by Means of Electron Spectroscopy* (Almquist and Wiksells, Upsala, 1967).
(43) T.A. Carlson, J.C. Carver, L.J. Saethre, F.G. Santibanez and G.A. Vernon, *J. Electron Spec. Related Phen., 5* 247 (1974).
(44) C.D. Wagner, *J. Electron Spectroscopy 10* 305 (1977).
(45) W.L. Baun, *Appl. Surface Sci., 1* 81 (1977).
(46) H.W. Werner in: *Advances in Applied Surface Analysis,* ASTM STP, to be published.
(47) J.A. McHugh, in *Methods of Surface Analysis,* Ed. by A.W. Czanderna (Elsevier, N.Y., 1975) Chapter 6.
(48) C.J. Powell, *Surface Sci.* 44 29 (1974).
(49) J.W. Coburn and E. Kay, *Crit. Reviews Solid State Sci.* 4 561 (1974).
(50) G.E. McGuire, *Surf. Sci.* 76 130 (1978).
(51) J.S. Williams in: *Ion Beam Surface Layer Analysis,* Ed. by O. Meyer, G. Linker, and F. Kappeler (Plenum Press, N.Y., 1976) p. 223.
(52) P.H. Holloway, *J. Electron Spectrosc.* 7 215 (1975).
(53) R.J. Baird, C.S. Fadley, S.K. Kawamoto, M. Mehta, R. Alvarez and J.A. Silve, *Anal. Chem.* 48 843 (1976).
(54) G.C. Nelson, *J. Appl. Phys.* 47 1253 (1976).

Index

A15, 296, 302
Abrasive wear, 11, 155
Abrupt discontinuity in composition, 8
Absorbers, 503
Acetal, 504
Acrylics, 498, 499, 502
Activated evaporation, 127
Activated reactive evaporation (ARE)
    process, 4, 7, 17, 122, 128, 259,
    290, 319, 320, 327
  using a resistance heated source, 126
Activation, 402
  of plastics, 423
Activation—surface
  metals, 79
  oxides, 76
  polymers, 79
Active coatings, 503
Adatom, 212
  mobility, 7, 221, 225
  surface diffusion, 212, 213, 217, 219
Addition (see Polymerization)
Addition agents, 397
Additives, 491, 498, 499, 503
Adhesion, 63, 193, 221, 228, 231, 506
  criterion for good, 63
  deposit to substrate, 12
  effect of contamination, 229
  failure of, 63, 70
  "good" adhesion, 63
  sputtered coatings, 193
  testing, 71
  use of bead blasting, 229
Adhesive wear, 11
Advantages of PVD processes, 89
Age, 503

Air and airless spraying, 16
Air pollution, 501
Air-to-air seals, 92
Al-alloy films, 544
Al + Cu, 542, 544
Al + Cu + Si, 544
Alkyd, 498
Alloy deposition, 14, 404
Alloy evaporation, 118
Alloy films, 520
Alloying, 534
Alloys and compounds, 183
  deposits, 227
  sputtering of, 183
Al_2O_3, 293, 298, 302, 303, 306-308, 310
Alumina, 338, 361
Aluminum, 306, 347, 514, 524, 534, 538,
  542
Aluminum films, 534
Aluminum powder (also see Filler), 503
Amide, 501
Amine, 501, 507
Amorphous, 289, 290, 292, 323, 330,
  332, 333
Amorphous deposits, 89
Amorphous materials, 9
  incorporation of entrapped gas, 221
  structure of sputtered coatings, 214
Analyzers
  cylindrical mirror, 551, 552, 553
  magnetic sector, 554, 556
  quadrupole, 554
  resolution, 551
  spherical sector, 553
Angle of incidence
  coating flux, 177, 217, 222, 224

Index

influence on stress in sputtered coatings, 223
ions impacting sputtering target, 178
Anions, 385, 493
Anode reactions, 386
Anodes, 31, 33, 42, 170, 172
  auxiliary for bias sputtering, 193
  magnetrons, 194
  sheath, 32
Anodic oxidation of Ta, 527
Anodization (see also Conversion coatings), 7, 15, 385, 419, 452, 507
  of aluminum, 419
  of magnesium, 422
  sealing, 419
Antistatic coatings, 504
Applications, 158, 174
  sputtering, 174, 197, 200, 225
Applications of coatings, 10
Arc evaporation, 108
Arc plasma spraying, 17
ARE process (see Activated reactive evaporation)
Artificial organs, 12
Assembly, 514, 544
Atomic deposition processes, 8
Atomistic deposition, 6, 7
Atomize, 509
Au, 293, 538
Auger, 295
  AES, 551, 562, 564, 566
  spectroscopy, 548
  transition, 551

Barrier height, 537
Barrier layer, 538
Barrier metal, 545
Batch processing, 197, 198, 236
  using cylindrical magnetrons, 197, 198
Beeswax (see Binders)
Beryllium foil, 84
Beryllium oxide, 358
Bias sputtering, 6, 172, 173, 220, 221, 229
  effect on coating composition (also see Ion bombardment), 230
  magnetrons, 230
  nonconducting substrates, 230
Biased activated reactive evaporation (BARE), 17, 128
Bifunctional, 492
Bi-metal, 538
Binders, 490, 491, 502
  beeswax, 490
  casein, 490
  egg albumen, 490
  gelatin, 490
  gum arabic, 490

Biocides, 503
Biomedical devices, 12
Blistering, 221, 506
Blowing agents, 502
Blush, 499
Bombardment of substrate/deposit by inert gas ions, 14
Bonding, 507, 514, 534
  thermocompression (TC), 544
  ultrasonic, 544
Boron, 337, 345
Boron nitride, 338, 361
Boron trichloride, 347
Boundary layer, 395
Brush, 508
Brush painting, 16
Bulk coatings, 7
Bulk properties, 4
By-product disposal, 345

Calomel electrode, 389
Capacitors, 514, 523
  tantalum film, 526
  thin film, 524
Carbon, 342, 347
  amorphous, 375
  pyrolytic, 360
Carbon black (also see Filler), 503
Carrier gas, 339
Casein (see Binders), 490
Catalysts, 492, 498, 499, 500, 501
Cathode reactions, 386
Cathodes, 31, 34, 170
  dark space (also see Plasma, sheaths), 32, 33, 34, 35, 42, 43, 190
  shape, 224
Cations, 385, 493
Cavity, 291, 295-298, 301-304, 306, 311, 312, 314, 320-325, 328, 329, 333
  cracks, 320
  microcavity, 320, 323, 324, 325
  porosity, 320
  voids, 302, 305, 306, 308, 333
Cellulose, 498
Cellulose acetate, 498, 504
Cermet films, 520, 522
Chain reaction (also see Polymerization), 493
Chalk (also see Filler), 490
Characterization of chemical composition and microstructure, 3
Charcoal (also see Filler), 490
Charge, 509
Charge exchange processes, 122
Chemical analysis, 294, 295
Chemical conversion, 7
Chemical interaction, 95

Chemical ion plating, 18
Chemical processes, 5
Chemical reduction, 6
Chemical shifts
  auger parameter, 562
  multiplet splitting, 562
  shake-up, 562
Chemical transport, 335, 336
Chemical vapor, 7
Chemical vapor deposition (CVD) (see Plasma-assisted CVD), 2, 5, 6, 8, 16
Chemically functional, 10, 158
Child-Langmuir law, 33
Chromate (also see Conversion coatings), 507
Chromate conversion coatings, 404
Chromium, 338, 347
Chromizing, 343
Cladding, 7
Classification of coating processes, 5
Classification of evaporation processes, 127
"Clean" environment, 72
"Clean" surface, 72
Cleaning processes, 74, 402
Cleanliness, 506
Close cathode gun, 103
Cluster ion beam deposition, 18
Coalesce, 499, 511
Coatings, 404 ff
  alloys, 404
  composite, 408
  diffusion, 408
  dispersion, 408
  electrophoretic, 410
  mechanical, 412
  selection, 404, 406
Coating-substrate interface, 2
Co-deposition, 227, 229
Cold cathode plasma electron beam, 106
Colligative (see Molecular weight)
Collision frequency, 24, 25, 27
  electron-electron, 25
  electron-ion, 25
  temperature, 25
Collisions, 20
  attachment, 20
  charge exchange, 27, 34, 35, 191, 192
  collisional scattering, 224
  cross sections, 20, 21
  elastic, 20
  electron-atom, 20, 46, 47, 196
  electron-molecule, 48
  excitation, 20, 48
  ionization, 20, 48
  momentun exchange, 20
Color, 502

Columnar
  boundaries, 219
  defects, 140, 219
  grains, 4, 217
  microstructure, 8
  morphology, 132
  structure, 215
  voids, 217
Complex ions, 386
Complex shapes, 197
  coating by sputtering, 197, 220
Complexing agents, 390
Composite material, 83
Compound semiconductors, 213, 218, 360
  sputtering of, 213
Compounds
  sputtering of, 183
Computer modeling
  sputtering process, 180
Condensation (see Polymerization)
  adatom, 64
Condensation of vapors followed by film nucleation and growth, 85
Conditioning, 402
Conduction and diffusion processes, 5, 15
Conduction in thin films, 517
Conductive carbon, 504
Conductivity (also see Electrical properties), 504
Cone formation
  during sputtering, 184, 185
Contact angle, 505
Contact resistance, 538
Contacts, 528-534
  metal-insulator, 528
  metal-metal, 528
  metal-semiconductor, 529, 531
  ohmic, 514, 533
  platinum silicide, 536
  rectifying, 514
  Si-Al, 535, 536
Contaminants, 498, 503
  surface, 72
Continuous film, 129, 131
Convection, 395
Conversion coatings, 507
  anodize, 507
  chromate, 507
  phosphate, 507
Conversion and conversion/diffusion coating, 15
Coordination complexes, 493
Copolymerized, 501
Corona discharge, 507
Corrosion, 503, 506, 507
Corrosion/erosion-resistant coatings, 12
Corrosion resistance, 155

Index 573

Corrosion-resistant CVD coatings, 360
Cosine distribution, 188
  sputtered species, 188
Cosine law, 91
Cost of CVD coatings, 346, 347
Cottonseed oil, 498
Coulomb-dominated plasma, 25, 30
Coulometer, 388
Creation of vapor phase specie, 5
Crosslinked, 493, 495, 498, 499, 500, 501, 502, 511
Crossovers, 541
Cr-SiO films, 521, 523
Crucible liners, 102
Cryoscopy (see Molecular weight determination)
Crystal oscillators, 110
Crystalline, 495
Cu, 293, 313-316, 318, 319
Curing (also see Crosslinked), 493, 499, 511
Current distribution, 389
Current-voltage characteristic sputtering discharges, 201
Curtain coating, 510
Cutting tools, 11
CVD (see Chemical vapor deposition)
CVD number, 354
Cylindrical magnetrons, 174, 176, 197, 210
  hollow-type, 198, 199
  operating conditions, 197
  post type, 194
  scaling, 197
Cylindrically symmetric sputter-coating system, 89

Debye length, 30, 31, 36
Decorative, 10, 158
Defect concentration, 9
Defect structure, 122
"Defects" found in vapor-deposited materials, 140
Degreasing, 506
Delta (δ) (see Solubility parameter), 495
Dendrites, 7
Deposition, 7
  of alloys, 115
  of intermetallic compounds, 119
  of refractory compounds, 122
  onto the surface and film growth, 4
Deposition flux profile, 202
  cylindrical-post magnetron, 202
  hollow-type, 197
  rectangular planar magnetron, 204, 205
  ring-type planar magnetron, 204, 205

Deposition mechanism, 392
Deposition rate, 14, 191, 192, 193, 200
  monitors, 94, 108
  for PVD processes, 89
Detectors
  electron multiplier, 551
  surface barrier, 556
Detergent cleaners, 506
Detonation gun coatings, 17, 454-489
  coating process, 461
  coating structure, 464
  equipment, 459
  equipment-related limitations, 461
  powder temperature, velocity, 459
  properties, 464
  structure, 464
  technology, 3
D-gun (see Detonation gun coatings), 7
Diacid, 492
Dialcohol, 492
Dielectric (also see Electrical properties), 504, 514
Dielectric constant (also see Electrical properties), 499, 505, 523
Dielectric films, 524
Dielectric strength (also see Electrical properties), 505
Differentiation
  beam blanking, 552
  potential modulation, 552
Diffusion, 495
  adatom diffusion, 212, 213, 217, 219
  ambipolar, 35, 36
  barrier, 11, 539
  bulk diffusion, 7, 225
  interfaces, 229
  limitation, 351 ff
  plasma, 28, 29
  rate, 395
Diffusion coating or chemical conversion coating, 1, 3, 15
Diffusion or compound interfacial region, 8
Diisocyanates, 500
Dilatency, 499
Diode ion-plating, 87
Dip coating, 16, 508, 509
Direct evaporation, 127
Direction evaporation, 122
Dirt, 506
Dislocations
  sputtered coatings, 214
Dispersion-strengthened, 135
Dispersion-strengthened alloy, 115
Dispersion-strengthened alloy deposits, 151
Disproportionation, 16

Dissipation factor (also see Electrical properties), 505
Distribution (see Molecular weight)
Driers, 499
Drift velocity
  electric and magnetic fields, 29, 36, 43, 194, 196
  electric field, 27
Dry film lubricant coatings, 11
"Dry plating," 245
Drying, 499
Drying oils, 499
Ductility, 385, 443
  of copper deposits, 431
  of nickel deposits, 427, 432

Ebulliometry (see Molecular weight determination)
Ecological considerations, 13
Economics, 161
Egg albumen (see Binders)
Electric-arc spraying, 17
Electric discharge (see Plasma discharge), 170
  abnormal negative glow, 170, 190
  capacitive discharges, 37
  cold cathode discharges, 41, 43
  collective phenomena, 30
  glow discharge, 19, 37, 170, 189
  positive column, 42
  RF discharges, 37, 204, 208, 209
Electrical double layer, 392
Electrical properties, 504, 505
  conductivity, 504
  dielectric, 504
  dielectric constant, 495, 505
  dielectric strength, 505
  dissipation factor, 505
  insulation resistance, 505
  resistivity, 504
Electrically functional, 10, 158
Electrode efficiency, 388
Electrode potential, 389
Electrodeposition, 3, 385, 491
Electrodeposits, 426 ff
  electroforming, 385, 415
  electroless deposition, 3, 16, 385, 412
  properties, 426, 431, 440, 441
  reducing agents, 413
  solution formulations, 450
  stress, 414
  structures, 426, 428, 431
Electroless plating, 7
Electrolyte, 395
Electrolytic deposition, 15
Electromigration, 514, 543
Electromotive force (EMF) series, 389
Electron, 22, 23, 25, 28, 43
  acceleration in plasma electric field, 22
  confinement in magnetron discharge, 43
  cyclotron frequency, 28
  energy distribution function, 23, 24, 42
  energy loss in collision, 22
  motion in magnetic field, 28
  motion in static magnetic fields, 196
  stimulated absorption, 53
  stimulated desorption, 53
  temperature, 22, 23
  thermal velocity, 25
Electron beam heated sources, 102
Electron spectroscopy for chemical analysis (ESCA), 295
Electron volt, 22, 172
Electrophoretic, 6, 7
Electrophoretic coating, 15
Electrophoretic deposition, 508, 511
Electroplating (see Electrodeposition), 6, 7, 511
Electropolishing, 292, 293
Electrostatic, 508, 509
Electrostatic deposition, 15
Electrostatic spraying, 7
Elimination of pollutants and effluents, 89
Emulsion, 499, 509
Enameling, 7
Encapsulation, 505
Energetic neutrals, 122
Energy band diagrams, 529, 530
Energy of deposit-species, 14
Enhanced ARE process, 128
Entrapped working gas
  sputter process, 182, 212, 221, 224, 230
Environmental corrosion, 11
Epitaxial growth, 217, 227
Epitaxy, 7
Epoxy, 498, 501, 502, 506, 507, 511
Equiaxed grains, 217
  morphology, 132
"Equilibrium" properties, 2
Equivalent weight, 387
Ester, 492
Etching, 402, 506
Ethylcellulose, 498
Evaporation, 2, 6, 17, 85, 86, 218, 498, 499, 534
  activated reactive, 37, 39
  three-temperature method of, 218
Evaporation coefficient α_v, 91
Evaporation process and apparatus, 92
Evaporation process control, 113
Evaporation sources, 94
Evaporation technologies, 84
Evolution of the microstructure, 84
Explosive, 7

Fabrication, 503
Fabrication of microelectronic devices, 3

Faceted surfaces, 217, 227
Failure-adhesion, 63
  fracture, 63, 70, 71
  static fatigue, 71
  stress corrosion, 71
  time-dependent, 71
Faraday, 387
Faraday's laws, 387
Fast cycle coaters, 92
Fasteners, 11
Fatigue-crack initiation, 140
Fatty acids, 498
Fibrous grains, 215
Filler, 498, 499, 502
  aluminum powder, 503
  carbon black, 503
  mica, 503
  minerals, 490
  titanium dioxide, 502
  zinc oxide, 503
Film, 499
Film deposition, 514
Film growth on the substrate, 5
Films and coatings, 83
Fingerprints, 506
"First Wall" of thermonuclear reactor vessels, 12
Flake formation, 140
Flame spraying, 6, 17
Flash evaporation, 119
Flash vaporizer, 339
Flexibilizer, 504
Floating potential, 31, 32
Flow coating, 510
Flow properties, 498, 499
Fluid bed coating, 511
Fluidized bed, 341, 360
Fluorocarbon, 503
Fluoropolymers, 511
Formaldehyde, 501
Free radicals, 493, 502
Friction, 492, 504
Friction and wear, 11
"Frontier Areas," 12
Full-density coatings, 84
Functional groups, 492
Fused salt electrolysis, 7

Gallium arsenide, 338
Gallium arsenide-phosphide, 360
Galvanic couple, 503
Gamma (γ) (see Hydrogen bonding index)
Gas discharge processes, 246
Gas dynamics, 341
Gas scattering, 14, 86
Gaseous anodization, 6, 15
Gasless ion plating, 5

Gel, 495
Gel coatings, 500
Gel permeation chromatography, 494
Gelatin (see Binders)
Gelatinous, 499
Geometrical shadowing, 8
Getter pumping, 232
Glass, 506
Glazing, 6
Gloss, 502
Glow discharge, 86, 511
Glow discharge (also see Plasma)
  decomposition, 369
Glow discharge evaporation and sputtering, 17
Glycerine, 498
Glycidyl, 507
Glycols, 500
"Gold-colored" wear-resistant coating of titanium nitride, 10
Graded interface, 140
Graft polymerization, 510
Grain boundaries, 219
Grain boundary sliding, 308
Grain refinement in CVD, 358
Grain size, 397, 428
Graphite, 347
  pyrolytic, 360
Grounded shield, 41, 190, 191
Growth boundaries
  flake defects, 219
  fracture cross sections, 215
  in vacuum deposited coatings, 219, 220
Growth interface perturbation, 14
Growth stresses, 139
Gum arabic (see Binders), 490
Gun-type magnetrons, 173, 174, 176, 194, 200

Hall-Petch relationship, 150, 227
Hard anodizing, 420, 453
Hardener, 501
Hardness, 385, 442, 498
  chromium, 439
  copper, 431
  electroless nickel, 413
  hard anodize, 420
  metals and alloy deposits, 145
  nickel, 432
  scales, 406
  sputtered coatings, 227
Harmonic electrical spraying, 17
Heart valves, 12
High deposition rate PVD processes, 84
High solids, 491
High strength-high toughness ceramics, 12
High temperature phases, 225

Hillocks, 541, 546
  growth, 221
Hollow cathode, 43, 89, 199, 214, 220
  sputtering, 88, 220
Hot hollow cathode discharge beam, 106
Humidity, 506
Hydrogen bonding, 495
Hydrogen bonding index (γ) (also see Solubility), 495, 497
Hydrogen electrode, 389
Hydrogen embrittlement, 392, 404
Hydrophilic method, 16
Hydrophilicity (also see Wettability), 505
Hydrophobic property (also see Wetting), 504, 505
Hygroscopicity, 504

Ideal creep-resistant material, 152
Impact erosion wear conditions, 155
Impact plating, 7
Impingement rate of gas molecules, 90
Impurities, 398
  carbon, 415, 438
  effect on coating growth, 220
  incorporation of, 67-68
  metallic, 399, 438
  organic, 399
  sulfur, 415, 438
Induction heated sources, 101
Inhibitors, 503
Initiation (also see Polymerization), 493, 502
Initiator (also see Polymerization), 493, 511
Insulation resistance (also see Electrical properties), 505
Insulator films, 541
Insulators
  multilevel, 542
Integrated circuits (IC), 514, 532, 538
Interconnections, 514, 538
  multilevel, 542
  two-level, 540
Intercrystalline boundaries, 217
Interdiffusion, 140
Interface
  abrupt, 229
  bond, 221
  graded, 231
  region, 228, 229, 231
  structure formed during sputtering, 213
  voids, 230
Interfacial interaction, 8
Interfacial regions, 63-65
  microstructure, 65, 77
  types of, 65
Interlayers, 345
Intermediate layers
  "glue" layers, 64
  reactive deposition, 68
Intermetallic compounds, 119, 229
Internal (residual) stress, 221, 229, 231, 385, 441
  atomic peening model, 224
  in chromium deposits, 440
  compressure stresses, 224
  effect of neutralized and reflected ions, 224
  in electroless nickel deposits, 414
  evaporated coatings, 221
  intrinsic stress, 221
  in nickel deposits, 432
  sputtered coatings, 224
  thermal, 221
Intrinsic stresses, 8
Ion
  mobility, 27, 33
  negative, 48, 49, 187
  neutralization at sputtering target, 178, 182
Ion beam deposition, 7, 18, 210, 211
Ion beam etching, 211
Ion beam milling, 292, 293, 330
Ion bombardment, 213, 217, 224, 235
Ion bombardment effects (also see Bias sputtering)
  coating growth, 213, 217
  coating stress, 224
  film growth, 255-257
  interface formation, 254
  surface preparation, 251-253
Ion bombardment methods
  DC diode, 249
  enhanced sources of, 259
  ion gun, 245
  magnetron, 250
  RF diode, 250-251
  triode, 250
Ion carburizing, 15
Ion confinement
  magnetron sputtering sources, 197
Ion current monitor for electron beam heated source, 110
Ion implantation, 1, 2, 7, 18
Ion induced x-rays (IIX), 548, 558, 562, 565, 566
  escape depth, 564
Ion milling, 38
Ion nitriding, 15
Ion plating, 2, 6, 7, 18, 37, 39, 77, 78, 86, 244 ff
  benefits, 244
  definition, 244

Index

types of, 245
Ion plating apparatus, 260
  gas discharge chamber, 261
  gas handling system, 262
  high voltage power supplies, 264
  high voltage substrate, 263
  vacuum system, 260
  vapor sources, 258
Ion plating applications, 267
Ion plating—process control, 264
  deposition variables, 265
  film properties (sensitive), 265
  flow rate (gas), 265
  gas composition monitoring, 266
  pressure monitoring, 266
  rate/deposition monitoring, 266
Ion plating sources (deposition), 258
  chemical gases, 258
  electron beam, 258
  flash evaporation, 258
  resistively heated, 249, 250, 258
  rf heating, 258
  sputtering targets, 258
Ion plating—surface preparation, 267
Ion scattering spectroscopy (ISS), 548
  560, 562, 566
Ion sources
  duoplasmatron, 210, 554, 564
  hollow cathode, 210
Ionic contaminants, 506
Ionic migration, 394
Ionic residues, 506
Ion-induced surface chemistry, 52
Ionization, 20, 40, 190, 201
  potential, 20
  processes, 201
  production of, 190
  rate, 40
Ionization gauge rate monitor, 109
Ionize, 504
Ionized cluster beam deposition, 259
Iron, 352
Iron carbonyl, 345, 352
Island growth, 129, 130
Isocyanate, 500

Josephson tunnel devices, 546
Joules law, 390

Kinetic control, 351 ff
Kinetic theory of gases, 90
Kinetics, chemical, 356
Kirkendall effect, 545
Kirkendall porosity, 65
Knife coater, 508

Lacquer, 498, 499
Laminate composites, 152
Lamination, 504, 510
Langmeir Blodgett technique, 16
Large scale integrated circuits (LSI)
  514, 525, 536, 540, 546
Larmor radius, 29
Laser beam evaporation, 108
Laser interference monitoring, 346
Laser mirrors, 12
Laser trim, 519, 523
Latent heat of evaporation (see Solubility parameter)
Latex paints, 499
Leak hunting, 94
Level, 499
Leveling agents, 499
Light scattering (also see Optical properties), 491, 494
Limitations of PVD processes, 89
Line-of-sight, 5
Line-of-sight coverage, 14
Linseed, 498
Liquid phase epitaxy, 7
Low pressure plasma deposition (LPPD), 129
L-shaped resistor, 515, 516
Lubricant, 504

Macro and micro porosity, 8
Magnetic bubble memory, 546
Magnetic coatings, 385, 444
Magnetic field, 29, 193, 194, 195
  application to triodes, 193
  coil system for cylindrical magnetrons, 195
  effect of gradient, 29
  magnetron sputtering sources, 194
  strength for glow discharge devices, 29
Magnetic materials
  sputtering of, 200
Magnetic moment, 29
Magnetron sputtering sources, 174, 193, 194, 195, 199, 200
  cylindrical hollow type, 194
  cylindrical-post type, 195
  definition of, 194
  sputtering, 209, 220
Maleic anhydride, 498
Marine oils, 498
Masonry, 504
Mass transport, 351 ff
Matching network
  rf sputtering sources, 208
Materials conservation, 11
Matthiessen's Rule, 518
Maxwellian velocity distribution, 21, 23
M-D diagram, 256
Mean free path, 20
Meandering resistor, 516

Mechanical properties, 301, 302, 307, 311, 313, 314, 319, 330
  creep, 290, 301, 302, 307-310
  hardness, 302, 307, 320, 324, 329
Mechanical properties—film, 66, 70
  determining factors, 70
Mechanical properties of thick condensates, 145
Mechanically functional, 10, 158
Mechanism of production of depositing species, 14
Medium (also see Binders), 491
Melamine, 501
Melt extrusion, 504, 510
Melting, 491, 492, 494, 495
Metal, 514
Metal cleaning, 78
  abrasive, 78
  chemical etching, 78
  glow discharge, 78
  hydrogen firing, 78
  oxidation, 78
  reactive plasma, 79
  solvent, 78
  synthetic, 78
  vacuum firing, 78
Metal coverage, 539, 540
Metal deposition, 14
Metal-semiconductor junction, 530
Metallic powder, 504
Metalliding, 15
Metastable phases, 8
Metastable species, 49, 50
Methane, 347
Mica (also see Filler), 503
Microbalances, 110
Microballoons, 502
Microelectronics, 505, 514, 545
Microhardness testing, 441
Microstructure, 7, 131, 288-290, 293, 296, 297, 301, 313, 319, 320
Microstructure evolution, 129
Microstructure—film, 66
  effects of, 70
  M-D Diagram, 66-67
  modification of, 70
Microstructure of PVD condensates, 129
Microvoids (also see Optical properties), 502
Microwave discharges, 38
Microwelding, 11
Minerals (also see Filler), 490
Mo-An, 540
Modulated current forms, 399
  asymmetrical AC, 400
  interrupted DC (IC), 400
  periodic reverse (PR), 400
  pulsed, 400
  superimposed AC, 400
Moisture, 506
Molar volume (see Solubility parameter)
Molded, 503
Molecular adsorption, 52
  effect of ion bombardment, 52
Molecular beam, 7
Molecular beam epitaxy, 2, 17, 93
Molecular collisions, 48
  association efficiency, 48
  electrons (see Collisions), 48
Molecular flow conditions, 5
Molecular weight, 492, 493, 497, 498
  distribution, 494
  number average, 494
  weight average, 494
Molecular weight determination
  colligative, 494
  cryoscopy, 494
  ebulliometry, 494
  osmometry, 494
Monitoring of deposited mass, 110
Monitoring surface cleaning
  direct, 80
  indirect, 80
Monolithic material, 10
Monomer, 492, 498, 499, 502, 511
Morphology of thick single phase films, 131
Multilayered deposit, 115
Multilevel interconnections, 540
Multiple sources, 115

Naphthenates, 500
Nb, 296, 313
Nb_3Ge, 295
Nb_3Sn, 293-302, 304, 307, 308, 311, 314-319
$Nb_3Sn-Al_2O_3$, 302-306, 314
Nb_3Sn-Cu, 294, 311, 313-315, 317-319
Negative glow, 23, 42, 172, 190
Nernst equation, 389
  complex ion modification, 390
Nernst thickness, 395
Neurological electrodes, 12
Neutralized and reflected ions, 182, 212, 224
  effect on coating stress, 224
  effect on substrate heating, 212
New crystallographic modifications, 89
Newtonian viscosity (see Viscosity)
Ni, 301, 302, 306-308, 310, 311
$Ni-Al_2O_3$, 307-309
$Ni-ZrO_2$, 311, 312
Nichrome films, 520, 521
Nickel, 338, 340, 341, 345, 347, 360

Index   579

Nickel carbonyl, 339, 345, 347
Niobium, 348, 349, 350, 358
Niobium carbide, 348, 349, 350, 357
Niobium germanide, 350
Nitrocellulose, 498
Nodular defects, 219
Non-aqueous dispersions, 491
"Non-equilibrium" properties, 2
Non-Newtonian viscosity (see Viscosity)
Nuclear reaction analysis, 548, 561, 562, 566
Nucleation
  density during flux growth, 230
  density of film, 64
  kinetics, 221
  sites during sputter deposition, 213, 220
Nucleation and growth, 6, 8, 212
  coatings during sputtering, 212
  effects of ion bombardment (see Bias sputtering), 213
  effect of substrate roughness, 214, 219, 220
Number average (see Molecular weight)
Nylon, 504, 511

Oblique coating flux
  effect on coating stress, 224
  effects on structure, 214, 219, 220
Offset printing, 508, 510
Ohmic contacts, 534, 538
Ohm's law, 388
Oils, 506
Oleates, 500
Oleoresinous coatings, 498
Oleoresins, 499
Opacifiers (also see Optical properties), 502
Optical coatings, 2
Optical monitors, 110
Optical properties
  light scattering, 491, 494
  microvoids, 502
  opacifiers, 502
  reflect, 502, 505
  refract, 502, 505
  refractive index, 502, 503
Optically functional, 10, 158
Orange-peel texture, 499
Oriented microstructures, 8
Osmometry (see Molecular weight determination)
Outgassing flux, 177, 236
  water vapor, 236
Overlay coating, 1, 3, 9
Overpotential, 390
Overvoltage, 390
  hydrogen overvoltage, 392

  oxygen overvoltage, 392
Oxides, 122, 507
  cleaning, 74-76
  in contact with metals, 100
Oxidized, 499, 508
Oxygen, 500

Pack cementation, 336
Pad, 508
Paints, 490
Partial pressure gage, 94
Particle impingement rate monitors, 109
Particle size, 502
Particles, 503
Particulate deposition processes, 7
Parts handling, 508
Pascal, 170
  unit of pressure measurement, 170
Passive coatings, 503
Peelable coatings, 504
Peening, 70
Penning ionization processes, 50
Pentaerythritol, 498
Phenolics, 498, 501
Phosphate (also see Conversion coatings), 507
Photoelectron
  ESCA, 549
  XPS, 548, 551, 562, 564, 566
Photolithography, 511
Photomasks, 546
Photosensitizer, 502
Photovoltaic devices, 12
Physical vapor deposition (PVD), 2, 4, 5, 6, 83, 85, 288, 301, 302, 311, 320
Pickling, 402
Pigments (also see Filler), 490, 491, 498, 499, 502, 503, 506
Pinhole, 140
Planar diode, 37, 42, 172, 174, 189, 190, 191, 192, 193, 204
  sputtering device (also see Electric discharge), 172, 220, 229
  typical DC operating conditions, 42, 191
Planar magnetrons, 44, 174, 176, 185, 194, 200, 202
  deposition rates, 200
  deposition uniformity, 200
  magnetic field production (also see Gun-type magnetrons), 202
  scaling, 200
Planatery or nutating devices, 94
Plasma, 170, 366, 511
  Debye length, 30, 31, 36
  diffusion, 28, 29, 35, 36
  electrical conductivity, 28
  floating potential, 31, 32
  frequency, 30

580 Deposition Technologies for Films and Coatings

glow discharge (see Electric discharge), 37
oscillations, 36
polymerization, 50
potential, 31, 32, 207
quasi-neutral region, 34
sheaths, 30, 33, 34
temperature, 22
wall losses, 40
Plasma-assisted chemical vapor deposition, 16, 37, 50, 335, 365
Plasma-assisted etching, 37, 38, 50
Plasma-assisted thermal growth, 6
Plasma coating, 454 ff
  equipment, 454, 460
  equipment-related limitations, 460
  gas properties, 455
  operating parameters, 454
  powder characteristics, 456, 457, 458
  structure, 464
  torches, 454
Plasma coating processes
  coating, 462
  finishing, 463
  masking, 462
  powder, 461
  substrate preparation, 461
Plasma coating properties
  bond strength, 472
  corrosion, 486
  density, 474
  electrical, 487
  friction, 479
  hardness, 477
  mechanical, 475
  residual stress, 473
  thermal, 485
  wear, 479
Plasma discharge, 37, 41, 190, 194, 201
  conditions for sustaining, 190
Plasma electron beam gun, 103, 106
Plasma-induced polymerization, 6
Plasma oxidation, 15
Plasma polymerization, 7, 18, 381
Plasma reaction, 6
Plasma reduction, 6
Plasma SiO_2, 525
Plasma spraying, 3, 7
Plasmas in deposition processes, 3, 19
Plastic substrates, 225
  coating by sputtering, 192, 225
Plastisols, 504
Plated coatings, 385
Plating (see Electrodeposition), 534
Plating on plastics, 423
  preparatory steps, 423
Platinum silicide (see PtSi), 514, 533
Point defect concentration, 8

Polarization, 390
  concentration polarization, 391
Polyacetal, 492
Polyacids, 498
Polyacrylics, 498, 499, 502
Polyamide, 492
Polybutadiene-styrene, 499
Polybutyrates, 498
Polydispersity (see Molecular weight)
Polyester, 492, 498, 500, 501, 506, 507
Polyethylene, 511
Polyhydric alcohols, 498
Polymer, 492
Polymer characteristics, 493
Polymer cleaning, 79
Polymer surfaces
  crosslinking, 80
  roughening, 79
Polymeric coatings, 3
Polymerization, 6, 498, 499, 500, 501
  addition, 492, 493
  condensation, 492
Polymers
  plasma polymerization, 7, 18, 381
Polymethyl methacrylate, 504
Polymorphic phases, 225
Polyols, 500
Polyphenylene sulfide, 504
Polytetrafluoroethylene, 504
Polyurethane, 493, 498, 500
Polyvinyl acetate, 504
Polyvinyl butyral, 498
Polyvinyl chloride, 498, 504
Porosity
  film, 71
  interfacial, 65
Positive column, 42
Post cathode, 89
Powder, 503
Powder coating, 491, 504
Power density, 95
Preferential sputtering, 221
Preferred orientation, 132, 360
Prepolymer, 501
Pressure plating, 86
Primary electrons, 42, 43, 190, 192, 211
Primer, 506
Printed circuit board (PCB), 385, 424, 505
  laminates, 424
  preparation steps, 424
  types of PCB's, 424
Printed wiring boards, 424
Printing, 508
Printing process, 16
Process control, 108
Propagation (also see Polymerization), 493

Properties, 491
Propionates, 498
Protection, 490, 503, 506
Protrusions in CVD coatings, 355
Pseudodiffusion, 254
  "pseudodiffusion" type of interface, 8
PtSi (see Platinum silicide), 536, 537, 538, 542
Pt_2Si, 536
Pulse plating, 400
Purification of metals by evaporation, 157
Purple plague, 544
Pyrolysis, 16
Pyrolytic carbon, 342, 347, 360
Pyrolytic graphite, 347, 360

Radiation chemistry, 49
Radiation curing, 491
Radiation damage, 211
Radio frequency (rf) sputtering, 44, 45, 204
  bias sputtering, 209
  double ended, 209
  electrode area ratio, 209
  equivalent circuit, 207
  floating potential, 204, 208
  magnetrons, 209, 210
  matching network, 207
  operating frequency, 44, 204
  planar diode, 45
  plasma potential, 206
  rf shielding (see Electric discharge), 205
  self-bias voltage, 44, 45, 204, 206
  single ended, 209, 210
Raschen relation, 40
Rate control, 113
Rate of deposition, 12
Reactive deposition, 68
Reactive evaporation, 17, 122, 127
Reactive ion etching, 37, 38
Reactive ion plating, 4, 5, 18, 128
Reactive plasma cleaning, 79
Reactive sputtering, 4, 172, 187, 231, 232, 233, 234, 235
  cathode poisoning, 233, 237
  cathode surface compounds, 232
  coating stoichiometry, 232, 234
  effects of ion bombardment, 235
  magnetrons, 234
  metal to compound transition, 232
  target processes, 187
Reciprocating brush coater, 508
Recrystallization, 217, 219
  sputtered coating, 217
Rectifying contacts, 528

Reflective surfaces, 12
Refract (also see Optical properties), 502, 505
Refractive index (also see Optical properties), 492, 502, 503
Refractory compound deposition, 14
Refractory compounds, 154
Reinforcement (also see Filler), 502
Reliability, 543
Residual stress (see Internal stress), 139
Resinates, 500
Resins, 498
Resistance, 515
  temperature coefficient, 518
Resistance heated sources, 95
Resistance monitors, 112
Resistivity (also see Electrical properties), 504
  residual, 518
Resistors, 514
  design, 519
  materials, 520
  meandering, 515
  trim, 519
Resonance neutralization, 553
Reverse-roll coater, 508
RF reactive ion plating, 129
Ripple, 399
Roll bonding, 7
Roller coating, 508, 510
Room temperature vulcanization (RTV), 500
Rotting, 503
Roughness, 506
Rutherford backscattering (RBS), 548, 554, 561, 562, 566

Schottky barrier, 529
Schottky barrier diode (SBD), 531, 532
Schottky barrier heights, 532
Schottky clamped transistor, 533
Schottky diode, 536, 537
Screen printing, 509
Secondary electrons, 190, 194
Secondary emission coefficient, 42
Secondary ion
  negative, 554
  positive, 554
  SIMS, 548, 560, 562, 565, 566
  yield, 554
Selection criteria, 12
Self-accelerated gun, 102
Self-shadowing
  effects in deposition, 213, 214, 217, 219
Self-supported shapes, 83, 84
Semiconductor, 505
Semiconductor coatings

deposited by sputtering, 204, 215, 217, 221
Semicontinuous in-line systems, 92
Shear rate (also see Viscosity), 499
Sheet resistance, 515
Shellac, 498
Shot peening, 7
Shrinking, 497
Si/Al contacts, 535, 541
Silane, 506
Silicon, 344, 345, 347, 348, 349, 352, 353, 354, 356, 360
Silicon, amorphous, 335, 374
Silicon carbide, 346, 355, 360, 378
Silicon dioxide, 525
Silicon monoxide, 525
Silicon nitride, 335, 376
  films, 526
Silicon oxide, 335, 378
Silicon oxynitride, 378
Silicon tetrachloride, 339, 347
Silicone, 493, 498, 500, 504
Single rod-fed electron beam source, 118
Sintering, 503
SiO, 525
SiO_2, 524, 525, 542
Sn, 296
Solar energy, 503
Solids, 498
Solubility, 494
Solubility parameter (also see Solubility), 495, 497
  delta (δ), 497
Solvent, 491, 495, 497
Solvent maps, 495
Solventless coatings, 491, 501, 511
Source-container reactions, 90
Soybean, 498
Space charge region, 31
Spark-hardening, 15
Specific contact resistance, 531, 533, 537
Spin-on, 511
"Spit," 140
Spray coating, 509
Spray pyrolysis, 7
Spraying processes, 5, 16
Sputter cleaning, 172, 173, 229, 230, 231
  nonconducting substrates, 230
  typical conditions, 230
Sputter deposition, 7, 17, 290, 330
Sputter deposition flux
  calculation of, 202
Sputter etching, 38
Sputtered atoms
  average velocities, 187
  backscattered by working gas, 185
  contribution to substrate heating, 212
  energies, 172, 187, 212, 224, 229
  energy thermalization during transport, 192
Sputtered coatings
  composition of, 183
Sputtered species, 187, 188
  atom clusters, 187
  atoms, 187
  average velocity, 187
  energy distribution, 187, 188
  kinetic energy, 187
  molecular fractions, 187
  negative ions, 187
Sputtered species ejection, 188
  amorphous targets, 188
  effect of target roughness, 188
  oblique ion incidence, 188
  polycrystalline targets, 188
Sputtering, 2, 6, 7, 85, 88, 178, 180, 182, 247
  chemical, 252
  diode, 249-250
  effect of bombarding particle energy, 181
  efficiency, 182, 183
  erosion rate, 178
  ion beam, 245
  magnetron, 250
  molecules, 178
  momentum exchange, 180, 181, 182
  nonconducting materials, 204, 220
  RF, 251
  triode, 250
Sputtering apparatus
  geometry effects on coating properties, 222
  in-line type, 174, 200
  scale-up, 177, 225
Sputtering target, 170, 171, 172, 174, 175, 183, 184, 190
  altered layer, 184
  bonding, 228
  composite, 183, 185
  conditioning, 177
  cone formation, 184
  cooling, 228
  cracking, 185, 210
  cylindrical magnetrons, 201, 202
  fabrication of, 183, 202, 228
  hot pressed, 183, 185, 228
  planar magnetrons, 201, 202
  planar type, 193
  surface composition, 186
  surface compounds, 186
  surface topography, 184, 186

Sputtering yield, 178, 180, 181, 186, 187, 554, 560, 564
  chemisorbed species, 187
  compounds, 179 (table)
  effect on ion species, 186
  ion angle of incidence effect, 180
  metals, 179 (table)
  reactive sputtering, 186
  theoretical prediction, 181
Stabilize, 499
Staining, 502
Stainless steel (304 SS), 330-333
Standard electrode potential, 389
Static electricity, 504
Steps in the formation of a deposit, 4
Sticking coefficient, 232, 233
Stoichiometric range, 122
Stoichiometry (anion:cation ratio), 122
Storage—clean surfaces, 81
Stress—film, 68-70
  compressive stress, 69, 70
  effects of, 70
  modification of, 69
  origin of, 68
  "peening," 70
  tensile stress, 69
  thermal expansion effects, 68, 69
Strikes, 403
  chromium, 403
  cyanide copper, 403
  gold, 403
  silver, 403
  Wood's nickel, 403
Structural imperfections, 6
Structural morphology, 132
Structural zones in condensates, 131
Structure of CVD coatings, 357
Structure zone models, 213
  of Movchan and Demshishin, 8
Styrenated oils, 498
Styrene, 498, 501
Styrene-butadiene, 498
Sublimation source, 99
Submicron microelectronic devices, 12
Substrate, 170, 171, 172, 173, 193, 506
  plasma bombardment, 193
  preparation, 404, 449
  sputtering, 171
  thermally sensitive, 173
Substrate heating, 14, 182, 193, 211, 212, 343
  fundamental processes, 212
  magnetrons, 212

  sputtering, 182, 193
  triode devices, 212
Substrate holders, 94
Substrate ion bombardment (see Bias sputtering), 218, 220, 229
Substrate surface roughness, 213, 214, 220
  deposited coatings (see Nucleation and growth), 213
  effect on structure of vacuum, 213
Substrate temperature, 4, 212, 218, 225, 229
  during sputtering (see Nucleation and growth), 212
  influence of, in sputtering, 218
Sunlight, 491
Superconducting materials, 12
Superconducting properties, 290, 295, 296, 299, 301, 302, 311, 313, 314
Supersaturation
  effect on structure of CVD, 357
Supported discharge ion-plating process, 87
Surface chemistry
  metals, 78
  oxides, 74
  polymers, 79
Surface diffusion, 213
  during growth of sputtered films (see Adatom), 213
Surface enrichment, 7
Surface mobility, 132
Surface modification, 7
Surface preparation, 63, 72, 345
  activation, 72
  "cleaning," 72
  "non-cleaning," 81
Surface properties, 4
Surface roughness, 8, 224
  of coatings, 224
Surface tension, 498
Surface treatment, 498
Surfactant, 499
Suspensions, 498, 499
Swell, 495
Synthesis or creation of the depositing species, 4
Synthesis of material to be deposited, 85
Synthetic polymers, 498

Tantalum, 338, 361
Tantalum films, 520, 521, 526
Tantalum oxide, 524, 525, 526
Tantalum oxide films, 527
Ta_2O_5, 524, 527
Tapered crystallites, 132

584 Deposition Technologies for Films and Coatings

TC bonding, 544, 545
TCC, 525, 528
TEM (see Transmission electron microscopy)
Temperature coefficient of resistance (TCR), 517, 520, 521
Temporary coatings, 504
Tensile properties, 145
Tensile strength, 443
  of copper deposits, 431
  of nickel deposits, 427, 432
Termination, 493
Tetrafluoroethylene, 503
Textiles, 504
Texture of evaporated deposits, 139
Thermal barrier coatings, 12
Thermal decomposition, 493
Thermal fatigue resistance, 155
Thermal growth, 6
Thermal spraying, 7
Thermal stability of refractory, 100
Thermionic emission (TE) model, 530
Thermionic gun, 103
Thermocompression (TC) bonding, 544, 545
Thermodynamic compositional constraints, 9
Thermodynamic effect, 495
Thermodynamics, 348
Thermoplastic, 493
Thermoset, 493, 495
Thick films, 4
Thickening, 499
Thickness control, 113
Thickness distribution, 92
Thin films, 4, 514, 518
Thin self-standing shapes, 9
Thixotropy (also see Viscosity), 499
Throwing power, 4, 12, 14, 245, 392
  macrothrowing power, 392
  microthrowing power, 395
Throwing power of CVD, 355
Ti, 326, 327, 328
TiC, 293, 320-323, 325-327
TiN, 320, 327, 329, 330
Ti_2N, 327-330
Titanium boride, 361
Titanium carbide, 338, 345, 347, 361
Titanium dioxide (also see Filler and Optical properties), 502
Titanium nitride, 361
Titanium sheets, 84
Titanium tetrachloride, 339, 347
Ti:W, 542
Ti:W/Al-Cu, 543
Ti:W/Au, 540
Topcoat, 506
Torr (unit of pressure measurement), 170
Transition temperatures between various structural zones, 133
Transmission electron microscopy (TEM), 288-296, 301, 303, 306, 308, 311, 313, 314, 327-330
  electron diffraction, 291, 294
  imaging (BF) (DF), 290, 291, 324, 330
  out-of-focus, 291
  selected area diffraction (SAD), 291, 296-298, 302, 304, 314, 315, 317-321, 323, 325-329, 331, 333
TEM/STEM, 290, 295, 306
Transparent conductive coatings, 12
Transport from source to substrate, 4, 5, 85
Transverse linear cathode guns, 104
Triisobutylaluminum, 347
Tri-metal, 538
Triode, 170, 175, 193
  hot cathode type, 193
  sputtering device, 170, 173
Tungsten, 338, 340, 346, 347, 357, 358, 360
Tungsten hexachloride, 360
Tungsten hexafluoride, 339, 347
Tungsten—rhenium, 359
Two-source evaporation, 115

Ultimate electrons, 42, 197
Ultracentrifugation (also see Molecular weight determination), 494
Ultrafine powders. 12
Ultrafine (<1 μm) powders of metals, 159
Ultra-high molecular weight polyethylene (UHMWPE), 492, 504
Ultra-high vacuum operation, 93
Ultrasonic agitation, 397
Ultrasonic bonding, 544, 545
Ultraviolet light, 501, 502, 503
  stabilizers, 503
Ultraviolet photoelectron spectroscopy (UPS), 295
Unsaturated fatty acid, 499
Unusual microstructures, 89
Ureaformaldehyde, 501
Urethanes, 500, 502
UV/O_3 cleaning, 76

Vacuum deposition, 6
Vacuum evaporation, 7
Vapor phase, 6
Vapor phase nucleation, 249
Varnishes, 499, 502
VC, 320, 323, 327
VC-TiC, 323
Vegetable oils (also see Binders), 498
Vias, 541
Vinyl, 493, 507
Vinylidene chloride, 504
Viscoelasticity, 491
Viscometry, 494
Viscosity, 497, 498, 502
 Newtonian, 499
 non-Newtonian, 499
Volatility, 495, 497
Voltage coefficient of resistance (VCR), 517
V-Ti, 323
(V,Ti)C, 320, 321, 324, 325

Water repellence, 504
Water vapor transmission rate (WVTR), 504
Water-borne coatings, 491
Water-break-free surface, 506
Water-breaks, 506
Waterproofing, 490

Wear resistance, 155, 385
 chromium, 439
 electroless nickel, 414
 hard anodize, 420, 421
Wear-resistant CVD coatings, 360
Weathering, 503
Weld coating, 7
Welding processes, 16
Wetting, 498, 505, 506
Wetting processes, 5, 7, 16
Wire and metal-foil sources, 95
Work-accelerated, 102
Working gas
 argon, use of in sputtering, 178
 entrapped in sputtered films, 182, 212, 221, 224, 230

x-ray sources
 bremsstrahlung, 567
 monochromatic radiation, 549
 monochrometer, 550
 synchrotron, 567

Zinc chromate (also see Primer, Corrosion), 503, 506
Zinc oxide (also see Filler), 503
Zinc sulfide, 357
Zirconium carbide, 361
ZrO_2, 293, 302, 311